The Yeast Nucleus

Frontiers in Molecular Biology

SERIES EDITORS

B. D. Hames

*Department of Biochemistry
and Molecular Biology
University of Leeds, Leeds LS2 9JT, UK*

D. M. Glover

*Department of Genetics,
University of Cambridge UK*

TITLES IN THE SERIES

The Yeast Nucleus

EDITED BY

Peter Fantes and Jean Beggs

Institute of Cell and Molecular Biology
University of Edinburgh
Edinburgh, UK

UNIVERSITY PRESS

OXFORD

UNIVERSITY PRESS

Great Clarendon Street, Oxford OX2 6DP

Oxford University Press is a department of the University of Oxford
It furthers the University's objective of excellence in research, scholarship,
and education by publishing worldwide in

Oxford New York

Athens Auckland Bangkok Bogotá Buenos Aires Calcutta
Cape Town Chennai Dar es Salaam Delhi Florence Hong Kong Istanbul
Karachi Kuala Lumpur Madrid Melbourne Mexico City Mumbai
Nairobi Paris São Paulo Singapore Taipei Tokyo Toronto Warsaw

with associated companies in
Berlin Ibadan

Oxford is a registered trade mark of Oxford University Press
in the UK and in certain other countries

Published in the United States
by Oxford University Press Inc., New York

A catalogue record for this book is available from the British Library

Library of Congress Cataloging in Publication Data
(Data applied for)

ISBN 0 19 963773 3 (Hbk)
ISBN 0 19 963772 5 (Pbk)

Typeset by Footnote Graphics, Warminster, Wilts

Printed in Great Britain on acid free paper by
The Bath Press Ltd, Avon

Preface

Two yeasts—*Saccharomyces cerevisiae* and *Schizosaccharomyces pombe*—have in the past few decades proved excellent model eukaryotes for a wide range of cellular processes. The genetic amenability of yeasts has permitted the identification of hitherto unsuspected cellular mechanisms, many of which were initially defined only by conditional lethal mutations. The development of powerful molecular genetical techniques has allowed the physical isolation of the corresponding wild-type genes and their analysis. In many cases the predicted protein products of these genes have shown sequence similarities to proteins isolated directly from cell types more suited to classical biochemical procedures. Often it has turned out that the underlying processes are highly conserved. Indeed, one of the major insights into biology over the past decade is that there is far more unity than diversity as regards the molecular bases of fundamental cellular processes.

Of crucial importance has been the determination of the complete DNA sequence of the *S. cerevisiae* nuclear genome. This has revealed the presence of many genes whose predicted products show structural homology to proteins identified in higher cells. Molecular genetical methods have made it a relatively simple matter to determine whether a yeast gene has an important cellular role, and to shed light on the nature of that role. Genomic sequence determination has also revealed the types and numbers of particular protein families present. The impending completion of the *S. pombe* genomic sequence will allow a detailed comparison of the genomes of these two somewhat unrelated yeasts, and will lead to insights about what constitutes the essential eukaryotic gene set.

This book is about cellular systems rather than individual genes or gene products, and, where appropriate, our aim has been to have authors who are expert in their fields write comparative chapters dealing with both budding and fission yeasts. Our hope has been that in this way we would not only provide useful reviews of the various topics, but also provoke thought about the similarities and differences between the yeasts, and whether parallel mechanisms might operate in other cell types.

We chose to concentrate on activities of the yeast nucleus, because it is in nuclear processes that most progress has been made. For example, cell cycle mechanisms and cell cycle checkpoints have been studied in depth in both yeasts, and there are interesting and important similarities and differences between them. Likewise, chromosome structure and the mechanisms of DNA replication and nuclear division have been targets for study in both yeasts. Intron splicing has been studied much more extensively in *S. cerevisiae* than in *S. pombe*, but it is clear that the mechanism of splicing is very highly conserved from budding yeast to humans. Nevertheless, introns are more abundant in *S. pombe* than in *S. cerevisiae*, and splicing regulation in *S. pombe* may turn out to be more similar to that in higher cells. Other processes such as nuclear transport and transcriptional regulation have, for largely historical reasons, been studied far more in *S. cerevisiae*, and the content of those chapters reflects this bias.

We hope this book will facilitate a comparison of different nuclear processes as well as highlighting observed links between them. For example, nucleocytoplasmic transport processes determine the intracellular localization of many proteins and RNAs, and thus might potentially affect all nuclear processes. Along the same lines, transcription is linked with various aspects of mRNA processing, while cell cycle checkpoints monitor DNA replication and repair. A striking and poorly understood example is that in both yeasts there are mutations that affect pre-mRNA splicing and cell cycle progress, two apparently unrelated processes. In view of these developments we hope that a book dealing with various activities of the nucleus in the two yeasts will be useful and stimulating for a wide readership, experts and students alike.

Edinburgh
April 2000

P. F.
J. B.

Nomenclature

This book is concerned primarily with the cell and molecular biology of two yeasts: *Saccharomyces cerevisiae* and *Schizosaccharomyces pombe*. These are genetically well studied organisms, and the functions of many genes have been defined by the cellular consequences of mutations. Many gene names were originally chosen on the basis of mutant phenotypes, and consequently bear no obvious relation to the product encoded by the gene. More confusingly, some gene symbols refer to unrelated genes in the two yeasts (e.g. *rad9*), while homologous genes often have different names (e.g. *CDC28* and *cdc2$^+$*). We decided that, as genes and gene products from both yeasts are discussed in this book, we would try to minimize confusion while conforming as far as possible to established conventions.

S. cerevisiae wild-type genes are written as (e.g.) *CDC28*; mutant alleles as *cdc28-1*. The corresponding designations for *S. pombe* are *cdc2$^+$* and *cdc2-33*. It is generally clear from these conventions and from the context which yeast is intended. For gene products the conventions are less strict, and there is more possibility of confusion. For this book we have chosen to describe *S. cerevisiae* gene products as, for example, Cdc28p, and those of *S. pombe* as, for example, Cdc2. Note the absence of italicization. We hope that the added 'p' for *S. cerevisiae* gene products will help discrimination between the yeasts. Much of Chapter 4 deals with a subset of 'checkpoint *rad*' genes in both yeasts, and it is particularly unfortunate that in several cases the same symbol is used for unrelated genes in the two yeasts. For this chapter only, we have spelled out the gene names unambiguously by the use of suffixes indicating which yeast is meant.

For further information on nomenclature see the *Trends in Genetics* genetic nomenclature guide (*Trends Genet.*, 1998, **14**, issue 11).

Contents

3 The cell cycle   58

MIKE TYERS AND PAUL JORGENSEN

4 Cell cycle checkpoints

KRISTI CHRISPELL FORBES AND TAMAR ENOCH

5 Mechanisms of nuclear division 143

HIROHISA MASUDA AND YASUSHI HIRAOKA

6 RNA polymerase II transcriptional machinery 176

MICHAEL HAMPSEY

7 The structure of yeast centromeres and telomeres and the role of silent heterochromatin 212

ALISON L. PIDOUX AND ROBIN C. ALLSHIRE

8 Splicing pre-mRNA introns 246

ANDREW E. MAYES, JUDITH A. POTASHKIN AND JEAN D. BEGGS

9 Yeast nuclear transport 276

CHARLES N. COLE

Contributors

ROBIN C. ALLSHIRE
MRC Human Genetics Unit, Western General Hospital, Crewe Road, Edinburgh EH4 2XU, UK.

JEAN D. BEGGS
Institute of Cell and Molecular Biology, University of Edinburgh, Mayfield Road, Edinburgh EH9 3JR, UK.

PETER M. J. BURGERS
Department of Biochemistry and Molecular Biophysics, Washington University School of Medicine, BOX 8231, 660 South Euclid, Saint Louis, MO 63110, USA.

CHARLES COLE
Department of Biochemistry, Dartmouth Medical School, 7200 Vail Building, Room 413, Hanover, NH 03755, USA.

DANIELA DELNERI
School of Biological Sciences, University of Manchester, 2.205 Stopford Building, Oxford Road, Manchester M13 9PT, UK.

TAMAR ENOCH
Department of Genetics, Harvard Medical School, Warren Alpert Building, 200 Longwood Avenue, Boston, MA 02115, USA.

KRISTI CHRISPELL FORBES
Department of Developmental Biology, Stanford University, Beckman Center B351, 279 Campus Drive, Palo Alto, CA 94304-5329, USA.

MICHAEL HAMPSEY
Department of Biochemistry, University of Medicine and Dentistry of New Jersey, Robert Wood Johnson Medical School, 675 Hoes Lane, Piscataway, NJ 08854, USA.

YASUSHI HIRAOKA
Structural Biology Section and CREST Research Project, Kansai Advanced Research Center, Communications Research Laboratory Iwaoka-cho, Nishi-ku, Kobe, 651-2492, Japan.

PAUL JORGENSEN
Samuel Lunenfeld Research Institute, Room 1078, Mount Sinai Hospital, 600 University Avenue, Toronto, Ontario, Canada M5G 1X5.

STUART A. MACNEILL
Wellcome Centre for Cell Biology, Institute of Cell and Molecular Biology, University of Edinburgh, Mayfield Road, Edinburgh EH9 3JR, UK.

HIROHISA MASUDA
Structural Biology Section and CREST Research Project, Kansai Advanced Research Center, Communications Research Laboratory Iwaoka-cho, Nishi-ku, Kobe, 651-2492, Japan.

ANDREW E. MAYES
Unilever Research Colworth, Colworth House, Sharnbrook, Bedford MK44 1LQ, UK.

STEPHEN G. OLIVER
School of Biological Sciences, University of Manchester, 2.205 Stopford Building, Oxford Road, Manchester M13 9PT, UK.

ALISON L. PIDOUX
MRC Human Genetics Unit, Western General Hospital, Crewe Road, Edinburgh EH4 2XU, UK.

JUDITH A. POTASHKIN
The Chicago Medical School, 3333 Green Bay Road, North Chicago, IL 60064, USA.

MICHAEL TYERS
Samuel Lunenfeld Research Institute, Room 1080, Mount Sinai Hospital, 600 University Avenue, Toronto, Ontario, Canada M5G 1X5.

1 | Functional genomics with yeast

DANIELA DELNERI and STEPHEN G. OLIVER

1. Introduction

DNA sequencing and other biological data are accumulating at a vertiginous rate. Within the last 5 years, sequences of 23 prokaryotic genomes, the unicellular eukaryote *Saccharomyces cerevisiae* (1), and the metazoan *Caenorabditis elegans* (2) were completed (Entrez Genomes at http://www.ncbi.nlm.nih.gov/Entrez/Genome/org.html). The fission yeast *Schizosaccharomyces pombe* and the fruitfly *Drosophila melanogaster* will be the next eukaryotes to have their genomes sequenced and released by the end of this year. There are about 70 microbial genomes (with sizes between 1.5–5 Mb) whose sequences are in progress. In autumn 1999, the first sequence of a human chromosome arm, 22p, was completed by an international consortium of sequencing centres (3). The Human Genome Project (HGP), which started at the beginning of the 1990s, should be completed in the next 5 years (4). By 2005, about 50 years after the landmark paper of Watson and Crick on the structure of DNA (5), the complete description of all the genes contained in human chromosomes will be available. In parallel with the HGP, the mouse genome project has also been undertaken and is expected to be completed by 2008 (6). Although these are major achievements, in reality they are only the tip of the iceberg. DNA sequencing data provide an accurate genetic blueprint that needs to be matched to a functioning organism, in which the genes are expressed in specific temporal, spatial, and functional contexts. Thus, the sequencing era is leading to a phase of genome-wide functional analysis, in which attention will shift to the study of the biochemical and biological function of the novel genes and to the determination of the overall organization and evolution of these genomes. This chapter will provide an overview of different systematic approaches that have been employed in the yeast *S. cerevisiae* to elucidate the functions of the 6000 protein-coding genes predicted by the nuclear DNA sequence (7).

2. Organization of the yeast genome and the problem of redundancy

Owing to its intrinsic genetic characteristics, the brewers' and bakers' yeast *S. cerevisiae* is a model organism for the study of eukaryotic molecular genetics. The complete sequence of this organism revealed the extent of our knowledge of yeast genetics. In fact, less than 45% of the protein-encoding genes predicted by the genome sequence had had their functions determined previously (8). Unlike more complex eukaryotes, *S. cerevisiae* has a small genome of about 13 Mb (less than four times the size of the genome of the prokaryote *Escherichia coli* (9)) and a high frequency of open reading-frames (ORFs), one every 2 kb, with almost 70% of the total sequence consisting of genes. (By contrast, the human genome contains, on average, one potential ORF every 30 kb.) The analysis of the yeast genome revealed the presence of approximately 6000 ORFs, 140 ribosomal RNA genes in a large tandem array on chromosome XII, 275 transfer RNA genes, and 40 genes encoding small nuclear RNAs (snRNAs) distributed between the 16 chromosomes. In addition, there are a number of small ORFs whose protein products are less than 100 amino acids long. The sequencing project set an arbitrary cut-off of 300 bp to define ORFs, thus excluding small ORFs, such as some ribosomal protein genes, whose length is less than 100 amino acids. Special software packages have now been developed (10–12) in order to identify small ORFs that are likely to be real genes and to discriminate between genuine incidences of pseudogenes and ORFs which have been artificially terminated due to sequencing errors. None the less, it is probable that the veracity of the small ORFs as genuine coding sequences can be established unambiguously only by studies at the proteomic level.

The data arising from the yeast genome-sequencing project revealed, not only that gene numbers were higher than expected from conventional or 'function first' genetic analyses (13), but also that many genes were members of families that comprised two or more members whose predicted protein products were at least 50% identical at the amino-acid sequence level (7). This remarkably high level of apparent genetic redundancy (Plate 1) seems to be a common feature among eukaryotes: for example, in *D. melanogaster*, it was once believed that the total number of genes corresponded to the number of polytene chromosome bands (ca. 5000), while it is now estimated that the number of ORFs is around 20 000, most of which are not essential for viability (14, 15).

Part of the explanation for genetic redundancy arises from the presence of gene sets with overlapping functions; in yeast, most of the duplicated genes are members of families with just two or three members, but some gene family are significantly larger. Blocks of duplicated ORFs, called cluster homology regions (CHR), are found both in the telomeric regions and at internal sites within chromosome arms (1). For example, the *SUC*, *MAL*, and *MEL* gene families, which are involved, respectively, in the fermentation of sucrose, maltose, and melibiose (16–18), are predominantly associated with the chromosome ends, flocculation genes (*FLO*) constitute a new subtelomeric gene set (19), and there are at least 15 genes belonging to the family of the hexose transporter (*HXT*), spread around the yeast genome (20).

The problem of redundancy poses questions about the true meaning of the gene duplications within the yeast genome. Are the various members of these gene sets truly redundant? Are all the genes belonging to a family expressed, each making a small contribution to the functional activity? Are they expressed under different physiological conditions? Do the products of these genes have different characteristics in terms of their subcellular localization, kinetic properties, or substrate specificity? For instance, the *CIT2* gene, present on chromosome III, encodes a peroxisomal form of the citrate synthase enzyme, whereas *CIT1* on chromosome XIV, and *CIT3* on chromosome XVI, encode mitochondrially located forms of the same enzyme (20, 21). The yeast ribosomal proteins Cry1p and Cry2p represent, instead, a case of differential regulation in which the second copy of the gene is expressed only when the first copy has been deleted (23). The aryl alcohol dehydrogenase gene family (*AAD*) contains six telomere-associated members, and a seventh gene encoding a similar protein product is found at an internal chromosomal location. However, only two (*AAD4* and *AAD6*) of these seven genes were found to be induced by chemicals that cause oxidative stress by inactivating the glutathione reservoir in the cell. Of these two genes, only *AAD4* is likely to be functional since *AAD6* is interrupted by a stop codon (24). Thus, in terms of the function in response to oxidative stress, the apparent sevenfold redundancy of the *AAD* gene set disappears completely.

Genetic redundancy appears to be the rule at the ends of the yeast chromosomes, and many duplicate genes seem to be phenotypically redundant. The 'single gene duplication' mechanisms, due to recombination events between either repeat sequences, or yeast transposons or their long terminal repeats (LTRs) (25, 26), are insufficient to account for the full extent of redundancy in the *S. cerevisiae* genome. Wolfe and Shields (27) have provided a more radical view of this phenomenon. Inspired by Ohno's theories on genome evolution (28), they proposed that the yeast genome, at some stage in its evolutionary history, underwent a complete duplication and has been subsequently reduced to its present size via a series of deletion and reciprocal translocation events (29).

Whatever the origin of this redundancy, understanding its true nature, and developing experimental strategies to tackle it, will be a major conceptual and technical challenge. For instance, the problem of genetic redundancy could, in a sense, be an artificial problem arising from our inability either to measure subtle fitness changes or to mimic the range of environments in which that particular species has evolved. One approach, undertaken by Oliver and Louis (personal communication) in order to elucidate gene function, is to construct a 'minimalist' yeast in which all the non-essential genes are deleted from the genome. This strategy, leaving the eukaryotic cell with only the basic set of genes required for its survival, enables the contribution of non-essential yeast genes to be assessed by adding them back to this minimal set. Recently, a similar approach has been taken to study the 517 genes of *Mycoplasma genitalium* (the smallest genome of any independently replicating cell so far identified) leading to the identification of all non-essential genes, and making a significant progress towards the creation of the first true minimal genome under laboratory growth conditions (30).

A great variety of physiological challenges needs to be analysed to reveal the functions of novel genes and, even though obvious stresses such as heat, cold, starvation and osmotic shock can be examined, a vast number of conditions relevant to the natural history of the yeast might be unknown. In order to uncover the functions of the 'orphan' genes (unknown ORFs that show no homology with any other gene of any organism in the available databases), systematic methods of analysis will be required. Oliver (13) has proposed a hierarchical taxonomy of gene function that will allow us to perform only the tests relevant for genes of a given class. The system must be hierarchical in order to limit the number of phenotypic tests needed to elucidate the function of a gene. Informatic analysis, large-scale gene deletions, systematic phenotype screening, transcriptome analysis, proteome studies and top-down metabolic control analysis are some of the approaches that can be applied to uncover gene function on a genome-wide scale.

3. Informatic analysis

As a result of the improvements of large-scale experimental methods, such as the DNA sequencing technology, a new science has emerged, able to process and analyse the wealth of biological data generated. In fact, the first task of bioinformatics, a fusion of biology, mathematics, and computer science, has been to organize and manage the vast amount of data accumulating from ongoing genome projects: to find the start and stop codons of reading-frames, to look for promoters and terminators of transcription, introns, splice sites, and the binding sites for ribosomes and transcriptional activators. A second major, and more challenging, task for bioinformatics is to make biological sense of the data: to assign functions to uncharacterized coding regions, or study the evolution of gene families within and between species. Sequence alignments provide a powerful way to compare related ORFs and assign provisional functions to novel genes; the alignment of two sequences could uncover common structural features of the gene products and/or reveal their evolutionary history by describing the most common mutations (31–33).

At present, for *Saccharomyces*, there are three major databases, the Munich Information Center for Protein Sequences (MIPS), the Saccharomyces Genome Database (SGD) and the Yeast Protein Database (YPD), while several other useful yeast web sites are growing rapidly (34). All these databases maintain web servers and offer a fast, easy, and reliable method for the yeast community to acquire information about the *S. cerevisiae* genome. Moreover they provide links with pertinent information in other systems as the complete genome sequences of new organisms become available. More recently, a series of informatic programmes was developed for analysis of the regulatory sequences in yeast, allowing a more precise study of the regulation of gene expression (35, 36). With this tool, it is possible to search for binding sites of known transcriptional activators in novel genes or predict potential binding sites for novel activators in the regulatory regions of genes that are co-regulated.

Although the homology-based predictions of gene functions have proved to be extremely powerful in exploring genomic information (37), it is important to be aware of the limitations of such analyses in assigning an unequivocal function to orphan genes. In fact, they sometimes provide only a general or superficial description of the predicted protein products, but give no clear indication about their biological role (38). For instance, the *AAD* homologues in the yeast genome show about 90% of similarity to the aryl alcohol dehydrogenase enzyme of the lignin-degrading fungus *Phanerochaete chrysosporium* (39). As *S. cerevisiae* is not a lignin degrader, the sequence alignments data retrieved in this case did not make a great deal of sense even though it is possible that yeast has to deal with the aromatic monomers of the lignocellulose catabolism generated by filamentous fungi and bacteria in its immediate environment (40–42). Another example is given by the NifS protein of nitrogen-fixing bacteria that has homologues in *S. cerevisiae*, *Bacillus* and *E. coli*. Neither yeast nor either of these bacteria is able to fix molecular nitrogen, and the set of NifS homologues is probably a set of transaminases, which, like NifS itself, employ pyridoxal phosphate as a cofactor (43). These examples demonstrate that, powerful as the computational approach is, its results need to be treated with caution and intelligence; wet experiments are necessary to elucidate the functions of the 'orphans' discovered by systematic sequencing.

4. Large-scale functional analysis

The small size of the yeast nuclear genome, and the facility with which it can be manipulated using recombinant DNA techniques, encouraged the yeast research community to embark upon a systematic deletion of the ORFs, expecting to uncover the functions of unknown novel genes. A large European research network, called EUROFAN (European Functional Analysis Network), was created with the intention to produce a collection of deletion mutants covering all the 6000 ORFs revealed by the *S. cerevisiae* genome sequence (44). An effective approach to gene disruption has been developed using a polymerase chain reaction (PCR)-mediated gene replacement strategy, which takes advantage of the efficiency and precision of yeast's mitotic recombination system (45, 46). This technique consists of the PCR amplification of a gene deletion cassette, containing a selectable marker flanked by approximately 40 bp of DNA homologous to the upstream and downstream regions of the gene of interest (short flanking homology PCR, SFH-PCR), and the transformation of yeast with the resulting PCR products. Recombination occurs between the flanking regions of the disruption cassette and the homologous sites on the chromosome, resulting in the replacement of the gene with the selectable marker. The first markers used in the PCR-mediated gene replacement method were homologous nutritional markers (e.g. *HIS3*) (45), which had the limitation of a rather low efficiency of correct integration due to the incomplete deletion of the marker locus in the yeast genome.

An important improvement of this technique was achieved with the introduction of *kan*MX as a heterologous dominant resistance marker (46). The *kan*MX selectable module consists of the coding sequence of the kanamycin resistance gene from the bacterial transposon Tn903, coding for an aminoglycoside phosphotransferase (47), the activity of which confers resistance to the drug geneticin in the host yeast cell (48). Since *kan*MX is expressed in yeast by using promoter and terminator sequences from the filamentous ascomycete fungus, *Ashbya gossypii*, the only regions of homology to the yeast genome that the disruption cassette contains are those sequences complementary to the flanking regions of the target ORF. Compared with the nutritional markers, the heterologous *kan*^r gene has proved to be a 'safer' module to use in functional analysis, exhibiting a negligible effect on specific growth rate (49). Moreover it is particularly useful in the genetic manipulation of industrial brewing yeasts, which do not have any auxotrophic markers (50). To improve the recombination efficiency and the site specificity of the replacement event in laboratory and industrial strains of *S. cerevisiae*, longer flanking regions (200–300 bp) could be generated using a nested PCR reaction (long flanking homology PCR, LFH-PCR) (51).

Because of the redundancy problem, it might be necessary to create deletions of all members of a gene family in the same strain, in order to define a correct phenotype and to investigate gene interactions. Thus, new methods that would permit the recycling of the selectable marker become necessary, and deletion cassettes in which the nutritional or resistance markers are flanked by two site-specific DNA regions such as FRT (52), I-*SceI* (53), or *loxP* (54) have been constructed, allowing the pop-out of the marker through *in vivo* excision. For gene families with a large number of members, these marker rescue methods are time consuming because they require that the marker gene is excised from the genomic locus in order to proceed with subsequent deletions. Moreover, an additional transformation step, with a plasmid carrying the corresponding site-specific recombinase (cre for *loxP*, Flp for FRT) or endonuclease (I-*SceI*) is essential. This plasmid needs to be lost once excision of the marker has occurred.

To overcome this problem, a new series of deletion cassettes with different selectable genes placed between two *loxP* loci were constructed, and used to generate sequential gene replacements of four homologues, followed by the simultaneous excision of all different markers, performed as the final step (55). Moreover, disruption cassettes conferring resistance to aureobasidin A (56), hygromycin B, nourseothricin and bialaphos (57) were recently released, providing a wider choice of replacement markers to use in the multiple gene deletions.

A genomic-scale approach to determine the biological function of newly identified ORFs has been developed, using the PCR targeting strategy. With this technique, each deletion mutant is labelled with a unique 20-base tag sequence, called a 'molecular bar code' (58), which can be detected by hybridization to a high-density oligonucleotide array (59). This method, allowing whole genome analysis with a single selective growth condition and a single hybridization, should greatly accelerate the systematic analysis of gene function. Bar-coded deletants for all 6000 yeast ORFs are now being created as part of a transatlantic collaboration between a number of US and Canadian laboratories and the EUROFAN network in the EU (60).

More recently, Ross-Macdonald and collaborators (61) developed a transposon tagging strategy for genome-wide analysis of disruption phenotypes, gene expression, and protein localization. Their approach utilizes a multipurpose minitransposon, mTn-3xHA/*lacZ*, which contains a *lacZ* reporter gene flanked by two *loxP* loci and three copies of a haemagglutinin epitope tag, to allow the analysis of the gene products by indirect immunofluorescence. Introduction of the transposon into yeast results in production of β-galactosidase, while the expression of cre recombinase results in recombination of *loxP* sites reducing the mTn to a 93-codon ORF (including 3xHA), which allows the production of a full-length, epitope-tagged protein from that gene. This study, with the generation of 11 000 mutant strains, each carrying a transposon inserted within a region of the genome expressed during different growth conditions, constitutes the largest single study ever undertaken of the functionality of the yeast genome.

5. Quantitative phenotypic analysis

One reason that classical genetic approaches have failed to identify such a large proportion of genes within the yeast genome might be that phenotype is usually assessed using qualitative, rather than quantitative, tests. The high level of apparent genetic redundancy in the *S. cerevisiae* genome may mean that, for some gene families (at least), each individual member may have a fractional quantitative impact on the final phenotype. For these reasons, a quantitative approach to the analysis of the phenotypes produced by specific gene deletions is necessary. Metabolic control analysis (MCA) aims to exploit such mutants in order to place unknown genes in particular functional domains (62), starting with global assessments of fitness and metabolic profiles in a 'top-down' approach (13).

Differences in growth rates are most easily revealed in competition experiments because these are very sensitive to small effects (63). Competition experiments using mutant strains generated by PCR-mediated gene replacement protocols have been performed using quantitative PCR analysis to determine the proportions of the different competing strains (64). Competition analysis has recently been used to uncover drug target sites by exploiting haplo-insufficiency in diploid strains heterozygous for single-ORF deletions (65, 66).

An attractive alternative approach to competition analysis, called 'genetic footprinting', has been developed by Smith and co-workers (67). Yeast mutants, generated by Ty1 transposon insertion, were grown under different selective conditions and the changing proportions of the mutant strains with time were monitored by cycle sequencing from gene-specific fluorescently labelled primers. The genetic footprinting technique was employed for the quantitative phenotypic analysis of yeast chromosome V (68), revealing a phenotype for approximately 60% of the chromosome's 268 predicted genes.

In terms of MCA theory, these competition experiments assess the flux control coefficient of the different gene products. It is expected that such coefficients will usually be low, and that single gene deletions will have only a small impact on specific growth rate. An alternative approach is to measure the change in concentration of metabolic intermediates as a result of the deletion or overexpression of a gene. Such measurements allow the assessment of concentration control coefficients, which (unlike flux control coefficients) can adopt very high values, i.e. it may be expected that large changes in the concentration of one or more metabolites will result from the deletion or overexpression of single genes. An analytical strategy, termed FANCY (functional analysis by co-responses in yeast (69), based on the theoretical work by Rohwer and collaborators (70), was suggested in order to take advantage of the fact that the number of different metabolites found in yeast is an order of magnitude lower than the number of genes, and that the functions of 40% of these genes are already known.

Some enzymes can be grouped into 'monofunctional units', in which any perturbations within them will always produce the same co-responses (coordinate changes in a number of metabolites) outside the unit, regardless of which enzyme was affected or of the magnitude of the perturbation. This property could be used for functional analysis purposes. For example, when two mutants have the same co-responses, they will affect the same monofunctional unit and, if the function of one of the genes is known, it is possible to infer that of the other. To achieve a fast and effective way of sampling the concentration of as many metabolites as possible (and thereby take a 'metabolic snapshot' of each deletant strain), a two-stage strategy was developed by Oliver and co-workers (44, 71). In the first phase, mutants of novel genes, grown under different ranges of conditions, are grouped with genes of known functions by comparing their infrared spectra (72, 73). These data then determine the type of metabolic snapshot to be taken in the second phase, using tandem mass spectrometry (74) or proton nuclear magnetic resonance (NMR) spectroscopy (75).

6. The transcriptome

Powerful methods have been developed in the last 5 years that allow a quantitative and comprehensive analysis of gene expression at the transcriptional level. As the characteristics of an organism are determined by the genes expressed within it, a way to uncover the functions of novel genes is to determine the physiological conditions or developmental and disease states in which those orphan genes are expressed and to compare their expression profiles with those of genes whose function is well known. During the 1980s and the early 1990s, techniques based on complementary DNA (cDNA) subtraction or differential display (76, 77) were used successfully to compare gene expression differences, even though, with these methods, no direct information about transcript abundance was retrieved. A more recent approach for the rapid and detailed analysis of transcription on a genome-wide scale is the use of 'serial analysis of gene expression' (SAGE), in which short diagnostic sequence tags, isolated under specific physiological conditions, are concatenated, cloned, and sequenced, revealing the distinctive gene expression pattern (78). The availability of the complete *S. cerevisiae* genome sequence, and the possibility of uniquely identifying an ORF with a sequence of just 12 nucleotides, encouraged Velculescu and co-workers to employ the SAGE technology for studying the 'yeast transcriptome', the complete set of genes expressed under a given set of conditions (79). In this work, they analysed 60 633 transcripts revealing a total of 4665 genes (about 75% of the predicted ORFs), most of which were expressed at a low level.

A more efficient high-throughput approach to transcriptome analysis is the use of hybridization array technologies. Oligonucleotides can now be synthesized *in situ* on

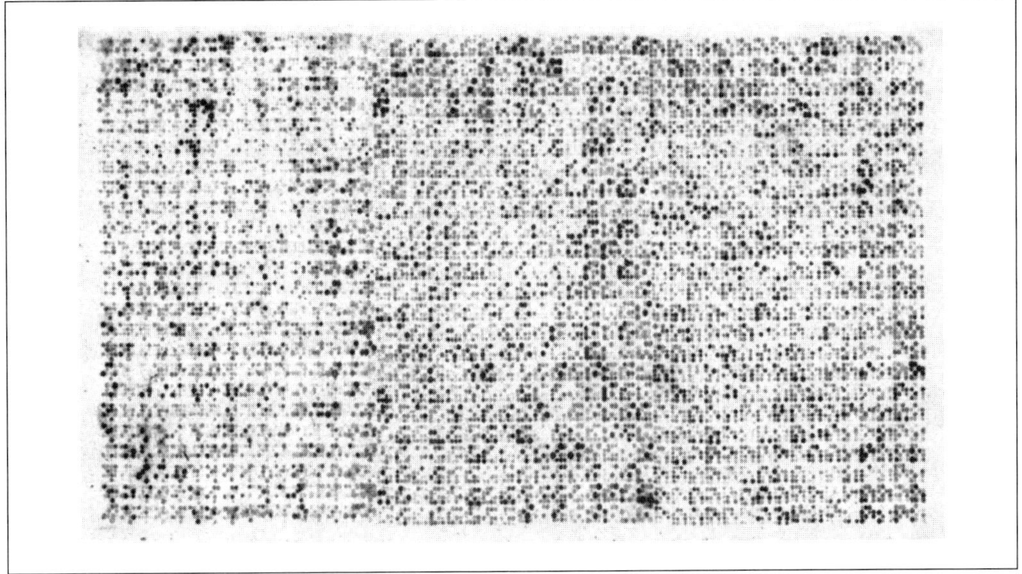

Fig. 1 Typical yeast microarray (82) hybridized with a complex probe from a glucose-grown yeast culture.

a silicon chip using a combination of photolithography and oligonucleotide chemistry (80, 81). Thus each yeast gene can be represented by a number of specific oligonucleotides at a defined location on the chip. With the development of automated techniques for PCR amplification, it is possible to create DNA microarrays, consisting of PCR products corresponding to each of 6000 yeast ORFs, arrayed on a positively charged glass support (i.e. slide) by a high-speed robot (59, 82). In addition, DNA macroarray technologies have been developed, where the PCR products are spotted on to nylon or polypropylene membranes (typical size 22 × 11cm) using various robotic devices (Fig. 1) (83). All these arrays allow the expression of all protein-encoding genes to be monitored in parallel, and differential expression measurements may be performed by means of simultaneous two-colour fluorescent hybridization.

An important use of transcriptome analysis in elucidating the biological functions of orphan genes is to group genes into co-expressed clusters that may contain some members with well defined functions. For example, transcriptome analysis of mutant strains, in which the gene encoding a specific transcription factor is either deleted or overexpressed, can identify novel genes under the control of the given transcriptional activator and provide important clues about the functions of these genes. In this context, DeRisi and collaborators (84) have applied DNA microarrays to identify genes whose expression was affected by either the deletion of the gene specifying the transcriptional co-repressor Tup1p, or the overexpression of the transcriptional activator gene *YAP1*. More recently, a comprehensive catalogue of yeast genes whose transcript levels vary periodically during either the cell cycle (85) or sporulation (86) was created using DNA microchip technology. It should be noted that the micro- and macro-array procedures are unlikely to be able to discriminate between nearly identical members of a gene family, and it requires very careful oligonucleotide design to achieve this with DNA chips. In such instances, additional genetic evidence is required to establish whether there is differential expression of the different members of such a family (24). To organize all the data generated from transcriptome analysis, a system of clustering genes according to similarity in pattern of gene expression was developed. The method used to group genes is a standard statistical algorithm and the output is displayed graphically, linking the clustering and the expression data simultaneously in a form accessible to biologists (87).

On the threshold of the genome-wide research era, classical Northern blot techniques are likely to disappear. However, the availability of data sets, obtained using this classical technology (88), that are sufficiently large to permit the application of the clustering algorithms (A. J. Brown *et al.*, unpublished results), will provide a useful bridge between the old approach to transcript analysis and the array technologies. At present, there is intense competition to establish the worth of the different array formats in terms of efficiency, time-saving, throughput, and accuracy in the quantitation of transcript levels.

7. The proteome

The proteome is the complete set of proteins found in a cell under given conditions and it could be considered the protein-level correlate of the transcriptome (89). However, protein molecules, unlike the majority of transcripts, are functional entities within the cell and therefore proteome analysis approaches the determination of gene function much more directly than does transcriptome analysis. The drawback is that it is much more technically demanding to perform. Mass spectrometry has had a long-standing part in the study of protein sequence and structure, and has recently become an important technique for correlating proteins with genes. In fact, improvements in ionization and mass analysis techniques concurrently with large-scale DNA sequencing of whole genomes have provided a powerful resource for protein identification. Matrix-assisted laser desorption ionization time-of-flight mass spectrometry (MALDI-TOF-MS), electrospray mass spectrometry (ESI-MS), and tandem mass spectrometry (MS-MS) are the most common techniques applied to the identification of electrophoretically separated protein mixtures. To correlate the mass spectra of peptide fragments to sequences in the database, computer algorithms have been developed that search the protein sequence database for multiple peptides of individual proteins that match the measured masses (90). The yeast proteome has been defined by mass spectrometry from two dimensional gels (91, 92); using MALDI-MS on protein spots digested with trypsin, 32 novel proteins were matched to previously uncharacterized ORFs in the yeast genome.

Since, for identification of a protein, it is necessary to identify only a protein fragment whose mass is unique to that particular protein, it should be possible to analyse mixtures of unknown proteins. A new method for the direct identification of proteins in macromolecular complexes uses multidimensional liquid chromatography and tandem mass spectrometry to separate and fragment peptides (Fig. 2).

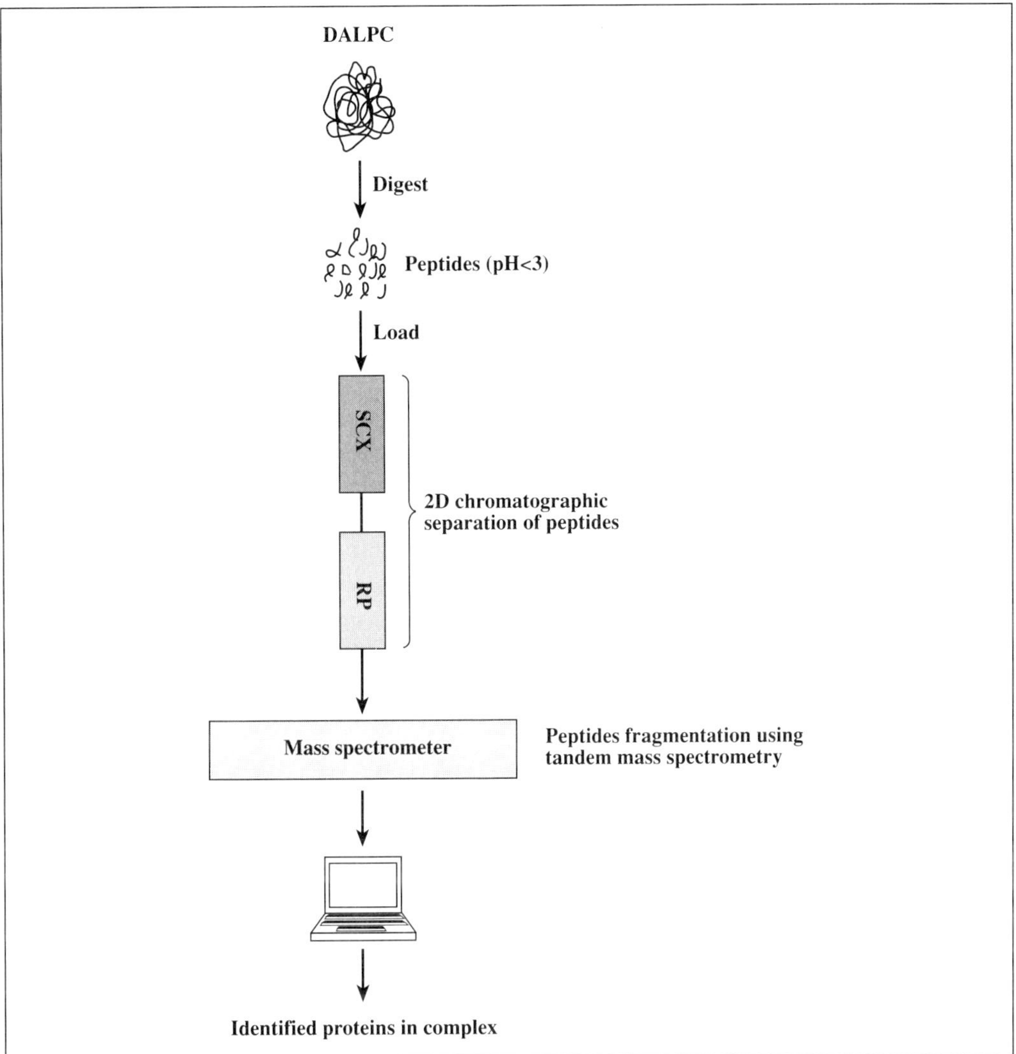

Fig. 2 Direct analysis of large protein complexes (DALPC). A denatured protein complex is digested with trypsin and loaded on to the strong cation exchange column (SCX). Discrete peptides then flow on to the reverse phase (RP) liquid chromatography column, which is attached to the mass spectrometer. This iterative process is repeated, obtaining the fragmentation patterns of peptides in the original peptide mixture. An algorithm is used to correlate the tandem mass spectra of fragmented peptides to amino acids sequences using nucleotide databases. From Link *et al.* (Reference 92).

This technique allows the detection of more than 100 proteins in a single run and the identification of the components of a protein mixture without the need to purify each member to homogeneity. Link and co-workers (93) applied direct analysis of large protein complexes (DALPC) to the *S. cerevisiae* 80S ribosome, leading to the identification of a novel protein component of the yeast 40S subunit (Ymr116p). The homologous human protein (*RACK1*) was also shown to be part of the 40S subunit.

In the near future, the information acquired on protein complexes by mass spectrometry will be correlated with *in vivo* data obtained using the yeast two-hybrid system (94), thus integrating our knowledge of the role of protein interactions in the biology of the yeast cell (95). Protein–protein interaction maps of the subunits of RNA polymerase III and of members of a novel set of Sm-like proteins were recently investigated in *S. cerevisiae*, by exhaustive two-hybrid screening (96, 97). A comprehensive approach has also been undertaken for the *E. coli* bacteriophage T7 genome (encoding 55 proteins), where a complete protein linkage map was provided by Bartel and collaborators (98). Fields and co-workers have recently attempted an exhaustive two-hybrid analysis of all potential protein–protein interactions in the yeast cell (99), uncovering novel interactions involved in both cell cycle control and meiotic recombination.

Recently, relationships between transcript and protein expression levels were determined under a given set of physiological conditions. Spots from a two-dimensional protein gel were analysed by liquid chromatography (LC)–MS-MS and the proteins quantified by radioisotopic labelling and scintillation counting. The levels of the corresponding mRNA molecules were calculated from the published SAGE analysis (100). While Futcher and co-workers (101) found a good correlation between protein abundance, mRNA abundance, and codon bias (measured in glucose and ethanol media), Gygi and collaborators claimed that, in the mid-exponential phase, the correlation between mRNA and protein levels was insufficient to predict protein expression levels from quantitative mRNA data. In fact, for some genes, the relative mRNA and protein levels varied by more than 20-fold. According to this second study, the limits of the current approaches for quantitative analysis of protein expression prevent a global view of the yeast proteome from the simple deduction of the mRNA transcript levels.

8. Conclusion and prospects

The availability of complete genome sequences has dramatically altered the scale of biological experiments. New data from transcriptome and proteome experiments quickly accumulate, and even conceptually simple experiments yield vast amounts of information. In the post-sequencing era, the pattern of research for many biologists will become more similar to that of the particle physicist. Experiments will be designed with enormous care because of the quantity of data that they will generate and the expense involved in their execution and analysis. Relatively little time will be spent actually performing these experiments and the subsequent period of data analysis and evaluation will become the rate-limiting step. These analyses will be undertaken by exploiting the data-mining and model-building approaches of information technologists, physical scientists, mathematicians, and engineers. It is to be hoped that the collaborations between these disciplines will enable the construction of computer tools and visualization methods to allow experimental biologists to continue to bring their biological insight and intellectual creativity to bear on an increasingly integrative view of the living world.

Acknowledgements

Andy Hayes is thanked for useful discussions and for always locating the missing reference. Work on functional genomics in the authors' laboratory is supported by the European Commission (in the frame of EUROFAN), Biotechnology & Biological Sciences Research Council, Wellcome Trust, the BBSRC/Engineering & Physical Sciences Research Council Bioinformatics Initiative, and by Nycomed-Amersham.

References

1. Goffeau, A., Barrell, B. G., Bussey, H., Davis, R. W., Dujon, B., Feldmann, H., Galibert, F., Hoheisel, J. D., Jacq, C., Johnston, M., Louis, E. J., Mewes, H. W., Murakami, Y., Philippsen, P., Tettelin, H., and Oliver, S. G. (1996) Life with 6000 genes. *Science*, **274**, 546.
2. Wilson, R. K. (1999) How the worm was won. The *C. elegans* genome sequencing project. *Trends Genet.*, **15**, 51.
3. Dunham, I., Shimizu, N., Roe, B. A., Chissoe, S., Hunt, A. R., Collins, J. E., Bruskiewich, R., Beare, D. M., Clamp, M., Smink, L. J., Ainscough, R., Almeida, J. P., Babbage, A., Bagguley, C., Bailey, J., Barlow, K., Bates, K. N., Beasley, O., Bird, C. P., Blakey, S., Bridgeman, A. M., Buck, D., Burgess, J., Burrill, W. D., and O'Brien, K. P. (1999) The DNA sequence of human chromosome 22. *Nature*, **402**, 489.
4. Collins, F. S. (1995) Ahead of schedule and under budget: the Genome Project passes its fifth birthday. *Proc. Natl. Acad. Sci. U.S.A.*, **92**, 10821.
5. Watson, J. D. and Crick, F. H. (1974) Molecular structure of nucleic acids: a structure for deoxyribose nucleic acid. Published in Nature, no. 4356, 25 April 1953. *Nature*, **248**, 765.
6. Collins, F. S., Patrinos, A., Jordan, E., Chakravarti, A., Gesteland, R., and Walters, L. (1998) New goals for the U.S. Human Genome Project: 1998–2003. *Science*, **282**, 682.
7. Mewes, H. W., Albermann, K., Bahr, M., Frishman, D., Gleissner, A., Hani, J., Heumann, K., Kleine, K., Maierl, A., Oliver, S. G., Pfeiffer, F., and Zollner, A. (1997) Overview of the yeast genome. *Nature*, **387**, 7.
8. Dujon, B. (1996) The yeast genome project: what did we learn? *Trends Genet.*, **12**, 263.
9. Blattner, F. R., Plunkett III, G., Bloch, C. A., Perna, N. T., Burland, V., Riley, M., Collado-Vides, J., Glasner, J. D., Rode, C. K., Mayhew, G. F., Gregor, J., Davis, N. W., Kirkpatrick, H. A., Goeden, M. A., Rose, D. J., Mau, B., and Shao, Y. (1997) The complete genome sequence of *Escherichia coli* K-12. *Science*, **277**, 1453.
10. Barry, C., Fichant, G., Kalogeropoulos, A., and Quentin, Y. (1996) A computer filtering method to drive out tiny genes from the yeast genome. *Yeast*, **12**, 1163.
11. Andrade, M. A., Daruvar, A., Casari, G., Schneider, R., Termier, M., and Sander, C. (1997) Characterization of new proteins found by analysis of short open reading frames from the full yeast genome. *Yeast*, **13**, 1363.
12. Termier, M. and Kalogeropoulos, A. (1996) Discrimination between fortuitous and biologically constrained open reading frames in DNA sequences of *Saccharomyces cerevisiae*. *Yeast*, **12**, 369.
13. Oliver, S. G. (1996) From DNA sequence to biological function. *Nature*, **379**, 597.
14. Judd, B. H., Shen, M. W., and Kaufman, T. C. (1972) The anatomy and function of a segment of the X chromosome of *Drosophila melanogaster*. *Genetics*, **71**, 139.
15. Brookfield, J. F. Y. (1997) Genetic redundancy: screening for selection in yeast. *Curr. Biol.*, **7**, R366.

16. Carlson, M., Celenza, J. L., and Eng, F. J. (1985) Evolution of the dispersed *SUC* gene family of *Saccharomyces* by rearrangements of chromosome telomeres. *Mol. Cell. Biol.*, **5**, 2894.
17. Charron, M. J., Read, E., Haut, S. R., and Michels, C. A. (1989) Molecular evolution of the telomere-associated *MAL* loci of *Saccharomyces*. *Genetics*, **122**, 307.
18. Naumov, G., Turakainen, H., Naumova, E., Aho, S., and Korhola, M. (1990) A new family of polymorphic genes in *Saccharomyces cerevisiae*: alpha-galactosidase genes *MEL1–MEL7*. *Mol. Gen. Genet.*, **224**, 119.
19. Teunissen, A. W. and Steensma, H. Y. (1995) Review: the dominant flocculation genes of *Saccharomyces cerevisiae* constitute a new subtelomeric gene family. *Yeast*, **11**, 1001.
20. Andre, B. (1995) An overview of membrane transport proteins in *Saccharomyces cerevisiae*. *Yeast*, **11**, 1575.
21. Lalo, D., Stettler, S., Mariotte, S., Slonimski, P. P., and Thuriaux, P. (1993) Two yeast chromosomes are related by a fossil duplication of their centromeric regions. *C. R. Acad. Sci. III*, **316**, 367.
22. Jia, Y. K., Becam, A. M., and Herbert, C. J. (1997) The *CIT3* gene of *Saccharomyces cerevisiae* encodes a second mitochondrial isoform of citrate synthase. *Mol. Microbiol.*, **24**, 53.
23. Li, Z., Paulovich, A. G., and Woolford, J. L., Jr. (1995) Feedback inhibition of the yeast ribosomal protein gene *CRY2* is mediated by the nucleotide sequence and secondary structure of CRY2 pre-mRNA. *Mol. Cell. Biol.*, **15**, 6454.
24. Delneri, D., Gardner, D. C., and Oliver, S. G. (1999) Analysis of the seven-member *AAD* gene set demonstrates that genetic redundancy in yeast may be more apparent than real. *Genetics*, **153**, 1591.
25. Melnick, L. and Sherman, F. (1993) The gene clusters *ARC* and *COR* on chromosomes 5 and 10, respectively, of *Saccharomyces cerevisiae* share a common ancestry. *J. Mol. Biol.*, **233**, 372.
26. Wicksteed, B. L., Collins, I., Dershowitz, A., Stateva, L. I., Green, R. P., Oliver, S. G., Brown, A. J., and Newlon, C. S. (1994) A physical comparison of chromosome III in six strains of *Saccharomyces cerevisiae*. *Yeast*, **10**, 39.
27. Wolfe, K. H. and Shields, D. C. (1997) Molecular evidence for an ancient duplication of the entire yeast genome. *Nature*, **387**, 708.
28. Ohno, S. (1970) *Evolution by gene duplication*. George Allen and Unwin, London.
29. Keogh, R. S., Seoighe, C., and Wolfe, K. H. (1998) Evolution of gene order and chromosome number in *Saccharomyces*, *Kluyveromyces* and related fungi. *Yeast*, **14**, 443.
30. Hutchison, C. A., Peterson, S. N., Gill, S. R., Cline, R. T., White, O., Fraser, C. M., Smith, H. O., and Venter, J. C. (1999) Global transposon mutagenesis and a minimal *Mycoplasma* genome. *Science*, **286**, 2165.
31. Altschul, S. F., Gish, W., Miller, W., Myers, E. W., and Lipman, D. J. (1990) Basic local alignment search tool. *J. Mol. Biol.*, **215**, 403.
32. Pearson, W. R. and Lipman, D. J. (1988) Improved tools for biological sequence comparison. *Proc. Natl. Acad. Sci. U.S.A.*, **85**, 2444.
33. Page, R. D. (1996) TreeView: an application to display phylogenetic trees on personal computers. *Comput. Appl. Biosci.*, **12**, 357.
34. Dolinski, K., Ball, C. A., Chervitz, S. A., Dwight, S. S., Harris, M. A., Roberts, S., Roe, T., Cherry, J. M., and Botstein, D. (1998) Expanding yeast knowledge online. *Yeast*, **14**, 1453.
35. van Helden, J., André, B., and Collado-Vides, J. (1998) Extracting regulatory sites from the upstream region of yeast genes by computational analysis of oligonucleotide frequencies. *J. Mol. Biol.*, **281**, 827.

36. van Helden, J., André, B., and Collado-Vides, J. (2000) A web site for the computational analysis of yeast regulatory sequences. *Yeast*, **16**, 177.

37. Bork, P., Ouzounis, C., and Sander, C. (1994) From genome sequences to protein function. *Curr. Opin. Struct. Biol.*, **4**, 393.

38. Oliver, S. G. (1996) From DNA sequence to biological function: the new 'Voyage of the Beagle'. *Biochem. Soc. Trans.*, **24**, 291.

39. Delneri, D., Gardner, D. C., Bruschi, C. V., and Oliver, S. G. (1999) Disruption of seven hypothetical aryl alcohol dehydrogenase genes from *Saccharomyces cerevisiae* and construction of a multiple knock-out strain. *Yeast*, **15**, 1681.

40. Kersten, P. J., Tien, M., Kalyanaraman, B., and Kirk, T. K. (1985) The ligninase of *Phanerochaete chrysosporium* generates cation radicals from methoxybenzenes. *J. Biol. Chem.*, **260**, 2609.

41. Delneri, D., Degrassi, G., Rizzo, R., and Bruschi, C. V. (1995) Degradation of *trans*-ferulic and *p*-coumaric acid by *Acinetobacter calcoaceticus* DSM 586. *Biochim. Biophys. Acta*, **1244**, 363.

42. Vicuna, R. (1988) Bacterial degradation of lignin. *Enzyme Microb. Technol.*, **10**, 646.

43. Leong-Morgenthaler, P., Oliver, S. G., Hottinger, H., and Soll, D. (1994) A *Lactobacillus* nifS-like gene suppresses an *Escherichia coli* transaminase B mutation. *Biochimie*, **76**, 45.

44. Oliver, S. G., Winson, M. K., Kell, D. B., and Baganz, F. (1998) Systematic functional analysis of the yeast genome. *Trends Biotechnol.*, **16**, 373.

45. Baudin, A., Ozier-Kalogeropoulos, O., Denouel, A., Lacroute, F., and Cullin, C. (1993) A simple and efficient method for direct gene deletion in *Saccharomyces cerevisiae*. *Nucleic Acids Res.*, **21**, 3329.

46. Wach, A., Brachat, A., Pohlmann, R., and Philippsen, P. (1994) New heterologous modules for classical or PCR-based gene disruptions in *Saccharomyces cerevisiae*. *Yeast*, **10**, 1793.

47. Oka, A., Sugisaki, H., and Takanami, M. (1981) Nucleotide sequence of the kanamycin resistance transposon Tn903. *J. Mol. Biol.*, **147**, 217.

48. Jimenez, A. and Davies, J. (1980) Expression of a transposable antibiotic resistance element in *Saccharomyces*. *Nature*, **287**, 869.

49. Baganz, F., Hayes, A., Marren, D., Gardner, D. J. C., and Oliver, S. G. (1997) Suitability of replacement markers for functional analysis studies in *Saccharomyces cerevisiae*. *Yeast*, **13**, 1563.

50. Hutter, A. and Oliver, S. G. (1998) Ethanol production using nuclear petite yeast mutants. *Appl. Microbiol. Biotechnol.*, **49**, 511.

51. Amberg, D. C., Botstein, D., and Beasley, E. M. (1995) Precise gene disruption in *Saccharomyces cerevisiae* by double fusion polymerase chain reaction. *Yeast*, **11**, 1275.

52. Storici, F., Coglievina, M., and Bruschi, C. V. (1999) A 2-micron DNA-based marker recycling system for multiple gene disruption in the yeast *Saccharomyces cerevisiae*. *Yeast*, **15**, 271.

53. Fairhead, C., Llorente, B., Denis, F., Soler, M., and Dujon, B. (1996) New vectors for combinatorial deletions in yeast chromosomes and for gap-repair cloning using 'split-marker' recombination. *Yeast*, **12**, 1439.

54. Guldener, U., Heck, S., Fielder, T., Beinhauer, J., and Hegemann, J. H. (1996) A new efficient gene disruption cassette for repeated use in budding yeast. *Nucleic Acids Res.*, **24**, 2519.

55. Delmeri, D., Tomlin, G. C., Wilson, J. L., Hutter, A., Sefton, M., Louis, E. J., Olivet, S. G. (2000) Exploring redundancy in the yeast genome: an improved strategy for use of the ore/loxP System, *Gene*, **252**, 127–135.

56. Hashida-Okado, T., Ogawa, A., Endo, M., Yasumoto, R., Takesako, K., and Kato, I. (1996) *AUR1*, a novel gene conferring aureobasidin resistance on *Saccharomyces cerevisiae*: a study of defective morphologies in Aur1p-depleted cells. *Mol. Gen. Genet.*, **251**, 236.

57. Goldstein, A. L. and McCusker, J. H. (1999) Three new dominant drug resistance cassettes for gene disruption in *Saccharomyces cerevisiae*. *Yeast*, **15**, 1541.

58. Shoemaker, D. D., Lashkari, D. A., Morris, D., Mittmann, M., and Davis, R. W. (1996) Quantitative phenotypic analysis of yeast deletion mutants using a highly parallel molecular bar-coding strategy. *Nat. Genet.*, **14**, 450.

59. Schena, M., Shalon, D., Davis, R. W., and Brown, P. O. (1995) Quantitative monitoring of gene expression patterns with a complementary DNA microarray. *Science*, **270**, 467.

60. Winzeler, E. A., Shoemaker, D. D., Astromoff, A., Liang, H., Anderson, K., Andre, B., Bangham, R., Benito, R., Boeke, J. D., Bussey, H., Chu, A. M., Connelly, C., Davis, K., Dietrich, F., Dow, S. W., El Bakkoury, M., Foury, F., Friend, S. H., Gentalen, E., Giaever, G., Hegemann, J. H., Jones, T., Laub, M., Liao, H., and Davis, R. W. (1999) Functional characterization of the *S. cerevisiae* genome by gene deletion and parallel analysis. *Science*, **285**, 901.

61. Ross-Macdonald, P., Coelho, P. S., Roemer, T., Agarwal, S., Kumar, A., Jansen, R., Cheung, K. H., Sheehan, A., Symoniatis, D., Umansky, L., Heidtman, M., Nelson, F. K., Iwasaki, H., Hager, K., Gerstein, M., Miller, P., Roeder, G. S., and Snyder, M. (1999) Large-scale analysis of the yeast genome by transposon tagging and gene disruption. *Nature*, **402**, 413.

62. Brand, M. D. (1996) Top down metabolic control analysis. *J. Theor. Biol.*, **182**, 351.

63. Danhash, N., Gardner, D. J. C., and Oliver, S. G. (1991) Heritable damage to yeast caused by transformation. *Biotechnology (N. Y.)*, **9**, 179.

64. Baganz, F., Hayes, A., Farquhar, R., Butler, P. R., Gardner, D. C., and Oliver, S. G. (1998) Quantitative analysis of yeast gene function using competition experiments in continuous culture. *Yeast*, **14**, 1417.

65. Giaever, G., Shoemaker, D. D., Jones, T. W., Liang, H., Winzeler, E. A., Astromoff, A., and Davis, R. W. (1999) Genomic profiling of drug sensitivities via induced haplo-insufficiency. *Nat. Genet.*, **21**, 278.

66. Oliver, S. G. (1999) Redundancy reveals drugs in action. *Nat. Genet.*, **21**, 245.

67. Smith, V., Botstein, D., and Brown, P. O. (1995) Genetic footprinting: a genomic strategy for determining a gene's function given its sequence. *Proc. Natl. Acad. Sci. U.S.A.*, **92**, 6479.

68. Smith, V., Chou, K. N., Lashkari, D., Botstein, D., and Brown, P. O. (1996) Functional analysis of the genes of yeast chromosome V by genetic footprinting. *Science*, **274**, 2069.

69. Teusink, B., Baganz, F., Westerhoff, H. V., and Oliver, S. G. (1998) Metabolic control analysis as a tool in the elucidation of the function of novel genes. *Methods Microbiol.*, **26**, 297.

70. Rohwer, J. M., Schuster, S., and Westerhoff, H. V. (1996) How to recognize mono-functional units in a metabolic system. *J. Theor. Biol.*, **179**, 213.

71. Oliver, S. G. (1997) Yeast as a navigational aid in genome analysis. 1996 Kathleen Barton-Wright Memorial Lecture. *Microbiology*, **143**, 1483.

72. Naumann, D., Shultz, C., and Helm, D. (1996) What can infrared spectroscopy tell us about the structure and composition of intact bacterial cells? In *Infrared spectroscopy of biomolecules* (ed H. H. Mantsch and D. Chapmann), p. 279. Wiley-Liss, New York.

73. Goodacre, R., Timmins, E. M., Rooney, P. J., Rowland, J. J., and Kell, D. B. (1996) Rapid identification of *Streptococcus* and *Enterococcus* species using diffuse reflectance–absorbance Fourier transform infrared spectroscopy and artificial neural networks. *FEMS Microbiol. Lett.*, **140**, 233.

74. Rashed, M. S., Ozand, P. T., Bucknall, M. P., and Little, D. (1995) Diagnosis of inborn errors of metabolism from blood spots by acylcarnitines and amino acids profiling using automated electrospray tandem mass spectrometry. *Pediatr. Res.*, **38**, 324.

75. Brindle, K. M., Fulton, S. M., Gillham, H., and Williams, S. P. (1997) Studies of metabolic control using NMR and molecular genetics. *J. Mol. Recog.*, **10**, 182.

76. Hedrick, S. M., Cohen, D. I., Nielsen, E. A., and Davis, M. M. (1984) Isolation of cDNA clones encoding T cell-specific membrane-associated proteins. *Nature*, **308**, 149.

77. Liang, P. and Pardee, A. B. (1992) Differential display of eukaryotic messenger RNA by means of the polymerase chain reaction. *Science*, **257**, 967.

78. Velculescu, V. E., Zhang, L., Vogelstein, B., and Kinzler, K. W. (1995) Serial analysis of gene expression. *Science*, **270**, 484.

79. Velculescu, V. E., Zhang, L., Zhou, W., Vogelstein, J., Basrai, M. A., Bassett, D. E., Jr., Hieter, P., Vogelstein, B., and Kinzler, K. W. (1997) Characterization of the yeast transcriptome. *Cell*, **88**, 243.

80. Lockhart, D. J., Dong, H., Byrne, M. C., Follettie, M. T., Gallo, M. V., Chee, M. S., Mittmann, M., Wang, C., Kobayashi, M., Horton, H., and Brown, E. L. (1996) Expression monitoring by hybridization to high-density oligonucleotide arrays. *Nat. Biotechnol.*, **14**, 1675.

81. Wodicka, L., Dong, H., Mittmann, M., Ho, M. H., and Lockhart, D. J. (1997) Genome-wide expression monitoring in *Saccharomyces cerevisiae*. *Nat. Biotechnol.*, **15**, 1359.

82. Schena, M., Shalon, D., Heller, R., Chai, A., Brown, P. O., and Davis, R. W. (1996) Parallel human genome analysis: microarray-based expression monitoring of 1000 genes. *Proc. Natl. Acad. Sci. U.S.A.*, **93**, 10614.

83. Hauser, N. C., Vingron, M., Scheideler, M., Krems, B., Hellmuth, K., Entian, K. D., and Hoheisel, J. D. (1998) Transcriptional profiling on all open reading frames of *Saccharomyces cerevisiae*. *Yeast*, **14**, 1209.

84. DeRisi, J. L., Iyer, V. R., and Brown, P. O. (1997) Exploring the metabolic and genetic control of gene expression on a genomic scale. *Science*, **278**, 680.

85. Spellman, P. T., Sherlock, G., Zhang, M. Q., Iyer, V. R., Anders, K., Eisen, M. B., Brown, P. O., Botstein, D., and Futcher, B. (1998) Comprehensive identification of cell cycle-regulated genes of the yeast *Saccharomyces cerevisiae* by microarray hybridization. *Mol. Biol. Cell*, **9**, 3273.

86. Chu, S., DeRisi, J., Eisen, M., Mulholland, J., Botstein, D., Brown, P. O., and Herskowitz, I. (1998) The transcriptional program of sporulation in budding yeast. *Science*, **282**, 699.

87. Eisen, M. B., Spellman, P. T., Brown, P. O., and Botstein, D. (1998) Cluster analysis and display of genome-wide expression patterns. *Proc. Natl. Acad. Sci. U.S.A.*, **95**, 14863.

88. Planta, R. J., Brown, A. J., Cadahia, J. L., Cerdan, M. E., de Jonge, M., Gent, M. E., Hayes, A., Kolen, C. P., Lombardia, L. J., Sefton, M., Oliver, S. G., Thevelein, J., Tournu, H., van Delft, Y. J., Verbart, D. J., and Winderickx, J. (1999) Transcript analysis of 250 novel yeast genes from chromosome XIV. *Yeast*, **15**, 329.

89. Wilkins, M. R., Pasquali, C., Appel, R. D., Ou, K., Golaz, O., Sanchez, J. C., Yan, J. X., Gooley, A. A., Hughes, G., Humphery-Smith, I., Williams, K. L., and Hochstrasser, D. F. (1996) From proteins to proteomes: large scale protein identification by two-dimensional electrophoresis and amino acid analysis. *Biotechnology (N. Y.)*, **14**, 61.

90. Yates, J. R., III (1998) Mass spectrometry and the age of the proteome. *J. Mass Spectrom.*, **33**, 1.

91. Shevchenko, A., Jensen, O. N., Podtelejnikov, A. V., Sagliocco, F., Wilm, M., Vorm, O., Mortensen, P., Shevchenko, A., Boucherie, H., and Mann, M. (1996) Linking genome and proteome by mass spectrometry: large-scale identification of yeast proteins from two dimensional gels. *Proc. Natl. Acad. Sci. U.S.A.*, **93**, 14440.

92. Sagliocco, F., Guillemot, J. C., Monribot, C., Capdevielle, J., Perrot, M., Ferran, E., Ferrara, P., and Boucherie, H. (1996) Identification of proteins of the yeast protein map using genetically manipulated strains and peptide-mass fingerprinting. *Yeast*, **12**, 1519.

93. Link, A. J., Eng, J., Schieltz, D. M., Carmack, E., Mize, G. J., Morris, D. R., Garvik, B. M., and Yates, J. R., III (1999) Direct analysis of protein complexes using mass spectrometry. *Nat. Biotechnol.*, **17**, 676.

94. Fromont-Racine, M., Rain, J. C., and Legrain, P. (1997) Toward a functional analysis of the yeast genome through exhaustive two-hybrid screens. *Nat. Genet.*, **16**, 277.

95. Oliver, S. (2000) Guilt-by-association goes global. *Nature*, **403**, 601.

96. Flores, A., Briand, J. F., Gadal, O., Andrau, J. C., Rubbi, L., Van, M., V, Boschiero, C., Goussot, M., Marck, C., Carles, C., Thuriaux, P., Sentenac, A., and Werner, M. (1999) A protein–protein interaction map of yeast RNA polymerase III. *Proc. Natl. Acad. Sci. U.S.A.*, **96**, 7815.

97. Mayes, A. E., Verdone, L., Legrain, P., and Beggs, J. D. (1999) Characterization of Sm-like proteins in yeast and their association with U6 snRNA. *EMBO J.*, **18**, 4321.

98. Bartel, P. L., Roecklein, J. A., SenGupta, D., and Fields, S. (1996) A protein linkage map of *Escherichia coli* bacteriophage T7. *Nat. Genet.*, **12**, 72.

99. Uetz, P., Giot, L., Cagney, G., Mansfield, T. A., Judson, R. S., Knight, J.R., Lockshon, D., Narayan, V., Srinivasan, M., Pochart, P., QureshiEmili, A., Li, Y., Godwin, B., Conover, D., Kalbfleisch, T., Vijayadamodar, G., Yang, M. J., Johnston, M., Fields, S., Rothberg, J. M. (2000) Protein–protein interactions in *Saccharomyces cerevisiae*. *Nature* , **403**, 623.

100. Gygi, S. P., Rochon, Y., Franza, B. R., and Aebersold, R. (1999) Correlation between protein and mRNA abundance in yeast. *Mol. Cell. Biol.*, **19**, 1720.

101. Futcher, B., Latter, G. I., Monardo, P., McLaughlin, C. S., and Garrels, J. I. (1999) A sampling of the yeast proteome. *Mol. Cell. Biol.*, **19**, 7357.

2 | Chromosomal DNA replication in yeast: enzymes and mechanisms

STUART A. MacNEILL and PETER M. J. BURGERS

1. Introduction

Chromosomal DNA replication in eukaryotic cells is a remarkable process requiring the complex interplay of a variety of distinct essential factors in a spatially and temporally coordinated manner. Each replication fork supports a large multiprotein complex containing as many as 30 to 40 proteins, the vast majority of which are essential for the successful duplication of the genome. In the yeasts *Saccharomyces cerevisiae* and *Schizosaccharomyces pombe*, the molecular basis of the replication process has been studied in great detail. Indeed much of our current understanding of replication in higher eukaryotes is founded on work performed on these two distantly related ascomycetes. Each has a genome size of approximately 15 Mb and a doubling time in the range 110–240 min depending on strain characteristics and the precise growth conditions employed. S phase occupies approximately 10–15% of the cell cycle, or 15–20 min at optimal growth rates.

In both yeasts, components of the replication apparatus have been identified by a variety of approaches. Genetic analysis of the original collections of cell cycle mutants in both organisms (1–5) led to the identification of a series of mutants with specific defects in the S and G2 phases of the cell cycle, suggestive of possible defects in DNA replication. On further analysis, a subset of mutants independently isolated as being sensitive to DNA-damaging agents also showed replication defects (for examples see references 6–8). Subsequent cloning and sequencing of the corresponding wild-type genes led to the identification of many key replication factors. For instance, genes encoding the catalytic subunits of all three essential replicative DNA polymerases in *S. pombe* were identified in this manner (9–12), although not before two of the three genes had been identified by other methods (13–15). In addition to these genetic approaches, a number of essential proteins have been identified as a result of biochemical purification of replication enzymes or as novel factors physically associating with previously identified proteins. In many cases the determination of partial protein sequences by peptide sequencing and the subsequent identification of the corresponding gene sequence has been followed by reverse genetic analysis of *in vivo* function, particularly through the generation of conditional lethal or null alleles. For example, analysis of the function of the non-catalytic subunits of DNA polymerase δ (Pol δ) in *S. cerevisiae* has followed this pattern (16, 17). More recently, cloning by homology and by whole-genome sequence analysis has led to the identification of additional replication genes.

As a result of the multiplicity of approaches that can be taken to identify new replication factors, individual genes have often been assigned a variety of different names by different workers. For instance, the *S. cerevisiae* gene encoding the FEN-1 exonuclease (see section 6 for a discussion of the function of this enzyme) has been designated *RAD27*, indicating that *rad27* mutants are sensitive to DNA-damaging agents (18), *RTH1*, indicating homology to *RAD2* (*RTH* is an acronym for *RAD two homologue*, see section 6) (19), and *YKL510*, indicating open reading frame 510 on the left arm of chromosome 12 (20). In this chapter we have tried wherever possible to use the most commonly used gene and protein names, or those that we believe, for whatever reason, will be the most widely used in the future. Where alternative designations are given, these are indicated as, for example, *POL2/CDC17*.

In this chapter we review our current state of knowledge of the molecular machinery of DNA replication in *S. cerevisiae* and *S. pombe* based on genetic and biochemical studies of the two yeasts. We begin with a brief overview of the principles underlying the formation of the replication fork, the complex molecular assembly at the heart of the replication apparatus. This is followed with a detailed discussion of different stages of the replication process, from origin recognition through to the final processing of Okazaki fragments.

2. Biogenesis of the replication fork: an overview

The mechanism of initiation of chromosomal DNA replication in eukaryotic cells is believed to conform to the model originally proposed in the 1960s for replication of the bacterial chromosome (21), reviewed in references 22 and 23. In this model, a DNA sequence element termed a replicator was proposed to be the primary determinant of DNA sequence-specific replication initiation. The replicator was proposed to be a binding site for a specific initiator protein that in turn acted as a landing pad for other replication proteins, such as the helicase required to unwind the DNA ahead of the moving replication fork (21). Biochemical studies on model DNA replication systems, such as bacteriophage T4 or *Simian virus 40* (SV40), have subsequently shown that replication initiation also involves the partial DNA unwinding of replication origin DNA mediated by the initiator protein, a process that is facilitated by local destabilization of the DNA and by the presence of a single-stranded DNA-binding protein (22). In yeasts the initiator protein is believed to be the origin recognition complex (ORC) (24). The ORC binds to chromosomal replication origins (replicator sequences) and is required for origin firing (see section 3 below) but whether the ORC promotes local DNA unwinding remains to be seen. The eukaryotic single-stranded DNA-binding protein is replication protein A (RP-A) (25).

The next step in origin activation involves expansion of the unwound DNA at the replication origin (sometimes termed the replication bubble) by the replicative DNA helicase (22). The helicase must first be loaded on to the DNA, either by the action of accessory proteins or by virtue of its interactions with the initiator complex. In some prokaryotic and eukaryotic viral systems the initiator and helicase activities are present in the same protein, but this is not universal. A likely candidate for the eukaryotic chromosomal replicative helicase is the MCM (mini-chromosome maintenance) protein complex (26) (see section 3 below); association of the MCM complex with origin DNA is dependent upon ORC, and a DNA helicase activity associated with a sub-complex of human MCM proteins has been reported (27). The helicase responsible for initial expansion of the replication bubble probably also remains as an active component of the replication fork during the highly processive synthesis of non-origin DNA. This has been shown to be the case for the MCM complex (28, 29).

Following unwinding of the origin DNA, a complex referred to as the primosome synthesizes short RNA–DNA primers on either side of the expanded replication bubble. These mark the starting points of the two leading strands. In eukaryotic cells primer synthesis is catalysed by the Pol α–primase complex. The primers are then elongated by a holoenzyme complex consisting of a replicative polymerase (Pol δ or Pol ε; see section 5 below), a sliding clamp (proliferating cell nuclear antigen; PCNA) and a clamp loader, replication factor C (RF-C). The sliding clamp is a toroidal molecule that encircles the DNA (30), thereby acting to tether the moving polymerase.

The switch from the primosomal polymerase to the replicative polymerase occurs on the leading strand after synthesis of the initial primer at the replication origin, and also in the synthesis of each Okazaki fragment on the lagging strand. Thus, the Pol δ and/or Pol ε holoenzymes carry out efficient continuous elongation of the leading strand, whereas Pol α and either the Pol δ or Pol ε holoenzymes cycle in primer synthesis and elongation of Okazaki fragments on the lagging strand. The final processing of Okazaki fragments is achieved through the action of multiple enzymes, most likely including another DNA helicase, the Dna2 protein (see section 6). Ligation of processed Okazaki fragments, formally the final step of the replication process, is carried out by DNA ligase I.

Model replication studies have shown that the leading and lagging strands of the fork are replicated coordinately (22). This coordination is likely to be achieved by interactions between the leading and lagging strand replication machineries. To explain coordinate replication of both strands in the bacteriophage T4 system, a model was proposed in which looping of the lagging strand allowed both strands to be replicated in the same direction, allowing protein–protein interactions between the leading and lagging strand complexes to be maintained during synthesis (31). Recent electron micrographic studies of the bacteriophage T7 replication fork (32) are consistent with this model, which is illustrated in Fig. 1.

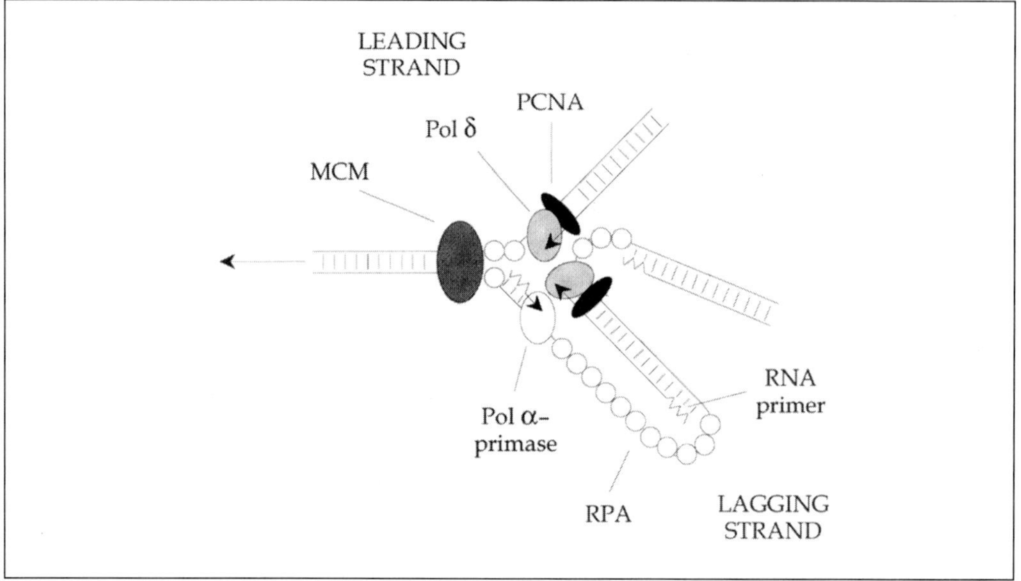

Fig. 1 Schematic representation of the eukaryotic replication fork. The DNA ahead of the fork is believed to be unwound by the MCM complex, a putative DNA helicase, most likely in association with Cdc45p (not shown). Unwound single-stranded DNA is coated with RP-A. Synthesis of the leading strand is carried out by Pol δ in association with the toroidal processivity factor PCNA. PCNA is loaded on to the DNA by the clamp loader RF-C (not shown). For lagging strand (Okazaki fragment) synthesis, Pol α–primase and Pol δ–PCNA complexes cycle: Pol α-primase synthesises a short RNA-DNA primer to initiate each Okazaki fragment, while Pol δ synthesizes the remainder of the Okazaki fragment. For clarity the enzymes required for processing of the Okazaki fragment have been omitted (see Fig. 9 and section 6). Coordinated synthesis of the leading and lagging strands is thought to be achieved by looping of the lagging-strand DNA. The direction of fork movement is indicated by the arrow. See text for details and references.

3. Replication origin structure, recognition, and activation

3.1 Overview

Replication of the large genomes of eukaryotic cells is made possible by the use of multiple bidirectional replication origins on each chromosome. In *S. cerevisiae* it is estimated that there is an active replication origin every 20–40 kb, making a total of approximately 500 active origins per genome. In this section we review what is known of the structure and function of replication origins in *S. cerevisiae* and *S. pombe* and of the specific protein complexes that bind either directly or indirectly to them to bring about the initiation of bidirectional replication.

3.2 Origins of chromosomal replication

Yeast replication origins were first identified as sequences that conferred on plasmid DNA the ability to replicate extrachromosomally (33–35). For this reason they were named autonomously replicating sequences (ARSs). Subsequent physical analysis using two-dimensional gel electrophoresis methods demonstrated that many but not all of the ARS sequences identified in *S. cerevisiae* were indeed chromosomal replication origins (36, 37), reviewed in reference 38, although it was not until very recently that the origin of bidirectional replication associated with an ARS element was actually shown to lie within the ARS itself (39).

Studies on *S. cerevisiae* have shown that whenever origins are removed from their usual chromosomal location and reinserted elsewhere, origin activity at the original location is lost and is frequently established at the new location, although the efficiency and timing of origin firing at the new location can vary greatly (for examples see references 40–44). This indicates that replication origin sequences are both necessary and sufficient for localized replication initiation *in vivo* but also shows that sequences surrounding the origin can exert an influence on origin function (38).

Typically, deletion of an *S. cerevisiae* replication origin from its authentic chromosomal location does not greatly affect the ability of the surrounding sequences to be efficiently replicated. In one study (45), all the known replication origins within a 200-kb region of *S. cerevisiae* chromosome III were deleted, with only minimal effects on the stability of the chromosome. This result suggests that there are many more replication origins present on the chromosome than are required to replicate the genome under normal circumstances, and that many origins are cryptic, becoming active only when the function of a nearby active origin is impaired by deletion or mutation. Indeed the presence of a strong or early firing origin generally suppresses the activity of a nearby weak or late firing origin. The mechanism underlying this phenomenon is not understood (43, 44).

Although there are significant differences in replication origin structure between the two yeasts, they share common features. *S. cerevisiae* origins are generally 150–200 bp in length and comprise two distinct domains, termed A and B (46), both of which are essential for origin function. In *S. pombe*, the minimum origin size is considerably larger than in *S. cerevisiae* (650–850 bp in two well studied examples), and origins appear to consist of a greater number of modular elements, each of which is stimulatory rather than essential for origin activity (47, 48).

Figure 2.2A shows the arrangement of the A and B elements at *ARS1*, the best studied origin in *S. cerevisiae*. Domain A contains a conserved, though rather degenerate, 11-bp sequence 5'-(A/T)TTTA(T/C)(A/G)TTT(A/T)-3' known as the ACS (ARS consensus sequence, shown in Fig. 2), deletion of which abolishes origin firing at chromosomal replication origins. In contrast, elimination of the B1, B2, or B3 domains reduces origin function but does not abolish it (46, 49–51). The B3 element is a binding site for the general transcription factor Abf1p (52); *abf1*Δ strains display an increased frequency of plasmid loss, suggesting that Abf1p binding to the B3 element may contribute to ARS function (53), reviewed in reference 38. Another conserved feature of yeast chromosomal replication origins is a DNA-unwinding element (DUE), located within the B2 domain (54–56). It is believed that the A-T-rich DUE element promotes opening of the duplex DNA at replication initiation.

Recently, a novel method has been developed to map precisely the sites of rep-

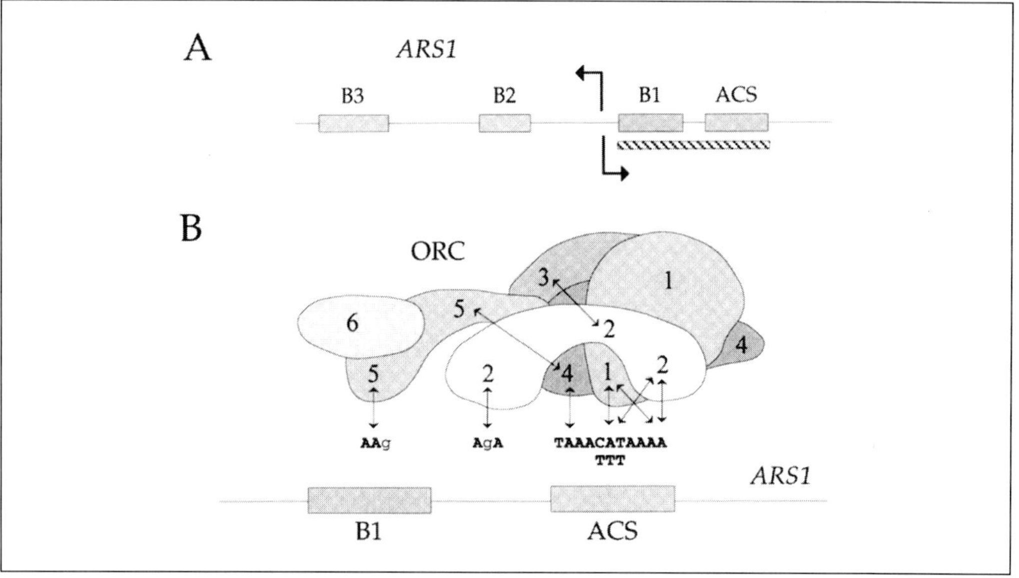

Fig. 2 (A) Structure of the *S. cerevisiae* replication origin *ARS1* showing the conserved ARS. B1, B2 and B3 regions and the origin of bidirectional replication determined by RIP mapping (39). (B) Interactions between ORC proteins (numbered 1–6) and *ARS1*. Residues of *ARS1* whose modification interferes with ORC binding are shown. Matches to the ACS consensus are shown in bold, as are four residues located outside the ACS (in the B1 element and in the spacer between B1 and the ACS) that are conserved in some but not all origins. Redrawn with minor modifications from reference 67.

lication initiation at origin sequences. This method (replication initiation point mapping, or RIP mapping) has been applied to both *S. cerevisiae* and *S. pombe* replication origins in their correct chromosomal context to demonstrate that leading-strand replication initiates at single distinct sites on each strand (39, 57, 58). In the case of *S. cerevisiae ARS1*, the initiation sites on the top and bottom strands, which are located between the B1 and B2 elements, are 1 bp apart, so that the origin of bidirectional replication (OBR) spans only 2 bp (Fig. 2A). As noted above, the B2 element has been shown to a be stretch of easily unwound DNA (54–56). It remains to be seen whether single initiation sites will be a feature of all yeast chromosomal replication origins.

The use of two-dimensional gel technologies to analyse replication origin firing in populations of cells undergoing synchronous passage through S phase has also allowed the identification of origins that are predominantly early or late firing (40, 42, 43, 59, 60). Exactly how the timing of origin firing is regulated remains unclear, as early and late origins do not differ in the timing of their association with origin-binding proteins (see below). Instead, chromosomal context plays an important role. For example, if *ARS1*, which normally fires early in S phase, is moved near to a late-firing origin, *ARS1* itself becomes late firing, and also somewhat less efficient (42).

3.3 Origin recognition complex

A breakthrough in the understanding of replication origin function came with the identification and subsequent purification of the origin-binding complex (ORC) from *S. cerevisiae* (24). ORC is a six-subunit protein complex that binds specifically to origin DNA and which is required for replication origin function both *in vivo* and *in vitro* (24, 61, 62). ORC appears to be a conserved component of the eukaryotic replication machinery, as ORC-related proteins have been identified in a range of eukaryotic species, including *S. pombe*, and in certain cases ORC complexes have been purified (for example, see reference 63). In addition to its essential role in DNA replication in *S. cerevisiae*, ORC is also involved in the process of transcriptional silencing of genes at the mating type locus (reviewed in reference 64).

S. cerevisiae ORC appears to be bound to origins of replication throughout most, if not all, of the cell cycle (61) and is thought to act as a landing pad for Cdc6p, which in turn is required for loading of the MCM protein complex on to DNA. The functions of Cdc6p and the MCM proteins are discussed in detail below (see sections 3.4 and 3.5). ORC binds to two separate elements within replication origins: the ACS in domain A and the B1 element in domain B (see Fig. 2). Mutations in the ACS or the B1 element that reduce origin function *in vivo* also reduce ORC DNA binding *in vitro* and *in vivo* (24, 65, 66). A recent protein–DNA crosslinking study (67) has shown that ORC binds preferentially to one strand of the origin DNA at *ARS1* close to, but not overlapping with, the origin of bidirectional replication defined by RIP mapping (discussed above; Fig. 2B) (39, 57).

The six subunits of ORC in *S. cerevisiae* are the Orc1p–Orc6p proteins, encoded by *ORC1–ORC6* (68–73) (Table 1). Cells carrying temperature-sensitive mutants in *ORC2* initially arrest with unreplicated DNA following shift to the restrictive temperature. This is followed by a slow progression through S phase accompanied by a marked drop in cell viability, indicating that passage through S phase in the absence of a fully functional ORC complex is highly detrimental to the cells. Similar phenotypes have been reported for cells carrying temperature-sensitive mutations in *ORC1* (Aparicio and Bell, unpublished results, cited in reference 74) and *ORC5*, although in the latter case the mutant *orc5* cells arrest with largely replicated DNA (70). Analysis of the frequency of replication initiation at defined chromosomal origins shows that initiation is reduced in *orc2* and *orc5* mutants, although the extent of the reduction appears to depend on the origin tested. In addition, *S. cerevisiae* plasmids that have only a single origin of replication, and which are therefore particularly sensitive to defects in the initiation of replication, show a marked decrease in stability in *orc2* and *orc5* mutants at the permissive temperature. This instability can be reversed by addition of extra replication origins to the plasmids. The added origins presumably provide extra opportunities for successful initiation (70, 73).

In *S. pombe*, genes encoding four putative ORC subunits have been characterised: *orc1$^+$/cdc30$^+$* (75–77), *orc2$^+$* (78), *orc4$^+$* (79), and *orc5$^+$* (80) (Table 1). Cells carrying temperature-sensitive mutations in *orc1$^+$* arrest with unreplicated DNA at the restrictive temperature (81). One striking difference between the ORC proteins of the

Table 1 Proteins at the replication fork: replication origin recognition and activation[a]

	S. cerevisiae	*S. pombe*	**Comments**
ORC	Orc1p	Orc1/Orp1/Cdc30	Binds and hydrolyses ATP
	Orc2p	Orc2/Orp2	
	Orc3p	Orc3	
	Orc4p	Orc4	Binds to DNA via AT-hook motifs
	Orc5p	Orc5	ATP binding
	Orc6p	Orc6	
	Cdc6p	Cdc18	ATP binding
	Cdc7p	Hsk1	Catalytic subunit of Cdc7–Dbf4 protein kinase
	Dbf4p	Dfp1/Him1/Rad35	Regulatory subunit of Cdc7–Dbf4 protein kinase
MCM	Mcm2p	Nda1/Cdc19	
	Mcm3p	Mcm3	
	Mcm4p/Cdc54p	Cdc21	
	Mcm5p/Cdc46p	Nda4	
	Mcm6p	Mis5	
	Mcm7p/Cdc47p		
	Cdc45p	Snu41	Cdc45 physically associates with the MCM proteins, possibly acting as a loading factor for Pol α–primase
	Mcm10p/Dna43p	Cdc23	Essential factor implicated in fork movement

[a] See text and Reference 74 for references.

two yeasts is seen with *S. cerevisiae* Orc4p and *S. pombe* Orc4. The *S. pombe* protein is larger than its *S. cerevisiae* counterpart, having a extended N-terminal domain consisting of nine copies of the A-T hook motif found in a number of DNA-binding proteins, including the HMG-I(Y) family of chromatin proteins. Structural studies have shown that A-T hooks are responsible for binding the minor groove of A-T-rich DNA. As *S. pombe* origins have extended A-T-rich regions, the hooks may serve to stabilize ORC binding to the origin (79).

With the exception of Orc6p, each subunit of *S. cerevisiae* ORC is required for origin DNA binding, yet none of the subunits is capable of sequence-specific DNA recognition on its own (67). Instead, sequence-specific DNA binding of origin DNA requires the participation of several subunits of the complex. Protein–DNA crosslinking has shown that the Orc2p, Orc3p, and Orc5p proteins are located close to a region of *ARS1* between the ACS and the B1 element, whilst Orc1p, Orc4p, and Orc6p are located close to the DNA within a region of *ARS1* that is relatively insensitive to nuclease cleavage (67) (see Fig. 2B).

Stable DNA binding by purified ORC also requires the presence of adenosine triphosphate (ATP) (24). At least two of the six subunits of *S. cerevisiae* ORC (Orc1p and Orc5p) are able to bind ATP; Orc1p also hydrolyses ATP (82). Interestingly ATP and origin binding by ORC have been shown to be interdependent, that is, ATP binding by Orc1p requires sequence-specific DNA binding and sequence-specific DNA binding requires ATP binding by Orc1p. However, although ATP binding is essential for ORC DNA recognition, ATP hydrolysis is not, leading to the suggestion that ATP acts as a co-factor to lock ORC on the origin DNA (82). This raises the intriguing possibility that regulated ATP hydrolysis may drive transitions between different functional states of ORC, but whether this will prove to be the case awaits further investigation (discussed in Reference 82). As noted above, ORC appears to remain bound to origin DNA throughout the cell cycle and does not appear to move with the replication fork from origin to non-origin DNA (29). This raises interesting questions about how ORC is duplicated following replication of the origin sequences.

3.4 Cdc6p

Although ORC appears to remain bound to origin DNA throughout the cell cycle in *S. cerevisiae*, genomic footprinting of replication origins has revealed that the origin chromatin exists in at least two distinct states during the cell cycle. In G1, replication origins are said to exist in a prereplicative state, while from S phase until the end of mitosis they exist in a postreplicative state. Several components of the prereplicative complex (pre-RC) have been identified, including ORC, the Cdc6p and Cdc45p proteins, and members of the MCM protein family discussed in the following section (83). That all these proteins are required for replication initiation suggests that the pre-RC itself is an important structure.

The *S. cerevisiae* Cdc6p protein (a homologue of Cdc18 in *S. pombe*) is an essential ORC-interacting factor. Cells carrying mutations in the *CDC6* gene display a range of phenotypes that are consistent with an impairment of DNA replication, such as a decrease in plasmid stability that can be suppressed by addition of extra replication origins (84). *CDC6* interacts genetically with ORC in a number of ways; for instance, overexpression of *CDC6* can rescue *orc5* mutant cells (85), *cdc6* mutants are synthetically lethal with *orc5* mutants (70, 85), and overexpression of *ORC6* is lethal in *cdc6* cells (86). Similarly, in *S. pombe*, *cdc18* mutants are synthetically lethal with *orc1*/*orp1* mutants (75). Physical interactions between Cdc6p and ORC have been demonstrated (85) and on the basis of studies in yeast and in higher eukaryotes, notably in *Xenopus*, it is widely believed that ORC acts as a landing pad on chromatin for Cdc6p. Certainly the association of Cdc6p with chromatin is ORC dependent.

Studies in *S. cerevisiae* have shown that Cdc6p must be synthesized during the G1 of each cell cycle to allow entry into S phase (87). Synthesis of Cdc6p is required for formation of pre-RC complexes and several lines of evidence point to Cdc6p being a crucial target for the regulatory mechanism that prevents pre-RC assembly outside of G1 (see chapter by Tyers, this volume, for further details). Cdc6p is distantly related in structure to Orc1 and Orc4 and contains a nucleotide binding motif, suggesting that nucleotide binding and possibly hydrolysis may be important for its function. Consistent with this, mutations within the presumptive nucleotide binding site abolish Cdc6p function *in vivo* (88).

In addition to sharing sequence similarity with the Orc1p and Orc4p proteins, Cdc6p is also distantly related to the prokaryotic and eukaryotic clamp loader proteins that are responsible for the loading on to DNA of the ring-shaped DNA polymerase processivity factors, known as sliding clamps, during the elongation stage of replication (88) (see sections 5.2 and 5.3). In the absence of Cdc6p, the MCM proteins fail to assemble on to chromatin. This has led to the suggestion that Cdc6p might be responsible for catalysing the loading of the MCM proteins on to DNA and that this process might be mechanistically related to the loading of the sliding clamp on to DNA by the clamp loader complex. There are certainly significant similarities between the two processes (discussed in reference 88).

3.5 MCM protein complex

Although unwinding of the origin DNA constitutes a key step in the process of replication initiation, the identities of the cellular factors responsible for catalysing this unwinding remain unclear. At present the only DNA helicase known to be essential for chromosomal replication in the yeasts is the Dna2 protein, but this appears to be required only for lagging-strand synthesis and/or Okazaki fragment process-ing and not for replication initiation (discussed further in section 6.3). In the SV40 system, unwinding of the origin and of the template DNA ahead of the replication fork is accomplished by the virally encoded initiator protein T antigen. In common with the *Escherichia coli* replicative DNA helicase DnaB, and with those from the bacteriophages T4 and T7, T antigen is hexameric in structure and it is tempting to speculate that the cellular replicative helicase will also be a hexamer (reviewed in reference 89). However, the identification of possible helicases amongst the existing repertoire of factors essential for chromosomal replication is hampered by the fact that different classes of helicases share very little primary sequence similarity be-yond their nucleotide-binding motifs. At present the prime candidate for the role of the eukaryotic replicative helicase is the MCM protein complex, as weak *in vitro* heli-case activity has been found associated with a complex of the human Mcm4, Mcm6 and Mcm7 proteins (27). In addition, studies in *S. cerevisiae* (described in greater detail below) show that the MCM proteins move with the replication fork from origin to non-origin DNA as replication proceeds (29).

Genes encoding the yeast MCM proteins were first identified in screens for muta-tions defective in cell cycle progression or mini-chromosome maintenance (MCM is an acronym for mini-chromosome maintenance) and were initially grouped together on the basis that they interacted with one another genetically and encoded proteins of related sequence (reviewed in reference 26). It is now apparent that the MCM proteins from any one species can be grouped into six distinct classes (Fig. 3) and the *S. cerevisiae* proteins have been renamed Mcm2–Mcm7 to reflect this (see Table 1 for the former names of the *S. cerevisiae* MCM proteins). All the proteins share a conserved domain of approximately 200 amino acids that has similarity to DNA-dependent ATPases and includes a nucleotide-binding motif (Fig. 3). Despite their sequence similarity, each of the *S. cerevisiae* and *S. pombe* MCM proteins is independ-ently required for DNA replication, indicating that each performs a distinct essential function in the cell.

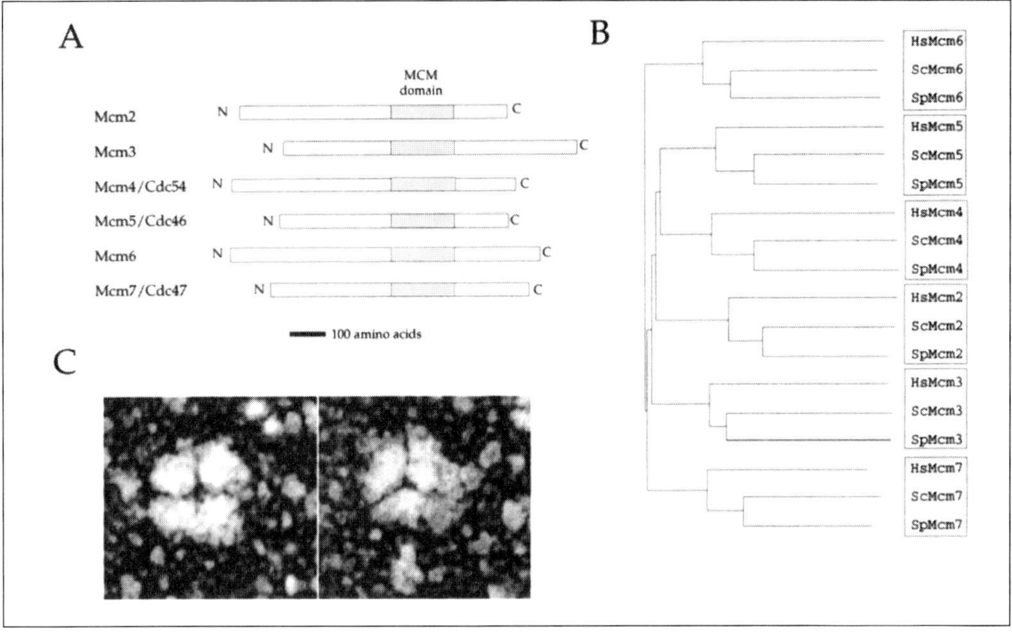

Fig. 3 Structure and relatedness of MCM proteins. (A) Schematic representation of the structures of the *S. cerevisiae* MCM proteins Mcm2p through Mcm7p, showing the location of the highly conserved 200 amino acid MCM domain containing the ATP-binding consensus sequences (shaded box). The original names of Mcm4p (Cdc54p), Mcm5p (Cdc46p) and Mcm7p (Cdc47p) are given. (B) Phylogenetic tree of MCM protein families in humans (Hs), *S. cerevisiae* (Sc) and *S. pombe* (Sp). Six clear groupings (corresponding to Mcm2 through Mcm7) can be seen within the MCM protein family. (C) Rotary-shadowing electron micrographic images of MCM protein complexes purified from *S. pombe* (images kindly supplied by Dr Y. Adachi, University of Edinburgh).

In addition to interacting genetically with one another, extensive physical interactions between different MCM proteins have been reported. In *S. cerevisiae*, for example, Mcm3p binds tightly to Mcm5p, and these proteins each bind weakly to Mcm2p (90), whereas in *S. pombe* all six proteins have been shown to co-purify in equimolar ratios in a 560-kDa multiprotein complex (91). Rotary-shadowing electron microscopy of the purified MCM complex (91) reveals globular particles with surface pits or cavities dividing the structures into several possibly symmetrical domains (Fig. 3C).

Using a chromatin crosslinking and immunoprecipitation assay, it has been shown that the MCM proteins move with the replication fork following origin activation (29). These observations have led to the model for the early steps in origin activation shown in Fig. 4 and discussed below (section 3.8).

3.6 Cdc7p–Dbf4p protein kinase

The *S. cerevisiae* Cdc7p protein kinase and its regulatory subunit Dbf4p (Fig. 4, Table 1) are required for the initiation of DNA replication (reviewed in references

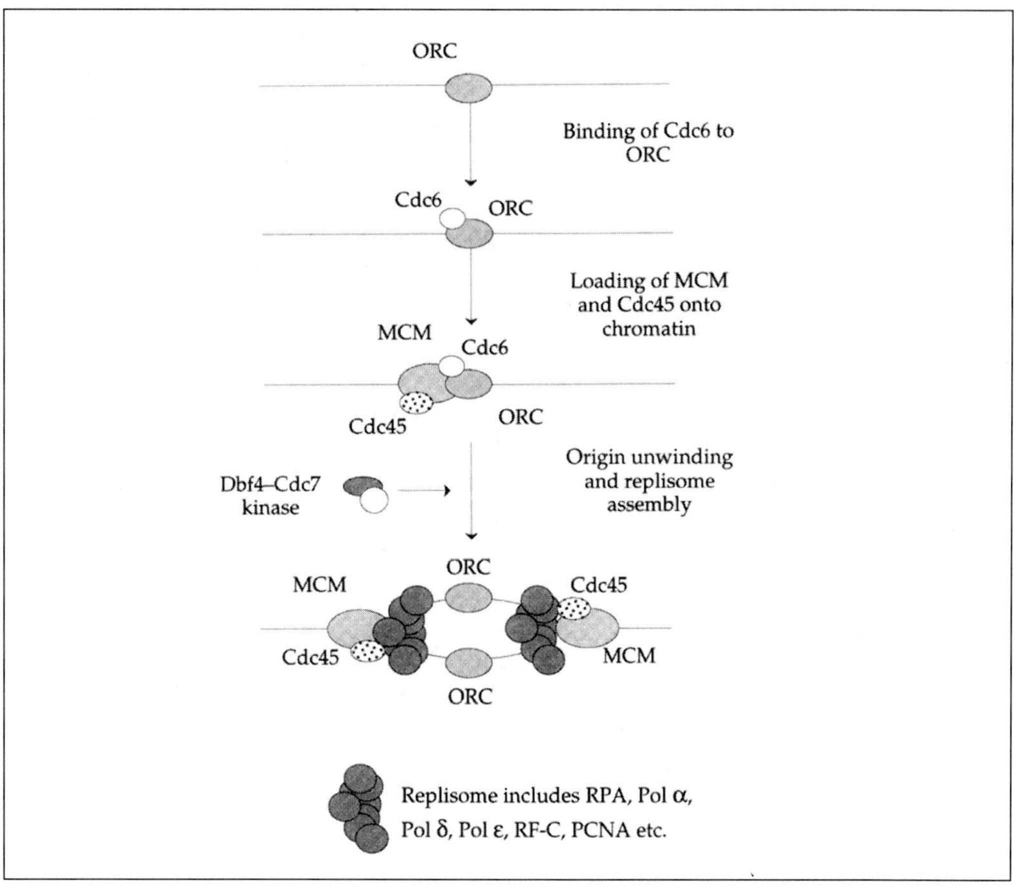

Fig. 4 Schematic representation of the events at the initiation of DNA replication in *S. cerevisiae*. The ORC, which is bound to origin DNA throughout the cell cycle, associates in G1 with Cdc6p to promote loading of the putative helicase, the MCM complex. Cdc45p, an MCM-associated factor, also loads on to the chromatin at this time. Following origin activation, the MCM proteins and Cdc45p appear to move with the replication fork from origin to non-origin DNA, while ORC remains bound to origins. How ORC is duplicated is not known.

26,74). The Cdc7p–Dbf4p enzyme is believed to act by phosphorylating one or more of the six subunits of the MCM complex to bring about its activation, but the precise details of this remain to be fully worked out. In *S. pombe*, the Hsk1 and Dfp1/Him1 proteins are homologues of Cdc7p and Dbf4p respectively (8, 92–94). As in *S. cerevisiae*, the Hsk1 and Dfp1/Him1 proteins are essential for replication initiation and Dfp1/Him1 levels are cell-cycle regulated (26, 74).

3.7 Cdc45p

The *S. cerevisiae* Cdc45p protein is an essential replication factor that physically associates with Mcm5p, and *CDC45* displays genetic interactions with genes encoding five of the six MCM family members (all except *MCM6*) and also with *ORC2* (74). The

Cdc45p protein is present at replication forks and, like the MCM proteins RP-A and Pol α–primase, appears to move with the fork following origin activation from origin DNA to neighbouring sequences (29). In *S. pombe*, a Cdc45p homologue is encoded by the *sna41*$^{+}$ gene, mutations in which were initially isolated as suppressors of a cold-sensitive *nda4* (*mcm5*) mutation (95).

3.8   Dynamic protein complexes at the replication fork

Figure 2.4 shows a current model for the assembly of protein factors at the replication fork based on the results of analysis in *S. cerevisiae* and to a lesser extent, *Xenopus laevis*. During the G1 phase of the cell cycle, the pre-RC complex comprising ORC, Cdc6p, MCMs, and Cdc45p, binds origin DNA. In fact, ORC appears to remain bound to the origin throughout the cell cycle. Once replication is initiated, Cdc6p is degraded, and the MCM proteins and Cdc45p move with the replication fork into non-origin sequences. The moving replication apparatus also contains Pol α, Pol δ, and Pol ε (29). After mitosis is complete, Cdc6p again becomes associated with ORC. This in turn leads to the incorporation of the MCM proteins and Cdc45p into the pre-RC in preparation for the next S phase.

4.   Initiation of replication: from origin unwinding to primer synthesis

4.1   Overview

In the SV40 system, origin unwinding can be considered as a two-step process (22). First, the viral initiator protein T antigen brings about a transient opening of the duplex DNA at the origin and in so doing facilitates initial binding of RP-A to the single-stranded DNA (ssDNA). Next, a more extensive region of the DNA is stably unwound as a consequence of replication protein A (RP-A, also known as RF-A or SSB) binding to ssDNA generated by the further action of T antigen. RP-A and T antigen then recruit the Pol α–primase complex which acts to synthesize the RNA–DNA primers required for synthesis of the leading strand and for each Okazaki fragment on the lagging strand. As noted above, the identity of the cellular factor that brings about origin unwinding is unclear but it is not unreasonable to expect that the events following the initial melting of the chromosomal DNA will mirror those in the SV40 system. In this section we review what is known of the structure and function of RP-A and Pol α–primase in yeast chromosomal replication.

4.2   Replication protein A (RP-A)

RP-A is a single-stranded DNA-binding protein complex that was first identified as an essential factor for the *in vitro* replication of SV40 viral DNA (96). In this system RP-A acts to promote unwinding of origin DNA catalysed by T antigen. RP-A also functions to stimulate Pol α–primase activity and is required for PCNA- and

RF-C-dependent DNA synthesis catalysed by Pol δ (see section 5). Studies in *S. cerevisiae* have shown that RP-A is associated with origin DNA in early S phase (i.e. in cells arrested in early S phase with the replication inhibitor hydroxyurea) but appears to move with the replication fork into neighbouring sequences at later time points, a process that is dependent upon Pol α–primase function (97). The origin association of RP-A is also dependent upon Cdc7p/Dbf4p and on MCM function, consistent with the notion that the MCM proteins may be responsible for initial unwinding of the origin DNA. Similar behaviour to RP-A is shown by the Pol α–primase complex (see below) but, significantly, origin association of Pol α–primase is dependent upon the presence of functional RP-A (97), suggesting that RP-A acts directly or indirectly to recruit the Pol α–primase complex to the unwound origin.

Like its mammalian counterparts, yeast RP-A is a heterotrimeric complex, with each subunit being encoded by an essential gene. In *S. cerevisiae* the three subunits of the complex are Rpa1p, Rpa2p, and Rpa3p, encoded by *RFA1*, *RFA2*, and *RFA3* respectively (Fig. 5); in *S. pombe*, the homologous subunits are encoded by *rpa1*$^+$, *rpa2*$^+$, and *rpa3*$^+$ (see Table 2 for alternative names) (80, 98). Cells carrying temperature-sensitive mutations in *RFA2* and *RFA3* undergo a rapid cessation of DNA synthesis when shifted to the restrictive temperature (99, 100), consistent with an absolute requirement for these functions for replication fork movement. In *S. cerevisiae*, genetic interactions have been detected between RP-A and both Pol α–primase and Pol δ, suggesting a close functional (and perhaps structural) association at the replication fork. Interestingly the middle subunit of RP-A in both *S. cerevisiae* and *S. pombe* (Rpa2) is phosphorylated as cells enter S phase, remains phosphorylated in G2, and is dephosphorylated in mitosis (98, 101). At present the significance of this phosphorylation for RP-A function is unclear but recent work with mammalian RP-A points to it having a potential role in regulating RP-A assembly and disassembly (102).

Recent structural analysis of RP-A suggests that it may contain up to four, and

Table 2 Proteins at the replication fork: origin unwinding and primer synthesis[a]

	S. cerevisiae	*S. pombe*	**Comments**
RPA	Rpa1p	Rpa1/Rad11/Ssb1	Large subunit of RPA; contains three distinct ssDNA-binding folds
	Rpa2p	Rpa2/Ssb2	Phosphorylated middle subunit of RPA; contains a single ssDNA-binding fold
	Rpa3p	Rpa3/Ssb3	Small subunit of RPA; contains a single putative ssDNA-binding fold
Pol α–primase	Pol1p/Cdc17p	Pol1/Cdc29/Swi7	Catalytic subunit of Pol α–primase complex
	Pol2p	Spb1	B subunit of unknown function but phosphorylated in a cell cycle-dependent manner suggesting a possible regulatory role
	Pri1p	Spp1	Small (catalytic) subunit of primase
	Pri2p	Spp2	Large subunit of primase

[a] See text and Reference 144 for references.

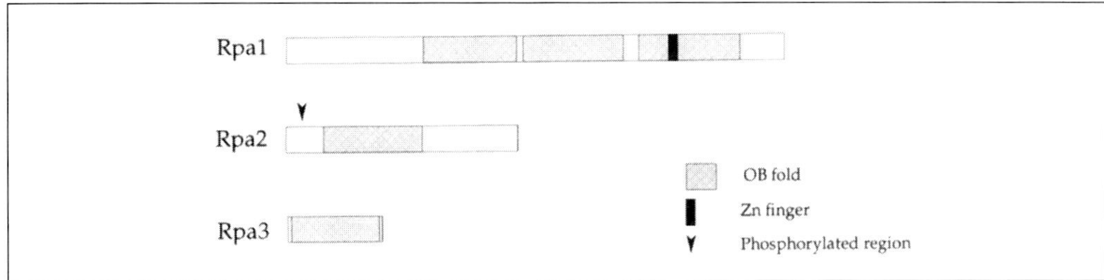

Fig. 5 Structure of RP-A subunits showing the relative size of the three subunits and the location of the five OB fold ssDNA-binding domains within them (103, 104). Note that DNA-binding activity has been demonstrated for all three OB fold domains in Rpa1p as well as the single OB fold domain in Rpa2p, but not for the OB fold in Rpa3p. The arrowhead indicates the phosphorylated domain of Rpa2p.

possibly five, distinct ssDNA-binding domains (103, 104). Three of these domains are located in the large subunit (Rfa1p), one in the middle subunit (Rfa2p), and possibly one in the small (Rfa3p) subunit (see Fig. 5). There is also evidence that RP-A may bind ssDNA in a higher-order structure; that is, that the ssDNA may be wrapped around the RP-A complex.

4.3 Pol α–primase complex

The Pol α–primase complex functions to synthesize the short RNA–DNA segment (sometimes referred to as the iDNA) that serves as a primer for continuous synthesis of the leading strand and for each Okazaki fragment on the discontinuously synthesized lagging strand. In the SV40 system, the primer is approximately 40 nucleotides in length, with around 10 nt of RNA. Unlike both Pol δ and Pol ε described in the following sections, Pol α–primase does not possess 3′–5′ proofreading activity, although the presence of certain protein sequence motifs in the Pol1 protein (the Pol α catalytic subunit) suggests that this activity has been lost during evolution (see reference 105 for further details). The absence of proofreading activity gives a strong indication that Pol α is not responsible for bulk DNA replication, as the accumulation of errors during Pol α-catalysed DNA replication would sooner or later prove fatal. Indeed it is likely that the Pol α-synthesized portion of the RNA–DNA primer is itself removed and resynthesized by one of the proofreading polymerases (see section 6, Fig. 9).

As in mammalian cells, *S. cerevisiae* Pol α–primase is a four-subunit complex, comprising Pol1p, Pol12p, Pri1p, and Pri2p, each of which is essential for Pol α–primase function (summarized in Table 2 and discussed below). In *S. pombe*, the genes encoding these four subunits have been identified (see legend to Table 2) but only *pol1*+ has been characterized in any detail (13, 15, 106–108). *S. cerevisiae* Pol1p and Pri1p are the catalytic subunits of Pol α and primase respectively, whereas Pol12p is the Pol α B subunit. Although the precise function of this protein is not yet known, Pol12p appears to play an important structural role in linking Pol1p with the dimeric primase, since Pol1p does not interact with either Pri1p or Pri2p directly, and Pol1p fails to co-immunoprecipitate with Pri1p and Pri2p in *pol12* mutant cells (109). In the

SV40 system, the Pol α–primase B subunit interacts directly with T antigen, suggesting a possible role in loading Pol α–primase at the fork, but whether this will prove to be the case in yeast remains to be seen. In addition to its structural role, there is increasing evidence that the Pol12p protein may play an important regulatory role, as Pol12p is phosphorylated and dephosphorylated in a cell cycle-dependent manner (109).

5. Polymerase switching: enzymes and mechanisms

5.1 Overview

Once Pol α–primase has synthesized the primer, polymerase switching occurs (Fig. 6). This process, which was first described in the SV40 system (110), sees the Pol α–primase complex being replaced by Pol δ and/or Pol ε in association with their processivity factor PCNA. PCNA is a ring-shaped homotrimeric molecule that encircles the DNA, forming what is known as a sliding clamp, that acts to tether Pol δ and Pol ε to the DNA, thus maximizing their processivity on the template. Loading of PCNA onto the DNA, which is believed to occur before it associates with Pol δ and/or Pol ε, is accomplished by the action of the clamp loader complex, replication factor C (RF-C). In this section we describe the components of the polymerase-switching apparatus and consider what is known of their biochemical properties. We also discuss recent evidence concerning the differing requirements for Pol δ and Pol ε in the replication process.

5.2 Replication factor C (RF-C): the eukaryotic clamp loader

RF-C was first identified as an essential factor in the SV40 system, where it was shown to bind in a structure-specific manner to the 3′ end of the nascent RNA–DNA primer to facilitate the loading of PCNA on to the template DNA (111, 112) (Fig. 6). RF-C possesses DNA-dependent ATPase activity that is stimulated both by primer–template junctions and by PCNA, and ATP hydrolysis is required for PCNA loading (111, 113, 114). Mammalian RF-C is a five-subunit complex, comprising one large and four small subunits, and the same is true of RF-C purified from either *S. cerevisiae* (114–117) or *S. pombe* (118). Genes encoding all five subunits of the complex have now been identified in both yeasts (119–128) (see Table 3) and the encoded proteins have been shown to be related not only to their higher eukaryotic counterparts but also to one another, suggesting that they originated from a common ancestor (128). As shown in Fig. 7, the RF-C proteins are also distantly related in sequence to the subunits of the *E. coli* clamp loader, the γ complex, which functions to load the PCNA-like β sliding clamp on to DNA during chromosomal replication by Pol III. The three-dimensional structure of one of the subunits of the γ complex, the δ′ subunit, has been determined and reveals a protein with a C-shaped structure comprising three distinct domains: an N-terminal domain (domain 1, indicated as D1 in Fig. 7) that has greatest similarity to the RF-C proteins, a short central domain (D2)

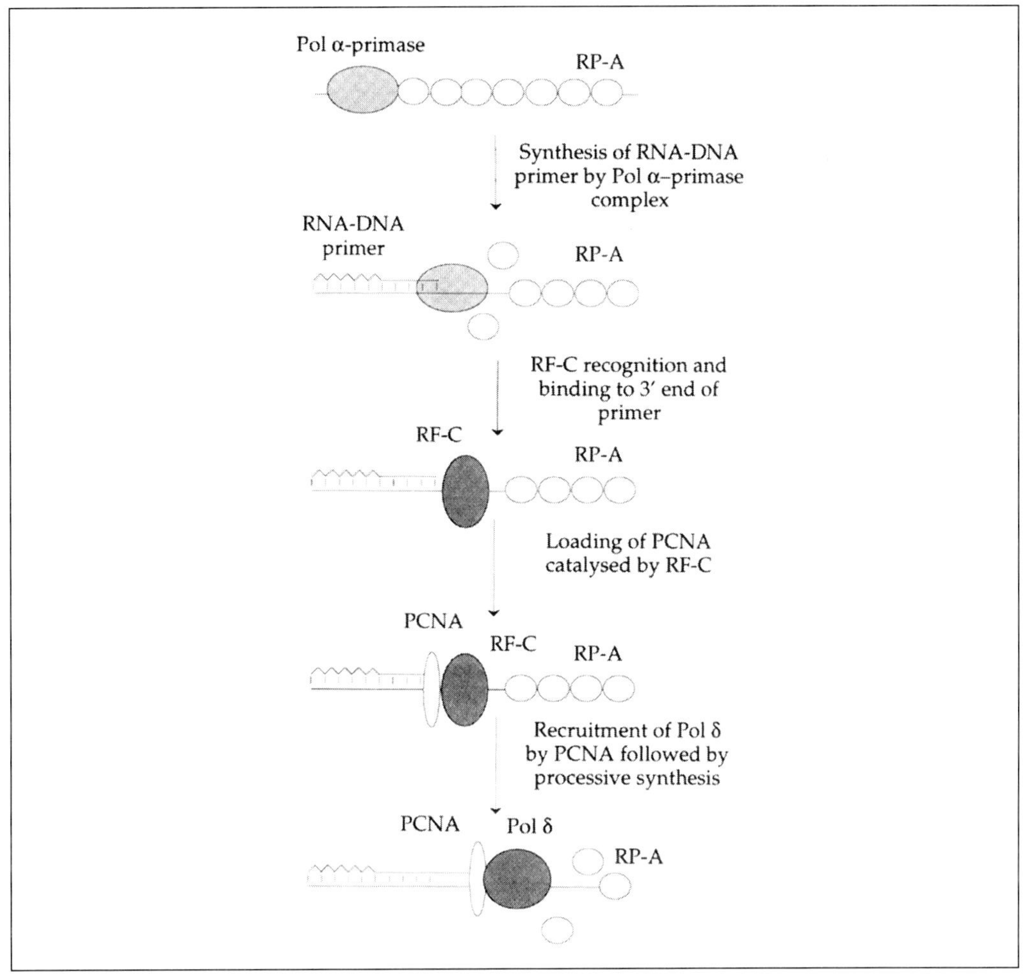

Fig. 6 Polymerase switching. Pol α–primase synthesizes short RNA–DNA (iDNA) molecules that act as primers for leading-strand synthesis and for each Okazaki fragment on the lagging strand. Following iDNA synthesis, RF-C binds to the 3' end of the primer and catalyses the loading of PCNA on to the DNA. This is followed by recruitment of the replicative polymerase (most likely Pol δ) and processive synthesis. See text for details and references.

that may act as a hinge, and an α-helical C-terminal domain (D3). On the basis of protein sequence alignments across the kingdoms, it seems likely that the structure of the RF-C proteins will broadly conform to that of δ' but confirmation of this awaits determination of the structure of an RF-C subunit.

Each of the RF-C subunits contains a number of conserved protein sequence motifs (called the RF-C boxes; see Fig. 7) (128) within a region of the protein corresponding to domains D1 and D2 of the δ' structure. The conserved regions include the Walker A and B motifs (RF-C boxes III and V respectively), characteristic of nucleotide-binding proteins. However, it appears that not all the RF-C subunits actually bind

Table 3 Proteins at the replication fork: polymerase switching and processive synthesis[a]

	S. cerevisiae	*S. pombe*	Comments
RF-C	Rfc1p/Cdc44p	Rfc1	Large subunit of RF-C interacts with DNA and with PCNA
	Rfc2p	Rfc2	Small RF-C subunit: binds DNA
	Rfc3p	Rfc3	Small RF-C subunit: DNA-dependent ATPase
	Rfc4p	Rfc4	Small RF-C subunit: ATP binding
	Rfc5p	Rfc5	Small RF-C subunit
PCNA	Pol30p	Pcn1	PCNA: toroidal heterotrimeric Pol δ/Pol ε processivity factor
Pol δ	Pol3p/Cdc2p	Pol3/Cdc6	Catalytic subunit of Pol δ complex, processes both polymerase and 3′-5′ proofreading exonuclease activity: interacts directly with Pol31p/Hys2p
	Pol31p/Hys2p	Cdc1	B subunit of Pol δ complex: interacts directly with Pol32p but function unknown
	Pol32p	Cdc27	PCNA binding C subunit of Pol δ complex: essential in *S. pombe* only
		Cdm1	Non-essential D subunit of *S. pombe* Pol δ: no known equivalent in *S. cerevisiae*
Pol ε	Pol2p	Cdc20/Pol2	Catalytic subunit of Pol ε complex, processes both polymerase and 3′-5′ proofreading exonuclease activity, although polymerase activity is not absolutely required for *in vivo* function
	Dpb2p		Essential B subunit of Pol ε complex: function unknown
	Dpb3p		Non-essential third subunit of Pol ε: *S. pombe* homologue not yet characterized
	Dpb4p		Fourth subunit of Pol ε: *S. pombe* homologue not yet characterized
	Dpb11p	Cut5/Rad4	Essential Pol ε-interacting factor: Dpb11 interacts directly with Dpb2
	Sld2p		Essential Pol ε-interacting factor
	Drc1p		Essential Pol ε-interacting factor

[a] See text and Reference 144 for references.

ATP and, of those that do, not all are capable of ATP hydrolysis, echoing the situation with ORC (described in section 3.3).

In *S. cerevisiae* the subunits of RF-C are encoded by *CDC44/RFC1* (122, 127, 128), *RFC2* (123, 128), *RFC3* (120, 128), *RFC4* (121, 128), and *RFC5* (119, 128). The small subunits (Rfc2 through Rfc5) range in size from 36 to 40 kDa and share a high degree of sequence similarity to one another and to the large subunit (Fig. 7) (128). The large subunit (Cdc44p/Rfc1p) is a 94-kDa protein that differs from the small subunits in having lengthy (200–300 amino acids) N- and C-terminal extensions (Fig. 7). This subunit interacts directly with DNA, and also with PCNA via a p21^{Cip1}-like canonical PCNA-binding motif at its extreme N-terminus (129) (see Fig. 8). The N-terminal domain also contains a region with similarity to the prokaryotic NAD-dependent DNA ligase enzymes that is distantly related to the BRCT protein–protein interaction motif (128). Studies of the mammalian Rfc1 homologue, hRFC140, have shown that

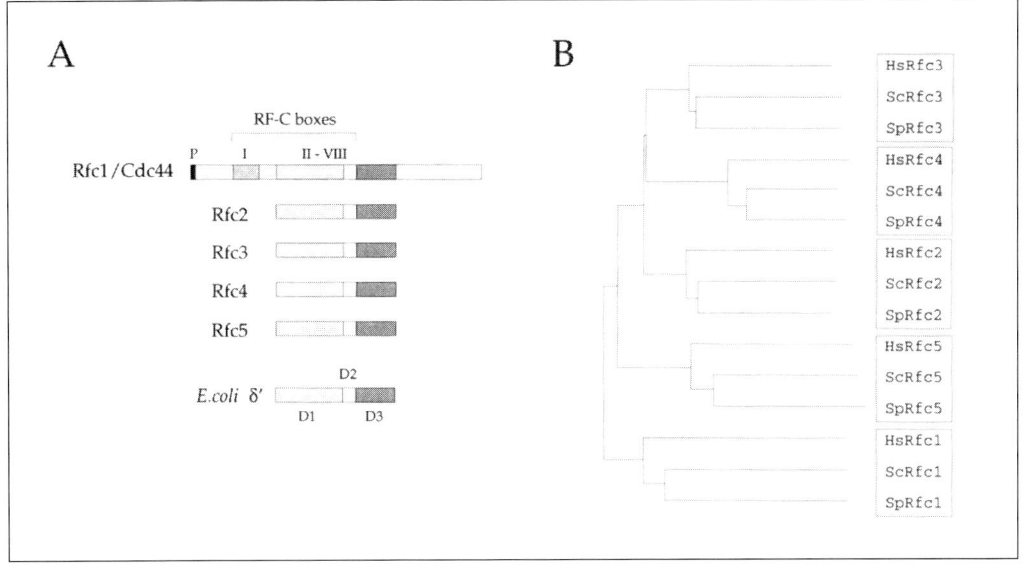

Fig. 7 Structure and relatedness of RF-C proteins. (A) Schematic representation of the structural features of the eukaryotic RF-C subunits alongside the *E. coli* clamp loader γ complex δ′ subunit whose tertiary structure has been determined by crystallography. The large subunit (Rfc1p/Cdc44p in *S. cerevisiae*, Rfc1 in *S. pombe*) contains a PCNA-binding motif at its N-terminus (black box labelled P), a region of homology to prokaryotic DNA ligases (RF-C box I, unique to the large subunit), a central portion containing RF-C boxes II–VIII that is related in structure to the small subunits and to the δ′ protein, and a 200 amino acid C-terminal domain of unknown function. The small subunits are closer in size to the δ′ protein and are likely to have a simple three-domain structure (shown as D1, D2, D3). (B) Phylogenetic tree of RF-C subunits (128, 133). Five clear groupings (corresponding to Rfc1 through Rfc5) can be seen within the RF-C protein family. The phylogenetic tree was constructed with njplot using an alignment generated by Clustal_X.

the protein binds specifically to the primer–template junction and it is thought that this specificity will be shared by the yeast proteins. *S. cerevisiae* cells carrying temperature-sensitive mutations in any of the *CDC44, RFC2*, or *RFC5* genes display a terminal phenotype consistent with a failure to complete DNA replication, although in the case of *RFC2* (124) and *RFC5* (130) this is somewhat obscured by the failure to activate the DNA replication checkpoint that occurs in these mutants. Various genetic interactions between *CDC44* and *POL30* (encoding PCNA, see below) have been observed, consistent with a direct interaction between their products; for example, mutations have been isolated in *POL30* that suppress cold-sensitive mutants in *CDC44* but not a *cdc44Δ* strain (131, 132). In *S. pombe*, *rfc1*[+], *rfc2*[+] and *rfc3*[+] have been shown to encode essential functions (125, 126, 133). To our knowledge, *rfc4*[+] and *rfc5*[+] have yet to be analysed (see Table 3).

5.3   PCNA, the eukaryotic sliding clamp

PCNA is a homotrimeric ring-shaped protein complex (Plate 2), structural and functional homologues of which have been identified in many eukaryotes, in *E. coli* (see below), and in bacteriophage T4. There is also evidence for PCNA-like proteins in the archaea (134). The *S. cerevisiae* and *S. pombe* PCNA proteins (encoded by *POL30* and *pcn1$^+$* respectively) share approximately 50% amino acid sequence identity to their mammalian PCNA counterparts. Both proteins have been demonstrated to be essential for DNA replication and cells carrying mutant *pol30* (*S. cerevisiae*) or *pcn1* (*S. pombe*) alleles display a range of phenotypes consistent with defects in either DNA replication or DNA repair (131, 135–140).

Remarkably, *S. cerevisiae* PCNA is the only component of the yeast chromosomal replication apparatus whose three-dimensional structure has been solved crystallographically (141). The resulting structure (shown in Plate 2) is strikingly similar to that of the β sliding clamp of *E. coli*, despite there being little or no primary sequence identity between the two proteins. The β sliding clamp is a homodimer, with each monomer consisting of three structurally similar domains giving the structure of the dimer an overall sixfold symmetry (142). In contrast, each monomer of the PCNA complex consists of only two structurally similar domains but, as PCNA is a trimer, sixfold symmetry still ensues. The centre of the PCNA ring is large enough to accommodate double-stranded DNA (dsDNA) and it is believed that the arrangement of the α helices that line this central cavity (see Plate 1) effectively restricts protein–DNA interactions to non-specific contacts with the phosphate backbone of the DNA helix (141).

The toroidal structure of the PCNA trimer is consistent with its biochemical properties and provides an immediate explanation for the requirement for a clamp-loading activity. PCNA can load on to linear DNA molecules only by diffusion on to the dsDNA end, consistent with it having a ring-like structure (143). However, this process is considerably less efficient than enzymatic loading of PCNA onto dsDNA by RF-C, which can be accomplished on a circular DNA molecule. PCNA loaded on linear DNA by RF-C is also likely to slip off the DNA via the ends (143).

5.4   DNA polymerase δ

Pol δ is an essential component of the SV40 *in vitro* replication system where it is responsible for replicating both the leading and lagging strands from primers laid down by Pol α–primase (reviewed in reference 144). The question of whether Pol δ played a similar role in replicating both strands of chromosomal DNA in yeast was, until recently, complicated by the fact that in both *S. cerevisiae* and *S. pombe* Pol ε is also an essential enzyme, and cells carrying mutations in the catalytic subunit of this enzyme arrest with unreplicated DNA when shifted to their restrictive temperature (9, 10, 145, 146). However, this issue appears now to have been resolved, at least in part, with the demonstration (147, 148) that, while Pol ε is indeed required for chromosomal DNA replication in *S. cerevisiae*, the catalytic activity of the enzyme is not, indicating that under certain conditions, Pol δ alone is able to replicate both DNA strands (discussed in detail below).

Mammalian Pol δ purifies as a heterodimeric enzyme, consisting of a large catalytic subunit and a smaller B subunit, although the existence of a third (C) subunit has recently been reported (149). In contrast, highly purified *S. cerevisiae* Pol δ comprises three distinct subunits (16, 17, 137, 150, 151), while that of *S. pombe* has four (118, 152). In both yeasts the catalytic subunit is the Pol3 protein and is encoded by *POL3/CDC2* (*S. cerevisiae*) and *pol3$^+$/cdc6$^+$* (*S. pombe*) (12, 14, 15, 145, 153–155). In *S. cerevisiae* the B and C subunits are Pol31p and Pol32p, encoded by *POL31/HYS2* and *POL32* (16,17,137,156,157), whereas in *S. pombe* the equivalent proteins are Cdc1 and Cdc27, encoded by *cdc1$^+$* and *cdc27$^+$* (118, 158). Cdm1, the fourth subunit of the *S. pombe* enzyme, which appears to have no counterpart in *S. cerevisiae*, is encoded by *cdm1$^+$* (152).

The catalytic subunit is a protein of about 125 kDa that possesses both polymerase and 3′–5′ exonuclease activities. Yeast cells carrying temperature-sensitive mutations in the *POL3* (*S. cerevisiae*) or *pol3$^+$* (*S. pombe*) genes undergo cell cycle arrest when shifted to the restrictive temperature, becoming arrested in S phase (12, 145, 159, 160). Mutational analysis of the 3′–5′ exonuclease domain of *S. cerevisiae* Pol3p has shown that this is required for accurate replication, consistent with a role in proofreading (146, 155, 161, 162).

The catalytic subunit interacts directly with the B subunit (either Pol31p/Hys2p or Cdc1) and in both yeasts the B subunits are also essential proteins (3, 16, 17, 156, 158). A range of genetic interactions has been demonstrated between the genes encoding these subunits (16, 17, 137, 158). In *S. pombe*, for example, overexpression of *pol3$^+$* suppresses temperature-sensitive *cdc1* alleles, and certain *pol3* and *cdc1* mutants are also synthetically lethal (158).

The third subunit of the Pol δ core enzymes from *S. cerevisiae* and *S. pombe*, Pol32p and Cdc27 respectively, show a low level of protein sequence identity only, at approximately 20%, and are similar to the recently identified putative third subunit of mammalian Pol δ (149). In *S. cerevisiae*, *POL32* is non-essential but *pol32Δ* cells display a range of phenotypes consistent with this protein playing an important role in the replication process, such as an increased sensitivity to the replication inhibitor hydroxyurea and synthetic lethality with *pol3*, *pol31*, and *pol30* mutations (16, 17). In contrast, *S. pombe cdc27$^+$* is an essential gene; *cdc27Δ* cells are incapable of completing DNA replication (5, 158, 163). The reason for this distinction is unclear as in all other respects these subunits appear to have similar properties. Interaction studies have shown that Pol32p and Cdc27 are able to interact with PCNA (16, 17, 137, 164) and that this interaction is mediated via a protein sequence motif similar to that first identified in the mammalian p21^{Cip1} protein (165) (see Fig. 8). In the case of Cdc27, binding to Pcn1 (PCNA) is essential for function (164). This is discussed in greater detail below.

The fourth subunit of the *S. pombe* Pol δ core enzyme, Cdm1, is a 160 amino acid protein ($M_r \sim 22$ kDa) with no significant sequence similarity to proteins in existing protein sequence databases. *In vitro* protein–protein interaction studies have failed to determine precisely how Cdm1 contacts the other subunits of the Pol δ complex but overexpression of *cdm1$^+$* is sufficient to rescue temperature-sensitive mutations in *pol3*, *cdc1*, and *cdc27*, as well as *cdc24* (discussed in section 6.3). Haploid *cdm1Δ* cells are indistinguishable from wild-type cells in every respect examined, so the function of this subunit remains a mystery (152).

In addition to interacting with itself, with Pol30p and with Pol31p, Pol32p also appears able to interact directly with Pol1p, the catalytic subunit of Pol α in *S. cerevisiae* (166). A similar interaction is seen between the *S. pombe* Cdc27 and Pol1 proteins, and their human counterparts (S. A. MacNeill, unpublished results). These observations represent the first indication of a direct interaction between components of the Pol α and Pol δ complexes. The site of interaction on Pol1 has been mapped to a conserved non-catalytic domain of previously unknown function. There are temperature-sensitive Pol1 mutants in both yeasts that map within this domain (167, 168), raising the possibility that the defect in these proteins could be a failure to interact with Pol32p or Cdc27. However, this argument assumes that the observed interaction is an essential one, which need not be the case.

5.5 DNA polymerase ε

In both *S. cerevisiae* and *S. pombe*, Pol ε is essential for chromosomal DNA replication. *S. cerevisiae* cells carrying mutations in *POL2*, encoding the catalytic subunit of the Pol ε complex, or *DPB2*, which encodes the B subunit of the enzyme, are unable to replicate their chromosomal DNA successfully (154, 169, 170). The same is true of cells carrying mutations in the gene encoding the catalytic subunit of *S. pombe* Pol ε, $cdc20^+/pol2^+$ (9, 10). Furthermore, Pol ε is associated with *S. cerevisiae* replication origins in early S phase and moves with the replication fork from origin to non-origin sequences as S phase progresses (29). Remarkably, however, two recent reports (147, 148) have shown that neither the exonuclease nor polymerase activities of Pol2p are required for its function in *S. cerevisiae*; cells expressing mutated forms of the Pol2p protein in which the polymerase and exonuclease domains have been completely deleted are viable. Instead it is the C-terminal protein–protein interaction domain of Pol2p that has an essential role (147, 148). Whether this result indicates that Pol ε plays no significant role in nucleotide polymerisation in wild-type cells is unclear at present. However, it is noteworthy that the duration of S phase in cells expressing mutant forms of Pol2p lacking catalytic activity is twice that seen in wild-type cells, suggesting that Pol ε does indeed play a role in the latter. Presumably, in its absence, Pol δ replicates both DNA strands, but with reduced efficiency. Interestingly, cells carrying exonuclease-deficient *pol2* mutants accumulate strand-specific lesions during replication that are thought to be indicative of a role for the Pol2p exonuclease in proofreading on one DNA strand only, most likely the leading strand (146, 171).

Purified *S. cerevisiae* Pol ε consists of at least four subunits: Pol2p, Dpb2p, Dpb3p, and Dpb4p (172) (Table 3). Dpb3p is non-essential but *dpb3Δ* cells display an increased frequency of spontaneous mutagenesis (173); Dpb4p is not yet character-ized. Several putative Pol ε-interacting factors have been identified, encoded by *DPB11*, *SLD2*, and *DRC1* (*DPB11* is equivalent to $cut5^+/rad4^+$ in *S. pombe*; see Table 3) (174–178). The precise functions of these proteins are not understood at present.

5.6 PCNA binding via a conserved protein sequence motif

Several of the replication proteins that interact with PCNA appear to do so by similar means, utilizing a canonical PCNA-binding motif that was first identified in the mammalian replication inhibitor p21[Cip1] (165). The consensus PCNA-binding sequence is Q--I--FF, although not all the proteins have the Q (Fig. 8). Five budding yeast proteins that interact with PCNA appear to do so through a motif of this type: Rfc1p/Cdc44p, Pol32p (Table 3), Rad27p/Rth1p, Cdc9p (Table 4), and Rad2p (see legend to Fig. 8), although the latter is not involved in chromosome replication, only in the repair of UV-damaged DNA. The *S. pombe* homologues of these five proteins (Rfc1, Cdc27, Rad2, Cdc17, and Rad13 respectively) also contain the conserved motif (Fig. 8).

Genetic analysis has shown that, despite being conserved across evolution, the motif is not always required for protein function in S phase. In the case of *S. cerevisiae* Cdc44p/Rfc1p, for example, deletion of a large N-terminal region, including the

Table 4 Proteins at the replication fork: Okazaki fragment processing[a]

S. cerevisiae	S. pombe	Comments
Dna2p	Dna2	DNA helicase with additional single-stranded endonuclease activity
	Cdc24	Essential Dna2 interacting factor in *S. pombe*: no apparent homologues in *S. cerevisiae*
Rad27p/Rth1p	Rad2	FEN-1 exonuclease, implicated in Okazaki fragment processing
Cdc9p	Cdc17	DNA ligase I: ATP-dependent enzyme essential for ligation of processed Okazaki fragments

[a] See text and Reference 144 for references.

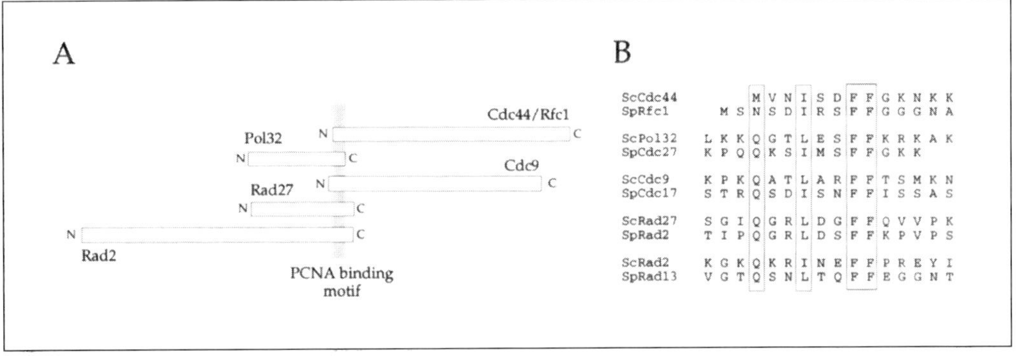

Fig. 8 PCNA binding via a conserved sequence motif. (A) Schematic representation of the structures of five *S. cerevisiae* proteins that bind to PCNA via the canonical PCNA-binding motif first identified in the mammalian p21[Cip1] DNA replication inhibitor. The location of the PCNA-binding motif within each protein is indicated by the shaded box. Rad2p is a homologue of the mammalian XP-G nuclease that functions in nucleotide excision repair. (B) Sequence alignment of conserved PCNA-binding motifs in *S. cerevisiae* (Sc) and *S. pombe* (Sp) proteins. See text for further details.

entire PCNA-binding motif, does not appear to affect the replication function of the protein (X. V. Gomes, S. L. Gary, and P. M. J. Burgers, unpublished results). This is in sharp contrast to the effects of deleting the PCNA-binding motif from the C-terminus of the *S. pombe* Pol δ subunit Cdc27; in this case the truncated protein is completely non-functional (164). Interestingly, the *S. pombe* Cdc24 protein (which has no apparent homologue in *S. cerevisiae*) also binds PCNA and contains a sequence that resembles the canonical PCNA-binding sequence, but deletion of this sequence does not prevent PCNA binding, suggesting that these residues are not required for the observed interaction (126).

6. Okazaki fragment processing

6.1 Overview

The joining of adjacent Okazaki fragments requires that they are first processed to remove the RNA segment of the RNA–DNA primer. The DNA segment of the primer is also likely also be replaced at this time; as this was originally synthesized by Pol α, which lacks proofreading activity, it is likely to contain mutations at an unacceptably high frequency. Resynthesis of the DNA segment of the primer by either of the proofreading polymerases, Pol δ or Pol ε, would leave the final product error-free.

In the SV40 system, in addition to Pol δ, processing of lagging-strand Okazaki fragments requires three distinct factors: the FEN-1 nuclease, DNA ligase I, and RNaseH1 (see reference 179 and references therein). In the yeasts, structural and functional homologues of FEN-1 and DNA ligase I have been extensively characterized and, together with the Dna2 helicase (discussed below), are believed to function in Okazaki fragment processing. In contrast, although the identity of the large subunit of RNaseH1 is known in both yeasts (180), whether this enzyme plays a role in the processing of Okazaki fragments during chromosomal DNA replication in yeast is unclear.

6.2 FEN-1 nuclease

The FEN-1 nuclease is required for Okazaki fragment processing in the SV40 system and its yeast homologues (Rad27p/Rth1p in *S. cerevisiae*, Rad2 in *S. pombe*) are believed to play a similar role. *S. cerevisiae rad27Δ* cells are viable, but grow slowly and are temperature-sensitive (arresting in S phase at the restrictive temperature), suggesting that at low temperatures the essential function of Rad27p/Rth1p can be performed by a different enzyme (18, 19, 181); see also reference 6. In this regard it may be significant that the Rad27p/Rth1p (*S. cerevisiae*) and Rad2 (*S. pombe*) proteins are members of a family of related endonucleases that includes the *S. cerevisiae* Rad2p and Exo1p proteins (homologues of *S. pombe* Rad13 and Exo1) implicated in the repair of UV-damaged DNA and recombination respectively (182, 183). It is conceivable that either of these proteins could contribute to Okazaki fragment processing in the absence of Rad27p/Rth1p (*S. cerevisiae*) or Rad2 (*S. pombe*). Alternatively, the

Dna2p helicase, which also has endonuclease activity, may substitute for Rad27p/Rth1p. In common with DNA ligase I (see below), the *S. cerevisiae* Rad27p/Rth1p and *S. pombe* Rad2 proteins also interact with PCNA via a p21^{Cip1}-like PCNA-binding motif (129) (see Fig. 8).

6.3 Dna2

Mutations in *S. cerevisiae DNA2* were originally identified in a screen for replication-deficient mutants using a permeabilized cell DNA replication assay (184–186). *DNA2* encodes an essential 172-kDa protein with protein sequence motifs characteristic of DNA helicase proteins. Biochemical analysis of immunoprecipitated Dna2p protein led to the conclusion that Dna2p possessed 3′–5′ DNA helicase activity (184, 185), although this latter point (specifically, the direction of translocation of the helicase on its DNA substrate) is now in dispute (187). It is possible that this discrepancy may be due to the helicase activity of Dna2p being modulated by interaction with other protein factors. The identity of these factors is unknown, but one candidate is Rad27p/Rth1p which appears to co-precipitate with Dna2p (186).

Although the precise role of the Dna2p helicase in the replication process remains unclear, there is increasing evidence that Dna2p functions in Okazaki fragment processing (187) (S.-H. Bae and Y.-S. Seo, personal communication; see Fig. 9 and section 6.5). In addition to possessing helicase activity, it has also been shown that highly purified *S. cerevisiae* Dna2p possesses ssDNA-specific endonuclease activity.

In *S. pombe* one additional essential factor has been identified that may have a role to play in either Okazaki fragment synthesis or processing. Cells carrying temperature-sensitive mutations in the *cdc24⁺* gene undergo cell cycle arrest in late S phase when shifted to the restrictive temperature (5, 126, 188). However, whilst bulk DNA replication is possible under the restrictive conditions, the nascent DNA in *cdc24* cells is approximately one-twentieth of the size of a chromosome (126, 188). The Cdc24 protein interacts physically with both Pcn1 and Rfc1 (126), while the *cdc24⁺* gene interacts genetically with *dna2⁺* (188) and with *cdm1⁺* which encodes the smallest subunit of Pol δ (152). Whether the latter two interactions are an indication of direct physical interactions between Cdc24 and either Dna2 or Pol δ remains to be seen. At present no homologues of Cdc24 are apparent in protein sequence databases.

6.4 DNA ligase I

Higher eukaryotes contain at least three distinct ATP-dependent DNA ligases, two of which (DNA ligases I and IV) have homologues in the yeasts (see reference 189 for a recent review). DNA ligase I plays an essential role in the replication process, being required for the ligation of Okazaki fragments on the lagging strand and for the completion of leading-strand synthesis. Cells carrying temperature-sensitive mutant alleles of *CDC9* (*S. cerevisiae*) or *cdc17⁺* (*S. pombe*) undergo cell cycle arrest in S phase when shifted to the restrictive temperature, accumulating nascent DNA of a size consistent with a failure of Okazaki fragment ligation (2, 4). In contrast, DNA ligase

IV (encoded by *LIG4/DNL4* in *S. cerevisiae*) has no part to play in chromosomal DNA replication. Instead, Lig4p/Dnl4p is a key component of the non-homologous end-joining apparatus implicated in the repair of double-strand breaks in DNA caused by ionizing radiation (190–192).

The *S. cerevisiae* Cdc9p and *S. pombe* Cdc17 proteins share approximately 45% primary sequence identity (193–195). Structurally, DNA ligase I proteins can be considered as being composed of at least four distinct protein domains (189): an N-terminal domain of variable length and limited cross-species sequence conservation (with the exception that all contain the canonical PCNA-binding motif discussed above), a central domain of about 200 amino acids that is conserved in all eukaryotic DNA ligase I homologues and also in archaeal DNA ligase proteins, and two C-terminal domains that are related to prokaryotic ATP-dependent ligases (such as T7 ligase). The active site of the enzyme lies within the T7-related domains, centering on an invariant lysine residue that is absolutely required for DNA ligase I function. Sequence analysis of *cdc17* mutant alleles of *S. pombe* has shown that these encode proteins with mutations within the central and C-terminal catalytic domains of the Cdc17 protein (I. Martin and S. A. MacNeill, unpublished results).

As noted above, in common with their higher eukaryotic counterparts, the N-terminal domains of the yeast DNA ligase I proteins contain a p21^{Cip1}-like Pcn1-binding motif (reviewed in reference 129) (see Fig. 8). Recent experiments have confirmed that the *S. pombe* Cdc17 and Pcn1 proteins do indeed interact with one another *in vitro* and that the PCNA-binding motif plays an important role in Cdc17 function (I. Martin and S. A. MacNeill, unpublished results). Whether DNA ligase I interacts with components of the replication machinery other than PCNA is not known.

6.5 A model for the processing of Okazaki fragments in yeast

Figure 2.9 shows one current model for Okazaki fragment processing in yeast, based on recent biochemical analysis of the properties of the *S. cerevisiae* Dna2p protein on a series of artificial substrates intended to resemble unprocessed Okazaki fragments (S.-H. Bae and Y.-S. Seo, personal communication) (187). In this model, Pol δ first displaces the 5′ end of the previous Okazaki fragment, forming an extended RNA–DNA flap structure. Dna2p then recognizes and cleaves the RNA–DNA primer, completely removing the RNA segment. Following dissociation of Pol δ from the DNA, the displaced strand reanneals to the template strand, and is cleaved by Rad27p/Rth1p (FEN-1). Finally the nicked DNA is sealed by Cdc9p (DNA ligase I). Whether this model presents a true picture of the events of Okazaki fragment processing remains to be seen, but it appears to correlate well with the biochemical properties of Dna2p and Rad27p/Rth1p. For example, cleavage by Dna2p of the DNA segment of the RNA–DNA primer is significantly stimulated by the presence of the RNA segment, and occurs efficiently in conjunction with displacement synthesis by Pol δ (S.-H. Bae and Y.-S. Seo, personal communication).

Plate 1 View of 53 clustered gene duplications between the 16 chromosomes of yeast. The chromosomes are represented by black horizontal lines and are aligned according to their centromeres. Gene clusters are represented by coloured boxes. A cluster is detected if, within a window of 25 kb, at least five blocks of coding DNA (block size 500 bp) are duplicated in a syntenic way. Ty elements and telomeres are filtered out because these sequences are known to be highly repetitive (taken from [7]).

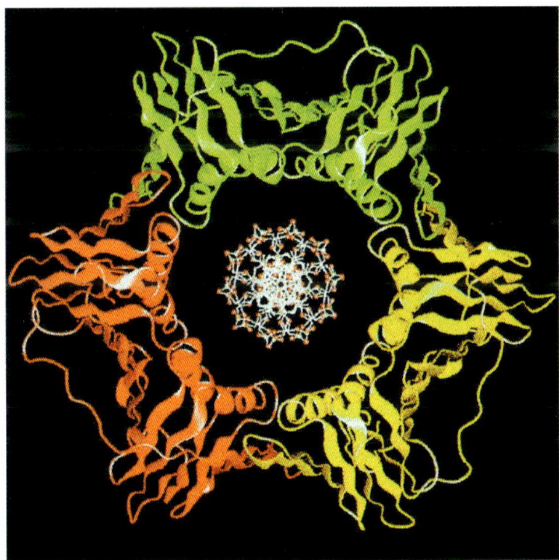

Plate 2 Ribbon representation of the polypeptide backbone of *S. cerevisiae* PCNA, with hypothetical duplex DNA. Strands of β sheet are shown as flat ribbons; α helices are shown as spirals. Individual monomers within the ring are distinguished by different colours. A hypothetical model of duplex DNA is placed in the geometrical centre of the ring, to illustrate the hypothesis that PCNA encircles duplex DNA. Reproduced, with permission, from Reference 30. © Cell Press, Cambridge, Massachusetts, USA.

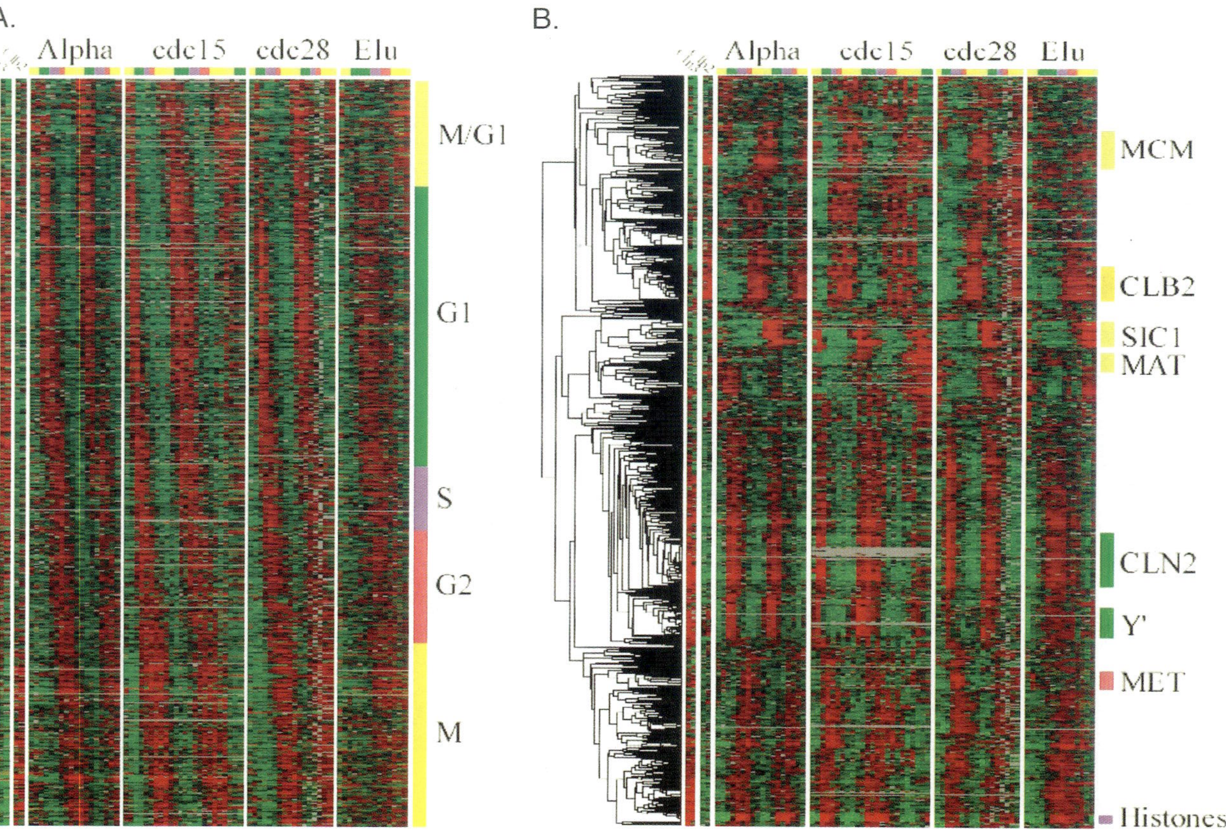

Plate 3 Genome-wide analysis of cell cycle-regulated transcription in *S. cerevisiae*. Cultures were synchronized by mating pheromone block and release (Alpha), *cdc15* or *cdc28* block and release, or elutriation of G1 cells (Elu), and progress was followed for one or two cell cycles, as represented by vertical columns advancing from left to right. Each cell cycle-regulated gene is represented by a horizontal line. Induced genes are red, repressed genes are green, and unchanged genes are black. The two columns at the extreme left represent genes that are induced or repressed by *CLN3* or *CLB2* overexpression. (A) Genes ordered according to expression pattern. (B) Clusters of co-regulated genes, as described in Table 3.3. Only 400 of the 800 cell cycle-regulated genes fit into the eight groups of tightly clustered genes. Reprinted with permission by the American Society for Cell Biology from Reference 77. For details see: http//cellcycle-www.stanford.edu

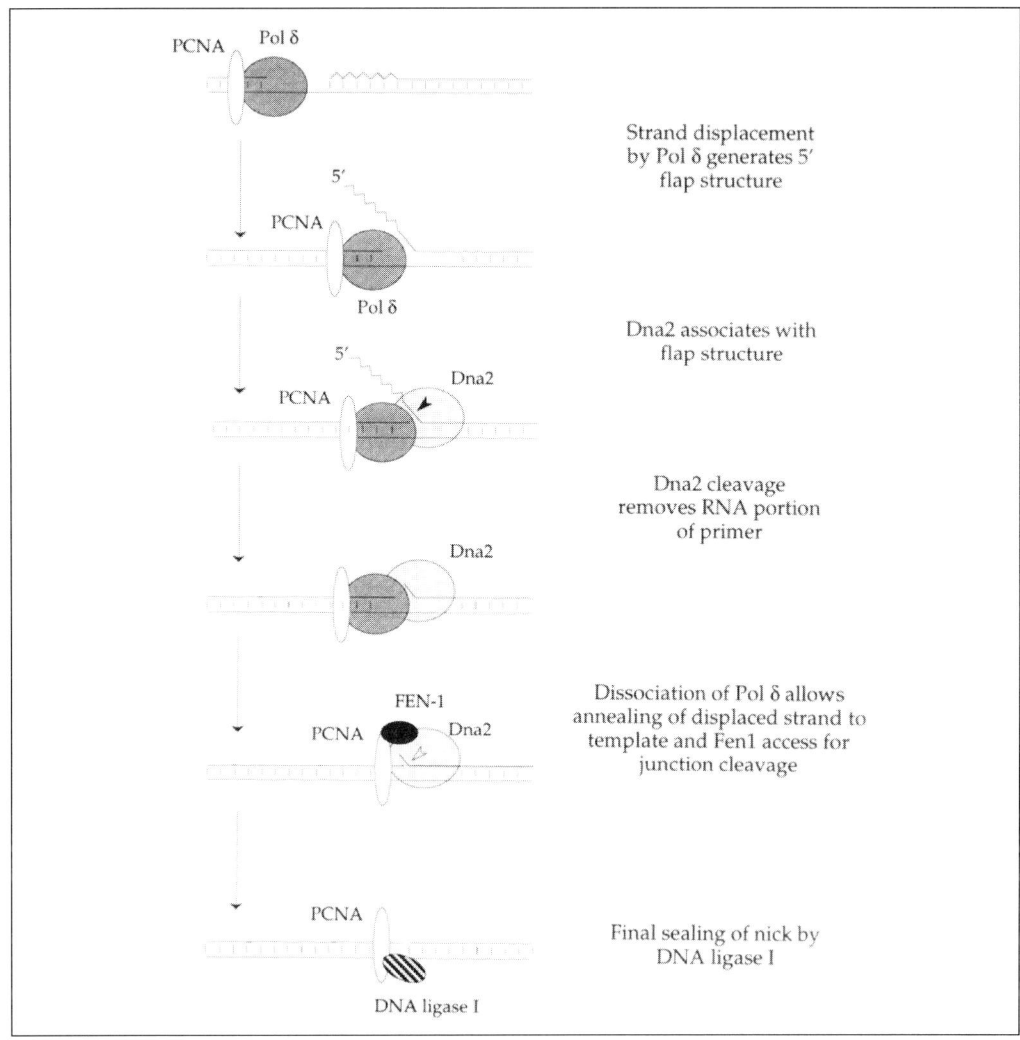

Fig. 9 A speculative model for Okazaki fragment processing in yeast based on *in vitro* studies of the *S. cerevisiae* Dna2p enzyme (S.-H. Bae and Y.-S. Seo, personal communication). Displacement synthesis by Pol δ–PCNA results in the formation of an extended RNA–DNA flap structure that is subsequently cleaved by Dna2p. Dissociation of Pol δ then allows junction cleavage by Rad2p (FEN-1) and sealing of the nick by Cdc9p (DNA ligase I). The arrowheads indicate the positions of cleavage by Dna2p (black arrowhead) and Rad2p (white arrowhead). Note that Pol δ, Rad2p, and Cdc9p all interact directly with PCNA (as in Fig. 8).

7. Conclusions

Recent years have seen rapid advances in understanding the molecular mechanisms of chromosomal DNA replication in the yeasts and of the complex multiprotein structures that are responsible for ensuring that the complete 15 Mb genomes of these unicellular organisms are replicated with such astonishing accuracy within the time available. Despite these advances, however, there are still a large number of questions that remain to be answered before the molecular events of the replication fork can be described in full. For example, while there have been a number of reports of interactions between different protein complexes present at the replication fork, such as Cdc45p and the MCM proteins (83, 196, 197) or RF-C and PCNA (122, 127, 131), little is known of the precise manner in which these proteins contact one another or of the effects on the replication process of specifically disrupting such interactions. Indeed it is striking that of the 30 or so proteins identified as being essential for chromosomal DNA replication in yeast, the three-dimensional structure of only one (*S. cerevisiae* PCNA, Pol30p) has been solved (30) (see Plate 1). Given the high degree of conservation apparent between the yeast replication fork and its higher eukaryotic counterpart, future work aimed at addressing these fundamental issues in *S. cerevisiae* and *S. pombe* will no doubt prove highly informative for understanding the molecular biology of chromosomal DNA replication in all eukaryotic cells.

Acknowledgements

The authors thank the members of their laboratories for reading and commenting on the manuscript, Y. Adachi for supplying the EM images of *S. pombe* MCM complexes, and S.-H. Bae and Y.-S. Seo for communicating unpublished results. Work in S. M.'s laboratory is supported by the Wellcome Trust. Work in P. B.'s laboratory is in part supported by NIH grants GM 32431 and GM 58534.

References

1. Hartwell, L. H. (1973) Three additional genes required for deoxyribonucleic acid synthesis in *Saccharomyces cerevisiae*. *J. Bacteriol.*, **115,** 966.
2. Hartwell, L. H. (1971) Genetic control of cell division in yeast. II. Genes controlling DNA replication and its initiation. *J. Mol. Biol.*, **59,** 183.
3. Nurse, P., Thuriaux, P., and Nasmyth, K. (1976) Genetic control of the cell division cycle in the fission yeast *Schizosaccharomyces pombe*. *Mol. Gen. Genet.*, **146,** 167.
4. Nasmyth, K. A. (1977) Temperature-sensitive lethal mutants in the structural gene for DNA ligase in the yeast *Schizosaccharomyces pombe*. *Cell,* **12,** 1109.
5. Nasmyth, K. and Nurse, P. (1981) Cell-division cycle mutants altered in DNA replication and mitosis in the fission yeast *Schizosaccharomyces pombe*. *Mol. Gen. Genet.*, **182,** 119.
6. Murray, J. M., Tavassoli, M., Alharithy, R., Sheldrick, K. S., Lehmann, A. R., Carr, A. M., and Watts, F. Z. (1994) Structural and functional conservation of the human homolog of the *Schizosaccharomyces pombe rad2* gene, which is required for chromosome segregation and recovery from DNA damage. *Mol. Cell. Biol.,* **14,** 4878.
7. Parker, A. E., Clyne, R. K., Carr, A. M., and Kelly, T. J. (1997) The *Schizosaccharomyces pombe rad11*+ gene encodes the large subunit of replication protein A. *Mol. Cell. Biol.,* **17,** 2381.
8. Takeda, T., Ogino, K., Matsui, E., Cho, M. K., Kumagai, H., Miyake, T., Arai, K., and Masai, H. (1999) A fission yeast gene, *him1*+/*dfp1*+, encoding a regulatory subunit for Hsk1 kinase, plays essential roles in S-phase initiation as well as in S-phase checkpoint control and recovery from DNA damage. *Mol. Cell. Biol.,* **19,** 5535.

9. D'Urso, G. and Nurse, P. (1997) *Schizosaccharomyces pombe cdc20*⁺ encodes DNA polymerase epsilon and is required for chromosomal replication but not for the S phase checkpoint. *Proc. Natl. Acad. Sci. U.S.A.*, **94**, 12491.

10. Sugino, A., Ohara, T., Sebastian, J., Nakashima, N., and Araki, H. (1998) DNA polymerase epsilon encoded by *cdc20*⁺ is required for chromosomal DNA replication in the fission yeast *Schizosaccharomyces pombe*. *Genes to Cells*, **3**, 99.

11. Murakami, H. and Okayama, H. (1995) A kinase from fission yeast responsible for blocking mitosis in S-phase. *Nature*, **374**, 817.

12. Iino, Y. and Yamamoto, M. (1997) The *Schizosaccharomyces pombe cdc6* gene encodes the catalytic subunit of DNA polymerase δ. *Mol. Gen. Genet.*, **254**, 93.

13. Damagnez, V., Tillit, J., Derecondo, A. M., and Baldacci, G. (1991) The *pol1* gene from the fission yeast, *Schizosaccharomyces pombe*, shows conserved amino-acid blocks specific for eukaryotic DNA polymerases α. *Mol. Gen. Genet.*, **226**, 182.

14. Pignede, G., Bouvier, D., Derecondo, A. M., and Baldacci, G. (1991) Characterization of the *pol3* gene-product from *Schizosaccharomyces pombe* indicates interspecies conservation of the catalytic subunit of DNA polymerase-delta. *J. Mol. Biol.*, **222**, 209.

15. Park, H., Francesconi, S., and Wang, T. S. F. (1993) Cell-cycle expression of 2 replicative DNA polymerases α and δ from *Schizosaccharomyces pombe*. *Mol. Biol. Cell*, **4**, 145.

16. Burgers, P. M. J. and Gerik, K. J. (1998) Structure and processivity of two forms of *Saccharomyces cerevisiae* DNA polymerase δ. *J. Biol. Chem.*, **273**, 19756.

17. Gerik, K. J., Li, X. Y., Pautz, A., and Burgers, P. M. J. (1998) Characterization of the two small subunits of *Saccharomyces cerevisiae* DNA polymerase δ. *J. Biol. Chem.*, **273**, 19747.

18. Reagan, M. S., Pittenger, C., Siede, W., and Friedberg, E. C. (1995) Characterization of a mutant strain of *Saccharomyces cerevisiae* with a deletion of the *rad27* gene, a structural homolog of the *rad2* nucleotide excision-repair gene. *J. Bacteriol.*, **177**, 364.

19. Sommers, C. H., Miller, E. J., Dujon, B., Prakash, S., and Prakash, L. (1995) Conditional lethality of null mutations in *rth1* that encodes the yeast counterpart of a mammalian 5′-exonuclease to 3′-exonuclease required for lagging-strand DNA synthesis in reconstituted systems. *J. Biol. Chem.*, **270**, 4193.

20. Jacquier, A., Legrain, P., and Dujon, B. (1992) Sequence of a 10.7 kb segment of yeast chromosome XI identifies the *apn1* and the *baf1* loci and reveals one transfer RNA gene and several new open reading frames including homologs to *rad2* and kinases. *Yeast*, **8**, 121.

21. Jacob, F., Brenner, S., and Cuzin, F. (1964) On the regulation of DNA replication in bacteria. *Cold Spring Harbor Symp. Quant. Biol.*, **28**, 329.

22. Kornberg, A. and Baker, T. A. (1991) *DNA replication*. Freeman, New York.

23. Stillman, B. (1996) Comparison of DNA replication in cells from prokarya and eukarya. In *DNA replication in eukaryotic cells* (ed . M. L. DePamphilis), p. 435. Cold Spring Harbor Laboratory Press, New York.

24. Bell, S. P. and Stillman, B. (1992) ATP-dependent recognition of eukaryotic origins of DNA replication by a multiprotein complex. *Nature*, **357**, 128.

25. Iftode, C., Daniely, Y., and Borowiec, J. A. (1999) Replication protein A (RPA): The eukaryotic SSB. *Crit. Rev. Biochem. Mol. Biol.*, **34**, 141.

26. Tye, B. K. (1999) MCM proteins in DNA replication. *Annu. Rev. Biochem.*, **68**, 649.

27. Ishimi, Y. (1997) A DNA helicase activity is associated with an MCM4, -6, and -7 protein complex. *J. Biol. Chem.*, **272**, 24508.

28. Aparicio, O. M., Stout, A. M., and Bell, S. P. (1999) Differential assembly of Cdc45p and DNA polymerases at early and late origins of DNA replication. *Proc. Natl. Acad. Sci. U.S.A.*, **96**, 9130.

29. Aparicio, O. M., Weinstein, D. M., and Bell, S. P. (1997) Components and dynamics of DNA replication complexes in *S. cerevisiae*: redistribution of MCM proteins and Cdc45p during S phase. *Cell*, **91**, 59.

30. Krishna, T. S. R., Kong, X. P., Gary, S., Burgers, P. M., and Kuriyan, J. (1994) Crystal structure of the eukaryotic DNA polymerase processivity factor PCNA. *Cell*, **79**, 1233.

31. Sinha, N.K., Morris, C.F., and Alberts, B.M. (1980) Efficient *in vitro* replication of doubled-stranded DNA templates by a purified T4 bacteriophage replication system. *J. Biol. Chem.*, **255**, 4290.

32. Lee, J., Chastain, P. D., Kusakabe, T., Griffith, J. D., and Richardson, C. C. (1998) Coordinated leading and lagging strand DNA synthesis on a minicircular template. *Mol. Cell*, **1**, 1001.

33. Stinchcomb, D. T., Struhl, K., and Davis, R. W. (1979) Isolation and characterisation of a yeast chromosomal replicator. *Nature*, **282**, 39.

34. Hsiao, C. L. and Carbon, J. (1979) High-frequency transformation of yeast by plasmids containing the cloned yeast *ARG4* gene. *Proc. Natl. Acad. Sci. U.S.A.*, **76**, 3829.

35. Maundrell, K., Hutchison, A., and Shall, S. (1988) Sequence analysis of ARS elements in fission yeast. *EMBO J.*, **7**, 2203.

36. Brewer, B. J. and Fangman, W. L. (1987) The localization of replication origins on ars plasmids in *Saccharomyces cerevisiae*. *Cell*, **51**, 463.

37. Huberman, J. A., Spotila, L. D., Nawotka, K. A., Elassouli, S. M., and Davis, L. R. (1987) The *in vivo* replication origin of the yeast 2 μm plasmid. *Cell*, **51**, 473.

38. Marahrens, Y. and Stillman, B. (1996) The initiation of DNA replication in the yeast *Saccharomyces cerevisiae*. In *Eukaryotic DNA replication* (ed J. J. Blow), p. 66. IRL Press, Oxford.

39. Bielinsky, A. K. and Gerbi, S. A. (1999) Chromosomal *ARS1* has a single leading strand start site. *Mol. Cell*, **3**, 477.

40. Ferguson, B. M., Brewer, B. J., Reynolds, A. E., and Fangman, W. L. (1991) A yeast origin of replication is activated late in S-phase. *Cell*, **65**, 507.

41. Deshpande, A. M. and Newlon, C. S. (1992) The ARS consensus sequence is required for chromosomal origin function in *Saccharomyces cerevisiae*. *Mol. Cell. Biol.*, **12**, 4305.

42. Ferguson, B. M. and Fangman, W. L. (1992) A position effect on the time of replication origin activation in yeast. *Cell*, **68**, 333.

43. Brewer, B. J. and Fangman, W. L. (1993) Initiation at closely spaced replication origins in a yeast chromosome. *Science*, **262**, 1728.

44. Marahrens, Y. and Stillman, B. (1994) Replicator dominance in a eukaryotic chromosome. *EMBO J.*, **13**, 3395.

45. Newlon, C. S., Collins, I., Dershowitz, A., Deshpande, A. M., Greenfeder, S. A., Ong, L. Y., and Theis, J. F. (1993) Analysis of replication origin function on chromosome III of *Saccharomyces cerevisiae*. *CSH Symp. Quant. Biol.*, **58**, 415.

46. Celniker, S. E., Sweder, K., Srienc, F., Bailey, J. E., and Campbell, J. L. (1984) Deletion mutations affecting autonomously replicating sequence *ARS1* of *Saccharomyces cerevisiae*. *Mol. Cell. Biol.*, **4**, 2455.

47. Dubey, D. D., Kim, S. M., Todorov, I. T., and Huberman, J. A. (1996) Large, complex modular structure of a fission yeast DNA replication origin. *Curr. Biol.*, **6**, 467.

48. Clyne, R. K. and Kelly, T. J. (1995) Genetic analysis of an ARS element from the fission yeast *Schizosaccharomyces pombe*. *EMBO J.*, **14**, 6348.

49. Palzkill, T. G. and Newlon, C. S. (1988) A yeast replication origin consists of multiple copies of a small conserved sequence. *Cell*, **53**, 441.

50. Bouton, A. H. and Smith, M. M. (1986) Fine-structure analysis of the DNA sequence

requirements for autonomous replication of *Saccharomyces cerevisiae* plasmids. *Mol. Cell. Biol.*, **6**, 2354.

51. Walker, S. S., Malik, A. K., and Eisenberg, S. (1991) Analysis of the interactions of functional domains of a nuclear origin of replication from *Saccharomyces cerevisiae. Nucl. Acids Res.*, **19**, 6255.

52. Diffley, J. F. X. and Stillman, B. (1988) Purification of a yeast protein that binds to origins of DNA replication and a transcriptional silencer. *Proc. Natl. Acad. Sci. U.S.A.*, **85**, 2120.

53. Rhode, P. R., Elsasser, S., and Campbell, J. L. (1992) Role of multifunctional autonomously replicating sequence binding factor I in the initiation of DNA replication and transcriptional control in *Saccharomyces cerevisiae. Mol. Cell. Biol.*, **12**, 1064.

54. Natale, D. A., Umek, R. M., and Kowalski, D. (1993) Ease of DNA unwinding is a conserved property of yeast replication origins. *Nucl. Acids Res.*, **21**, 555.

55. Huang, R. Y. and Kowalski, D. (1993) A DNA unwinding element and an ARS consensus comprise a replication origin within a yeast chromosome. *EMBO J.*, **12**, 4521.

56. Lin, S. and Kowalski, D. (1997) Functional equivalency and diversity of *cis*-acting elements among yeast replication origins. *Mol. Cell. Biol.*, **17**, 5473.

57. Bielinsky, A. K. and Gerbi, S. A. (1998) Discrete start sites for DNA synthesis in the yeast *ARS1* origin. *Science*, **279**, 95.

58. Gomez, M. and Antequera, F. (1999) Organization of DNA replication origins in the fission yeast genome. *EMBO J.*, **18**, 5683.

59. Friedman, K. L., Raghuraman, M. K., Fangman, W. L., and Brewer, B. J. (1995) Analysis of the temporal program of replication initiation in yeast chromosomes. *J. Cell Sci.*, 51.

60. Friedman, K. L., Diller, J. D., Ferguson, B. M., Nyland, S. V. M., Brewer, B. J., and Fangman, W. L. (1996) Multiple determinants controlling activation of yeast replication origins late in S phase. *Genes Devel.*, **10**, 1595.

61. Diffley, J. F. X. and Cocker, J. H. (1992) Protein–DNA interactions at a yeast replication origin. *Nature*, **357**, 169.

62. Pasero, P., Braguglia, D., and Gasser, S. M. (1997) ORC-dependent and origin-specific initiation of DNA replication at defined foci in isolated yeast nuclei. *Genes Devel.*, **11**, 1504.

63. Tugal, T., ZouYang, X. H., Gavin, K., Pappin, D., Canas, B., Kobayashi, R., Hunt, T., and Stillman, B. (1998) The Orc4p and Orc5p subunits of the *Xenopus* and human origin recognition complex are related to Orc1p and Cdc6p. *J. Biol. Chem.*, **273**, 32421.

64. Fox, C. A. and Rine, J. (1996) Influences of the cell cycle on silencing. *Curr. Opin. Cell Biol.*, **8**, 354.

65. Rao, H. and Stillman, B. (1995) The origin recognition complex interacts with a bipartite DNA binding site within yeast replicators. *Proc. Natl. Acad. Sci. U.S.A.*, **92**, 2224.

66. Rowley, A., Cocker, J. H., Harwood, J. and Diffley, J. F. X. (1995) Initiation complex assembly at budding yeast replication origins begins with the recognition of a bipartite sequence by limiting amounts of the initiator, ORC, *EMBO J.*, **14**, 2631.

67. Lee, D. G. and Bell, S. P. (1997) Architecture of the yeast origin recognition complex bound to origins of DNA replication. *Mol. Cell. Biol.*, **17**, 7159.

68. Bell, S. P., Kobayashi, R., and Stillman, B. (1993) Yeast origin recognition complex functions in transcription silencing and DNA replication. *Science*, **262**, 1844.

69. Bell, S. P., Mitchell, J., Leber, J., Kobayashi, R., and Stillman, B. (1995) The multidomain structure of orc1p reveals similarity to regulators of DNA replication and transcriptional silencing. *Cell*, **83**, 563.

70. Loo, S., Fox, C. A., Rine, J., Kobayashi, R., Stillman, B., and Bell, S. (1995) The origin recognition complex in silencing, cell-cycle progression, and DNA replication. *Mol. Biol. Cell*, **6**, 741.

71. Foss, M., McNally, F. J., Laurenson, P., and Rine, J. (1993) Origin recognition complex (ORC) in transcriptional silencing and DNA replication in *Saccharomyces cerevisiae*. *Science*, **262**, 1838.

72. Li, J. J. and Herskowitz, I. (1993) Isolation of *orc6*, a component of the yeast origin recognition complex by a one-hybrid system. *Science*, **262**, 1870.

73. Micklem, G., Rowley, A., Harwood, J., Nasmyth, K., and Diffley, J. F. X. (1993) Yeast origin recognition complex is involved in DNA replication and transcriptional silencing. *Nature*, **366**, 87.

74. Dutta, A. and Bell, S. P. (1997) Initiation of DNA replication in eukaryotic cells. *Annu. Rev. Cell Devel. Biol.*, **13**, 293.

75. Grallert, B. and Nurse, P. (1996) The *ORC1* homolog *orp1* in fission yeast plays a key role in regulating onset of S phase. *Genes Devel.*, **10**, 2644.

76. Muzi-Falconi, M. and Kelly, T. J. (1995) Orp1, a member of the Cdc18/Cdc6 family of S-phase regulators, is homologous to a component of the origin recognition complex. *Proc. Natl. Acad. Sci. U.S.A.*, **92**, 12475.

77. Gavin, K. A., Hidaka, M., and Stillman, B. (1995) Conserved initiator proteins in eukaryotes. *Science*, **270**, 1667.

78. Leatherwood, J., LopezGirona, A., and Russell, P. (1996) Interaction of Cdc2 and Cdc18 with a fission yeast ORC2-like protein. *Nature*, **379**, 360.

79. Chuang, R. Y. and Kelly, T. J. (1999) The fission yeast homologue of Orc4p binds to replication origin DNA via multiple AT-hooks. *Proc. Natl. Acad. Sci. U.S.A.*, **96**, 2656.

80. Ishiai, M., Dean, F. B., Okumura, K., Abe, M., Moon, K. Y., Amin, A. A., Kagotani, K., Taguchi, H., Murakami, Y., Hanaoka, F., Odonnell, M., Hurwitz, J., and Eki, T. (1997) Isolation of human and fission yeast homologues of the budding yeast origin recognition complex subunit ORC5: human homologue (ORC5L) maps to 7q22. *Genomics*, **46**, 294.

81. Grallert, B. and Nurse, P. (1997) An approach to identify functional homologues and suppressors of genes in fission yeast. *Curr. Genet.*, **32**, 27.

82. Klemm, R. D., Austin, R. J., and Bell, S. P. (1997) Coordinate binding of ATP and origin DNA regulates the ATPase activity of the origin recognition complex. *Cell*, **88**, 493.

83. Zou, L., Mitchell, J., and Stillman, B. (1997) *CDC45*, a novel yeast gene that functions with the origin recognition complex and MCM proteins in initiation of DNA replication. *Mol. Cell. Biol.*, **17**, 553.

84. Hogan, E. and Koshland, D. (1992) Addition of extra origins of replication to a minichromosome suppresses its mitotic loss in *cdc6* and *cdc14* mutants of *Saccharomyces cerevisiae*. *Proc. Natl. Acad. Sci. U.S.A.*, **89**, 3098.

85. Liang, C., Weinreich, M., and Stillman, B. (1995) ORC and Cdc6p interact and determine the frequency of initiation of DNA replication in the genome. *Cell*, **81**, 667.

86. Kroll, E. S., Hyland, K. M., Hieter, P., and Li, J. J. (1996) Establishing genetic interactions by a synthetic dosage lethality phenotype. *Genetics*, **143**, 95.

87. Cocker, J. H., Piatti, S., Santocanale, C., Nasmyth, K., and Diffley, J. F. X. (1996) An essential role for the Cdc6 protein in forming the pre-replicative complexes of budding yeast. *Nature*, **379**, 180.

88. Perkins, G. and Diffley, J. F. X. (1998) Nucleotide-dependent prereplicative complex assembly by Cdc6p, a homolog of eukaryotic and prokaryotic clamp-loaders. *Mol. Cell*, **2**, 23.

89. West, S. C. (1996) DNA helicases: new breeds of translocating motors and molecular pumps. *Cell*, **86,** 177.

90. Lei, M., Kawasaki, Y., and Tye, B. K. (1996) Physical interactions among Mcm proteins and effects of Mcm dosage on DNA replication in *Saccharomyces cerevisiae. Mol. Cell. Biol.*, **16,** 5081.

91. Adachi, Y., Usukura, J., and Yanagida, M. (1997) A globular complex formation by Nda1 and the other five members of the MCM protein family in fission yeast. *Genes to Cells*, **2,** 467.

92. Masai, H., Miyake, T., and Arai, K. (1995) Hsk1(+), a *Schizosaccharomyces pombe* gene related to *Saccharomyces cerevisiae cdc7*, is required for chromosomal replication. *EMBO J.*, **14,** 3094.

93. Brown, G. W. and Kelly, T. J. (1998) Purification of Hsk1, a minichromosome maintenance protein kinase from fission yeast. *J. Biol. Chem.*, **273,** 22083.

94. Brown, G. W. and Kelly, T. J. (1999) Cell cycle regulation of Dfp1, an activator of the Hsk1 protein kinase. *Proc. Natl. Acad. Sci. U.S.A.*, **96,** 8443.

95. Miyake, S. and Yamashita, S. (1998) Identification of *sna41* gene, which is the suppressor of *nda4* mutation and is involved in DNA replication in *Schizosaccharomyces pombe. Genes to Cells*, **3,** 157.

96. Brill, S. J. and Stillman, B. (1989) Yeast replication factor A functions in the unwinding of the SV40 origin of DNA replication. *Nature*, **342,** 92.

97. Tanaka, T. and Nasmyth, K. (1998) Association of RPA with chromosomal replication origins requires an Mcm protein, and is regulated by Rad53, and cyclin- and Dbf4-dependent kinases. *EMBO J.*, **17,** 5182.

98. Brill, S. J. and Stillman, B. (1991) Replication factor A from *Saccharomyces cerevisiae* is encoded by three essential genes coordinately expressed at S-phase. *Genes Devel.*, **5,** 1589.

99. Maniar, H. S., Wilson, R., and Brill, S. J. (1997) Roles of replication protein A subunits 2 and 3 in DNA replication fork movement in *Saccharomyces cerevisiae. Genetics*, **145,** 891.

100. Santocanale, C., Neecke, H., Longhese, M. P., Lucchini, G., and Plevani, P. (1995) Mutations in the gene encoding the 34 kDa subunit of yeast replication protein A cause defective S-phase progression. *J. Mol. Biol.*, **254,** 595.

101. Ishiai, M., Sanchez, J. P., Amin, A. A., Murakami, Y., and Hurwitz, J. (1996) Purification, gene cloning, and reconstitution of the heterotrimeric single-stranded DNA binding protein from *Schizosaccharomyces pombe. J. Biol. Chem.*, **271,** 20868.

102. Treuner, K., Findeisen, M., Strausfeld, U., and Knippers, R. (1999) Phosphorylation of replication protein a middle subunit (RPA32) leads to a disassembly of the RPA heterotrimer. *J. Biol. Chem.*, **274,** 15556.

103. Philipova, D., Mullen, J. R., Maniar, H. S., Lu, J. A., Gu, C. Y., and Brill, S. J. (1996) A hierarchy of SSB protomers in replication protein A. *Genes Devel.*, **10,** 2222.

104. Brill, S. J. and Bastin-Shanower, S. (1998) Identification and characterization of the fourth single-stranded-DNA binding domain of replication protein A. *Mol. Cell. Biol.*, **18,** 7225.

105. Burgers, P. M. J. (1996) Enzymology of the replication fork. In *Eukaryotic DNA replication* (ed J. J. Blow), p. 1. IRL Press, Oxford.

106. D'Urso, G., Grallert, B., and Nurse, P. (1995) DNA polymerase α, a component of the replication initiation complex, is essential for the checkpoint coupling S-phase to mitosis in fission yeast. *J. Cell Sci.*, **108,** 3109.

107. Bouvier, D., Pignede, G., Damagnez, V., Tillit, J., Derecondo, A. M., and Baldacci, G. (1992) DNA polymerase α in the fission yeast *Schizosaccharomyces pombe*—identification and tracing of the catalytic subunit during the cell-cycle. *Exp. Cell Res.*, **198,** 183.

108. Bouvier, D. and Baldacci, G. (1995) The N-terminus of fission yeast DNA polymerase α contains a basic pentapeptide that acts *in vivo* as a nuclear-localization signal. *Mol. Biol. Cell*, **6**, 1697.

109. Foiani, M., Marini, F., Gamba, D., Lucchini, G., and Plevani, P. (1994) The B-subunit of the DNA polymerase α-primase complex in *Saccharomyces cerevisiae* executes an essential function at the initial stage of DNA replication. *Mol. Cell. Biol.*, **14**, 923.

110. Waga, S. and Stillman, B. (1994) Anatomy of a DNA replication fork revealed by reconstitution of SV40 DNA replication *in vitro*. *Nature*, **369**, 207.

111. Tsurimoto, T. and Stillman, B. (1991) Replication factors required for SV40 DNA replication *in vitro*. 1. DNA structure-specific recognition of a primer–template junction by eukaryotic DNA polymerases and their accessory proteins. *J. Biol. Chem.*, **266**, 1950.

112. Tsurimoto, T. and Stillman, B. (1989) Purification of a cellular replication factor, RF-C, that is required for coordinated synthesis of leading and lagging strands during simian virus-40 DNA replication *in vitro*. *Mol. Cell. Biol.*, **9**, 609.

113. Lee, S. H., Kwong, A. D., Pan, Z. Q., and Hurwitz, J. (1991) Studies on the activator-1 protein complex, an accessory factor for proliferating cell nuclear antigen-dependent DNA polymerase δ. *J. Biol. Chem.*, **266**, 594.

114. Yoder, B. L. and Burgers, P. M. J. (1991) *Saccharomyces cerevisiae* replication factor C .1. Purification and characterization of its ATPase activity. *J. Biol. Chem.*, **266**, 22689.

115. Burgers, P. M. J. (1991) *Saccharomyces cerevisiae* replication factor C. 2. Formation and activity of complexes with the proliferating cell nuclear antigen and with DNA polymerase-delta and polymerase-epsilon. *J. Biol. Chem.*, **266**, 22698.

116. Fien, K. and Stillman, B. (1992) Identification of replication factor C from *Saccharomyces cerevisiae*—a component of the leading-strand DNA replication complex. *Mol. Cell. Biol.*, **12**, 155.

117. Gerik, K. J., Gary, S. L., and Burgers, P. M. J. (1997) Overproduction and affinity purification of *Saccharomyces cerevisiae* replication factor C. *J. Biol. Chem.*, **272**, 1256.

118. Zuo, S. J., Gibbs, E., Kelman, Z., Wang, T. S. F., O'Donnell, M., MacNeill, S. A., and Hurwitz, J. (1997) DNA polymerase δ isolated from *Schizosaccharomyces pombe* contains five subunits. *Proc. Natl. Acad. Sci. U.S.A.*, **94**, 11244.

119. Gary, S. L. and Burgers, P. M. J. (1995) Identification of the fifth subunit of *Saccharomyces cerevisiae* replication factor C. *Nucl. Acids Res.*, **23**, 4986.

120. Li, X. Y. and Burgers, P. M. J. (1994) Molecular-cloning and expression of the *Saccharomyces cerevisiae rfc3* gene, an essential component of replication factor C. *Proc. Natl. Acad. Sci. U.S.A.*, **91**, 868.

121. Li, X. Y. and Burgers, P. M. J. (1994) Cloning and characterization of the essential *Saccharomyces cerevisiae rfc4* gene encoding the 37 kDa subunit of replication factor C. *J. Biol. Chem.*, **269**, 21880.

122. McAlear, M. A., Tuffo, K. M., and Holm, C. (1996) The large subunit of replication factor C (Rfc1p/Cdc44p) is required for DNA replication and DNA repair in *Saccharomyces cerevisiae*. *Genetics*, **142**, 65.

123. Noskov, V., Maki, S., Kawasaki, Y., Leem, S. H., Ono, B. I., Araki, H., Pavlov, Y., and Sugino, A. (1994) The *rfc2* gene encoding a subunit of replication factor C of *Saccharomyces cerevisiae*. *Nucl. Acids Res.*, **22**, 1527.

124. Noskov, V. N., Araki, H., and Sugino, A. (1998) The *RFC2* gene, encoding the third largest subunit of the replication factor C complex, is required for an S-phase checkpoint in *Saccharomyces cerevisiae*. *Mol. Cell. Biol.*, **18**, 4914.

125. Reynolds, N., Fantes, P. A., and MacNeill, S. A. (1999) A key role for replication factor C

(RF-C) in DNA replication checkpoint function in fission yeast. *Nucl. Acids Res.*, **27,** 462.

126. Tanaka, H., Tanaka, K., Murakami, H., and Okayama, H. (1999) Fission yeast Cdc24 is a replication factor C- and proliferating cell nuclear antigen-interacting factor essential for S-phase completion. *Mol. Cell. Biol.*, **19,** 1038.

127. Howell, E. A., McAlear, M. A., Rose, D., and Holm, C. (1994) Cdc44—a putative nucleotide-binding protein required for cell-cycle progression that has homology to subunits of replication factor C. *Mol. Cell. Biol.*, **14,** 255.

128. Cullmann, G., Fien, K., Kobayashi, R., and Stillman, B. (1995) Characterization of the 5 replication factor C genes of *Saccharomyces cerevisiae. Mol. Cell. Biol.*, **15,** 4661.

129. Warbrick, E. (1998) PCNA binding through a conserved motif. *Bioessays*, **20,** 195.

130. Sugimoto, K., Shimomura, T., Hashimoto, K., Araki, H., Sugino, A., and Matsumoto, K. (1996) Rfc5, a small subunit of replication factor C complex, couples DNA replication and mitosis in budding yeast. *Proc. Natl. Acad. Sci. U.S.A.*, **93,** 7048.

131. McAlear, M. A., Howell, E. A., Espenshade, K. K., and Holm, C. (1994) Proliferating cell nuclear antigen (*pol30*) mutations suppress *cdc44* mutations and identify potential regions of interaction between the 2 encoded proteins. *Mol. Cell. Biol.*, **14,** 4390.

132. Beckwith, W. H., Sun, Q. A., Bosso, R., Gerik, K. J., Burgers, P. M. J., and McAlear, M. A. (1998) Destabilized PCNA trimers suppress defective Rfc1 proteins *in vivo* and *in vitro. Biochemistry*, **37,** 3711.

133. Gray, F. C. and MacNeill, S. A. (2000) The *Schizosacchromyces pombe rfc3*$^+$ gene encodes a homologue of the human hRFC36 and *Saccharomyces cerevisiae* Rfc3 subunits of replication factor C. *Curr. Genet.*, **37,** 159.

134. Edgell, D. R. and Doolittle, W. F. (1997) Archaea and the origin(s) of DNA replication proteins. *Cell*, **89,** 995.

135. Amin, N. S. and Holm, C. (1996) *In vivo* analysis reveals that the interdomain region of the yeast proliferating cell nuclear antigen is important for DNA replication and DNA repair. *Genetics*, **144,** 479.

136. Ayyagari, R., Impellizzeri, K. J., Yoder, B. L., Gary, S. L., and Burgers, P. M. J. (1995) A mutational analysis of the yeast proliferating cell nuclear antigen indicates distinct roles in DNA replication and DNA repair. *Mol. Cell. Biol.*, **15,** 4420.

137. Eissenberg, J. C., Ayyagari, R., Gomes, X. V., and Burgers, P. M. J. (1997) Mutations in yeast proliferating cell nuclear antigen define distinct sites for interaction with DNA polymerase δ and DNA polymerase ε. *Mol. Cell. Biol.*, **17,** 6367.

138. Arroyo, M. P., Downey, K. M., So, A. G., and Wang, T. S. F. (1996) *Schizosaccharomyces pombe* proliferating cell nuclear antigen mutations affect DNA polymerase δ processivity. *J. Biol. Chem.*, **271,** 15971.

139. Arroyo, M. P. and Wang, T. S. F. (1998) Mutant PCNA alleles are associated with *cdc* phenotypes and sensitivity to DNA damage in fission yeast. *Mol. Gen. Genet.*, **257,** 505.

140. Piard, K., Baldacci, G., and Tratner, I. (1998) Single point mutations located outside the inter-monomer domains abolish trimerization of *Schizosaccharomyces pombe* PCNA. *Nucl. Acids Res.*, **26,** 2598.

141. Krishna, T. S. R., Fenyo, D., Kong, X. P., Gary, S., Chait, B. T., Burgers, P., and Kuriyan, J. (1994) Crystallization of proliferating cell nuclear antigen (PCNA) from *Saccharomyces cerevisiae. J. Mol. Biol.*, **241,** 265.

142. Kong, X. P., Onrust, R., O'Donnell, M., and Kuriyan, J. (1992) Three-dimensional structure of the β-subunit of *Escherichia coli* DNA polymerase III holoenzyme—a sliding DNA clamp. *Cell*, **69,** 425.

143. Burgers, P. M. J. and Yoder, B. L. (1993) ATP-independent loading of the proliferating cell nuclear antigen requires DNA ends. *J. Biol. Chem.*, **268**, 19923.

144. Waga, S. and Stillman, B. (1998) The DNA replication fork in eukaryotic cells. *Annu. Rev. Biochem.*, **67**, 721.

145. Budd, M. E. and Campbell, J. L. (1993) DNA polymerases δ and ε are required for chromosomal replication in *Saccharomyces cerevisiae*. *Mol. Cell. Biol.*, **13**, 496.

146. Morrison, A. and Sugino, A. (1994) The 3'–5' exonucleases of both DNA polymerases δ and ε participate in correcting errors of DNA replication in *Saccharomyces cerevisiae*. *Mol. Gen. Genet.*, **242**, 289.

147. Kesti, T., Flick, K., Keranen, S., Syvaoja, J. E., and Wittenberg, C. (1999) DNA polymerase ε catalytic domains are dispensable for DNA replication, DNA repair, and cell viability. *Mol. Cell*, **3**, 679.

148. Dua, R., Levy, D. L., and Campbell, J. L. (1999) Analysis of the essential functions of the C-terminal protein–protein interaction domain of *Saccharomyces cerevisiae* pol ε and its unexpected ability to support growth in the absence of the DNA polymerase domain. *J. Biol. Chem.*, **274**, 22283.

149. Hughes, P., Tratner, I., Ducoux, M., Piard, K., and Baldacci, G. (1999) Isolation and identification of the third subunit of mammalian DNA polymerase delta by PCNA-affinity chromatography of mouse FM3A cell extracts. *Nucl. Acids Res.*, **27**, 2108.

150. Bauer, G. A., Heller, H. M., and Burgers, P. M. J. (1988) DNA polymerase III from *Saccharomyces cerevisiae*. 1. Purification and characterization. *J. Biol. Chem.*, **263**, 917.

151. Burgers, P. M. J. and Bauer, G. A. (1988) DNA polymerase III from *Saccharomyces cerevisiae*. 2. Inhibitor studies and comparison with DNA polymerases I and II. *J. Biol. Chem.*, **263**, 925.

152. Reynolds, N., Watt, A., Fantes, P. A., and MacNeill, S. A. (1998) Cdm1, the smallest subunit of DNA polymerase δ in the fission yeast *Schizosaccharomyces pombe*, is non-essential for growth and division. *Curr. Genet.*, **34**, 250.

153. Boulet, A., Simon, M., Faye, G., Bauer, G. A., and Burgers, P. M. J. (1989) Structure and function of the *Saccharomyces cerevisiae cdc2* gene encoding the large subunit of DNA polymerase III. *EMBO J.*, **8**, 1849.

154. Morrison, A., Araki, H., Clark, A. B., Hamatake, R. K., and Sugino, A. (1990) A third essential DNA polymerase in *Saccharomyces cerevisiae*. *Cell*, **62**, 1143.

155. Simon, M., Giot, L., and Faye, G. (1991) The 3' to 5' exonuclease activity located in the DNA polymerase δ subunit of *Saccharomyces cerevisiae* is required for accurate replication. *EMBO J.*, **10**, 2165.

156. Sugimoto, K., Sakamoto, Y., Takahashi, O., and Matsumoto, K. (1995) *Hys2*, an essential gene required for DNA replication in *Saccharomyces cerevisiae*. *Nucl. Acids Res.*, **23**, 3493.

157. Hashimoto, K., Nakashima, N., Ohara, T., Maki, S., and Sugino, A. (1998) The second subunit of DNA polymerase III (δ) is encoded by the *HYS2* gene in *Saccharomyces cerevisiae*. *Nucl. Acids Res.*, **26**, 477.

158. MacNeill, S. A., Moreno, S., Reynolds, N., Nurse, P., and Fantes, P. A. (1996) The fission yeast Cdc1 protein, a homolog of the small subunit of DNA polymerase δ, binds to Pol3 and Cdc27. *EMBO J.*, **15**, 4613.

159. Simon, M., Giot, L., and Faye, G. (1993) A random mutagenesis procedure—application to the *pol3* gene of *Saccharomyces cerevisiae*. *Gene*, **127**, 139.

160. Francesconi, S., Park, H., and Wang, T. S. F. (1993) Fission yeast with DNA polymerase δ temperature-sensitive alleles exhibits cell division cycle phenotype. *Nucl. Acids Res.*, **21**, 3821.

161. Morrison, A. and Sugino, A. (1992) Roles of *pol3*, *pol2* and *pms1* genes in maintaining accurate DNA replication. *Chromosoma*, **102,** S147.

162. Morrison, A., Bell, J. B., Kunkel, T. A., and Sugino, A. (1991) Eukaryotic DNA polymerase amino acid sequence required for 3′–5′ exonuclease activity. *Proc. Natl. Acad. Sci. U.S.A.*, **88,** 9473.

163. Hughes, D. A., Macneill, S. A., and Fantes, P. A. (1992) Molecular-cloning and sequence-analysis of *cdc27*$^+$ required for the G2-M transition in the fission yeast *Schizosaccharomyces pombe*. *Mol. Gen. Genet.*, **231,** 401.

164. Reynolds, N., Warbrick, E., Fantes, P. A., and MacNeill, S. A. (2000) Direct interaction between the fission yeast DNA polymerase δ subunit Cdc27 and Pcn1 (PCNA) mediated through a C-terminal p21^{Cip1}-like PCNA binding motif. *EMBO J.*, **19,** 1108.

165. Warbrick, E., Lane, D. P., Glover, D. M., and Cox, L. S. (1995) A small peptide inhibitor of DNA replication defines the site of interaction between the cyclin-dependent kinase inhibitor p21^{Waf1} and proliferating cell nuclear antigen. *Curr. Biol.*, **5,** 275.

166. Huang, M. E., LeDouarin, B., Henry, C., and Galibert, F. (1999) The *Saccharomyces cerevisiae* protein YJR043C (Pol32) interacts with the catalytic subunit of DNA polymerase α and is required for cell cycle progression in G2/M. *Mol. Gen. Genet.*, **260,** 541.

167. Francesconi, S., Copeland, W. C., and Wang, T. S. F. (1993) *In vivo* species specificity of DNA polymerase α. *Mol. Gen. Genet.*, **241,** 457.

168. Bhaumik, D. and Wang, T. S. F. (1998) Mutational effect of fission yeast Pol α on cell cycle events. *Mol. Biol. Cel.*, **9,** 2107.

169. Araki, H., Ropp, P. A., Johnson, A. L., Johnston, L. H., Morrison, A., and Sugino, A. (1992) DNA polymerase II, the probable homolog of mammalian DNA polymerase ε, replicates chromosomal DNA in the yeast *Saccharomyces cerevisiae*. *EMBO J.*, **11,** 733.

170. Araki, H., Hamatake, R. K., Johnston, L. H., and Sugino, A. (1991) *DPB2*, the gene encoding DNA polymerase II subunit B, is required for chromosome replication in *Saccharomyces cerevisiae*. *Proc. Natl. Acad. Sci. U.S.A.*, **88,** 4601.

171. Shcherbakova, P. V. and Pavlov, Y. I. (1996) 3′–5′-Exonucleases of DNA polymerase ε and DNA polymerase δ correct base analog induced DNA replication errors on opposite DNA strands in *Saccharomyces cerevisiae*. *Genetics*, **142,** 717.

172. Hamatake, R. K., Hasegawa, H., Clark, A. B., Bebenek, K., Kunkel, T. A., and Sugino, A. (1990) Purification and characterisation of DNA polymerase II from the yeast *Saccharomyces cerevisiae*. *J. Biol. Chem.*, **265,** 4072.

173. Araki, H., Hamatake, R. K., Morrison, A., Johnson, A. L., Johnston, L. H., and Sugino, A. (1991) Cloning *dpb3*, the gene encoding the third subunit of DNA polymerase II of *Saccharomyces cerevisiae*. *Nucl. Acids Res.*, **19,** 4867.

174. Kamimura, Y., Masumoto, H., Sugino, A., and Araki, H. (1998) Sld2, which interacts with Dpb11 in *Saccharomyces cerevisiae*, is required for chromosomal DNA replication. *Mol. Cell. Biol.*, **18,** 6102.

175. Araki, H., Leem, S. H., Phongdara, A., and Sugino, A. (1995) Dpb11, which interacts with DNA polymerase II (ε) in *Saccharomyces cerevisiae*, has a dual role in S-phase progression and at a cell-cycle checkpoint. *Proc. Natl. Acad. Sci. U.S.A.*, **92,** 11791.

176. Wang, H. and Elledge, S. J. (1999) *DRC1*, DNA replication and checkpoint protein 1, functions with *DPB11* to control DNA replication and the S-phase checkpoint in *Saccharomyces cerevisiae*. *Proc. Natl. Acad. Sci. U.S.A.*, **96,** 3824.

177. Saka, Y., Fantes, P., and Yanagida, M. (1994) Coupling of DNA replication and mitosis by fission yeast *rad4/cut5*. *J. Cell Sci.*, 57.

178. Saka, Y. and Yanagida, M. (1993) Fission yeast *cut5+*, required for S-phase onset and M-phase restraint, is identical to the radiation-damage repair gene *rad4+*. *Cell*, **74**, 383.

179. Murante, R. S., Henricksen, L. A., and Bambara, R. A. (1998) Junction ribonuclease: an activity in Okazaki fragment processing. *Proc. Natl. Acad. Sci. U. S. A.*, **95**, 2244.

180. Frank, P., Braunshofer-Reiter, C., Wintersberger, U., Grimm, R., and Busen, W. (1998) Cloning of the cDNA encoding the large subunit of human RNase HI, a homologue of the prokaryotic RNase HII. *Proc. Natl. Acad. Sci. U.S.A.*, **95**, 12872.

181. Johnson, R. E., Kovvali, G. K., Prakash, L., and Prakash, S. (1998) Role of yeast Rth1 nuclease and its homologs in mutation avoidance, DNA repair, and DNA replication. *Curr. Genet.*, **34**, 21.

182. Carr, A. M., Sheldrick, K. S., Murray, J. M., Alharithy, R., Watts, F. Z., and Lehmann, A. R. (1993) Evolutionary conservation of excision repair in *Schizosaccharomyces pombe*— evidence for a family of sequences related to the *Saccharomyces cerevisiae rad2* gene. *Nucl. Acids Res.*, **21**, 1345.

183. Harrington, J. J. and Lieber, M. R. (1994) Functional domains within FEN-1 and Rad2 define a family of structure-specific endonucleases—implications for nucleotide excision-repair. *Genes Devel.*, **8**, 1344.

184. Budd, M. E., Choe, W. C., and Campbell, J. L. (1995) *DNA2* encodes a DNA helicase essential for replication of eukaryotic chromosomes. *J. Biol. Chem.*, **270**, 26766.

185. Budd, M. E. and Campbell, J. L. (1995) A yeast gene required for DNA replication encodes a protein with homology to DNA helicases. *Proc. Natl. Acad. Sci. U.S.A.*, **92**, 7642.

186. Budd, M. E. and Campbell, J. L. (1997) A yeast replicative helicase, Dna2 helicase, interacts with yeast FEN-1 nuclease in carrying out its essential function. *Mol. Cell. Biol.*, **17**, 2136.

187. Bae, S. H., Choi, E., Lee, K. H., Park, J. S., Lee, S. H., and Seo, Y. S. (1998) Dna2 of *Saccharomyces cerevisiae* possesses a single-stranded DNA- specific endonuclease activity that is able to act on double-stranded DNA in the presence of ATP. *J. Biol. Chem.*, **273**, 26880.

188. Gould, K. L., Burns, C. G., Feoktistova, A., Hu, C. P., Pasion, S. G., and Forsburg, S. L. (1998) Fission yeast *cdc24+* encodes a novel replication factor required for chromosome integrity. *Genetics*, **149**, 1221.

189. Nash, R. and Lindahl, T. (1996) DNA ligases. In *DNA replication in eukaryotic cells* (ed M. L. DePamphilis), p. 575. Cold Spring Harbor Laboratory Press, New York.

190. Teo, S. H. and Jackson, S. P. (1997) Identification of *Saccharomyces cerevisiae* DNA ligase 4. Involvement in DNA double-strand break repair. *EMBO J.*, **16**, 4788.

191. Schar, P., Herrmann, G., Daly, G., and Lindahl, T. (1997) A newly identified DNA ligase of *Saccharomyces cerevisiae* involved in *RAD52*-independent repair of DNA double-strand breaks. *Genes Devel.*, **11**, 1912.

192. Wilson, T. E., Grawunder, U., and Lieber, M. R. (1997) Yeast DNA ligase IV mediates non-homologous DNA end joining. *Nature*, **388**, 495.

193. Barker, D. G. and Johnston, L. H. (1983) *Saccharomyces cerevisiae CDC9*, a structural gene for yeast DNA ligase which complements *Schizosaccharomyces pombe cdc17*. *Eur. J. Biochem.*, **134**, 315.

194. Barker, D. G., White, J. H. M., and Johnston, L. H. (1987) Molecular characterization of the DNA ligase gene, *cdc17*, from the fission yeast *Schizosaccharomyces pombe*. *Eur. J. Biochem.*, **162**, 659.

195. Johnston, L. H., Barker, D. G., and Nurse, P. (1986) Cloning and characterization of the *Schizosaccharomyces pombe* DNA ligase gene *cdc17*. *Gene*, **41**, 321.

196. Hopwood, B. and Dalton, S. (1996) Cdc45p assembles into a complex with Cdc46p/Mcm5p, is required for minichromosome maintenance, and is essential for chromosomal DNA replication. *Proc. Natl. Acad. Sci. U.S.A.*, **93**, 12309.
197. Dalton, S. and Hopwood, B. (1997) Characterization of Cdc47p-minichromosome maintenance complexes in *Saccharomyces cerevisiae*: identification of Cdc45p as a subunit. *Mol. Cell. Biol.*, **17**, 5867.

3 | The cell cycle

MIKE TYERS and PAUL JORGENSEN

1. Overview

The accurate replication and segregation of the genome during cell division is orchestrated by the cyclin-dependent protein kinases (CDKs) and their attendant regulatory partners, the cyclins and CDK inhibitors. At its most fundamental level, the cell division machinery is programmed by a state of low CDK activity in G1 phase and a state of high CDK activity in S, G2 and M phase. These two states dictate the once and only once replication and segregation of the genome per division cycle. Thus, low CDK activity triggers exit from mitosis and the establishment of competent origins of DNA replication, whereas high CDK activity activates firing of pre-replicative origins, prevents the refiring of spent origins and catalyses entry into mitosis. The alternation between the two CDK states is controlled by a switch that is flipped once in each direction per cell cycle. In late G1 phase, CDK activity is switched on while in late mitosis it is switched off. This switch is constructed from three major elements: CDK-dependent phosphorylation, cell cycle-regulated waves of transcription, and ubiquitin-dependent proteolysis of key regulatory proteins. In this chapter, we describe the mechanisms that generate the fundamental oscillation in CDK activity that lies at the heart of the cell cycle in the budding yeast *Saccharomyces cerevisiae* and the fission yeast *Schizosaccharomyces pombe*.

2. Background

2.1 Life cycles of budding yeast and fission yeast

Cell division in the two yeasts is organized into four phases, in much the same manner as in other eukaryotes. In G1 phase cells grow in size and, if environmental conditions are appropriate, commit to division; in S phase cells duplicate their genome; in G2 phase cells determine whether conditions are appropriate for mitosis; and in M phase cells segregate their duplicated genome, complete cytokinesis and re-enter the subsequent G1 phase. Both budding yeast and fission yeast can exist as either haploid or diploid cell types, through conjugation of haploids to form diploids and reductive meiotic divisions to form haploids. Because of their rather different life cycles, the comparative cell cycle genetics of *S. cerevisiae* and *S. pombe* have provided tremendous insight into the mechanisms of cell division (for comprehensive reviews see References 1–3).

In budding yeast, haploid cells mate under all growth conditions whereas diploid cells sporulate only under conditions of nutrient limitation, such that the default cell type is diploid. Cell cycle control occurs mainly in late G1 phase at a point or brief window called Start, at which cells commit to division and become insensitive to environmental inputs such as mating pheromones and nutrient availability (1). Landmark events that occur at or shortly after Start include bud emergence, polarization of the actin-based secretory network towards the apical tip of the new bud, initiation of DNA replication and duplication of the spindle pole body. At around the time DNA replication is completed, bud growth switches from a polarized apical pattern to an isotropic pattern, accounting for the more or less round shape of daughter cell. At the same time the spindle poles separate to form a short mitotic spindle that spans the nucleus. Following DNA replication the chromosomes become moderately condensed and the nucleus migrates to the bud neck as cells enter mitosis. At the onset of anaphase, sister chromatid separation and spindle elongation occurs, which drives the daughter chromosomes across the bud neck into the incipient daughter cell. Successful completion of anaphase signals nuclear division, spindle disassembly and cytokinesis, leaving the mother and daughter cell in early G1 phase.

In fission yeast, haploid cells conjugate only when starved for a nitrogen source. Because the resultant diploid sporulates immediately under such nutrient-poor conditions, the default cell type is haploid. Although the cell cycles of fission yeast and budding yeast have an overall similar organization there are also some notable differences (2). In order to maintain two copies of each chromosome as templates for repair of DNA strand breaks, fission yeast cells must replicate the genome soon after mitosis and thus the main control point for the mitotic cell cycle resides at the G2–M transition. The decision to either conjugate or enter meiosis is, however, necessarily retained at Start in G1 phase in order to maintain cell ploidy. Unlike budding yeast, but rather like most other eukaryotes, the fission yeast mitotic spindle does not develop until the early stages of mitosis. In addition, repolarization of the actin cytoskeleton to direct growth to the new end after fission, an event called NETO, for 'new end take off', does not occur until G2 phase. Despite these distinctions, the CDK engine that regulates cell division is remarkably conserved between the two yeasts, and indeed among all eukaryotes.

2.2 Foundations of the CDK paradigm

The seminal analysis of cell division cycle (*cdc*) mutants in budding yeast by Hartwell and colleagues (4) was based on the realization that conditional mutations in cell cycle processes cause cells to arrest with a uniform cellular and nuclear morphology. The prominent roles of numerous *CDC* gene products in cell cycle regulation attests to the spectacular insight of the original *cdc* concept. Of these, the most crucial was the observation that Cdc28p function is interdependent with the pheromone and nutrient execution points at Start (4). The subsequent isolation of *cdc28* mutants defective in the G2–M transition presaged the central role of Cdc28p in all phases of the cell cycle (5). A parallel search for *cdc* mutants in fission yeast, undertaken by Nurse and colleagues, identified an independent set of Cdc functions (2, 6). In fission yeast, *cdc⁻* mutants become large and highly elongated owing to continued growth in the absence of division. Importantly, a rarer class of mutants, termed Wee for their shorter than wild-type length, cause an acceleration of division with respect to growth. The *cdc2⁺* gene quickly emerged as a critical regulator because different alleles yield either a Cdc⁻ or Wee phenotype, and also because Cdc2 function is required in both G1 and G2 phases (2).

In the late 1980s, a convergence of research in several model systems led to a unified model of the eukaryotic cell cycle (reviewed in Reference 7). First, *cdc2⁺* and *CDC28* were found to encode similar protein kinases that are functionally interchangeable from yeast to humans. Second, a critical cell cycle regulator called cyclin was discovered because of its periodic destruction at mitosis in the embryonic cell cycles of sea urchins and clams. Subsequent studies in *Xenopus* revealed that cyclin is a necessary and sufficient activator of embryonic cell division. The product of the *cdc13⁺* gene, which is essential for mitosis in fission yeast, turned out to encode a cyclin homologue. Third, a succession of experiments in different model systems revealed that cyclin and Cdc2 form a heterodimeric kinase complex in which cyclin is an obligatory activator of the kinase subunit—hence the term cyclin-dependent kinase, abbreviated CDK. Most notably, the activity that stimulates meiotic progression in frog oocytes called maturation promoting factor (MPF) is a heterodimeric complex composed of cyclin and Cdc2. Taken together, these breakthroughs established the concept that CDK activity is the critical rate-limiting factor for each major cell cycle transition. The cyclin–CDK family now includes multiple kinase and cyclin subunits that regulate diverse aspects of cell division (reviewed in References 8, 9).

3. CDK regulation

CDK enzymatic activity is tightly controlled by three primary mechanisms (Fig. 1). First, different cyclin subunits confer substrate specificity, proper timing of activation and appropriate localization to the CDK kinase subunit. Second, phosphorylation of the CDK subunit couples CDK activity to both internal and external signals that modulate cell cycle progression. Third, the binding of inhibitory factors to CDK complexes attenuates CDK activity in response to yet other signals. Together, these mechanisms combine to create successive waves of different CDK activities from Start until the end of mitosis (Fig. 2).

3.1 CDK families

CDKs are serine/threonine kinases that tend to phosphorylate the nominal consensus sequence Ser/Thr-Pro-X-Lys/Arg. As opposed to the often drastic changes in cyclin abundance, CDK kinase subunits are expressed at constant levels through the cell

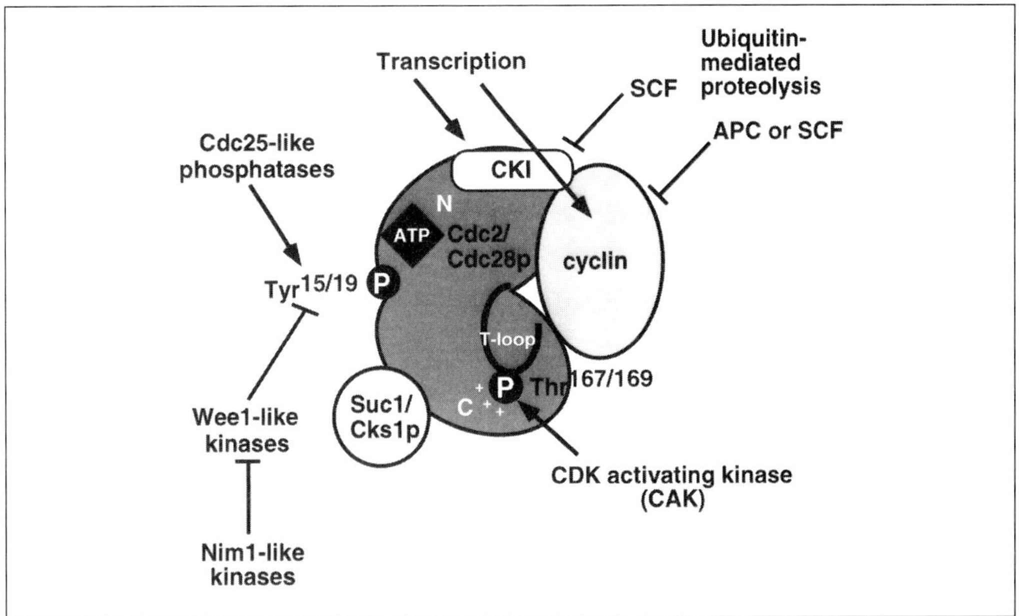

Fig. 1 Mechanisms that regulate CDK activity in *S. cerevisiae* and *S. pombe*. Arrows indicate activation; blunt arrows indicate inhibition. N and C refer to the N-terminus and C-terminus of Cdc2/Cdc28p. P, phosphorylation; + + + marks the arginine-rich pocket of Cdc2/Cdc28p.

cycle. The CDK and cyclin families of budding yeast and fission yeast are listed in Table 1. The completely defined *S. cerevisiae* genome encodes five different CDK subunits: Cdc28p, Pho85p, Kin28p, Srb10p, and Ctk1p. Cdc28p and Pho85p are specifically activated by large families of cyclin subunits, whereas Kin28p, Srb10p, and Ctk1p are each activated by a single cyclin partner. Cdc28p and its nine attendant cyclins are devoted entirely to cell cycle control (1). The versatile Pho85p kinase and its cohort of 10 cyclins regulate manifold processes, including phosphate metabolism, glycogen metabolism, the cytoskeleton and G1 phase progression (9). Kin28p, Srb10p, and probably Ctk1p are regulators of the RNA polymerase II holoenzyme (see Chapter 6).

In fission yeast, Cdc2 and its cyclins are dedicated to cell cycle control (2). The other known fission yeast cyclin–CDK pair, Mcs6–Mcs2, are homologues of Ccl1–Kin28, which, in addition to regulation of RNA polymerase II, also phosphorylate and activate Cdc2 (8). It seems likely that counterparts to the other budding yeast CDKs will be uncovered once the fission yeast genome sequence is completed.

3.2 Cyclins that activate Cdc2/Cdc28p

3.2.1 Fission yeast cyclins

All cyclins contain a conserved 100-residue domain, called the cyclin box, which mediates the interaction with the CDK subunit. The first cyclin discovered in yeast,

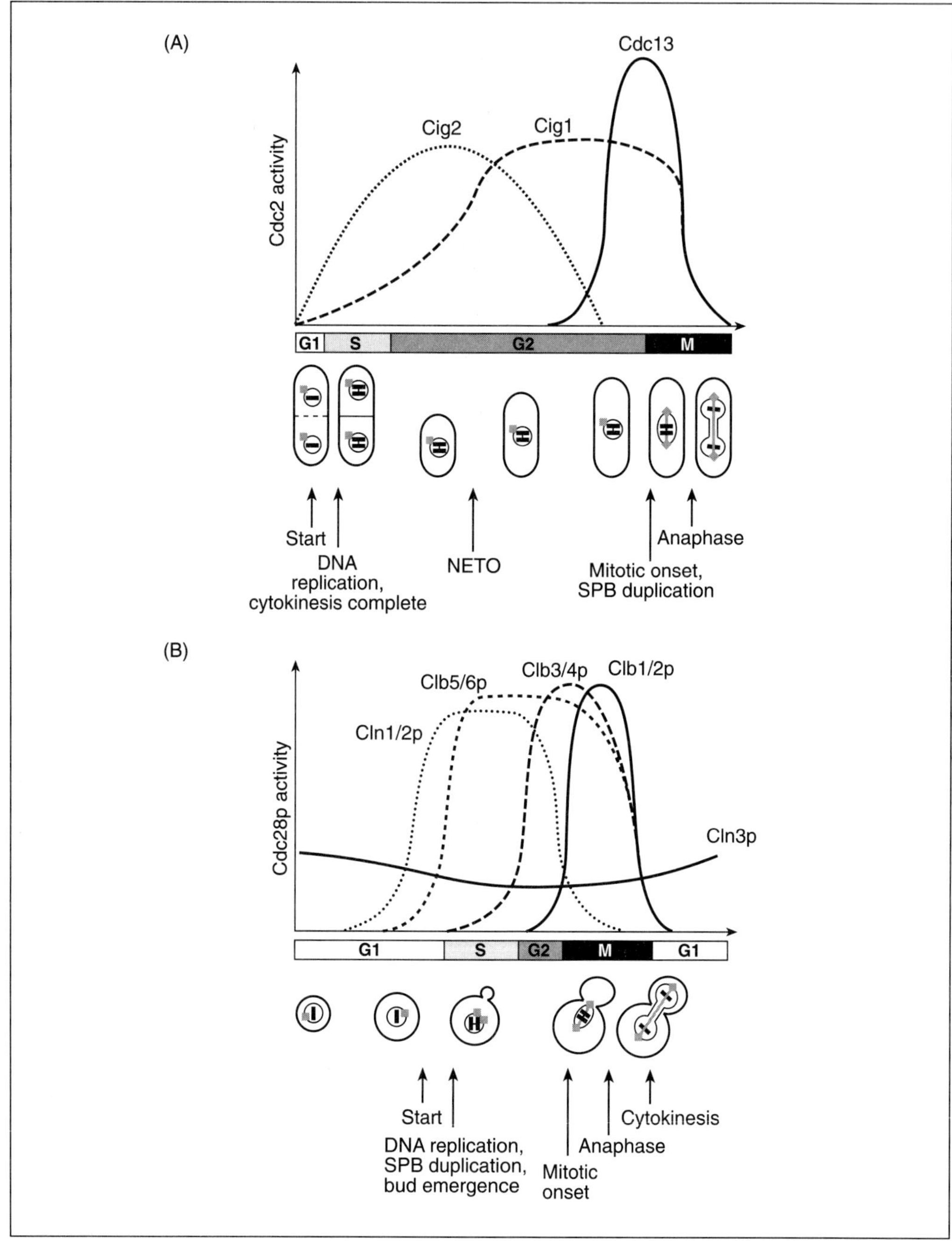

Fig. 2 Waves of cyclin–Cdc2/Cdc28p activities in (A) the *S. pombe* and (B) the *S. cerevisiae* cell cycles. Heights of the respective cyclin–Cdc2/Cdc28p peaks are not quantitative and differ only for the sake of clarity. Bars indicate the relative lengths of cell cycle phases. Schematics illustrate cell and nuclear morphology, as well as spindle structure.

Table 1 CDK and cyclin families in *S. cerevisiae* and *S. pombe*

CDK	Cyclin	Function
S. cerevisiae		
Cdc28p	Cln1p, Cln2p	G1 cyclins. Expressed at Start. Trigger spindle pole body duplication, bud emergence and polarized growth. Inhibit mating pathway. Phosphorylate and target Sic1p to SCF^{Cdc4p}. Phosphorylate Cdh1p, blocking its association with APC. *cln1 cln2 cln3* cells arrest at Start
	Cln3p	G1 cyclin. Slightly upregulated at M–G1. Potent activator of Start that drives entire programme of G1–S-specific transcription by stimulating SBF/MBF
	Clb1p, Clb2p	Mitotic cyclins. Expressed in G2–M. Promote isotropic growth, repression of SBF-dependent transcription, and onset of mitosis. *clb1 clb2* cells exhibit a near lethal delay at G2–M
	Clb3p, Clb4p	G2 cyclins. Expressed in early G2. Help initiate bipolar spindle formation
	Clb5p, Clb6p	S-phase cyclins. Expressed at Start. Initiate DNA replication once the Sic1p inhibitor is eliminated. *clb5 clb6* cells exhibit a severe delay in S phase
Pho85p	Pcl1p, Pcl2p	G1 cyclins. Expressed at Start. Facilitate bud emergence and polarized growth. *pcl1 pcl2 cln1 cln2* cells arrest at Start
	Pcl5p, Pcl6p, Pcl7p, Clg1p	Unknown function
	Pcl8p, Pcl10p	Cyclins that target Pho85p to phosphorylate and thereby inhibit glycogen synthase.
	Pcl9p	Expressed at M–G1. May have a role in polarized growth and septation.
	Pho80p	Represses *PHO5* (acid phosphatase) transcription in high phosphate concentrations by phosphorylation of Pho4p, a transcription factor
Srb10p	Srb11p	Cyclin–CDK pair that phosphorylates C-terminal repeats (CTD) on the largest subunit of RNA polymerase II (Pol II). A component of the Pol II mediator complex and a negative regulator of many genes.
Ctk1p	Ctk2p	Cyclin–CDK pair that phosphorylates CTD repeats on the largest subunit of Pol II.
Kin28p	Ccl1p	Cyclin–CDK pair that phosphorylates CTD repeats on the largest subunit of Pol II. A TFIIH component that is essential for Pol II transcription. Unlike its homologue in fission yeast, not a CDK-activating kinase (CAK)
S. pombe		
Cdc2	Cdc13	Mitotic cyclin. Expressed at G2–M. Essential for mitosis. Also required to prevent re-replication
	Cig1	G1 cyclin. Expressed at M–G1. Promotes G1 to S transition by phosphorylation of Rum1
	Cig2	S phase cyclin. Expressed at G1–S. Stimulates DNA replication
	Puc1	Similar to *S. cerevisiae* Clns in sequence. May promote entry into mitosis by positively regulating Cdc10–Res1 assocation. Transcription is not cell cycle regulated
	Pch1	C-type cyclin. Essential, but does not have an obvious cell cycle function. Transcription is not cell cycle regulated
Prk1	?	Homologue of Srb10p
Mcs6	Mcs2	C-type cyclin. Probable component of transcription factor TFIIH that is essential for PolII transcription. A partially redundant CAK for Cdc2

See text for references.

encoded by the *cdc13⁺* gene of fission yeast, was identified as a *cdc* mutant because of its essential role in mitotic entry (10). By sequence alignment within the cyclin box region, Cdc13 falls into the B-type cyclin subfamily, named after cyclin B of marine invertebrates (8). *cdc13⁺* exhibits allele-specific interactions with *cdc2⁺*, which reflects the physical association between the two proteins. While *cdc13⁺* transcription is moderately cell cycle regulated with a peak in mitosis, the Cdc13 protein is rapidly destroyed at the end of mitosis. Importantly, accumulation of Cdc13 is not the rate-limiting step for the G2–M transition. Rather, mitosis begins once the assembled Cdc13–Cdc2 complex is abruptly released from an inhibitory phosphorylation on the Cdc2 subunit (see Section 3.3).

Two other B-type cyclins, Cig1 and Cig2, are non-essential activators of Cdc2 that fine-tune G1- and S-phase progression (10). Cig1–Cdc2, whose activity peaks in G2, plays an important role at the G1–S transition and in the maintenance of the high CDK state by eliminating an inhibitor of the Cig2–Cdc2 and Cdc13–Cdc2 activities (11). $cig2^+$ expression peaks in S phase, and deletion of $cig2$ delays the onset of S phase, indicating that the Cig2–Cdc2 kinase activates DNA replication. A $cig1^-$ $cig2^-$ double mutant is delayed in S phase but otherwise viable, so the fission yeast cell cycle is capable of running on a single B-type cyclin, Cdc13 (12). Conversely, cells that lack Cdc13 function undergo successive rounds of DNA replication. This re-replication phenomenon depends on G1–S cyclin activity, since a $cdc13^-$ $cig1^-$ $cig2^-$ triple mutant permanently arrests in G1 phase (12). The pattern of Cdc2 activation by the three B-type cyclins is illustrated in Fig. 2A.

A fourth Cdc2-associated cyclin in fission yeast called Puc1 was isolated by complementation of a budding yeast strain defective in G1 cyclin function and is unrelated in sequence to the B-type cyclins (10). Although $puc1^+$ is non-essential, the Puc1–Cdc2 complex appears to promote S-phase entry by phosphorylation of Cdc10, a G1-specific transcription factor (13). Finally, an essential C-type cyclin called Pch1 may also interact with Cdc2, but as yet has no defined cell cycle role (14).

3.2.2 Budding yeast cyclins

The cyclins that activate Cdc28p fall into two groups: (1) the G1 cyclins, Cln1–3p, which function at Start, and (2) the B-type cyclins, Clb1–6p, which function in S, G2, and M phases (1). The first cyclin known to regulate G1 phase, now called Cln3p, was isolated by virtue of two dominant mutations, *WHI1-1* and *DAF1-1*, both of which reduce critical cell size at Start and confer resistance to mating pheromone-induced cell cycle arrest. Two other G1 cyclins, encoded by *CLN1* and *CLN2*, were isolated as high-copy suppressors of the Start defect caused by the *cdc28-4* mutation. The Clns are genetically redundant since a *cln1 cln2 cln3* strain arrests at Start, whereas as any one *CLN* gene suffices for viability. Complementation of the *cln1 cln2 cln3* strain has proven useful for the isolation of G1 cyclins from many organisms, notably cyclin D and cyclin E from humans (reviewed in Reference 15). Despite their genetic redundancy, the Cln proteins have quite different characteristics. Cln1p and Cln2p are 57% identical in sequence to each other but only 25% identical to Cln3p (1). *CLN1/2* expression is highly cell cycle periodic with a peak at Start, whereas *CLN3* oscillates only modestly with maximal expression at the M–G1 transition (16, 17). Finally, Cln3p is a far more potent activator of Start than Cln1/2p, despite its substantially lower abundance (16). In accord with these different characteristics, Cln1/2p and Cln3p serve distinct purposes and operate in a hierarchical manner at Start. Cln3p potently activates a broad programme of Start-specific gene expression, which includes *CLN1/2* (see section 4.1) In turn, Cln1/2p act as downstream effectors of Start that trigger bud emergence, spindle pole body duplication, and DNA replication (16, 18, 19). In support of this model, Cln3p by itself is not actually sufficient to activate events at Start, since a *cln1 cln2 pcl1 pcl2* strain arrests at Start (9). *PCL1/2* are also induced as part of the G1–S-specific programme initiated by Cln3p but are not normally rate limiting for Start in wild-type cells (9). Rather, the timing of Start is dictated largely through control of *CLN* transcription and translation (1). In contrast to their essential role in the mitotic cell cycle, Cln–Cdc28p activity is entirely dispensable for meiosis and sporulation, and may actually impede entry into premeiotic replication (20–22).

The budding yeast mitotic cyclins Clb1–4p were isolated both by sequence similarity to B-type cyclins in other species and as high-copy suppressors of the *cdc28-1N* mutation, while two other B-type cyclins, Clb5p and Clb6p, were isolated as a low-copy suppressor of the *cln⁻* arrest, and as a gene adjacent to *CLB1*, respectively (1). The Clbs act in partially redundant pairs that share similar primary sequences, expression patterns, and functions. The *CLB* genes are expressed in three separate waves of transcription: *CLB5/6* in late G1 phase, *CLB3/4* in early G2 phase, and *CLB1/2* in late G2 phase, just before mitosis (Fig. 2B). The function of the various Clbs coincides with their appearance. Thus, *clb5 clb6* double mutants are substantially delayed for entry into S phase, *clb3 clb4* double mutants in combination with other *clb* mutations have defects in formation of the mitotic spindle, and *clb1 clb2* double mutants are severely defective for mitotic entry (1).

Detailed genetic analysis suggests further specialization of Clb functions. With respect to DNA replication, Clb5p is the primary regulator of S-phase onset as it is capable of firing both early and late origins of replication, whereas Clb6p is specific for early origins (23). The specialized role of Clb5p in replication is not merely a matter of its early expression pattern because replacement of the *CLB5* coding sequence with *CLB2* fails to rescue the replication defect of a *clb5* strain (24). Nevertheless, any one Clb is sufficient to trigger S phase in the mitotic cell cycle, as elimination of all six *CLB* genes is necessary to cause cells to arrest with unreplicated DNA (25). Surprisingly though, *clb5 clb6* strains completely fail in premeiotic DNA replication and proceed into two lethal meiotic divisions (20, 21). An altered hierarchy of Clb function in meiosis is also suggested by the finding that, while Clb1/3/4p play only a supporting role to Clb2p in the mitotic cell cycle, the Clb1/4p pair is essential for the equational division of meiosis II (26). The precise functions of Clb3/4/5p in formation of the spindle are unknown, although Clb5p is necessary for proper positioning of the premitotic spindle (27).

3.2.3 Cdc2/Cdc28p substrates

Substrates of the various cyclin–Cdc2/Cdc28p complexes in fission yeast and budding yeast have proven difficult to identify. To date, most known substrates are themselves regulators of Cdc2/Cdc28p, rather than the anticipated downstream effectors of the kinases. As described in Section 5, in fission yeast Cdc2 substrates include the G1–S transcription factor Cdc10, the CDK inhibitor Rum1 and the replication protein Cdc18. Similarly, in budding yeast, Cln–Cdc28p kinases phosphorylate and inactivate the CDK inhibitors Sic1p and Far1p. The Cln–Cdc28p kinases also inhibit the mating pheromone pathway at Start, possibly through phosphorylation of the Ste20 kinase (28–30). As yet, there are no known substrates that link Cln–Cdc28p activity to the other major events at Start, such as bud emergence, polarization of the actin cytoskeleton, and spindle pole body duplication. The Cln and Clb forms of Cdc28p often phosphorylate the same substrate. For instance, two effectors of mitotic exit, Sic1p and Cdh1p, and the replication proteins Cdc6 and Mcm4, are all phosphorylated and inactivated by both forms of Cdc28p (see Sections 5 and 6).

The Clb–Cdc28 kinases undoubtedly phosphorylate numerous dedicated substrates in S, G2, and M phases. Known or suspected targets include the G1–S transcription factor Swi4p and various components of replication origins (see Chapter 2). An important morphogenetic function of Clb2p–Cdc28p is to depolarize the actin cytoskeleton once bud growth is underway. The transition from polarized to isotropic bud growth requires a Clb2p-associated protein called Nap1p, which allows Clb2p–Cdc28p to activate the Gin4p kinase, which in turn depolarizes the actin-based secretory network (31). However, the precise role of phosphorylation in cytoskeletal reorganization is not known. The means by which Clb–Cdc28p activity controls elaboration of the mitotic spindle and mitotic entry is also unknown. Evidence from other systems suggests that microtubule-associated proteins, Golgi-associated proteins, and histones are all substrates of mitotic CDK activity (32) and so it is likely that similar substrates will affect nuclear events in both budding and fission yeast. A common theme to emerge from both yeasts is that many CDK substrates are phosphorylated on multiple redundant sites, often making it difficult to ascribe a role for phosphorylation through mutation of one or even several sites.

3.3 Control of CDK activity by phosphorylation

3.3.1 Inhibition by tyrosine phosphorylation

As discovered through genetic analysis in fission yeast, Cdc2 activity is inhibited at the G2–M transition by phosphorylation of Tyr15, a conserved residue that lies within the adenosine triphosphate (ATP) binding loop of the kinase (Fig. 1; reviewed in References 7, 33). Control of Tyr15 phosphorylation directly links the cell cycle machinery to important signals, such as those emanating from DNA damage or nutrient limitation. A dual specificity kinase called Wee1 is the primary catalyst of Tyr15 phosphorylation on Cdc2 (Table 2). Cells that lack Wee1 activity divide at small size (the Wee phenotype) because Cdc2 is hyperactive and hence division is accelerated with respect to growth (7). Mutation of Tyr15 to a non-phosphorylatable residue causes a similar Wee phenotype. A Wee1 homologue called Mik1 plays an ancillary role in Tyr15 phosphorylation. Although *mik1⁻* cells have no size defect, at the restrictive temperature, a *wee1-50 mik1⁻* double mutant undergoes a lethal premature mitosis referred to as mitotic catastrophe (34). Wee1 is in turn directly

Table 2 Cell cycle regulators in *S. cerevisiae* and *S. pombe*

S. cerevisiae	*S. pombe*	Function
CDK regulators		
Cks1p	Suc1	Component of Cdc2/Cdc28p complexes that binds C-terminal lobe of kinase. Has a role in exit from mitosis. Assembly factor and/or targeting subunit of Cdc2/Cdc28p?
Sic1p	Rum1	Inhibitors of S phase and mitotic cyclin–CDK activity. Targeted for degradation upon phosphorylation by G1 cyclin–CDK kinases. *sic1* strains show high rates of chromosome loss. *rum1⁻* strains are sterile, while overproduction of *rum1⁺* causes re-replication
Far1p	?	Inhibitor of Cln–Cdc28p kinase activity. Activated by mating pheromone. Also mediates polarization of cytoskeleton in pheromone response. Unknown whether an analogous inhibitor imposes G1 arrest in response to pheromone in *S. pombe*
Swe1p	Wee1, Mik1	Kinases that inhibit Cdc2/Cdc28p by phosphorylation of Tyr15/Tyr19 and thereby block entrance into mitosis. Wee1 mediates G2–M size control in response to nutrients and blocks cell cycle progression in response to DNA damage and replication checkpoints. Swe1p inhibits mitotic entry in response to morphogenesis checkpoint
Mih1p	Cdc25	Phosphatases that activate Cdc2/Cdc28p by dephosphorylation of Tyr15/Tyr19. Cdc25 required for mitotic entry in a normal cell cycle and is a target of checkpoint kinases. Mih1p required for recovery from morphogenesis checkpoint arrest
Hsl1p, Kcc4p, Gin4p	Nim1/Cdr1, Cdr2	Kinases that bind and inhibit Wee1-like kinases. Nim1/Cdr1 and Cdr2 may transduce nutrient signals into size control by inhibiton of Wee1. Hsl1p, Kcc4p, Gin4p localize to the septin ring in the bud neck and inhibit Swe1p when there are no defects in bud and septin morphology
Transcription factors		
Swi6p, Swi4p, Mbp1p	Cdc10, Res1/Sct1, Res2/Pct1	Transcription factors that activate G1–S programme at Start. All six proteins contain ankryin repeats. Swi4p, Mbp1p, Res1, and Res2 contain a conserved DNA-binding domain. Only Cdc10 is essential, but others have overlapping essential functions. In *S. pombe*, Cdc10 heterodimerizes with either Res1 or Res2 to form MBF. In *S. cerevisiae*, Swi4p and Swi6p combine to form SBF, while Mbp1p and Swi6p combine to form MBF
Fkh1p, Fkh2p	ACC#, CAA19034?	Redundant components of SFF. In complex with Mcm1p, SFF activates expression of the *CLB2* cluster in G2–M
Swi5p/Ace2p	ACC#, CAB11298?	In complex with Mcm1p, activates transcription of the *SIC1* cluster in late mitosis. Nuclear entry blocked by Clb–Cdc28p kinases
Mcm1	?	Transcription factor required for activating several clusters (*CLB2, SIC1,* and *MCM* clusters) of cell cycle-regulated genes
SCF components		
Cdc53p	Pcu1	Core component of SCF–ubiquitin ligase complexes. Scaffold protein that links Skp1/F-box protein subcomplexes to the E2 enzyme Cdc34p
Skp1p	p19/Skp1	Core component of SCF–ubiquitin ligase complexes. Binds F-box proteins to Cdc53p. In *S. cerevisiae*, Skp1p has several SCF-independent functions, including an essential role in kinetochore assembly
Rbx1p	Acc#, CAB11672?	Core component of SCF–ubiquitin ligase complexes. RING-H2 finger protein that interacts directly with F-box proteins, Cdc53p and Cdc34p. Stimulates Cdc34p–Cdc53p interaction and allosterically activates Cdc34p

Table 2 Continued

S. cerevisiae	S. pombe	Function
SCF components		
Cdc4p	Pop1, Pop2/Sud1	F-box proteins with WD40 repeats. Recruit phosphorylated substrates for ubiquitination, including Sic1p/Rum1 and Cdc6p/Cdc18. Cdc4p is essential; Pop1/2 required to prevent re-replication
Grr1p	?	F-box protein with leucine-rich repeats (LRR). Recruits phosphorylated Cln1/2p and Gic1/2p for ubiquitination
Met30p	?	F-box protein with WD40 repeats. Recruits Swe1p and Met4p for ubiquitination. *met30* strains arrest before Start in a Met4p-dependent manner
Cdc34p	?	E2 ubiquitin-conjugating enzyme. Provides catalytic activity for SCF–ubiquitin ligase complexes. Interacts with Rbx1p and Cdc53p
Cdc6p	Cdc18	Required for pre-replicative complex assembly on origins of replication. Expressed as part of G1–S programme driven by SBF/MBF. Overexpression of $cdc18^+$, but not *CDC6*, causes re-replication. Targeted for SCF-dependent degradation upon phosphorylation by CDK kinases
APC components		
Cdc16p, Cdc23p, Cdc27p	Cut9, Nuc2	APC subunits that contain TPR repeats.
Cdc26p	Hcn1	APC subunit
Apc1p	Cut4	APC subunit
Apc2p	Genbank Z99162?	APC subunit that is a distant member of the Cdc53p/cullin family
Apc10/Doc1p	Apc10	APC subunit that contains a DOC domain
Apc11p	?	APC subunit that contains a RING-H2 domain. Similar to Rbx1
?	UbcP4	E2 ubiquitin-conjugating enzyme required for anaphase in *S. pombe*. No known combination of E2 enzymes is essential for anaphase in *S. cerevisiae*
Cdc20p	Slp1	Rate-limiting activator of APC at metaphase. Pds1p and Clb5p are the essential targets of APC^{Cdc20p}
Cdh1/Hct1p	Ste9/Srw1	Rate-limiting activators of APC from anaphase to G1. Cdh1p targets Clbs, Ase1p, Cdc5p, and Cdc20p for degradation. Ste9 is required for G1 arrest in response to pheromone
Regulators of anaphase onset and/or mitotic exit		
Pds1p	Cut2	Securin that restrains Cut1/Esp1p before anaphase, but that also has a positive role in Cut1/Esp1p function. Cut2/Pds1p destroyed in an APC-dependent manner at anaphase onset. Pds1p stabilized by DNA damage and spindle checkpoints
Esp1p	Cut1	Separin required for sister chromatid separation. Once released from Pds1p, Esp1p stimulates limited proteolysis of Scc1p, a cohesin protein
Cdc5p	Plo1	Polo-like kinases required for cytokinesis. Cdc5p required for anaphase B and mitotic exit. Plo1 required early in mitosis to form bipolar spindle and to form medial ring for septation
Cdc14p	?	Protein phosphatase required for mitotic exit. Dephosphorylates Sic1p, Swi5p, and Cdh1p at the end of mitosis
Cdc15p	Cdc7	Protein kinases required for cytokinesis. Cdc15p required for mitotic exit. Cdc7 required for septum formation but not nuclear division
Dbf2p, Dbf20p	Sid2	Protein kinases required for cytokinesis. Dbf2p and Dbf20p are highly similar with overlapping functions required for mitotic exit. Sid2 required for septum formation but not nuclear division

Table 2 Continued

S. cerevisiae	*S. pombe*	Function
Regulators of anaphase onset and/or mitotic exit		
Lte1p	?	Guanine nucleotide exchange factor that facilitates mitotic exit. Essential at low temperatures
Tem1p	Spg1	Ras-like GTPases required for cytokinesis. Tem1p required for mitotic exit. Spg1 required for septum formation but not nuclear division
Mob1p	ACC#, CAA22288?	Protein of unknown function required for mitotic exit. Interacts with Dbf2p and components of the Mad–Mps1 checkpoint pathway
?	Cdc14	No similarity to any known proteins. Required for septum formation but not nuclear division
Byr4p, Bub2p	Byr4, Cdc16	Inhibitors of cytokinesis. Byr4/Cdc16 and Byr4p/Bub2p are two component GTPase-activating complexes that convert Spg1/Tem1p into the inactive GDP-bound form

See text for references.

inhibited by two related dual-specificity kinases, Nim1/Cdr1 and Cdr2, which convey nutritional signals that accelerate entry into mitosis under conditions of nitrogen deprivation (7, 33, 35, 36). Nim1 phosphorylates Wee1 *in vitro* and thereby diminishes Wee1 kinase activity towards Tyr15 of Cdc2 (33). The predominantly cytosolic localization of Nim1 suggests that its access to the nuclear localized Wee1 may be regulated, an idea supported by the finding that forced nuclear localization of Nim1 accelerates entry into mitosis (37). Many subtleties in Wee1 regulation remain to be resolved, however, as overexpression of Nim1 causes a Wee phenotype, whereas overexpression of Cdr2 unexpectedly produces elongated cells with multiple septa (35, 36). The only known upstream regulator of Nim1 is a leucine zipper containing protein called Nik1, an apparent antagonist of Nim1 function that impedes mitotic entry (38).

Dephosphorylation of Tyr15 on Cdc2 is catalysed by a dual-specificity phosphatase encoded by the $cdc25^+$ gene, which is essential for mitotic entry in wild-type cells (7). A second dual-specificity phosphatase, Pyp3, also makes a modest contribution to Cdc2 activation (39). Entry into mitosis is dictated by a dynamic balance between Wee1 and Cdc25 activity, as was illustrated by the suppression of a *cdc25* defect by the *wee1-50* mutation (7). Cdc25 accumulates continuously as cells grow in size and becomes phosphorylated and increasingly active as cells approach the G2–M transition (40). Cdc25 phosphorylation and activation depend on Cdc2 function, and it seems likely that mitotic onset may be dictated by a positive feedback loop in which Cdc25 activates Cdc2, which in turn further activates Cdc25, and so on (40). A large body of evidence from higher species is consistent with such a positive feedback model for entry into mitosis (33). Cdc25 activity is limited in part through its turnover by the ubiquitin system, via an E3 enzyme of the HECT domain family called Pub1 (41). Intriguingly, a 'non-wee' suppressor of *cdc25* mutations called *stf1-1* is a gain of function mutation in the spindle pole body protein Cut12, suggesting that the primary role of Cdc2 in mitosis is to establish a bipolar spindle (42, 43). As befits its crucial role in activating mitosis, Cdc25 is a principal target of DNA damage and replication checkpoints (see Chapter 4).

Quite unlike fission yeast, the budding yeast homologues of Wee1 and Cdc25, called Swe1p and Mih1p respectively, do not normally regulate mitosis (Table 2). Cells bearing non-phosphorylatable mutants of Cdc28p that lack Tyr19 (the equivalent of Tyr15 in Cdc2) have wild-type cell cycle kinetics and an intact DNA damage response (1, 33). Despite these findings, phosphorylation of Tyr19 can be detected in wild-type cells, and indeed overexpression of *SWE1* causes a G2–M arrest, indicating that the pathway is operable (1). The enigma of Swe1p/Mih1p function in budding yeast was solved with the finding that phosphorylation of Tyr19 mediates a checkpoint delay when bud morphogenesis is perturbed (44). If cells fail to form a bud, either through mutational inactivation of the budding pathway or through depolarization of the actin-based secretory system by osmotic stress, nuclear division is delayed. This delay depends on Swe1p-mediated phosphorylation of Cdc28p, since deletion of *SWE1* causes the accumulation of multinucleate cells under osmotic stress, whereas deletion of *MIH1* blocks recovery from the checkpoint (44). Although mutation of Tyr19 partially overrides the checkpoint, it appears that at least a component of the response derives from a phosphorylation-independent effect of Swe1p, perhaps through sequestration of Cdc28p (45). Like its counterpart Wee1, Swe1p is inhibited by upstream kinases of the Nim1/Cdr1 family, called Kcc4p, Gin4p, and Hsl1p, all of which co-localize with the bud neck, where they are presumably activated once the bud is properly assembled (46). Nim1-like kinases participate in other, less well defined, pathways that regulate cell cycle progression through control of tyrosine phosphorylation on Cdc28p, such as in the filamentous growth response (47).

3.3.2 Activation by CAK

The catalytic activity of most CDK enzymes depends on phosphorylation of a conserved threonine residue, Thr167 in Cdc2 and Thr169 in Cdc28p, which resides in the kinase activation loop, or T loop (reviewed in Reference 8). A CDK-activating kinase, or CAK, was first identified in mammalian cells as a complex of cyclin H, CDK7, and a protein called MAT1 which contains a Zn^{2+} binding domain termed a RING finger. Unexpectedly, cyclin H–CDK7 also turned out to be a component of the RNA polymerase II holoenzyme subcomplex TFIIH (see Chapter 6). CAK activity in fission yeast is provided by the homologues of cyclin H–Cdk7, Mcs2–Mcs6 (Table 1). Mutations in the *mcs* genes were isolated as suppressors of the *cdc2-3w wee1-50* mitotic catastrophe phenotype, presumably because such mutations attenuate Cdc2 activity (8). A second kinase called Csk1 also contributes to CAK activity in fission yeast extracts (48).

However, the cyclin H–CDK7 orthologue in budding yeast, Ccl1p–Kin28p, does not exhibit CAK activity *in vitro* and is not directly required for Cdc28p function *in vivo* (8). Rather, CAK activity resides with an essential, novel protein kinase termed Cak1p, a finding that has raised some uncertainty as to whether cyclin H–CDK7 is the physiologically relevant CAK in metazoans (49, 50). Strains harbouring conditional alleles of *CAK1* have defects in both Cln–Cdc28p- and Clb–Cdc28p-dependent processes. Derivation of CAK-independent versions of Cdc28p demonstrates that the only essential function of Cak1p is to activate Cdc28p, although deletion of *CAK1* in such mutants indicates that Cak1p may also have additional non-essential functions (51). In both fission and budding yeast, CAK activity plays only a constitutive, non-regulatory role in the maturation of Cdc2/Cdc28p kinase activity.

3.4 CDK inhibitors

3.4.1 Sic1p and Rum1

A protein called Sic1p, originally identified as a tightly bound substrate of Cdc28, is a potent antagonist of Clb–Cdc28p kinases, and is the archetype of an important general class of regulators termed CDK inhibitors, or CKIs (Table 2) (52, 53). *SIC1* was also isolated as a high-copy suppressor of mutants defective in mitotic exit (54). Sic1p is present from the end of mitosis until the end of G1 phase, where it acts as a stoichiometric inhibitor of Clb–Cdc28p kinases, but not Cln–Cdc28p kinases. Deletion of *SIC1* causes DNA replication to initiate well in advance of bud emergence, and, conversely, overexpression of *SIC1* imposes a severe replication delay (53). The onset of Clb5/6–Cdc28p kinase activity at the end of G1 phase occurs only once Sic1p is eliminated, an event triggered by Cln–Cdc28p-dependent phosphorylation of Sic1 (25). As described in Section 5, Sic1p plays a key role in inactivation of Clb–Cdc28p activity at the end of mitosis and maintenance of the CDK-free window throughout G1 phase.

In fission yeast, a Sic1p analogue called Rum1 maintains cell ploidy by enforcing the alternation of S phase with mitosis (55). Overproduction of $rum1^+$ uncouples replication from mitosis, causing cells to accumulate more than 16 N DNA content. The physiological role of Rum1 is to repress Cdc13–Cdc2 and Cig2–Cdc2 kinase activity in G1 phase, but, because fission yeast normally spend so little time in G1, deletion of $rum1^+$ has little effect on cycling populations. However, $rum1^-$ mutants fail to arrest in G1 phase under low nitrogen conditions and, furthermore, if $rum1^-$ cells are blocked in G1 phase by a *cdc10* mutation, a lethal reductive mitosis ensues. Like Sic1p, Rum1 specifically inhibits the S phase and mitotic CDK activities of Cig2–Cdc2 and Cdc13–Cdc2, but not the M–G1 activity of Cig1–Cdc2 (11, 56). Rum1 and Sic1p share limited sequence similarity in their C-terminal regions, which contain the CDK inhibitory domain and are functionally interchangeable in the two yeasts (57). Finally, both Sic1p and Rum1 are eliminated in G1 phase by analogous phosphorylation-dependent degradation pathways (see Section 5).

3.4.2 Far1p

A screen for mutants defective in the G1 arrest component of the mating pheromone response in budding yeast yielded a gene called *FAR1* (reviewed in Reference 28). Far1p is induced by mating pheromones and is post-translationally activated by the pheromone-activated mitogen-activated protein (MAP) kinase Fus3p (58, 59). Far1p binds all Cln–Cdc28p complexes in a pheromone-stimulated manner (58, 60) and potently represses *CLN1/2* transcription, probably by inhibition of Cln3p–Cdc28p activity (61). Whether or not Far1p directly inhibits Cln1/2–Cdc28p activity is a matter of some dispute. Although Far1 can inhibit Cln2–Cdc28p *in vitro* (62), if *CLN2* is expressed from a constitutive promoter, levels of Cln2p–Cdc28p kinase activity are not altered by pheromone treatment, even though Far1p is recruited to the Cln2–Cdc28p complex (59). Interpretation of Cln–Far1p interactions is further complicated by the fact that Far1p is targeted for degradation by Cln–Cdc28p-dependent phosphorylation, in a similar manner as Sic1p (63). In addition to its role as a CDK inhibitor, Far1p has an entirely separate function in polarizing the actin cytoskeleton during the mating response (64). While haploid fission yeast arrest in G1 phase in response to mating pheromones and nitrogen starvation, it remains to be seen whether this response depends on a Far1p-like activity.

3.4.3 Other CKIs

Yeast contain several other factors that have CKI-like functions. A third CKI in budding yeast called Pho81p positively regulates *PHO5* gene expression by inhibiting the Pho80p–Pho85p kinase, but not other Pcl–Pho85p complexes (9). Pho81p contains an ankryin repeat domain like the p16[INK4A] class of CKIs in mammalian cells, but bears no similarity to other yeast CKIs. In fission yeast, an S-phase inhibitor called Spd1 physically associates with Cdc2 but it is not known whether Spd1 inhibits Cdc2 kinase activity *per se* (65). Finally, the replication protein Cdc6p/Cdc18 might also function as a CKI. The N-terminal 47 amino acids of Cdc6p mediate an interaction with Clb–Cdc28p complexes and, at least when present in stoichiometric excess, Cdc6p appears to inhibit Clb–Cdc28p kinase activity (66). Overexpression of N-terminal portions of Cdc6p or Cdc18 causes arrest phenotypes consistent with CDK inhibition *in vivo* (66, 67).

3.5 Other factors required for CDK activity

3.5.1 Suc1/Cks1p

A small, highly conserved, protein called Suc1 binds avidly to CDKs and is often exploited as an affinity reagent for the purification of CDK complexes (reviewed in Reference 2). *suc1*[+] is a high-copy suppressor of conditional alleles of *cdc2*[+] in fission yeast and appears to serve multiple functions, including facilitating exit from mitosis. The budding yeast homologue of Suc1, called Cks1p, is required at both the G1–S and G2–M transitions, and also interacts physically and genetically with Cdc28p (1). Cks1p is required to form active recombinant Cln–Cdc28p complexes, but not Clb–Cdc28p complexes, at least as produced in insect cells (68). The structure of the human CKS1–CDK2 complex indicates that Cks1p binds at the bottom of the C-terminal lobe of the kinase, at a site that overlaps the mitosis-specific *cdc28-1N* mutation (69). A number of unexpected genetic and biochemical interactions suggest a connection between Cks1p, Cdc28p, and the 26S proteasome, the protease particle that degrades ubiquitinated proteins (70). As Clb2p is stabilized in *cks1* mutants, and

Cdc13 accumulates in *suc1⁻* mutants, it has been suggested that Cks1p is a recycling factor that strips Cdc28p monomers away from the proteasome to clear the way for subsequent undegraded Clb–Cdc28p complexes (70).

3.5.2 Cdc37p

The genesis of active Cdc28p kinases requires at least one bona fide assembly factor, Cdc37p, which was originally isolated as a Start-specific *cdc* mutant (1). Cdc37 is required for the efficient interaction of the Cln and Cdc28p subunits, thus explaining the arrest phenotype of *cdc37* mutants (71). The assembly of Clb2p–Cdc28p complexes is also partially defective in *cdc37* mutants. Indeed, Cdc37 is a general assembly factor that acts in concert with the chaperonin Hsp90 to mediate the proper folding of numerous different kinases (72).

3.6 CDK structure

To provide a structural framework for the basis of CDK regulation we briefly describe the salient features of the human cyclin A–CDK2 complex, which has been crystallized in its active and inactive states (reviewed in Reference 73). As members of the eukaryotic protein kinase superfamily, CDKs conform to the standard bi-lobed kinase architecture in which the catalytic cleft is sandwiched between the N- and C-terminal lobes (Fig. 1). The monomeric CDK subunit is catalytically inert because active site residues are misaligned and substrate access is blocked by the T loop. At the cyclin A–CDK2 interface, the conserved cyclin box motif forms a compact α-helical domain that abuts an α helix derived from the conserved PSTAIRE sequence motif found in all CDKs. Cyclin binding causes large-scale rearrangements in the N-terminal lobe that rectify the catalytic defects of the CDK monomer. First, the PSTAIRE helix moves towards the catalytic centre and repositions the scissile β–γ phosphate bond of ATP for in-line attack by the substrate hydroxyl group. Second, the cyclin interaction melts a short α helix, Lα12, which in turn forces the T loop out of the active site towards the C-terminal lobe. Reorientation of the T loop allows CAK to phosphorylate efficiently Thr 160 of CDK2 to active the kinase fully. The phospho-Thr160 residue packs into an arginine-rich pocket on the C-terminal lobe, and in so doing pulls the T loop against the C-terminal lobe and stabilizes the overall structure of the kinase. The structural basis for CDK inhibition by tyrosine phosphorylation is not known. Systematic mutagenesis of key residues in Cln2p, based on the cyclin A–CDK2 interface, suggests that its interaction with Cdc28p is analogous to that of cyclin A with CDK2 (74). Amazingly, reiterative mutagenesis of Cdc28p has allowed for selection of a Cln-independent version of Cdc28p, which is both active as a monomer and able efficiently to catalyse Start in *cln⁻* cells (75). With respect to CKI interactions, structural analysis of the mammalian CKI p27[Kip1] in a complex with cyclin A–CDK2 reveals that p27[Kip1] interacts with both the N-terminal lobe and cyclin to distort and inhibit the kinase subunit. However, given the sequence divergence within the CKI family, it is difficult to predict whether similar mechanisms are employed by the yeast CKIs.

4. Transcriptional control in the cell cycle

The timely expression and repression of cell cycle genes in both yeasts is essential for cell cycle progression. The broad alterations in gene expression that occur at each major cell cycle transition are driven by changes in CDK activity. In budding yeast, genome-wide analysis of transcription by DNA microarray hybridization reveals that more than 800 of the approximately 6300 predicted open reading frames in *S. cerevisiae* are transcribed in a cell cycle-dependent manner (76, 77). Genome-wide microarray data can be assembled *de novo* into gene clusters by an algorithm that pairs genes, and then progressive sets of genes, according to similarity of expression pattern. Using this method, at least eight different clusters of cell cycle-regulated genes can be identified (Plate 3 and Table 3). Major clusters of cell cycle-regulated gene expression are switched on by Cln3p–Cdc28p activity at the G1–S transition, by Clb1/2–Cdc28p activity in G2 phase, and by the collapse of Clb–Cdc28p activity at the end of mitosis. Here, we focus on mechanisms of cell cycle-regulated transcription, with emphasis on budding yeast, which to date has been studied in greater depth than in fission yeast.

4.1 Cell cycle-regulated transcription in budding yeast

4.1.1 The G1–S programme

At around the time of Start, cells express a massive suite of some 120 genes, referred to as the *CLN2* cluster, after one of its more prominent members (77). The *CLN2* cluster is governed by two heterodimeric transcription factor complexes, called SBF and MBF (Plate 3, Fig. 3). The two complexes share a common regulatory subunit, Swi6p, which is tethered to DNA by its associated partners, Swi4p in SBF and Mbp1p in MBF (reviewed in Reference 78). All three proteins contain ankryin repeats, while Swi4p and Mbp1p share similar N-terminal DNA-binding domains and C-terminal Swi6p interaction domains. The Swi4p–Swi6p and Mbp1p–Swi6p heterodimers preferentially bind to SCB and MCB promoter elements respectively, which lie upstream of most genes induced at Start (77).

The *CLN2* cluster includes sets of genes involved in DNA synthesis (*CLB5*, *CLB6*, *CDC6*, *POL1*, *RNR1*), sister chromatid cohesion (*SMC1*, *SMC3*), spindle pole body duplication (*TUB4*, *SPC42*), and bud emergence (*SVS1*, *BNI4*) (77). Despite the elaborate programme of gene expression at Start, the only essential function of SBF and MBF is to drive transcription of G1 cyclins, which catalyse the main events of Start (79). Although highly regulated transcription of other genes in this cluster is not essential, it is likely that provision of additional gene products at Start facilitates each of the associated processes and/or that misregulated expression of such genes is detrimental to the cell. The timely expression of *CDC6* and other genes required for DNA synthesis is a case in point. Cdc6p is an essential loading factor for the mini-chromosome maintencance (Mcm) proteins, which, together with the origin recognition complex (ORC), forms pre-replicative origins of replication (see Chapter 2). In wild-type cells, *CDC6* transcription peaks moderately at the M–G1 transition and

Table 3 Clusters of cell cycle-regulated genes in *S. cerevisiae**

Cell cycle cluster	Peak	Transcription factor	Example genes	Significance
CLN2 cluster	Mid-G1	SBF/MBF	*CLN2, CLN1, CLB5, CLB6, CDC9, RNR1, POL1, MSH1, PMS1, SMC3, SMC1, MCD1*	119 genes. Cln1/2p–Cdc28p activity promotes Sic1p proteolysis, bud emergence and spindle pole body duplication, whereas Clb5/6p–Cdc28p activity initiates DNA replication. Nutrient and pheromone signals regulate Start through transcriptional control of *CLN2* cluster. Repressed by Clb1/2p–Cdc28p activity at G2–M
Y' cluster	Mid-G1	Unknown	*YPR203, YPR204*	26 open reading frames. All have high sequence similarity and are located in subtelomeric regions. Unknown significance
Late G1 group	Late G1	SBF/MBF	*FKS1,GOG5, GAS1, ALG7 PMI40*	92 genes. Poorly clustered, but simultaneously upregulated after Start. Numerous genes involved in cell wall synthesis (*GOG5, PMI40*), so may reflect biosynthetic and cell wall remodelling requirements during bud emergence. Includes *FKH1*
Histone cluster	S	Uncertain	*HTA2, HTB2*	Nine genes. Histone transcripts peak just before DNA synthesis
MET cluster	S	Met4p	*MET1, MET6, MUP1*	20 genes. Required for methionine biosynthesis. Cell cycle-regulated transcription is of unknown signficance
CLB2 cluster	M	SFF–Mcm1p	*CLB2,CLB1, SWI5,TEM1, APC1, BUD3, CDC20*	33 genes. Numerous mitotic regulators and bud-site components. Expression is dependent on Clb1/2p–Cdc28p kinases, and is repressed by loss of Clb–Cdc28p activity at the end of mitosis
MCM cluster	M–G1	Mcm1p	*MCM2, CDC54, FAR1, BEM1*	34 genes. Expression of all six components of the MCM complex peaks as pre-replicative complexes form in early G1 phase. *FAR1* induction may facilitate G1 arrest in response to mating pheromone
SIC1 cluster	M–G1	Swi5p/Ace2p– Mcm1p	*SIC1, TEC1, ASH1, CTS1, PCL9*	27 genes. *SIC1* expression important for mitotic exit. *ASH1* mRNA transported into daughter cells to repress Swi5p-dependent *HO* transcription and mating type switching
MAT cluster	Early G1	α1–Ste12p–Mcm1p	*MATα1, STE3, FUS1*	13 genes. Predominantly mating genes upregulated in early G1 and repressed at Start. The early G1 timing of transcription may help restrict the mating response to G1 phase

See text for references. *Derived primarily from Reference 77.

more strongly at Start. While low-level expression of Cdc6 suffices for sluggish progression through S phase, proficient replication requires *CDC6* expression in G1 phase (80). The timely onset of replication also depends on two other products of the *CLN2* cluster, Clb5p and Clb6p, which, once liberated from the Sic1 inhibitor, activate origins of replication (25).

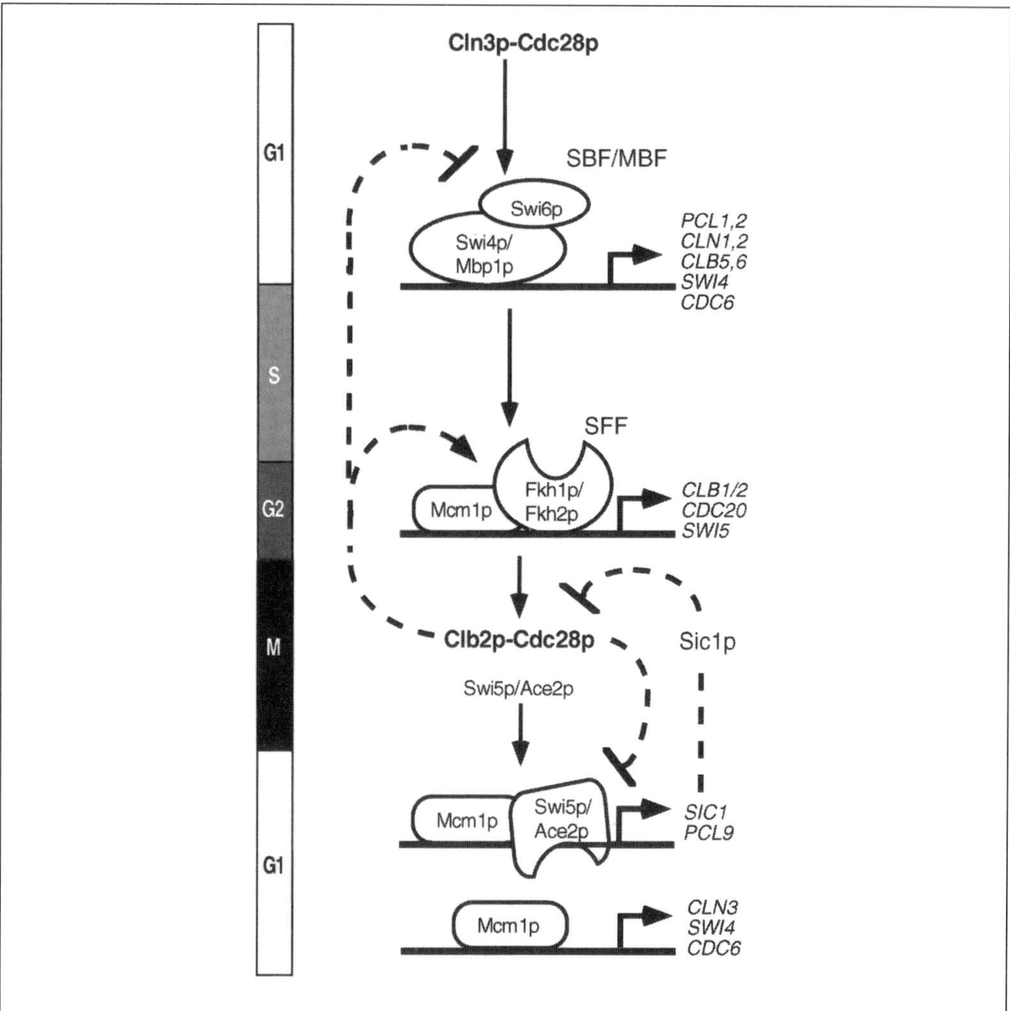

Fig. 3 Transcriptional hierarchies in the *S. cerevisiae* cell cycle. Arrows indicate activation, blunt arrows indicate inhibition.

SBF- and MBF-dependent gene expression is potently activated by the Cln3p–Cdc28p kinase (16). While it was originally posited that the interdependency of *CLN* transcription and Cln1/2–Cdc28 kinase activity might generate a positive feedback loop that activates Start in a switch-like manner, the timing of SBF/MBF activation under physiological conditions is determined solely by Cln3p–Cdc28p activity (16, 18, 19). Thus, *cln3* cells have a severe delay in the onset of both Start-specific transcription and Start, whereas *cln1cln2* cells activate SBF/MBF on schedule but are retarded for other events at Start (18, 19). Consistent with this view, overexpression of *CLN3* induces more than 100 Start-specific genes in the complete absence of *CLN1/2* (77). However, just what determines the timing of SBF/MBF activation by

Cln3p–Cdc28p is a mystery. It is possible that the level or activity of Cln3p–Cdc28p might increase in a growth-dependent manner. While there is evidence that nutrient signals do regulate the translation and stability of Cln3p, the level of Cln3p does not obviously increase as cells approach Start under constant nutrient conditions (81). Indeed, if anything, *CLN3* messenger RNA (mRNA) and protein peak around the M–G1 transition, and not at Start (16, 17).

The nature of the anticipated post-translational stimulus from Cln3p–Cdc28p to SBF/MBF has so far resisted all lines of investigation. SBF and MBF assemble on their cognate promoter elements in early G1 phase, probably through chromatin remodelling events, as in the case of the *HO* promoter (82–84). The Cln3p–Cdc28p-dependent signal must feed into such preassembled complexes. However, there is no evidence that Cln3p–Cdc28p directly phosphorylates Swi4p, Swi6p, or Mbp1p, and neither have ancillary factors that associate with Cln3p or SBF/MBF emerged as key activators of the Start programme. Genetic analysis has revealed several other, less well characterized, pathways that modulate SBF/MBF-dependent transcription, including a protein phosphatase, Sit4p, and a protein of unknown function, Bck2p (28).

4.1.2 The G2–M programme

In G2–M phase, expression of the *CLN2* cluster is extinguished and replaced by a suite of important mitotic regulators, referred to as the *CLB2* cluster (Plate 3, Fig. 3). The ascent of G2–M gene expression is driven through a positive feedback loop of *CLB2* transcription and Clb1/2–Cdc28 kinase activity (85). Conditional *clb1–4* mutants fail to express *CLB2* and other genes in the cluster, whereas overexpression of *CLB2* powerfully induces over 50 G2–M transcripts, including *CLB1/2, CDC20, CDC5, TEM1, MOB1, ACE2,* and *SWI5* (77, 85). The promoters of most genes in the *CLB2* cluster contain binding sites for a complex of the general transcription factor Mcm1 and its G2–M-specific partner, Swi Five Factor or SFF (77). SFF eluded identification until genome-wide analysis uncovered two related candidate genes called *FKH1* and *FKH2* that are expressed in S phase before the onset of G2–M transcription. *FKH1/2* appear to account for SFF activity as *fkh1 fkh2* double mutants are defective in activation of the G2–M programme[a]. Recent studies indicate that Mcm1p-Fkh1/2p probably serve as a platform for a cell cycle regulated transcription factor called Ndd1p[b]. In the simplest model, Clb-Cdc28 kinase might directly phosphorylate Ndd1p, thereby driving its association with Mcm1p-Fkh1/2p. The precise relationship between the onset of *CLB2* transcription and mitotic entry remains unclear.

The demise of the G1–S transcriptional programme occurs with the onset of Clb1/2–Cdc28p activity. Cells that lack Clb1/2p are unable to repress SBF-dependent genes and, moreover, approximately 100 genes in the *CLN2* cluster are repressed by overexpression of *CLB2* (77, 85). This repression may be direct as Clb2p interacts with the ankryin repeats of Swi4p, which could facilitate phosphorylation of Swi4p by Clb2p–Cdc28p (85, 86). Inhibitory phosphorylation of Swi4p may clear SBF from SCB

[a]Zhu, G., Spellman, P. T., Volpe, T., Brown, P. O., Botstein, D., Davis, T. N., and Futcher, B. (2000) Two yeast forkhead genes regulate the cell cycle and pseudohyphal growth. *Nature,* **406**, 90.

[b]Koranda, M., Schleiffer, A., Endler, L. and Ammerer, G. (2000) Forkhead-likje transcription factors recruit Ndd1 to the chromatin of G2-M-specific promoters. *Nature,* **406**, 94.

elements in G2 since the SBF footprint is lost just before mitosis (82, 83). A different mechanism appears to repress MBF because the decay of MBF-dependent transcription in G2 does not depend on Clb1/2–Cdc28p activity, nor does Mbp1p interact directly with Clb2p (85, 86).

4.1.3 The M–G1 programme

A burst of transcription at the end of mitosis, referred to as the *SIC1* cluster, is anticipated by the appearance of the transcription factors Ace2p and Swi5p as products of the G2–M cluster (Plate 3, Fig. 3). Swi5p and Ace2p are related zinc finger DNA-binding proteins that also act in conjunction with Mcm1p to activate distinct but partially overlapping sets of genes. For example, Ace2p preferentially activates *CTS1* transcription, which is required for efficient cytokinesis, whereas Swi5p induces *SIC1*, which facilitates exit from mitosis (87, 88). Ace2p and Swi5p are sequestered in an inactive state in the cytoplasm through Clb1/2–Cdc28p-dependent phosphorylation of residues within their nuclear localization sequences (89). After anaphase, Swi5p is dephosphorylated by the Cdc14p phosphatase, which allows Swi5p to enter the nucleus (90). As described in Section 4, *SIC1* is a critical target of Swi5p because Sic1p contributes to the collapse of Clb–Cdc28p activity as cells exit from mitosis (54, 90).

Two other clusters contribute to the M–G1 programme. One cluster is dominated by *MCM* genes such as *CDC46*, *MCM2*, and *MCM3*, which encode components of pre-replicative origin complexes that must assemble in G1 phase (77). Most genes in the *MCM* cluster contain an Mcm1p-binding site termed the ECB for Early Cell Cycle Box. *SWI4*, *CDC6*, and *CLN3* transcript levels also peak moderately at M–G1 due to ECB elements in their upstream regions (17). Finally, in haploids, a third aspect of the M–G1 programme is a cluster of mating-specific genes whose function is restricted to the G1 phase window (77). The three M–G1 clusters are downregulated at around the time of Start and, although many of the M–G1 genes are repressed by *CLN3* over-expression, the mechanism of repression is unknown.

4.1.4 Insights from genome-wide analysis

Three main themes emerge from genome-wide studies of cell cycle regulated transcription in budding yeast (Fig. 3). First, the successive transcriptional clusters that occur at the G1–S, G2–M, and M–G1 transitions are connected to each other by changes in Cdc28p activity that induce one cluster while at the same time repressing the previous one. In particular, the antagonism between Cln3p and Clb2p dominates the transcriptional landscape. Of 800 cell cycle-regulated genes, more than half are induced or repressed by Cln3–Cdc28p or Clb2–Cdc28p, usually in a reciprocal manner (77). Second, the expression of cell cycle-specific transcription factors within one cluster often anticipates the onset of the next cluster, as in the case of Swi4p, Fkh1/2p, and Swi5p. Third, some clusters contain elements that sustain their own expression until appropriate conditions are met to move on to the next cluster. Thus, the *CLB2*-positive feedback loop sustains the *CLB2* cluster, and the repression of Clb–Cdc28p activity by Sic1p following anaphase enforces the low CDK state required for the M–G1 programme.

4.2 Cell cycle-regulated transcription in fission yeast

The G1–S transcriptional programme in fission yeast is regulated by an activity called MBF or DSC1, which functions in an analogous manner to SBF and MBF in budding yeast (reviewed in References 2, 78). MBF is based on Cdc10, a Swi6p homologue, and two factors related to Swi4p/Mbp1p, called Res1/Sct1 and Res2/Pct1 (Table 2). Cdc10, Res1, and Res2 all contain ankryin repeats, while Res1 and Res2 contain N-terminal DNA-binding domains and C-terminal Cdc10-interaction domains. Cdc10 is essential for passage through Start, whereas Res1 is essential for Start at suboptimal growth temperatures, and Res2 is essential for premeiotic DNA replication. Genes regulated by MBF contain multiple upstream MCB elements and include $cig2^+$, which encodes an S-phase cyclin, $cdc18^+$, which encodes the ortholog of budding yeast Cdc6p, $cdc22^+$, which encodes the large subunit of ribonucleotide reductase, and $cdt1^+$, a gene of unknown function (2, 78). Despite the basic similarities, the wiring of Start-dependent transcription in fission yeast is quite different from that in budding yeast. First, in mitotic cells MBF is a heterotrimeric complex that contains Cdc10, Res1, and Res2, each of which contributes to proper regulation of MBF-dependent genes (91, 92). In meiotic cells, however, Res1 appears to be excluded from the MBF complex (2). Second, despite the apparent role of Puc1 and Cdc2 in activation of Cdc10 (13), Cdc2 activity is not required for MBF-dependent transcription because certain $cdc2^-$ mutations cause arrest before S phase with high levels of $cig2^+$ and $cdc18^+$ transcripts (91). Third, the limiting essential function of MBF is expression not of G1 cyclins but rather of $cdc18^+$ (93). Fourth, Res2 is dependent on Rep2, a co-factor that provides a transcriptional activation function (94). Finally, and more generally, it appears that the suite of genes expressed at Start in fission yeast may not be as large as in budding yeast. For example, many genes required for DNA synthesis are expressed constitutively in fission yeast (2). Little is known about the mechanisms that control transcriptional programmes in other parts of the fission yeast cell cycle, so at this point it is difficult to say whether a similar overall transcriptional hierarchy will operate as in the budding yeast cell cycle.

4.3 Environmental inputs into transcriptional programmes

The Cln3–Cdc28p kinase and the SBF/MBF transcription factors are targets of numerous environmental signals that modulate the timing of Start (78). Because the Cln proteins are highly unstable, alterations in rates of transcription are reflected by corresponding changes in protein levels and kinase activity, thus allowing effective transcriptional regulation of Start (95). Perhaps the best characterized connection between the cell cycle machinery and an environmental stimulus is the inhibition of Cln–Cdc28p activity by Far1p in response to mating pheromones (see Section 3.4). Pheromone stimulation rapidly terminates Start-specific transcription, probably through direct inhibition of Cln3–Cdc28p kinase activity by Far1p (61). Nutrient availability also modulates the timing of Start, as reflected in the different sizes at which cells pass through Start under different nutrient conditions, a correlation referred to as critical cell size. Glucose, the preferred carbon source of budding yeast, delays Start to impose a larger critical cell size, presumably to help cells acquire more biomass in rich nutrient conditions. Glucose activates the Ras/cyclic adenosine monophosphate (cAMP) pathway, which in turn preferentially represses *CLN1* transcription, thereby delaying Start (28). The effects of cAMP on cell division are complex since, in addition to its antagonistic role at Start upon nutrient shifts, the cAMP pathway is a general positive effector of growth that increases levels of all three Cln proteins in stationary phase cells (81).

Various stress conditions also feed into the Start transcriptional machinery. A pathway based on protein kinase C (Pkc1p) transduces signals in response to cell wall stress and activates SBF. At the bottom of the Pkc1p pathway, the MAP kinase Slt2p mediates hyperphosphorylation of Swi6p (96). Interestingly, some phenotypes caused by defects in the Pkc1p pathway are suppressed by overexpression of Pcl1/2p, but not Cln1/2p, consistent with the lower levels of *PCL1/2* but not *CLN1/2* mRNAs in *slt2* cells (96). How Slt2p differentially activates the expression of only some SBF-dependent genes is unclear. SBF is also a target for inhibitory signals, as in the case of the DNA damage checkpoint pathway, which represses *CLN1/2* transcription and imposes a G1 delay in response to DNA damage (97). In this instance, Swi6p is phosphorylated by the checkpoint kinase Rad53p, an event that may convert Swi6p from a transcriptional activator into a repressor.

In fission yeast, nutrient conditions regulate both the G1–S and the G2–M transitions. At Start, nutrient limitation alters the architecture of MBF complexes to induce commitment to meiosis. The switch into meiosis is controlled by a protein kinase called Ran1/Pat1, which represses sexual development (reviewed in Reference 98). Ran1/Pat1 appears to stimulate the Puc1–Cdc2-dependent association of Cdc10 and Res1, which together with Res2, promote commitment to the mitotic division cycle (13). In the absence of Ran1/Pat1 activity, Res1 does not assemble into Cdc10–Res2 complexes, which instead promotes meiosis-specific transcription (98). In addition to Wee1-dependent control of Cdc2, the G2–M transition is regulated by yet other nutrient and stress signals via a highly complex pathway based on the Spc1 MAP kinase, which probably acts by transcriptional regulation of as yet undefined mitotic regulators (reviewed in Reference 2).

5. Programmed proteolysis

The periodic degradation of cyclin in the embryonic cell cycles of marine invertebrates foreshadowed the crucial role of highly regulated or 'programmed' proteolysis in cell division (99). The ubiquitin proteolytic system mediates the precipitous collapse of cyclin–CDK activity at the end of mitosis, and an equally important collapse of CDK inhibitory activity at Start. In a complex regulatory web, the cell cycle ubiquitination machinery and its substrates are themselves tightly controlled by CDK activity.

5.1 The ubiquitin system

Many regulatory proteins are targeted for rapid intracellular proteolysis by conjugation to ubiquitin, a conserved 76-residue protein (reviewed in Reference 100). A cascade of enzymes, generically termed E1, E2, and E3, activates and transfers ubiquitin to its many substrates (Fig. 4A). An E1 or ubiquitin-activating enzyme first forms a thioester with the C-terminal glycine of ubiquitin in an ATP-dependent manner. Ubiquitin is subsequently transferred onto an E2 or ubiquitin-conjugating enzyme in a *trans*-thioesterification reaction. Finally, an E3 or ubiquitin ligase facilitates the nucleophilic attack of a substrate lysine residue on the ubiquitin thioester, resulting in an isopeptide linkage between the C-terminus of ubiquitin and the substrate. Known E3 enzymes fall into two classes: those that dock the E2 and the substrate for direct catalytic transfer of ubiquitin from the E2 to the substrate, and those that actually participate in catalytic transfer of ubiquitin through a third thioester intermediate. Importantly, reiterated cycles of ubiquitin conjugation on lysine residues of ubiquitin itself result in assembly of a polyubiquitination chain on the substrate. The polyubiquitin tag is recognized by an abundant protease particle, the 26S proteasome, which rapidly degrades the substrate into short peptides and recycles free ubiquitin (101). The ubiquitin–proteasome partnership is ruthlessly efficient, as substrates often have half-lives on the order of only seconds to minutes. Substrate degradation is further modulated by a host of deubiquitinating enzymes, processivity factors, and ubiquitin-related modifications (100). However, the critical regulatory step in ubiquitin-dependent proteolysis is that of substrate recognition, mediated by the ubiquitin ligases. Given the many hundreds of substrate of the ubiquitin system, the E3 enzyme family is necessarily diverse, with various members that fall into distinct subfamilies. Two related ubiquitin ligases, the anaphase-promoting complex (APC) and Skp1–Cdc53/F–box protein complexes (SCF), target cell cycle regulators for degradation at mitosis and Start respectively (Fig. 4B, Table 2; reviewed in References 102, 103).

5.2 SCF ubiquitin ligases and phosphorylation-dependent proteolysis

The identification of the budding yeast *CDC34* gene product as an E2 enzyme provided the first direct connection between the ubiquitin system and the cell cycle (reviewed in Reference 103). Conditional *cdc34* mutants, as well as *cdc4* and *cdc53* mutants, arrest in late G1 phase with multiple polarized buds, consistent with a defect in the degradation of an inhibitor of DNA replication. A vital clue as to the identity of such an inhibitor arose with the finding that cells depleted for all six Clb cyclins arrest with a phenotype indistinguishable from that of the *cdc34* arrest (25). Further, because Clb5p–Cdc28p complexes are present but inactive at the *cdc34* arrest point, it was hypothesized that Cdc34p was required for degradation of the CDK inhibitor Sic1p. This prediction is borne out by numerous results, most notably by-pass of the *cdc34* replication block by deletion of *SIC1* (25). The Cln–Cdc28p kinases also participate in Sic1p degradation, as Sic1p is stable in cells arrested at Start (25). Remarkably, deletion of *SIC1* is able to bypass the requirement for Cln–Cdc28p activity (104, 105). Sic1p is phosphorylated by Cln–Cdc28p kinases on multiple sites

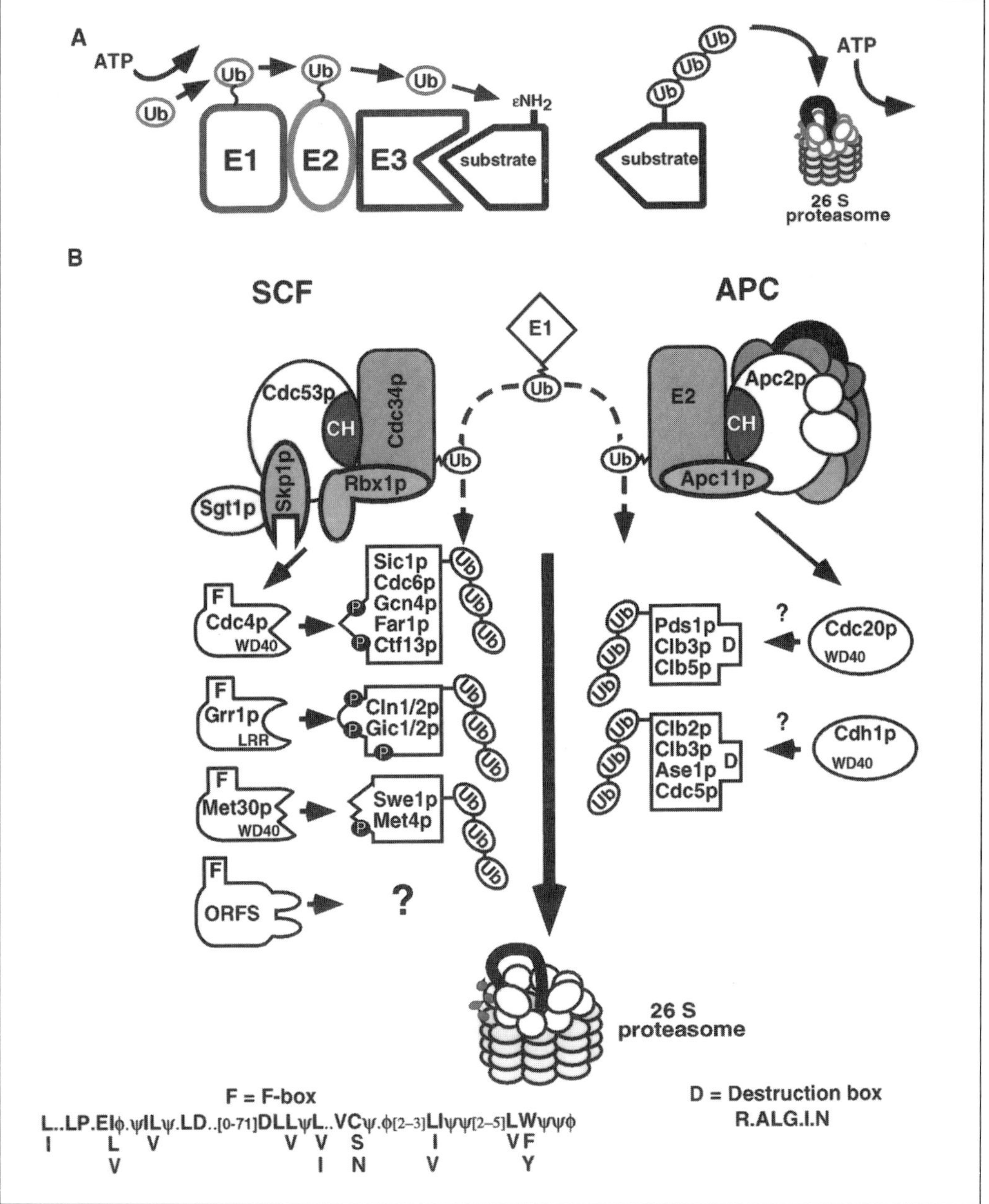

Fig. 4 (A) Schematic of the ubiquitin-proteasome system. (B) The SCF and APC ubiquitin ligases in *S. cerevisiae*. F-box proteins recruit phosphorylated substrates to SCF for ubiquitination by Cdc34p. Two related proteins, Cdc20p and Cdh1p/Hct1p, activate the APC towards distinct sets of substrates during mitosis. Uncertain or potentially indirect interactions are indicated by a question mark. P, phosphorylation; Ub, ubiquitin; WD40, WD40 repeats; LRR, leucine-rich repeats; CH, the cullin homology domain present in Cdc53p and Apc2p. Loose consensus sequences for the F box and the destruction box (D box) are provided. The F-box consensus is from 20 *S.cerevisiae* sequences and the *D*-box consensus is based on B-type cyclins from many species. φ and ψ denote hydrophobic and hydrophilic residues respectively, while a period denotes any residue. See Table 2 for *S. pombe* homologues.

in vivo and *in vitro*, and mutation of several such CDK phosphorylation sites stabilizes Sic1p *in vivo* (106). At least two other kinases, Pcl1p–Pho85p and the meiosis-specific kinase Ime2p, are also capable of targeting Sic1p for degradation (20, 107).

The instability of other proteins, including the Clns and Far1p, also depends on the Cdc34p pathway (103). In addition, a leucine-rich repeat-containing protein called Grr1p is specifically required for Cln1/2p degradation (108). Like Sic1p, Cln degradation depends on Cdc28p-dependent phosphorylation, as mutation or truncation of C-terminal CDK consensus sites in Cln2p causes its stabilization (109). A clue as to the identity of the E3 component for Cln degradation came from an analysis of proteins that interact physically with the Clns, which revealed that Cdc53p associates with all three Cln proteins in a phosphorylation-dependent manner (110). Moreover, Cdc53p physically associates with Cdc34p, consistent with an E3 ligase function (110, 111).

Description of a new class of ubiquitin ligase, the SCF family, fell into place with the identification of Skp1p, a small highly conserved protein that binds Cdc4 and is required for degradation of Sic1p, Cln2p, and, intriguingly, also Clb5p (112). In addition to Cdc4p, Skp1 interacts with many other proteins, including SKP2, as part of the cyclin A–CDK2 kinase complex in human cells, cyclin F, a mammalian cyclin of unknown function, and Ctf13p, in the yeast CBF3 kinetochore complex (112–114). Different conditional alleles of *SKP1* arrest in either late G1 phase with a *cdc34*-like phenotype or in metaphase, consistent with a dual function for Skp1p in Sic1p proteolysis and kinetochore function (112, 114). Alignment of the various Skp1p partners revealed a common sequence motif of approximately 40 amino acids, termed the F box, for cyclin F (Fig. 4B) (112). The F box is a binding site for Skp1p, and is found in many otherwise unrelated proteins, including Grr1p (103, 112).

These observations were amalgamated in a model sometimes referred to as the 'F-box hypothesis', in which numerous F-box proteins recruit a diverse set of substrates for ubiquitination by a core complex composed of Skp1p, Cdc53p, and Cdc34p (103, 112). The Skp1p–F box interaction links a host of F-box proteins to the core complex, in which Cdc53p serves as a platform for Cdc34p and Skp1p, while in turn the varied protein–protein interaction domains in different F-box proteins capture substrates for ubiquitination (103, 112). The physical interactions of Cdc34p, Skp1p, and other F-box proteins with Cdc53p in yeast extracts and purified complexes prepared from recombinant proteins supports the model shown in Fig. 4B (68, 110, 111, 115, 116). Unlike some other E3 enzymes, SCF complexes do not appear to participate directly in ubiquitin transfer, and instead rely on Cdc34p to provide catalytic activity (112, 116, 117). The various possible SCF complexes are designated by the F-box protein subunit, as in SCF^{Cdc4}, SCF^{Grr1}, and so on. Most known or suspected F-box protein–substrate interactions are driven by substrate level phosphorylation, thereby linking the degradation of many regulatory proteins to kinase-based signalling networks (Table 2).

Phosphorylation-dependent Sic1p ubiquitination has been completely reconstituted with recombinant proteins (68, 115). Ubiquitination of Sic1p by recombinant SCF^{Cdc4} requires the prior phosphorylation of Sic1p by Cln or Clb–Cdc28p kinases, which allows its capture by the WD40 repeat domain of Cdc4p. Importantly, Cdc4p selectively binds phospho-Sic1p but not phospho-Cln2p, whereas the converse specificity holds for Grr1p (115). Unlike the ubiquitination of Sic1p by SCF^{Cdc4}, an SCF complex composed solely of Skp1p, Cdc53p, and Grr1p is not sufficient to catalyse Cln2p ubiquitination, suggesting that such complexes lack a crucial co-factor (115). A small RING finger protein called Rbx1p (a.k.a. Roc1p or Hrt1p), which was first discovered as a binding partner of Cdc53p and its homologues in yeast and mammalian cells, turned out to be the missing component (117–120). Rbx1p potently stimulates phosphorylation-dependent ubiquitination of Cln1/2p by SCF^{Grr1} and also of Sic1p by SCF^{Cdc4}. Rbx1p interacts with Cdc34p, Cdc53p, and multiple F-box proteins, and so is at the hub of SCF complexes. As predicted from these interactions, *rbx1* mutants arrest with a *cdc34*-like phenotype. The Cdc34p–Cdc53p interaction is strengthened by Rbx1p, and, furthermore, it appears that the Rbx1p–SCF holocomplex allosterically stimulates the catalytic activity of Cdc34p (117, 119).

An SCF-like pathway controls the stability of Rum1 in fission yeast. Like Sic1p, Rum1 is targeted for degradation upon its phosphorylation by Cig1–Cdc2 (11). Unlike Sic1p, which is targeted by multiple phosphorylation events, mutation of one of two phosphorylation sites, Thr58 or Thr62, is sufficient to stabilize Rum1 *in vivo* and trigger re-replication (11). A hunt for mutants that cause polyploidization uncovered two non-essential Cdc4p homologues, Pop1 and Pop2 (121, 122). Cells that lack Pop1 or Pop2 function fail to eliminate Rum1, which thus accumulates to high levels and causes re-replication. Pop1 and Pop2 form a heterodimer that is necessary for Rum1 degradation *in vivo* (123, 124). An essential Cdc53p homologue called Pcu1 forms an SCF-like complex with Pop1/2 (123). Other SCF components in *S. pombe* have not yet been characterized (Table 2).

Although Sic1p is the essential substrate of SCF^{Cdc4} in G1 phase, a number of other cell cycle regulators are also targeted by SCF^{Cdc4}. In a dynamic regulatory balance that establishes the switch-like nature of the mate or divide decision, Far1p not only inhibits the Cln–Cdc28p kinases during the mating response, but is in turn targeted to SCF^{Cdc4} by Cln–Cdc28p-dependent phosphorylation (28, 63). Unlike Sic1p and Cln2p, mutation of a single phosphorylation site, Ser87, is sufficient to stabilize Far1p (59, 63). The replication protein Cdc6p is similarly targeted to SCF^{Cdc4} by Cdc28p-dependent phosphorylation, but puzzlingly stabilization of Cdc6p seems not to affect the timing or fidelity of DNA replication (125). In contrast, analogous phosphorylation site mutations in Cdc18 cause re-replication in fission yeast (126). Thus, the polyploid phenotype of *pop1⁻* or *pop2⁻* mutants derives from both direct effects on Cdc18 stability and on Rum1-mediated inhibition of Cdc2 activity, which would otherwise target Cdc18 to SCF$^{Pop1/2}$.

Additional cell cycle substrates are ubiquitinated by other SCF complexes (Table 2). Two proteins that are redundant activators of actin polarization during bud emergence, called Gic1p and Gic2p, are targeted to SCF^{Grr1} by as yet undetermined kinases (127). A third SCF complex based on the F-box protein Met30p mediates degradation of the Swe1p kinase (128). Numerous kinases may target Swe1p to SCF^{Met30}, including Clb2–Cdc28p, and one or more of the Nim1-related kinases Hsl1p,

Gin4p, and Kcc1p (46, 129). Another protein called Hsl7p, which has similarity to protein methytransferases, is also necessary for elimination of Swe1p (129). The mutual antagonism between Swe1p and its target Clb2–Cdc28p sets up another regulatory switch founded on phosphorylation and proteolysis. Finally, both Cdc4p and Grr1p are themselves degraded in an SCF-dependent manner, a pathway that may facilitate dynamic equilibrium of F-box proteins with a limiting amount of the core SCF complex (130).

As emphasized at the inception of the F-box hypothesis, the plethora of F-box proteins in sequence databases suggests that the tentacles of the SCF system reach far beyond cell cycle regulation (112). Indeed, recent genetic and biochemical analysis in plants, worms, flies, and humans has unearthed crucial roles for numerous SCF-like pathways in development and disease (103). In budding yeast, more than a dozen novel F-box proteins remain to be characterized, and in the relatively compact *C. elegans* genome more that 100 F-box proteins can be identified (103). While it is tempting to speculate that all F-box proteins will function within an SCF context, cautious interpretation is warranted since at least one Skp1p–F-box protein complex in yeast, the CBF3 kinetochore complex, is not an E3 enzyme (114). Other SCF components are also conserved through metazoan evolution, with a concomitant increase in diversity. The first identified homologue of Cdc53p, called CUL-1, is a negative regulator of proliferation in *Caenorhabditis elegans*, and is the namesake of the burgeoning cullin family (103). It is now evident that SCF complexes are the archetype for a conserved superfamily of E3 ubiquitin ligases that includes the VHL–ElonginBC–CUL2 complex in mammalian cells, and more distantly the APC (see Section 5.3).

5.3   The anaphase-promoting complex

The catastrophic destruction of cyclin–CDK activity at the end of mitosis is a *sine qua non* of cell cycle regulation (99). Mitotic cyclins are eliminated by the ubiquitin system via highly conserved, cell cycle-regulated E3 activity termed the anaphase-promoting complex (APC), also known as the cyclosome or APC/C (reviewed in Reference 102). The APC is turned on at the metaphase to anaphase transition and turned off at the G1 to S phase transition. Genetic screens for budding yeast mutants that stabilize Clb2p turned up three known *CDC* genes required for the metaphase to anaphase transition: *CDC16*, *CDC23*, and *CDC27* (131). At the same time, purification of the *Xenopus* APC particle yielded vertebrate homologues of Cdc16p, Cdc23p, and Cdc27p (132). Further purification revealed a total of 12 subunits in the yeast APC and at least eight corresponding subunits in the *Xenopus* APC particle (133, 134). Finally, parallel cytological screens for fission yeast mutants with altered nuclear structure (*nuc*) or with a septum through the undivided nucleus (*cut*, for cell untimely torn) identified Nuc2 and Cut9 as homologues of Cdc27p and Cdc16p respectively (reviewed in Reference 135). The sequence of most APC subunits does not provide any obvious insight into their function (Table 2). Cdc16p, Cdc23p, and Cdc27p all contain TPR repeats, known to mediate protein–protein interactions (135). Significantly though, two subunits, Apc2p and Apc11p, are homologues of Cdc53p and Rbx1p respectively, suggesting an ancestral relationship between the APC and SCF families (102). The E2 that functions in concert with the APC at anaphase has been identified as UbcP4 in fission yeast (136). Oddly, in budding yeast, neither the closest homologue of UbcP4 nor homologues of other E2s implicated in APC function in metazoans are required for mitosis, implying that two or more E2s must serve as redundant ubiquitin donors (137).

Substrate recognition by the APC depends on a loosely conserved nine-residue motif termed the destruction box (Fig. 4B). Mutation of the destruction box stabilizes mitotic cyclins, but, contrary to early expectations, cyclin degradation is not the trigger for the metaphase to anaphase transition. Thus, yeast mutants that arrest following anaphase, such as *cdc15* strains, retain substantial levels of Clb2p–Cdc28p activity (138). Rather, a protein responsible for sister chromatid cohesion is targeted for degradation by the APC prior to cyclin destruction. This factor, called Cut2 in fission yeast, was recovered in a *cut* mutant screen, while an analogous protein in budding yeast, Pds1p, emerged from a screen for mutants that undergo precocious dissociation of sister chromatids (139, 140). Cut2 and Pds1p both contain destruction box motifs that are required for their APC-dependent degradation at metaphase (139, 140). Deletion of *PDS1* bypasses the metaphase arrest of APC mutants and causes cells to arrest in late anaphase/telophase, suggesting that the essential role of the APC at metaphase is to eliminate Pds1p (140, 141).

Before anaphase, sister chromatids are bound together by cohesin complex, composed of Scc1p/Mcd1p, Scc3p, Smc1p, and Smc3p (reviewed in Reference 142; see Chapter 5). However, neither Cut2 nor Pds1p is a cohesion factor *per se*. Rather, Cut2 forms a physical complex with another protein isolated in the *cut* screens, called Cut1 (143). Likewise Pds1p binds the budding yeast homologue of Cut1, called Esp1p, which is also required for proper sister chromatid separation (141). Overexpression of Esp1p is sufficient to trigger sister separation, consistent with a model in which Cut2/Pds1p inhibits the activity of Cut1/Esp1p until anaphase onset (141). Esp1p in turn inactivates the cohesin complex through limited proteolytic cleavage of Scc1p (144). Cut2 and Pds1p also play a positive role in anaphase, perhaps by aiding the proper localization of Cut1 and Esp1p (141, 145). Somewhat surprisingly, *pds1* cells exhibit wild-type kinetics for mitotic entry and exit (146). Either the activating and inhibiting functions of Pds1p precisely counterbalance or, perhaps more likely, Pds1p serves to restrain anaphase only under conditions of DNA or spindle damage and simply plays no role in the unperturbed cell cycle.

While the degradation of both Pds1p and mitotic cyclins is mediated by the APC, Pds1p is eliminated slightly before Clb2p (140). This differential kinetics derives from activation of the APC by two distinct accessory factors, Cdc20p and Cdh1p/Hct1p, both of which contain WD40 repeat domains (147, 148). APC^{Cdc20} specifically targets Pds1p and is thus necessary for anaphase onset, while APC^{Cdh1} eliminates a number of proteins at the end of mitosis, including Clb2p, the spindle protein Ase1p, the polo-like kinase Cdc5p, and Cdc20p itself (reviewed in Reference 102). APC^{Cdc20} also targets Clb3p and Clb5p (146, 149). Cdc20p and Cdh1p are rate-limiting activators of the APC, since overexpression of either factor is sufficient to eliminate their respective substrates in a cell cycle-independent manner (147, 148). While it is assumed that Cdc20p and Cdh1p recruit substrates to the APC core particle, direct evidence for this idea has not been forthcoming. The basis for substrate discrimination by Cdc20p and Cdh1p is also unclear. In fission yeast, the orthologue of Cdc20p is called Slp1, which was isolated as a radiation-sensitive mutant that is unable to recover from cell cycle arrest after repair of DNA damage (150). The orthologue of Cdh1p, called Ste9 or Srw1, arose in a screen for sterile mutants and as a high-copy suppressor of a hyperactive allele of $cdc2^+$ (151–153). Like $rum1^-$ mutants, $ste9^-$ mutants are unable to curtail Cdc2 activity in G1 phase and hence are unable to arrest and mate. This phenotype also allowed the identification of the fission yeast counterpart of Apc10p (151).

The events that trigger APC^{Cdc20} activation at the metaphase to anaphase transition have not been unravelled. Although *CDC20* is expressed as part of the G2–M cluster driven by Clb2–Cdc28p, Cdc20p availability is apparently not a rate-limiting step for activation of the APC (154). Evidence from metazoan species suggests that APC activation depends on phosphorylation of the core particle by both Cdc2 and polo-family kinases (102). Conversely, cAMP-dependent protein kinase A (PKA) activity antagonizes the APC, as first shown through genetic analysis in fission yeast (135, 155). A plethora of factors identified in cytological screens in fission yeast is also required for sister chromatid separation, including at least two phosphatases, Dis2 and Sds21, and their associated regulatory factors (reviewed in Reference 3). Finally, the APC is a crucial target of mitotic checkpoint pathways that restrain mitosis until fully replicated chromosomes are condensed and properly aligned (see Chapter 4 and also below). The APC is thus a nexus for multiple regulatory inputs that control anaphase onset.

Unlike Cdc20p, Cdh1p abundance is not cell cycle regulated. Instead, Cdc28p-dependent phosphorylation of Cdh1p abrogates its interaction with the APC core particle (156, 157). In budding yeast, mutation of multiple CDK consensus sites in Cdh1p allows it constitutively to activate the APC, thereby preventing accumulation of Clb2p and Clb3p, and causing cell cycle arrest in G2 phase before bipolar spindle formation (156). In wild-type cells, APC^{Cdh1} activity is precisely reciprocal to that of Cdc28p, that is, APC^{Cdh1} is active from late anaphase/telophase until Start, whereas Cdc28p is active from Start until the end of anaphase. The mutual antagonism between CDK activity and APC^{Cdh1} activity lies at the crux of cell cycle control. The end of anaphase is marked by the transition from a state of high CDK activity and low APC^{Cdh1} activity to one of low CDK activity and high APC^{Cdh1} activity. This switch is necessary for cells to disassemble the mitotic spindle, undertake cytokinesis, and assemble pre-replicative origins in G1 phase, none of which can occur unless the CDK kinases are first turned off.

5.4   The mitotic exit/septation pathway

At the end of mitosis in budding yeast, the effects of Cdc28p-dependent phosphorylation are neutralized by the Cdc14p phosphatase. *cdc14* mutants arrest in telophase

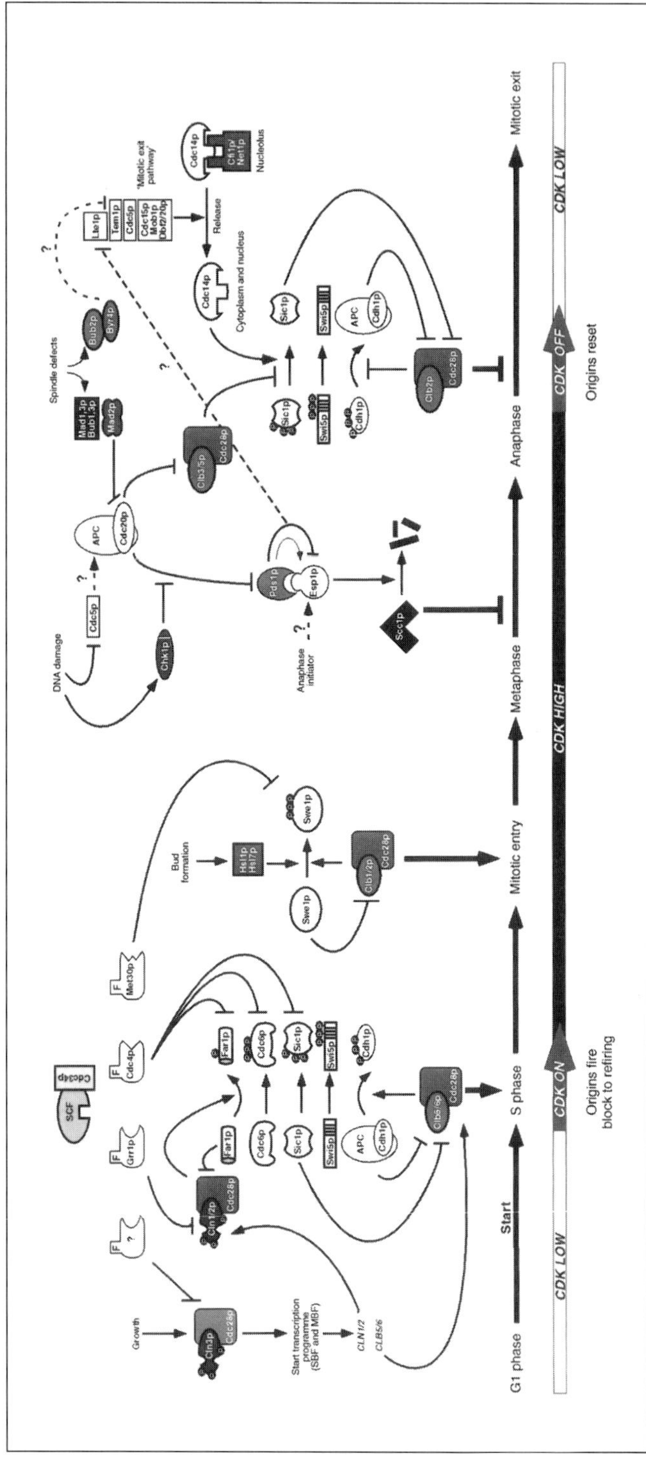

Fig. 5 The wiring of the *S. cerevisiae* cell cycle. Arrows indicate activation, blunt arrows indicate inhibition, and P indicates phosphorylation. An oscillation between a stable high CDK state and stable low CDK state determines the firing and resetting of replication origins. See text for details. See Table 2 for *S. pombe* homologues.

with substantial levels of Clb–Cdc28p kinase activity and highly elongated spindles, whereas overexpression of *CDC14* dramatically destabilizes Clb2p at all stages of the cell cycle (90). Conversely, Sic1p is extremely unstable in *cdc14* cells and is stabilized by overexpression of *CDC14*. Lastly, *SIC1* and other Swi5p-dependent genes are not expressed in *cdc14* mutants. This expression defect is partially rescued by intro- duction of a Swi5p mutant that lacks inhibitory Clb–Cdc28p phosphorylation sites. Consistent with these results, Cdc14p can specifically dephosphorylate Cdh1p, Sic1p, and Swi5p *in vitro* (90, 157). Thus, Cdc14p is a potent antagonist of Cdc28p- dependent phosphorylation and is crucial for elimination of Cdc28p activity at the end of mitosis (Fig. 5).

Cdc14p activity is controlled by an unusual mechanism, namely localization to the nucleolus through its association with an inhibitory factor called Cfi1p/Net1p (158, 159). Cdc14p is tightly localized to the nucleolus until the end of mitosis, at which point it is abruptly released into the nucleus and cytoplasm to dephosphorylate its targets. By the end of G1 phase, Cdc14p is re-sequestered in the nucleolus. In addition to directing the localization of Cdc14p, Cfi1p is both a substrate and inhibitor of Cdc14p *in vitro* (159). Numerous genetic interactions position Cdc14p as the down- stream effector of a network of proteins required for mitotic exit, including the Ras- like guanosine tripohosphatase (GTPase) Tem1p, an exchange factor Lte1p, the kinases Cdc5p, Cdc15p, Dbf2p, and Dbf20p, and Mob1p, a protein linked to the spindle checkpoint pathway (reviewed in Reference 160). The primary function of the mitotic exit pathway is to eliminate Clb–Cdc28p kinase activity, as illustrated by the fact that overexpression of *SIC1* also bypasses mutations in the mitotic exit pathway (160).

Most components of the budding yeast mitotic exit pathway are conserved in fission yeast (Table 2). However, the fission yeast version of the pathway appears dedicated to septum formation and is not required for nuclear division or spindle disassembly (reviewed in Reference 161). Thus, disruption of activators of the path- way such as Cdc7 abrogates septum formation without impeding the nuclear cycle, whereas elimination of inhibitory components such as Cdc16 results in multiple septa within the same cell. The sole exception to this theme is Plo1, a polo-like kinase and Cdc5p homologue that is required for both nuclear division and septum forma- tion. The rough order of the mitotic exit–septation pathway, shown for the budding yeast components in Fig. 5, can be deduced based on data from both budding and fission yeast. At the top of pathway in fission yeast, the Spg1 GTPase is inhibited by a complex of Cdc16 and Byr4, which form a two-component GTPase-activating com- plex that converts Spg1 into the inactive guanosine diphosphate (GDP)-bound form (162). Similarly, the analogue of the Cdc16–Byr4 complex in budding yeast, Bub2p– Byr4p, probably inhibits Tem1p (reviewed in Reference 163). The order of the kinases in the pathway is largely unknown, although Cdc5p is activated well before Dbf2p, consistent with its role in activation of both the APC and mitotic exit (160, 163). Cdc14p lies at the end of the pathway in budding yeast, since overexpression of *CDC14* bypasses all other mitotic exit mutants, and mutations in *NET1* bypass the deletion of either *CDC15* or *TEM1* (90, 159, 164). A Cdc14p homologue has just emerged from the *S. pombe* genome sequencing project (K. Gould, personal communication). While the role of the septation pathway in mitotic exit in *S. pombe* is uncertain, mitotic cyclin activity must nevertheless be eliminated for cells to complete nuclear division and cytokinesis (165).

5.5 Mitotic order and checkpoints

Events during mitosis have an intrinsic order in part because the activation of APC^{Cdh1} depends on the prior activation of APC^{Cdc20}. In addition to its role as an anaphase inhibitor, Pds1p also inhibits activation of the mitotic exit pathway, as deduced from the finding that expression of non-degradable forms of Pds1p prevents release from a *cdc15* block (166, 167). It is unclear whether Pds1p directly antagonizes the mitotic exit pathway, or whether Pds1p precludes Esp1p from activating mitotic exit (149, 166, 167). Regardless, elimination of Pds1p by APC^{Cdc20} is a necessary precondition for activation of APC^{Cdh1} and mitotic exit. A second barrier to mitotic exit is implied by the late anaphase/telophase arrest of *cdc20 pds1* double mutants. Surprisingly, Clb5p represents a second essential target of APC^{Cdc20}, as shown by the robust viability of *cdc20 pds1 clb5* triple mutants (149). Because Cdc14p is released from the nucleolus in *cdc20 pds1* double mutants, it appears that Clb5p–Cdc28p overwhelms the phosphatase activity of Cdc14p, rather than preventing its relocalization (149). Thus, APC^{Cdc20} plays a dual role in activation of APC^{Cdh1}: elimination of Pds1p, which allows the release of Cdc14p, and elimination of Clb5p, which would otherwise maintain phosphorylation of Swi5p, Sic1p, and Cdh1p, even in the presence of active Cdc14p. Together, these two targets of APC^{Cdc20} help to ensure that cytokinesis and spindle disassembly cannot occur before anaphase (Fig. 5). Clearly though, a redundant pathway must exist that is able to coordinate anaphase and mitotic exit in *cdc20 pds1 clb5* strains.

Important checkpoint pathways halt cell cycle progression in response to DNA and spindle damage by preventing activation of the APC (see Chapter 4). In both budding and fission yeast, unattached kinetochores activate the Mad2p checkpoint complex, which binds and inactivates APC^{Cdc20}, arresting cells at the metaphase to anaphase transition (168, 169). In a second branch of the spindle checkpoint pathway, which operates specifically in budding yeast, spindle defects cause the Bub2p–Byr4p complex to convert Tem1p into an inactive GDP-bound state and thus block mitotic exit (163). In another budding yeast specific pathway, DNA damage stimulates the Chk1p kinase to phosphorylate and stabilize Pds1p, simultaneously preventing both passage through anaphase and mitotic exit (170). A second branch of the DNA damage response appears to inhibit Cdc5p-dependent activation of APC^{Cdc20}, further locking cells in a pre-anaphase state until damage is repaired (170).

6. The cell cycle as a bistable oscillator

An underlying logic unifies cell division in all eukaryotes. From yeast to early embryos of marine invertebrates to human somatic cells, the bistable oscillation between a state of low CDK activity and a state of high CDK activity forms the core of every cell cycle. The inextricable link between CDK activity and the replication machinery mandates a once per cell cycle replication of the genome. Superimposed on top of this basic engine, dependency relationships coordinate replication, chromosome segregation, and other events necessary for high-fidelity transmission of the genome.

6.1 The CDK cycle and the replication–segregation cycle

The cell cycle is powered by a primal two-stroke engine that cycles between a state of high CDK activity and low CDK activity. The CDK oscillator depends on unidirectional changes of the CDK state, from low to high at Start and from high to low at the end of mitosis. Because each state perpetuates itself until all necessary events are complete, the CDK oscillator is intrinsically bistable, such that one round of division corresponds to one cycle of the oscillator. The crucial link between CDK oscillation and cell division resides in the requirements for CDK activity in DNA replication and chromosome segregation (reviewed in Reference 171). First, CDK activity both stimulates the firing of replication origins and prevents the refiring of spent origins (172). Each round of replication thus requires a complete cycle of CDK status: the high state fires origins, while the low state resets origins. Second, once replication is complete, CDK activity not only activates chromosome segregation, but also 'sows the seeds of its own destruction', because the activation of anaphase inevitably leads to activation of cyclin proteolysis. The key question then is how does the switch between CDK states occur? The principal elements of cell cycle control—CDK activity, transcriptional regulation, and programmed proteolysis—are wired together to create a toggle that is thrown into the high CDK state at G1–S, and back into the low CDK state at M–G1 (Fig. 5). In budding yeast, the switches are wired as follows.

6.1.1 The switch at G1–S

As cells approach Start and Cln1/2–Cdc28p activity rises, two events break the low CDK state. First, the Cln–Cdc28p kinases target Sic1p for ubiquitin-dependent proteolysis, thereby relieving inhibition of Clb5/6–Cdc28p kinase complexes. Second, the Cln–Cdc28p kinases phosphorylate Cdh1p, thereby abrogating its interaction with the APC and stabilizing the Clb proteins. Both events synergize to create permissive conditions for Clb–Cdc28p activity, which then initiates DNA replication. Importantly, the onset of Clb–Cdc28p activity entrains the high CDK state. That is, Clb–Cdc28p-dependent phosphorylation maintains repression of Sic1p and Cdh1p, while the *CLB1/2*-positive transcriptional feedback loop ensures an ongoing supply of Clb1/2–Cdc28p activity. In addition, Clb–Cdc28p-mediated phosphorylation blocks nuclear import of Swi5p, thereby repressing *SIC1* transcription and further locking in the high CDK state.

6.1.2 Entry into mitosis

Once DNA replication is complete, and checkpoint conditions are met, mitosis ensues. In neither fission yeast nor budding yeast is the exact trigger for mitotic onset known, although in each case it appears that a critical level of mitotic CDK activity is required, as opposed to S phase promoting CDK activity (reviewed in Reference 173). In fission yeast, and indeed most other species including humans, dephosphorylation of Tyr15 is the rate-determining step for entry into mitosis (33). As noted, several clues hint at the existence of a positive feedback loop between Cdc2 and Cdc25, although proof that such loops dictate the timing of mitosis under physiological conditions is difficult to obtain. None the less, it is reasonable to suppose that the gradual growth-dependent accumulation of Cdc25, measured against a nutrient-ependent setpoint of Wee1 activity, might determine when cells enter mitosis (40). Quite unlike other species, mitotic entry in budding yeast depends, at least as a pre-condition, upon the *CLB1/2*-positive transcriptional feedback loop, but again the trigger that sets this process in motion is unknown. Also in contrast to fission yeast and other species, the budding yeast checkpoint pathways that relay DNA damage signals do not modulate inhibitory tyrosine phosphorylation of Cdc2, but rather prevent chromosome segregation and cytokinesis through inhibition of the APC at metaphase. The rewiring of mitotic control in budding yeast may be a consequence of early bipolar spindle formation and the early commitment to cytokinesis through formation of the bud neck, both of which necessarily occur shortly after Start. This early formation of mitotic structures, well before DNA replication is complete, might simply preclude the use of CDK inhibition as a control mechanism.

6.1.3 The switch at M–G1

The Clb–Cdc28p stranglehold is broken by a direct reversal of the Cdc28p-dependent phosphorylation events that perpetuate the high CDK state. At the end of mitosis, Swi5p, Sic1p, and Cdh1p are each dephosphorylated by the Cdc14p phosphatase. As Swi5p is dephosphorylated, it enters the nucleus and activates *SIC1* transcription. In its stable dephosphorylated state, Sic1p is able to quell Clb-Cdc28p kinase activity effectively. In parallel, once Cdh1p is dephosphorylated, it associates with and activates the APC to degrade the Clbs and other positive regulators of mitosis. The catastrophic collapse in Clb–Cdc28p activity is entrained in a low CDK state by the active APC and by the stable expression of *SIC1* mRNA and protein, which persist until Cln–Cdc28p activity rises again in late G1 phase.

6.1.4 Throwing the switches

By definition, the signals that allow escape from the entrained CDK states must be independent of the mechanisms that sustain each state. Two characteristics of the G1–S switch meet this criterion. First, the Cln–Cdc28p complexes are immune to inhibition by Sic1p and degradation by APC[Cdh1]. Second, activation of the Start programme, including *CLN1/2*, is dictated by environmental inputs in pre-Start G1 phase. Growth to a critical cell size activates Cln3p–Cdc28p and subsequently SBF/MBF, although it is not known how the continuous increase in growth is converted into a switch-like event at Start (81). Once the switch is thrown, the initiating signal is extinguished because Clb–Cdc28p activity represses SBF-dependent transcription (85).

The M–G1 switch hinges on activation of the Cdc14p phosphatase, which defuses

Clb–Cdc28p activity. Cdc14p is activated by APC^{Cdc20}, first by elimination of Pds1p, which would otherwise block the mitotic exit pathway, and second by elimination of Clb5p, a potent antagonist of Cdc14p (149). Because Cdc20p is not inhibited by the high CDK state, it is a suitable catalyst for the onset of anaphase and mitotic exit. Again, once the switch is thrown, the initiator is defused as *CDC20* is not expressed in G1 phase and, further, both Cdc5p and Cdc20p are targeted for degradation by the APC in G1 phase (154, 174). The trigger that fires anaphase and the concomitant activation of APC^{Cdc20} remains elusive.

6.2 Genome stability and the cell cycle

6.2.1 CDK activity and DNA replication

The link between CDK activity and DNA replication was first realized in fission yeast, where certain alleles of *cdc2$^+$* can cause cells to re-replicate their genome (173). Re-replication appears to occur whenever CDK activity is lost, as in *cdc13* strains or when *rum1$^+$* is either overexpressed or mutationally stabilized. The re-replication effect is probably mediated through unregulated Cdc18 activity since both over-expression and mutational stabilization of Cdc18 trigger duplication of the genome in the absence of an intervening mitosis (93, 126). It is uncertain whether these re-replications require a cryptic oscillation in Cdc2 activity in order to reset replication origins.

The re-replication phenomenon does not occur quite so readily in budding yeast. Thus, *cdc4* cells, *clb1–6$^-$* cells, and cells expressing stabilized versions of Sic1p all arrest in G1 phase. However, this apparent difference may be more style than substance. Artificially imposed cycles of high and low Clb–Cdc28p activity do trigger re-replication, which elegantly demonstrates the link between CDK activity and origin firing (172, 175). Moreover, prolonged overexpression of *SIC1* causes some re-replication in *S. cerevisiae*, as does at least one dominant allele of *CDC6* (172, 176). For some reason, the replication machinery of budding yeast may be inherently more sensitive to CDK-mediated inhibition than in fission yeast. The exact mechanism whereby CDK activity both activates and inhibits replication origins is unresolved. In the simplest model, phosphorylation of factors such as Cdc6p might cause the pre-replicative complex to partially dissociate from origins, and, until CDK activity subsides, the phosphorylated components are unable to reassemble into new pre-replicative complexes. Indeed, recent evidence indicates that CDK dependent phos-phorylation of Mcm4, a component of pre-replicative origins, cause its exclusion from the nucleus (177).

6.2.2 Defective switches and genome instability

Any organism with a large linear genome requires literally thousands of origins to drive replication to completion in a timely manner, so if origins are not fired efficiently the risk of damage to the genome is great. The coupling of replication and segregation to the bistable CDK oscillator ensures that the genome is replicated once and only once per cell cycle. Because the mechanics of the replication requires a

CDK-free window to allow reloading of pre-replicative complexes, any event that perturbs the CDK-less window interferes with origin loading, and partially disables replication. As cases in point, *cdc14* strains, *sic1*strains, and strains that overexpress G1 cyclins all have extremely high rates of chromosome loss (53, 175, 178). In fission yeast, perturbation of the CDK oscillator leads to polyploidy, another route to genome instability.

6.2.3 Ploidy control and developmental alternatives

The sexual life cycle of budding yeast and fission yeast demands that conjugation and meiosis occur in G1 phase, and that the signals for such developmental alternatives act on pre-Start G1 cells. Importantly, switch-like mechanisms appear to control the choices between division and developmental alternatives. The mutual antagonism between the cell cycle machinery and the mating pheromone pathway in budding yeast is a good examples of such a switch. The self-perpetuating nature of the high CDK state also suggests a molecular basis for cell cycle commitment at Start; that is, there simply is no way back without first passing through mitosis. Because of its diploid G1 phase lifestyle, in budding yeast nutrient controls are integral to the Start decision (1). In contrast, fission yeast cells normally proceed directly into DNA replication before completion of cytokinesis unless severe nutrient depletion signals an arrest at Start (2). In accord with its haploid lifestyle, nutrient-sensing pathways in fission yeast pause the CDK oscillator at G2–M, which requires inhibitory phosphorylation on Cdc2. Some mechanism must therefore ensure that the loss of CDK activity at the G2–M transition is not perceived as re-entry in to G1 phase. A low level of residual Cdc2 activity in the tyrosine phosphorylated state and/or the absence of $cdc18^+$ expression may help prevent replication in G1 phase (173).

6.2.4 Dependency relationships

The CDK oscillator obviously does not explain the ordered series of events that occur in the high CDK state. Replication, bipolar spindle formation, mitosis, and cytokinesis must all be controlled in a precise temporal fashion. Such order is imposed by dependency relationships, the most important of which is the requirement for DNA replication before the onset of chromosome segregation. The dependency of segregation on replication must in some way be linked to replication origins. G1 cells that lack pre-replicative complexes but that do elaborate CDK activity undergo a lethal ectopic anaphase, suggesting that pre-replicative complexes may inhibit mitosis (80, 93). This mechanism might explain the dependency of mitosis on replication; that is, mitosis can occur only once all pre-replicative complexes have been discharged. The transcriptional hierarchies that make different cyclins available as cells progress towards mitosis may also contribute to order in the high CDK state by providing specific forms of CDK activity. Finally, as noted, multiple checkpoint pathways intercept both the CDK oscillator, dependency pathways, and downstream effectors as a means to cope with damage and other forms of stress that might interfere with faithful transmission of the genome.

6.3 Fundamental issues

The view that the cell cycle is primarily a CDK oscillator is supported by over-whelming evidence from budding yeast and fission yeast. The frontiers now lie at opposite poles of the cell cycle map, the upstream initiators of cell cycle events and the downstream effectors of CDK activity. Some key unanswered questions currently include:

- How is the continuous analogue signal of G1 phase growth converted into the digital output of Start?
- How does CDK activity at once discharge pre-replicative complexes and prevent their reformation?
- What is the signal that cues entry into mitosis and how is this signal connected to the completion of DNA replication?
- What is the signal that initiates anaphase and activates APC^{Cdc20}?
- What are the presumptive substrates that are effectors of CDK activity at the major cell cycle transitions?
- How has the cell cycle machinery of budding yeast and fission yeast been re-wired to accommodate their different lifestyles?

The comparative cell cycle genetics, biochemistry, and cell biology of budding yeast and fission yeast seems certain to resolve these central issues and, in the process, will no doubt reveal new and unanticipated facets of eukaryotic cell cycle control.

Acknowledgements

The authors are grateful to Bruce Futcher, Paul Spellman, and Gavin Sherlock for generously providing Plate 3. P. J. is supported by a predoctoral studentship award from the National Sciences and Engineering Research Council of Canada. Research in M.T.'s laboratory is supported by the National Cancer Institute of Canada, the Medical Research Council of Canada, and the Protein Engineering Centres of Excellence.

References

1. Lew, D., Weinert, T., and Pringle, J. R. (1997) Cell cycle control in *Saccharomyces cerevisiae*. In *The molecular and cellular biology of the yeast* Saccharomyces (ed J. R. Pringle, J. R. Broach, and E. W. Jones), Vol. 3, p. 607. Cold Spring Harbor Laboratory Press, New York.
2. MacNeill, S. and Nurse, P. (1997) Cell cycle control in fission yeast. In *The molecular and cellular biology of the yeast* Saccharomyces (ed J. R. Pringle, J. R. Broach, and E. W. Jones), Vol. 3, p. 697. Cold Spring Harbor Laboratory Press, New York.
3. Su, S.S.Y. and Yanagida, M. (1997) Mitosis and cytokinesis in the fission yeast, *Schizosaccharomyces pombe*. In *The molecular and cellular biology of the yeast* Saccharomyces (ed J. R. Pringle, J. R. Broach, and E. W. Jones), Vol. 3, p. 765. Cold Spring Harbor Laboratory Press, New York.

4. Hartwell, L. H., Culotti, J., Pringle, J. R., and Reid, B. J. (1974) Genetic control of the cell division cycle in yeast. *Science*, **183**, 46.

5. Piggott, J. R., Rai, R., and Carter, B. L. (1982) A bifunctional gene product involved in two phases of the yeast cell cycle. *Nature*, **298**, 391.

6. Nurse, P. (1975) Genetic control of cell size at cell division in yeast. *Nature*, **256**, 457.

7. Nurse, P. (1990) Universal control mechanism regulating onset of M-phase. *Nature*, **344**, 503.

8. Morgan, D. O. (1997) Cyclin-dependent kinases: engines, clocks, and microprocessors. *Annu. Rev. Cell. Devel. Biol.*, **13**, 261.

9. Andrews, B. and Measday, V. (1998) The cyclin family of budding yeast: abundant use of a good idea. *Trends Genet.*, **14**, 66.

10. Fisher, D. and Nurse, P. (1995) Cyclins of the fission yeast *Schizosaccharomyces pombe*. *Semin. Cell. Biol.*, **6**, 73.

11. Benito, J., Martin-Castellanos, C., and Moreno, S. (1998) Regulation of the G1 phase of the cell cycle by periodic stabilization and degradation of the p25[rum1] CDK inhibitor. *EMBO J.*, 17, 482.

12. Fisher, D. L. and Nurse, P. (1996) A single fission yeast mitotic cyclin B p34[cdc2] kinase promotes both S-phase and mitosis in the absence of G1 cyclins. *EMBO J.*, **15**, 850.

13. Caligiuri, M., Connolly, T., and Beach, D. (1997) Ran1 functions to control the Cdc10/Sct1 complex through Puc1. *Mol. Biol. Cell*, **8**, 1117.

14. Furnari, B. A., Russell, P., and Leatherwood, J. (1997) Pch1[+], a second essential C-type cyclin gene in *Schizosaccharomyces pombe*. *J. Biol. Chem.*, **272**, 12100.

15. Sherr, C.J. (1993) Mammalian G1 cyclins. *Cell*, **73**, 1059.

16. Tyers, M., Tokiwa, G., and Futcher, B. (1993) Comparison of the *Saccharomyces cerevisiae* G1 cyclins: Cln3 may be an upstream activator of Cln1, Cln2 and other cyclins. *EMBO J.*, **12**, 1955.

17. McInerny, C. J., Partridge, J. F., Mikesell, G. E., Creemer, D. P., and Breeden, L. L. (1997) A novel Mcm1-dependent element in the *SWI4*, *CLN3*, *CDC6*, and *CDC47* promoters activates M/G1-specific transcription. *Genes Dev.*, **11**, 1277.

18. Stuart, D. and Wittenberg, C. (1995) *CLN3*, not positive feedback, determines the timing of *CLN2* transcription in cycling cells. *Genes Dev.*, **9**, 2780.

19. Dirick, L., Bohm, T., and Nasmyth, K. (1995) Roles and regulation of Cln–Cdc28 kinases at the start of the cell cycle of *Saccharomyces cerevisiae*. *EMBO J.*, **14**, 4803.

20. Dirick, L., Goetsch, L., Ammerer, G., and Byers, B. (1998) Regulation of meiotic S phase by Ime2 and a Clb5,6-associated kinase in *Saccharomyces cerevisiae*. *Science*, **281**, 1854.

21. Stuart, D. and Wittenberg, C. (1998) CLB5 and CLB6 are required for premeiotic DNA replication and activation of the meiotic S/M checkpoint. *Genes Dev.*, **12**, 2698.

22. Colomina, N., Gari, E., Gallego, C., Herrero, E., and Aldea, M. (1999) G1 cyclins block the Ime1 pathway to make mitosis and meiosis incompatible in budding yeast. *EMBO J.*, **18**, 320.

23. Donaldson, A. D., Raghuraman, M. K., Friedman, K. L., Cross, F. R., Brewer, B. J., and Fangman, W. L. (1998) CLB5-dependent activation of late replication origins in *S. cerevisiae*. *Mol. Cell*, **2**, 173.

24. Cross, F. R., Yuste-Rojas, M., Gray, S., and Jacobson, M. D. (1999) Specialization and targeting of B-type cyclins. *Mol. Cell*, **4**, 11.

25. Schwob, E., Bohm, T., Mendenhall, M. D., and Nasmyth, K. (1994) The B-type cyclin kinase inhibitor p40[SIC1] controls the G1 to S transition in *S. cerevisiae*. *Cell*, **79**, 233.

26. Dahmann, C. and Futcher, B. (1995) Specialization of B-type cyclins for mitosis or meiosis in *S. cerevisiae*. *Genetics*, **140**, 957.

27. Segal, M., Clarke, D. J., and Reed, S. I. (1998) Clb5-associated kinase activity is required early in the spindle pathway for correct preanaphase nuclear positioning in Saccharomyces cerevisiae. *J. Cell Biol.*, **143**, 135.

28. Cross, F. R. (1995) Starting the cell cycle—what's the point? *Curr. Opin. Cell Biol.*, **7,** 790.

29. Oehlen, L. J. and Cross, F. R. (1998) Potential regulation of Ste20 function by the Cln1–Cdc28 and Cln2–Cdc28 cyclin-dependent protein kinases. *J. Biol. Chem.*, **273**, 25089.

30. Wu, C., Leeuw, T., Leberer, E., Thomas, D. Y., and Whiteway, M. (1998) Cell cycle- and Cln2p–Cdc28p-dependent phosphorylation of the yeast Ste20p protein kinase. *J. Biol. Chem.*, **273**, 28107.

31. Altman, R. and Kellogg, D. (1997) Control of mitotic events by Nap1 and the Gin4 kinase. *J. Cell. Biol.*, **138**, 119.

32. Nigg, E. A. (1993) Targets of cyclin-dependent protein kinases. *Curr. Opin. Cell Biol.*, **5,** 187.

33. Lew, D. J. and Kornbluth, S. (1996) Regulatory roles of cyclin dependent kinase phosphorylation in cell cycle control. *Curr. Opin. Cell Biol.*, **8,** 795.

34. Lundgren, K., Walworth, N., Booher, R., Dembski, M., Kirschner, M., and Beach, D. (1991) mik1 and wee1 cooperate in the inhibitory tyrosine phosphorylation of cdc2. *Cell*, **64,** 1111.

35. Breeding, C. S., Hudson, J., Balasubramanian, M. K., Hemmingsen, S. M., Young, P. G., and Gould, K. L. (1998) The $cdr2^+$ gene encodes a regulator of G2/M progression and cytokinesis in *Schizosaccharomyces pombe*. *Mol. Biol. Cell*, **9,** 3399.

36. Kanoh, J. and Russell, P. (1998) The protein kinase Cdr2, related to Nim1/Cdr1 mitotic inducer, regulates the onset of mitosis in fission yeast. *Mol. Biol. Cell*, **9,** 3321.

37. Wu, L., Shiozaki, K., Aligue, R., and Russell, P. (1996) Spatial organization of the Nim1–Wee1–Cdc2 mitotic control network in *Schizosaccharomyces pombe*. *Mol. Biol. Cell*, **7,** 1749.

38. Wu, L. and Russell, P. (1997) Nif1, a novel mitotic inhibitor in *Schizosaccharomyces pombe*. *EMBO J.*, **16,** 1342.

39. Millar, J. B., Lenaers, G., and Russell, P. (1992) Pyp3 PTPase acts as a mitotic inducer in fission yeast. *EMBO J.*, **11,** 4933.

40. Kovelman, R. and Russell, P. (1996) Stockpiling of Cdc25 during a DNA replication checkpoint arrest in *Schizosaccharomyces pombe*. *Mol. Cell. Biol.*, **16,** 86.

41. Nefsky, B. and Beach, D. (1996) Pub1 acts as an E6–AP-like protein ubiquitiin ligase in the degradation of cdc25. *EMBO J.*, **15,** 1301.

42. Hudson, J. D., Feilotter, H., and Young, P. G. (1990) *stf1*: non-wee mutations epistatic to *cdc25* in the fission yeast *Schizosaccharomyces pombe*. *Genetics*, **126,** 309.

43. Bridge, A. J., Morphew, M., Bartlett, R., and Hagan, I. M. (1998) The fission yeast SPB component Cut12 links bipolar spindle formation to mitotic control. *Genes Dev.*, **12,** 927.

44. Lew, D. J. and Reed, S. I. (1995) A cell cycle checkpoint monitors cell morphogenesis in budding yeast. *J. Cell Biol.*, **129,** 739.

45. McMillan, J. N., Sia, R. A., Bardes, E. S., and Lew, D. J. (1999) Phosphorylation-independent inhibition of Cdc28p by the tyrosine kinase Swe1p in the morphogenesis checkpoint. *Mol. Cell. Biol.*, **19,** 5981.

46. Barral, Y., Parra, M., Bidlingmaier, S., and Snyder, M. (1999) Nim1-related kinases coordinate cell cycle progression with the organization of the peripheral cytoskeleton in yeast. *Genes Dev.*, **13,** 176.

47. Edgington, N. P., Blacketer, M. J., Bierwagen, T. A., and Myers, A. M. (1999) Control of *Saccharomyces cerevisiae* filamentous growth by cyclin-dependent kinase Cdc28. *Mol. Cell. Biol.*, **19,** 1369.

48. Lee, K. M., Saiz, J. E., Barton, W. A., and Fisher, R. P. (1999) Cdc2 activation in fission yeast depends on Mcs6 and Csk1, two partially redundant Cdk-activating kinases (CAKs). *Curr. Biol.*, **9**, 441.

49. Kaldis, P., Sutton, A., and Solomon, M. J. (1996) The Cdk-activating kinase (CAK) from budding yeast. *Cell* 86, 553.

50. Thuret, J. Y., Valay, J. G., Faye, G., and Mann, C. (1996) Civ1 (CAK in vivo), a novel Cdk-activating kinase. *Cell*, **86**, 565.

51. Cross, F. R. and Levine, K. (1998) Molecular evolution allows bypass of the requirement for activation loop phosphorylation of the Cdc28 cyclin-dependent kinase. *Mol. Cell. Biol.*, **18**, 2923.

52. Mendenhall, M. D. (1993) An inhibitor of p34^{CDC28} protein kinase activity from *Saccharomyces cerevisiae*. *Science*, **259**, 216.

53. Nugroho, T. T. and Mendenhall, M. D. (1994) An inhibitor of yeast cyclin-dependent protein kinase plays an important role in ensuring the genomic integrity of daughter cells. *Mol. Cell. Biol.*, **14**, 3320.

54. Donovan, J. D., Toyn, J. H., Johnson, A. L., and Johnston, L. H. (1994) P40^{SDB25}, a putative CDK inhibitor, has a role in the M/G1 transition in *Saccharomyces cerevisiae*. *Genes Dev.*, **8**, 1640.

55. Moreno, S. and Nurse, P. (1994) Regulation of progression through the G1 phase of the cell cycle by the *rum1$^+$* gene. *Nature*, **367**, 236.

56. Correa-Bordes, J., Gulli, M. P., and Nurse, P. (1997) p25^{rum1} promotes proteolysis of the mitotic B-cyclin p56^{cdc13} during G1 of the fission yeast cell cycle. *EMBO J.*, **16**, 4657.

57. Sanchez-Diaz, A., Gonzalez, I., Arellano, M., and Moreno, S. (1998) The Cdk inhibitors p25^{rum1} and p40^{SIC1} are functional homologues that play similar roles in the regulation of the cell cycle in fission and budding yeast. *J. Cell Sci.*, **111**, 843.

58. Peter, M., Gartner, A., Horecka, J., Ammerer, G., and Herskowitz, I. (1993) FAR1 links the signal transduction pathway to the cell cycle machinery in yeast. *Cell*, **73**, 747.

59. Gartner, A., Jovanovic, A., Jeoung, D. I., Bourlat, S., Cross, F. R., and Ammerer, G. (1998) Pheromone-dependent G1 cell cycle arrest requires Far1 phosphorylation, but may not involve inhibition of Cdc28–Cln2 kinase, *in vivo*. *Mol. Cell. Biol.*, **18**, 3681.

60. Tyers, M., Futcher, B. (1993) Far1 and Fus3 link the mating pheromone signal transduction pathway to three G1-phase Cdc28 kinase complexes. *Mol. Cell. Biol.*, **13**, 5659.

61. Jeoung, D. I., Oehlen, L. J., and Cross, F. R. (1998) Cln3-associated kinase activity in *Saccharomyces cerevisiae* is regulated by the mating factor pathway. *Mol. Cell. Biol.*, **18**, 433.

62. Peter, M. and Herskowitz, I. (1994) Direct inhibition of the yeast cyclin-dependent kinase Cdc28–Cln by Far1. *Science*, **265**, 1228.

63. Henchoz, S., Chi, Y., Catarin, B., Herskowitz, I., Deshaies, R. J., and Peter, M. (1997) Phosphorylation- and ubiquitin-dependent degradation of the cyclin-dependent kinase inhibitor Far1p in budding yeast. *Genes Dev.*, **11**, 3046.

64. Butty, A. C., Pryciak, P. M., Huang, L. S., Herskowitz, I., and Peter, M. (1998) The role of Far1p in linking the heterotrimeric G protein to polarity establishment proteins during yeast mating. *Science*, **282**, 1511.

65. Woollard, A., Basi, G., and Nurse, P. (1996) A novel S phase inhibitor in fission yeast. *EMBO J.*, **15**, 4603.

66. Elsasser, S., Lou, F., Wang, B., Campbell, J. L., and Jong, A. (1996) Interaction between yeast Cdc6 protein and B-type cyclin/Cdc28 kinases. *Mol. Biol. Cell*, **7**, 1723.

67. Greenwood, E., Nishitani, H., and Nurse, P. (1998) Cdc18p can block mitosis by two independent mechanisms. *J. Cell Sci.*, **111**, 3101.

68. Feldman, R. M., Correll, C. C., Kaplan, K. B., and Deshaies, R. J. (1997) A complex of Cdc4p, Skp1p, and Cdc53p/cullin catalyzes ubiquitination of the phosphorylated CDK inhibitor Sic1p. *Cell*, **91**, 221.

69. Bourne, Y., Watson, M. H., Hickey, M. J., Holmes, W., Rocque, W., Reed, S. I., and Tainer, J. A. (1996) Crystal structure and mutational analysis of the human CDK2 kinase complex with cell cycle-regulatory protein CksHs1. *Cell*, **84**, 863.

70. Kaiser, P., Moncollin, V., Clarke, D. J., Watson, M. H., Bertolaet, B. L., Reed, S. I., and Bailly, E. (1999) Cyclin-dependent kinase and Cks/Suc1 interact with the proteasome in yeast to control proteolysis of M-phase targets. *Genes Dev.*, **13**, 1190.

71. Gerber, M. R., Farrell, A., Deshaies, R. J., Herskowitz, I., and Morgan, D. O. (1995) Cdc37 is required for association of the protein kinase Cdc28 with G1 and mitotic cyclins. *Proc. Natl. Acad. Sci. U.S.A.*, **92**, 4651.

72. Kimura, Y., Rutherford, S. L., Miyata, Y., Yahara, I., Freeman, B. C., Yue, L., Morimoto, R. I., and Lindquist, S. (1997) Cdc37 is a molecular chaperone with specific functions in signal transduction. *Genes Dev.*, **11**, 1775.

73. Pavletich, N. P. (1999) Mechanisms of cyclin-dependent kinase regulation: structures of Cdks, their cyclin activators, and Cip and INK4 inhibitors. *J. Mol. Biol.*, **287**, 821.

74. Huang, K. N., Odinsky, S. A., and Cross, F. R. (1997) Structure–function analysis of the *Saccharomyces cerevisiae* G1 cyclin Cln2. *Mol. Cell. Biol.*, **17**, 4654.

75. Levine, K., Kiang, L., Jacobson, M. D., Fisher, R. P., and Cross, F. R. (1999) Directed evolution to bypass cyclin requirements for the Cdc28p cyclin-dependent kinase. *Mol. Cell*, **4**, 353.

76. Cho, R. J., Campbell, M. J., Winzeler, E. A., Steinmetz, L., Conway, A., Wodicka, L., Wolfsberg, T. G., Gabrielian, A. E., Landsman, D., Lockhart, D. J., and Davis, R. W. (1998) A genome-wide transcriptional analysis of the mitotic cell cycle. *Mol. Cell*, **2**, 65.

77. Spellman, P. T., Sherlock, G., Zhang, M. Q., Iyer, V. R., Anders, K., Eisen, M. B., Brown, P. O., Botstein, D., and Futcher, B. (1998) Comprehensive identification of cell cycle-regulated genes of the yeast *Saccharomyces cerevisiae* by microarray hybridization. *Mol. Biol. Cell*, **9**, 3273.

78. Breeden, L. (1996) Start-specific transcription in yeast. *Curr. Topics Microbiol. Immunol.*, **208**, 95.

79. Nasmyth, K. and Dirick, L. (1991) The role of SWI4 and SWI6 in the activity of G1 cyclins in yeast. *Cell*, **66**, 995.

80. Piatti, S., Lengauer, C., and Nasmyth, K. (1995) Cdc6 is an unstable protein whose *de novo* synthesis in G1 is important for the onset of S phase and for preventing a 'reductional' anaphase in the budding yeast *Saccharomyces cerevisiae*. *EMBO J.*, **14**, 3788.

81. Polymenis, M. and Schmidt, E. V. (1999) Coordination of cell growth with cell division. *Curr. Opin. Genet. Dev.*, **9**, 76.

82. Harrington, L. A. and Andrews, B. J. (1996) Binding to the yeast Swi4,6-dependent cell cycle box, CACGAAA, is cell cycle regulated *in vivo*. *Nucleic Acids Res.*, **24**, 558.

83. Koch, C., Schleiffer, A., Ammerer, G., and Nasmyth, K. (1996) Switching transcription on and off during the yeast cell cycle: Cln/Cdc28 kinases activate bound transcription factor SBF (Swi4/Swi6) at start, whereas Clb/Cdc28 kinases displace it from the promoter in G2. *Genes Dev.*, **10**, 129.

84. Cosma, M. P., Tanaka, T., and Nasmyth, K. (1999) Ordered recruitment of transcription and chromatin remodeling factors to a cell cycle- and developmentally regulated promoter. *Cell*, **97**, 299.

85. Amon, A., Tyers, M., Futcher, B., and Nasmyth, K. (1993) Mechanisms that help the yeast cell cycle clock tick: G2 cyclins transcriptionally activate G2 cyclins and repress G1 cyclins. *Cell*, **74**, 993.

86. Siegmund, R. F., Nasmyth, K. A. (1996) The *Saccharomyces cerevisiae* Start-specific transcription factor Swi4 interacts through the ankyrin repeats with the mitotic Clb2/Cdc28 kinase and through its conserved carboxy terminus with Swi6. *Mol. Cell. Biol.*, **16**, 2647.

87. Dohrmann, P. R., Butler, G., Tamai, K., Dorland, S., Greene, J. R., Thiele, D. J., and Stillman, D. J. (1992) Parallel pathways of gene regulation: homologous regulators SWI5 and ACE2 differentially control transcription of *HO* and chitinase. *Genes Dev.*, **6**, 93.

88. Knapp, D., Bhoite, L., Stillman, D. J., and Nasmyth, K. (1996) The transcription factor Swi5 regulates expression of the cyclin kinase inhibitor $p40^{SIC1}$. *Mol. Cell. Biol.*, **16**, 5701.

89. Moll, T., Tebb, G., Surana, U., Robitsch, H., and Nasmyth, K. (1991) The role of phosphorylation and the CDC28 protein kinase in cell cycle-regulated nuclear import of the *S. cerevisiae* transcription factor SWI5. *Cell*, **66**, 743.

90. Visintin, R., Craig, K., Hwang, E. S., Prinz, S., Tyers, M., and Amon, A. (1998) The phosphatase Cdc14 triggers mitotic exit by reversal of Cdk-dependent phosphorylation. *Mol. Cell*, **2**, 709.

91. Baum, B., Wuarin, J., and Nurse, P. (1997) Control of S-phase periodic transcription in the fission yeast mitotic cycle. EMBO J 16, 4676.

92. Whitehall, S., Stacey, P., Dawson, K., and Jones, N. (1999) Cell cycle-regulated transcription in fission yeast: Cdc10–Res protein interactions during the cell cycle and domains required for regulated transcription. *Mol. Biol. Cell*, **10**, 3705.

93. Kelly, T. J., Martin, G. S., Forsburg, S. L., Stephen, R. J., Russo, A., and Nurse, P. (1993) The fission yeast *cdc18*$^+$ gene product couples S phase to START and mitosis. *Cell*, **74**, 371.

94. Nakashima, N., Tanaka, K., Sturm, S., and Okayama, H. (1995) Fission yeast Rep2 is a putative transcriptional activator subunit for the cell cycle 'start' function of Res2–Cdc10. *EMBO J.*, **14**, 4794.

95. Schneider, B. L., Patton, E. E., Lanker, S., Mendenhall, M. D., Wittenberg, C., Futcher, B., and Tyers, M. (1998) Yeast G1 cyclins are unstable in G1 phase. *Nature*, **395**, 86.

96. Madden, K., Sheu, Y. J., Baetz, K., Andrews, B., and Snyder, M. (1997) SBF cell cycle regulator as a target of the yeast PKC–MAP kinase pathway. *Science*, **275**, 1781.

97. Sidorova, J. M. and Breeden, L. L. (1997) Rad53-dependent phosphorylation of Swi6 and down-regulation of *CLN1* and *CLN2* transcription occur in response to DNA damage in *Saccharomyces cerevisiae*. *Genes Dev.*, **11**, 3032.

98. Yamamoto, M., Imai, Y., and Watanabe, Y. 1997. Mating and sporulation in *Schizosaccharomyces pombe*. In *The molecular and cellular biology of the yeast* Saccharomyces (ed J. R. Pringle, J. R. Broach, and E. W. Jones), Vol. 3, p. 1037. Cold Spring Harbor Laboratory Press, New York.

99. Evans, T., Rosenthal, E. T., Youngblom, J., Distel, D., and Hunt, T. (1983) Cyclin: a protein specified by maternal mRNA in sea urchin eggs that is destroyed at each cleavage division. *Cell*, **33**, 389.

100. Hershko, A. and Ciechanover, A. (1998) The ubiquitin system. *Annu. Rev. Biochem.*, **67**, 425.

101. Baumeister, W., Walz, J., Zuhl, F., and Seemuller, E. (1998) The proteasome: paradigm of a self-compartmentalizing protease. *Cell*, **92**, 367.

102. Zachariae, W. and Nasmyth, K. (1999) Whose end is destruction: cell division and the anaphase-promoting complex. *Genes Dev.*, **13**, 2039.

103. Patton, E. E., Willems, A. R., and Tyers, M. (1998) Combinatorial control in ubiquitin-dependent proteolysis: don't Skp the F-box hypothesis. *Trends Genet.*, **14**, 236.

104. Tyers, M. (1996) The cyclin-dependent kinase inhibitor p40^{SIC1} imposes the requirement for Cln G1 cyclin function at Start. *Proc. Natl. Acad. Sci. U.S.A.*, **93**, 7772.

105. Schneider, B. L., Yang, Q. H., and Futcher, A. B. (1996) Linkage of replication to start by the Cdk inhibitor Sic1. *Science*, **272**, 560.

106. Verma, R., Annan, R. S., Huddleston, M. J., Carr, S. A., Reynard, G., and Deshaies, R. J. (1997) Phosphorylation of Sic1p by G1 Cdk required for its degradation and entry into S phase. *Science*, **278**, 455.

107. Nishizawa, M., Kawasumi, M., Fujino, M., and Toh-e, A. (1998) Phosphorylation of Sic1, a cyclin-dependent kinase (Cdk) inhibitor, by Cdk including Pho85 kinase is required for its prompt degradation. *Mol. Biol. Cell*, **9**, 2393.

108. Barral, Y., Jentsch, S., and Mann, C. (1995) G1 cyclin turnover and nutrient uptake are controlled by a common pathway in yeast. *Genes Dev.*, **9**, 399.

109. Lanker, S., Valdivieso, M. H., and Wittenberg, C. (1996) Rapid degradation of the G1 cyclin Cln2 induced by CDK-dependent phosphorylation. *Science*, **271**, 1597.

110. Willems, A. R., Lanker, S., Patton, E. E., Craig, K. L., Nason, T. F., Mathias, N., Kobayashi, R., Wittenberg, C., and Tyers, M. (1996) Cdc53 targets phosphorylated G1 cyclins for degradation by the ubiquitin proteolytic pathway. *Cell*, **86**, 453.

111. Mathias, N., Johnson, S. L., Winey, M., Adams, A. E., Goetsch, L., Pringle, J. R., Byers, B., and Goebl, M. G. (1996) Cdc53p acts in concert with Cdc4p and Cdc34p to control the G1-to-S-phase transition and identifies a conserved family of proteins. *Mol. Cell. Biol.*, **16**, 6634.

112. Bai, C., Sen, P., Hofmann, K., Ma, L., Goebl, M., Harper, J. W., and Elledge, S. J. (1996) SKP1 connects cell cycle regulators to the ubiquitin proteolysis machinery through a novel motif, the F-box. *Cell*, **86**, 263.

113. Zhang, H., Kobayashi, R., Galaktionov, K., and Beach, D. (1995) p19^{Skp1} and p45^{Skp2} are essential elements of the cyclin A–CDK2 S phase kinase. *Cell*, **82**, 915.

114. Connelly, C. and Hieter, P. (1996) Budding yeast SKP1 encodes an evolutionarily conserved kinetochore protein required for cell cycle progression. *Cell*, **86**, 275.

115. Skowyra, D., Craig, K. L., Tyers, M., Elledge, S. J., and Harper, J. W. (1997) F-box proteins are receptors that recruit phosphorylated substrates to the SCF ubiquitin–ligase complex. *Cell*, **91**, 209.

116. Patton, E. E., Willems, A. R., Sa, D., Kuras, L., Thomas, D., Craig, K. L., and Tyers, M. (1998) Cdc53 is a scaffold protein for multiple Cdc34/Skp1/F-box protein complexes that regulate cell division and methionine biosynthesis in yeast. *Genes Dev.*, **12**, 692.

117. Seol, J. H., Feldman, R. M., Zachariae, W., Shevchenko, A., Correll, C. C., Lyapina, S., Chi, Y., Galova, M., Claypool, J., Sandmeyer, S., Nasmyth, K., and Deshaies, R. J. (1999) Cdc53/cullin and the essential Hrt1 RING-H2 subunit of SCF define a ubiquitin ligase module that activates the E2 enzyme Cdc34. *Genes Dev.*, **13**, 1614.

118. Kamura, T., Koepp, D. M., Conrad, M. N., Skowyra, D., Moreland, R. J., Iliopoulos, O., Lane, W. S., Kaelin, W. G., Jr., Elledge, S. J., Conaway, R. C., Harper, J. W., and Conaway, J. W. (1999) Rbx1, a component of the VHL tumor suppressor complex and SCF ubiquitin ligase. *Science*, **284**, 657.

119. Skowyra, D., Koepp, D. M., Kamura, T., Conrad, M. N., Conaway, R. C., Conaway, J. W., Elledge, S. J., and Harper, J. W. (1999) Reconstitution of G1 cyclin ubiquitination with complexes containing SCF^{Grr1} and Rbx1. *Science*, **284**, 662.

120. Ohta, T., Michel, J. J., Schottelius, A. J., and Xiong, Y. (1999) ROC1, a homolog of APC11, represents a family of cullin partners with an associated ubiquitin ligase activity. *Mol. Cell*, **3**, 535.

121. Kominami, K. and Toda, T. (1997) Fission yeast WD–repeat protein Pop1 regulates genome ploidy through ubiquitin–proteasome-mediated degradation of the CDK inhibitor Rum1 and the S-phase initiator Cdc18. *Genes Dev.*, **11**, 1548.

122. Jallepalli, P. V., Tien, D., and Kelly, T. J. (1998) *sud1*[+] targets cyclin-dependent kinase-phosphorylated Cdc18 and Rum1 proteins for degradation and stops unwanted diploidization in fission yeast. *Proc. Natl. Acad. Sci. U.S.A.*, **95**, 8159.

123. Kominami, K., Ochotorena, I., and Toda, T. (1998) Two F-box/WD-repeat proteins Pop1 and Pop2 form hetero- and homo-complexes together with cullin-1 in the fission yeast SCF (Skp1–Cullin-1–F-box) ubiquitin ligase. *Genes Cells*, **3**, 721.

124. Wolf, D. A., McKeon, F., and Jackson, P. K. (1999) F-box/WD-repeat proteins Pop1p and Sud1p/Pop2p form complexes that bind and direct the proteolysis of Cdc18p. *Curr. Biol.*, **9**, 373.

125. Drury, L. S., Perkins, G., and Diffley, J. F. (1997) The Cdc4/34/53 pathway targets Cdc6p for proteolysis in budding yeast. *EMBO J.*, **16**, 5966.

126. Jallepalli, P. V., Brown, G. W., Muzi-Falconi, M., Tien, D., and Kelly, T. J. (1997) Regulation of the replication initiator protein p65^{cdc18} by CDK phosphorylation. *Genes Dev.*, **11**, 2767.

127. Jaquenoud, M., Gulli, M. P., Peter, K., and Peter, M. (1998) The Cdc42p effector Gic2p is targeted for ubiquitin-dependent degradation by the SCF^{Grr1} complex. *EMBO J.*, **17**, 5360.

128. Kaiser, P., Sia, R. A., Bardes, E. G., Lew, D. J., and Reed, S. I. (1998) Cdc34 and the F-box protein Met30 are required for degradation of the Cdk-inhibitory kinase Swe1. *Genes Dev.*, **12**, 2587.

129. McMillan, J. N., Longtine, M. S., Sia, R. A., Theesfeld, C. L., Bardes, E. S., Pringle, J. R., and Lew, D. J. (1999) The morphogenesis checkpoint in *Saccharomyces cerevisiae*: cell cycle control of Swe1p degradation by Hsl1p and Hsl7p. *Mol. Cell. Biol.*, **19**, 6929.

130. Zhou, P. and Howley, P. M. (1998) Ubiquitination and degradation of the substrate recognition subunits of SCF ubiquitin–protein ligases. *Mol. Cell*, **2**, 571.

131. Irniger, S., Piatti, S., Michaelis, C., and Nasmyth, K. (1995) Genes involved in sister chromatid separation are needed for B-type cyclin proteolysis in budding yeast. *Cell*, **81**, 269.

132. King, R. W., Peters, J. M., Tugendreich, S., Rolfe, M., Hieter, P., and Kirschner, M. W. (1995) A 20S complex containing CDC27 and CDC16 catalyzes the mitosis-specific conjugation of ubiquitin to cyclin B. *Cell*, **81**, 279.

133. Zachariae, W., Shevchenko, A., Andrews, P. D., Ciosk, R., Galova, M., Stark, M. J., Mann, M., and Nasmyth, K. (1998) Mass spectrometric analysis of the anaphase-promoting complex from yeast: identification of a subunit related to cullins. *Science*, **279**, 1216.

134. Yu, H., Peters, J. M., King, R. W., Page, A. M., Hieter, P., and Kirschner, M. W. (1998) Identification of a cullin homology region in a subunit of the anaphase-promoting complex. *Science*, **279**, 1219.

135. Yanagida, M. (1998) Fission yeast cut mutations revisited: control of anaphase. *Trends Cell Biol.*, **8**, 144.

136. Osaka, F., Seino, H., Seno, T., and Yamao, F. (1997) A ubiquitin-conjugating enzyme in fission yeast that is essential for the onset of anaphase in mitosis. *Mol. Cell. Biol.*, **17**, 3388.

137. Townsley, F. M. and Ruderman, J. V. (1998) Functional analysis of the *Saccharomyces cerevisiae UBC11* gene. *Yeast*, **14**, 747.

138. Surana, U., Amon, A., Dowzer, C., McGrew, J., Byers, B., and Nasmyth, K. (1993) Destruction of the CDC28/CLB mitotic kinase is not required for the metaphase to anaphase transition in budding yeast. *EMBO J.*, **12**, 1969.

139. Funabiki, H., Yamano, H., Kumada, K., Nagao, K., Hunt, T., and Yanagida, M. (1996) Cut2 proteolysis required for sister-chromatid separation in fission yeast. *Nature*, **381**, 438.

140. Cohen-Fix, O., Peters, J. M., Kirschner, M. W., and Koshland, D. (1996) Anaphase initiation in *Saccharomyces cerevisiae* is controlled by the APC-dependent degradation of the anaphase inhibitor Pds1p. *Genes Dev.*, **10**, 3081.

141. Ciosk, R., Zachariae, W., Michaelis, C., Shevchenko, A., Mann, M., and Nasmyth, K. (1998) An ESP1/PDS1 complex regulates loss of sister chromatid cohesion at the metaphase to anaphase transition in yeast. *Cell*, **93**, 1067.

142. Nasmyth, K. (1999) Separating sister chromatids. *Trends Biochem. Sci.*, **24**, 98.

143. Funabiki, H., Kumada, K., and Yanagida, M. (1996) Fission yeast Cut1 and Cut2 are essential for sister chromatid separation, concentrate along the metaphase spindle and form large complexes. *EMBO J.*, **15**, 6617.

144. Uhlmann, F., Lottspeich, F., and Nasmyth, K. (1999) Sister-chromatid separation at anaphase onset is promoted by cleavage of the cohesin subunit Scc1. *Nature*, **400**, 37.

145. Kumada, K., Nakamura, T., Nagao, K., Funabiki, H., Nakagawa, T., and Yanagida, M. (1998) Cut1 is loaded onto the spindle by binding to Cut2 and promotes anaphase spindle movement upon Cut2 proteolysis. *Curr. Biol.*, **8**, 633.

146. Alexandru, G., Zachariae, W., Schleiffer, A., and Nasmyth, K. (1999) Sister chromatid separation and chromosome re-duplication are regulated by different mechanisms in response to spindle damage. *EMBO J.*, **18**, 2707.

147. Visintin, R., Prinz, S., and Amon, A. (1997) CDC20 and CDH1: a family of substrate-specific activators of APC-dependent proteolysis. *Science*, **278**, 460.

148. Schwab, M., Lutum, A.S., and Seufert, W. (1997) Yeast Hct1 is a regulator of Clb2 cyclin proteolysis. *Cell*, **90**, 683.

149. Shirayama, M., Toth, A., Galova, M., and Nasmyth, K. (1999) APC^{Cdc20} promotes exit from mitosis by destroying the anaphase inhibitor Pds1 and cyclin Clb5. *Nature*, **402**, 203.

150. Matsumoto, T. (1997) A fission yeast homolog of CDC20/p55^{CDC}/Fizzy is required for recovery from DNA damage and genetically interacts with p34^{cdc2}. *Mol. Cell. Biol.*, **17**, 742.

151. Kominami, K., Seth-Smith, H., and Toda, T. (1998) Apc10 and Ste9/Srw1, two regulators of the APC-cyclosome, as well as the CDK inhibitor Rum1 are required for G1 cell-cycle arrest in fission yeast. *EMBO J.*, **17**, 5388.

152. Yamaguchi, S., Murakami, H., and Okayama, H. (1997) A WD repeat protein controls the cell cycle and differentiation by negatively regulating Cdc2/B-type cyclin complexes. *Mol. Biol. Cell*, **8**, 2475.

153. Kitamura, K., Maekawa, H., and Shimoda, C. (1998) Fission yeast Ste9, a homolog of Hct1/Cdh1 and Fizzy-related, is a novel negative regulator of cell cycle progression during G1-phase. *Mol. Biol. Cell*, **9**, 1065.

154. Prinz, S., Hwang, E. S., Visintin, R., and Amon, A. (1998) The regulation of Cdc20 proteolysis reveals a role for APC components Cdc23 and Cdc27 during S phase and early mitosis. *Curr. Biol.*, **8**, 750.

155. Yamashita, Y. M., Nakaseko, Y., Samejima, I., Kumada, K., Yamada, H., Michaelson, D., and Yanagida, M. (1996) 20S cyclosome complex formation and proteolytic activity inhibited by the cAMP/PKA pathway. *Nature*, **384**, 276.

156. Zachariae, W., Schwab, M., Nasmyth, K., and Seufert, W. (1998) Control of cyclin ubiquitination by CDK-regulated binding of Hct1 to the anaphase promoting complex. *Science*, **282**, 1721.

157. Jaspersen, S. L., Charles, J. F., and Morgan, D. O. (1999) Inhibitory phosphorylation of the APC regulator Hct1 is controlled by the kinase Cdc28 and the phosphatase Cdc14. *Curr. Biol.*, **9**, 227.

158. Visintin, R., Hwang, E. S., and Amon, A. (1999) Cfi1 prevents premature exit from mitosis by anchoring Cdc14 phosphatase in the nucleolus. *Nature*, **398**, 818.

159. Shou, W., Seol, J. H., Shevchenko, A., Baskerville, C., Moazed, D., Chen, Z. W., Jang, J., Charbonneau, H., and Deshaies, R. J. (1999) Exit from mitosis is triggered by Tem1-dependent release of the protein phosphatase Cdc14 from nucleolar RENT complex. *Cell*, **97**, 233.

160. Morgan, D. O. (1999) Regulation of the APC and the exit from mitosis. *Nature Cell Biol.*, **1**, 47.

161. Gould, K. L. and Simanis, V. (1997) The control of septum formation in fission yeast. *Genes Dev.*, **11**, 2939.

162. Furge, K. A., Wong, K., Armstrong, J., Balasubramanian, M., and Albright, C. F. (1998) Byr4 and Cdc16 form a two-component GTPase-activating protein for the Spg1 GTPase that controls septation in fission yeast. *Curr. Biol.*, **8**, 947.

163. Taylor, S. S. (1999) Chromosome segregation: dual control ensures fidelity. *Curr. Biol.*, **9**, R562.

164. Jaspersen, S. L., Charles, J. F., Tinker-Kulberg, R. L., and Morgan, D. O. (1998) A late mitotic regulatory network controlling cyclin destruction in *Saccharomyces cerevisiae*. *Mol. Biol. Cell*, **9**, 2803.

165. Yamano, H., Gannon, J., and Hunt, T. (1996) The role of proteolysis in cell cycle progression in *Schizosaccharomyces pombe*. *EMBO J.*, **15**, 5268.

166. Tinker-Kulberg, R. L., Morgan, D. O. (1999) Pds1 and Esp1 control both anaphase and mitotic exit in normal cells and after DNA damage. *Genes Dev.*, **13**, 1936.

167. Cohen-Fix, O., Koshland, D. (1999) Pds1 of budding yeast has dual roles: inhibition of anaphase initiation and regulation of mitotic exit. *Genes Dev.*, **13**, 1950.

168. Kim, S. H., Lin, D. P., Matsumoto, S., Kitazono, A., and Matsumoto, T. (1998) Fission yeast Slp1: an effector of the Mad2-dependent spindle checkpoint. *Science*, **279**, 1045.

169. Hwang, L. H., Lau, L. F., Smith, D. L., Mistrot, C. A., Hardwick, K. G., Hwang, E. S., Amon, A., and Murray, A. W. (1998) Budding yeast Cdc20: a target of the spindle checkpoint. *Science*, **279**, 1041.

170. Sanchez, Y., Bachant, J., Wang, H., Hu, F., Liu, D., Tetzlaff, M., and Elledge, S. J. (1999) Control of the DNA damage checkpoint by Chk1 and Rad53 protein kinases through distinct mechanisms. *Science*, **286**, 1166.

171. Nasmyth, K. (1996) At the heart of the budding yeast cell cycle. *Trends Genet.*, **12**, 405.

172. Dahmann, C., Diffley, J. F. X., and Nasmyth, K. A. (1995) S-phase-promoting cyclin-dependent kinases prevent re-replication by inhibiting the transition of replication origins to a pre-replicative state. *Curr. Biol.*, **5**, 1257.

173. Stern, B., Nurse, P. (1996) A quantitative model for the cdc2 control of S phase and mitosis in fission yeast. *Trends Genet.*, **12**, 345.

174. Shirayama, M., Zachariae, W., Ciosk, R., and Nasmyth, K. (1998) The Polo-like kinase Cdc5p and the WD-repeat protein Cdc20p/fizzy are regulators and substrates of the anaphase promoting complex in *Saccharomyces cerevisiae*. *EMBO J.*, **17**, 1336.

175. Noton, E. and Diffley, J. F. X. (2000) CDK inactivation is the only essential function of the APC/C and the mitotic exit network proteins for origin resetting during mitosis. *Mol. Cell.* **5**, 85.

176. Liang, C. and Stillman, B. (1997) Persistent initiation of DNA replication and chromatin-bound MCM proteins during the cell cycle in cdc6 mutants. *Genes Dev.*, **11**, 3375.
177. Labib, K., Diffley, J. F. X., and Kearsey, S. E. (1999) G1-phase and B-type cyclins exclude the DNA-replication factor Mcm4 from the nucleus. *Nature Cell Biol.*, **1**, 415.
178. Hartwell, L. H. and Smith, D. (1985) Altered fidelity of mitotic chromosome transmission in cell cycle mutants of *S. cerevisiae*. *Genetics*, **110**, 381.

4 | Cell cycle checkpoints

KRISTI CHRISPELL FORBES and TAMAR ENOCH

1. Introduction

During each yeast cell cycle, chromosomes must be copied faithfully and segregated accurately between daughter cells. Surveillance systems termed 'checkpoints' ensure the fidelity of chromosome duplication by detecting abnormal structures and blocking cell cycle transitions while problems are corrected. Studies in *Saccharomyces cerevisiae* and *Schizosaccharomyces pombe* have identified a plethora of checkpoint genes and proteins. Remarkably, as was predicted in 1989 by Hartwell and Weinert (1), the same checkpoint proteins that maintain the yeast genomes play a crucial role in preventing cancer in mammals (2, 3). Thus studies in yeasts have helped to elucidate an important cause of genomic instability and cancer. Now that many checkpoint proteins have been identified, current investigations are aimed at understanding how these checkpoints work in molecular detail. As this review will detail, considerable progress has been made in this area in the last 10 years. Because many of the checkpoint proteins are conserved in higher eukaryotes, yeasts will continue to prove their value as model systems for the study of mechanisms that maintain the genome.

1.1 Cell cycle checkpoints

The eukaryotic cell cycle is composed of a sequence of phases: an initial gap phase (G1), DNA synthesis (S), a second gap phase (G2), and mitosis (M). (For additional information about the cell cycle, see Chapter 3). In order to maintain cell viability and genomic integrity, cells must ensure that the phases of the cell cycle occur correctly and in the proper order. Hartwell and Weinert proposed that the dependency of later cell cycle events on the completion of earlier events could have two possible explanations (1). First, the dependency could be due to substrate–product relationships. For example, duplicated and properly packaged chromosomes could be required as a substrate for the mitotic apparatus, or formation of a bud could be necessary for cytokinesis in budding yeast. Alternatively, dependency could be due to regulatory control mechanisms that Hartwell and Weinert named 'checkpoints'. Checkpoints may be thought of as surveillance mechanisms, monitoring the status of the cell, and causing a delay if something is amiss. If such inessential regulatory controls are present in the cell cycle, they argued, then it should be possible to find mutations, chemicals, or conditions that allow a later cell cycle event to take place when an earlier event has not occurred, a state they termed 'relief of dependence'. The substrate–product model would not allow relief of dependence mutations.

Hartwell and Weinert observed 'relief of dependence' in the case of the haploid *S. cerevisiae rad9(Sc)* loss of function mutant (1, 4). Although wild-type budding yeast cells delay in G2 following irradiation, resulting in a uniform population of large budded cells, the *rad9(Sc)* mutant cells continue to divide. The mutants eventually die as small microcolonies because they attempt to segregate damaged chromosomes (1, 4). Therefore, *rad9(Sc)* relieves the dependence of mitosis on repaired DNA. Since then, mutations in many other yeast genes that relieve dependence have been identified (reviewed in References 5, 6). The existence of these mutants demonstrates that wild-type cells have checkpoints in order to ensure that cell cycle events occur at the proper time and in the proper order. It has also become clear that checkpoints interface with the cell cycle machinery such as cyclin-dependent kinases (CDKs) and the anaphase-promoting complex (APC).

The *rad9(Sc)* gene participates in the DNA damage checkpoint, which arrests cell cycle progression in G2 when damaged DNA is present. Arrest in G2 is particularly advantageous for haploid cells because the cell has a duplicate copy of its genetic material available for use in recombinational repair. However, damage checkpoints can also delay the cell cycle at the G1–S transition, or during S. Some of the cell cycle checkpoints that have been identified in yeast are listed in Table 1. Here we will focus primarily on the surveillance system that monitors the integrity of chromosome structures and arrests cell cycle progression in response to damaged and unreplicated DNA. We will also consider the system that monitors attachment of chromosomes to the mitotic spindle.

We will be comparing the checkpoints of the yeasts *S. cerevisiae* and *S. pombe*. Although these two yeasts can be studied using the same molecular and genetic techniques, they are only distantly related. In fact, comparison of gene sequences of *S. cerevisiae* and *S. pombe* reveals that, in many cases, yeast proteins are as closely related to human proteins as they are to each other (7). Thus each yeast yields unique insights into cellular processes, and comparison of the two yeasts is particularly

Table 1 Yeast checkpoints

Name of Checkpoint	Defect	Arrest point(s)	References
S–M	Unreplicated DNA	S	6, 156
DNA damage	Damaged DNA	G1–S, G2–M	5, 6, 8
Intra-S damage	Damaged DNA	Extended S	5, 8
Re-replication	M not completed	Initiation of S	156, 157
Spindle assembly	Chromosomes misaligned, or defective spindle	Metaphase/anaphase	123, 125
Cytokinesis	Spindle misorientation	M	158
Morphogenesis	Actin cytoskeleton disorganized, no bud formation	G2	159
Cell size	Cell below size threshold, nutrient limitation	G0–G1	160
Meiotic prophase	Recombination incomplete	Meiosis I	5

Table 2 DNA damage and S–M checkpoint genes in *S. pombe* and *S. cerevisiae*

Budding yeast gene	Fission yeast gene	Potential human homologue(s)	Sequence motifs
Signals			
POL2 (64)	*cdc20⁺* (54)*	*POL ε**	DNA polymerase ε catalytic subunit
RFC2 (50)	*rfc2⁺* (44)	*hRFC37**	Subunit of RF-C complex
RFC5 (51, 53)		*hRFC38**	Subunit of RF-C complex
DPB11 (60)	*rad4⁺/cut5⁺* (45, 58)	XRCC1-related	BRCT domain
Sensors			
RAD9 (4, 23)	*rhp9⁺/ crb2⁺* (17, 59)		BRCT domain
RAD17 (13)	*rad1⁺* (15, 18)	*hRAD1* (161, 162)	3′→5′ exonuclease
RAD24 (12)	*rad17⁺* (163)	*hRAD17* (164)	RF-C related
DDC1 (19)	*rad9⁺* (15)	*hRAD9* (165)	
MEC3 (12, 63)			
NH	*hus1⁺* (14, 166)	*hHUS1* (166)	
NH	*rad26⁺* (16, 109)		
Transducers			
MEC1/ESR1/SAD3 (12, 167)	*rad3⁺* (15, 27)	*ATR* (27)	Kinase
TEL1 (20, 168)	*tel1⁺* (21)*	*ATM* (25)	
RAD53/SPK1/MEC2/ SAD1 (12, 62)	*cds1⁺* (69, 71)	*hCDS1/hCHK2* (149, 151)	Kinase, FHA domain
*CHK1**	*chk1⁺/rad27⁺* (28, 70)	*hCHK1* (150)	Kinase
BMH1 (169)*	*rad24⁺* (87)	Several 14-3-3s* (86)	14-3-3
BMH2 (169)*	*rad25⁺* (87)	Several 14-3-3s* (86)	14-3-3
Receivers			
G1–S			
SWI6 (100)	*cdc10⁺* (46)*		Transcription factor
S phase			
PRI1 (100, 101, 104, 105)		*PRIM1**	DNA primase
RFA1 (104)	*rad11⁺* (49)	*HRPA1**	Large subunit of single stranded binding protein (RP-A)
G2–M			
CDC28 (160)*	*cdc2⁺* (80, 157)	several CDK's (160)	Cyclin-dependent kinase
Metaphase/anaphase			
PDS1 (96, 143)	*cut2⁺* (144)		
Transcription			
DUN1 (113)			Kinase
CRT1 (113)			DNA-binding protein

*This gene is not known to play a role in checkpoint control, but is listed as a homologous gene for completeness. A blank space indicates that no homologue is known to date; NH, no homologue has been found in the completed genome sequence.

valuable for identifying features of control mechanisms that are likely to be common to all eukaryotic cells.

Several gene symbols for the genes discussed in this chapter are used in both yeasts, but for unrelated genes; see Table 2. To minimize confusion, suffixes indicating *S. cerevisiae* (Sc) or *S. pombe* (Sp) have been used throughout this chapter for genes and gene products.

1.2 Overview of checkpoint pathways

The function of a checkpoint is to act as a surveillance system, monitoring the status of the cell and arresting the cell cycle if conditions are not appropriate for progression. The checkpoint genes involved in each checkpoint fall into pathways. We will conceptually divide each checkpoint pathway into the following sections: signals, sensors, transducers, and receivers (or targets). Sensors are the components of the checkpoint pathway that monitor particular events or components of the cell during the cell cycle, such as cell size, the status of the chromosomes, or formation of a mitotic spindle. Exactly what signals these sensors detect is an interesting but poorly understood aspect of checkpoint pathways. The sensors then activate the transducers, but the mechanisms of activation are also unclear. The transducers are components, such as protein kinases, that transmit, and perhaps amplify, the checkpoint signal to the receivers. This part of each pathway may resemble more classical signal transduction pathways, in that it may involve a protein kinase cascade. The receivers are the downstream targets of each checkpoint and include the cell cycle machinery ultimately responsible for arresting the cell cycle at a particular point.

2. The checkpoint response to damaged and unreplicated DNA

One of the best studied cell cycle responses in yeasts is the response to unreplicated DNA and DNA damage. DNA damage can arrest the cell cycle during G1, S, or G2 (reviewed in References 5, 6, 8). Classically, checkpoints have been categorized by the point of the cell cycle at which arrest is observed. However this categorization is of limited usefulness for this checkpoint response, since many of the genes involved are required for cell cycle arrest at many different points in the cell cycle. It is probably more useful to consider these genes as components of a general surveillance system, which impinges on receivers with more specific cell cycle-specific functions. We begin this section with a discussion of checkpoint proteins that play general roles in the response to DNA damage, and also consider what the signals that activate the checkpoint response could be. In a later section we discuss specific mechanisms of cell cycle arrest. Overviews of the pathway in each yeast are presented in Figs 1, 2, and 3. It is also known that cells can 'adapt' to the presence of unrepaired damage, which must mean that the checkpoint signal decays or it is no longer able to act on targets (9, 10). Several genes required for this process have been identified (10, 11); however, because of space limitations, this aspect of the checkpoint response is not considered further here.

Very specific information on the function of most of the checkpoint components is not yet available. Our discussion thus includes some conjectures about their interactions and functions. We intend these speculations to stimulate discussion and experimentation, and we caution the reader that these ideas are not necessarily shared by others in the field. We encourage the interested reader to consult the many primary research papers and reviews cited throughout this chapter for more information.

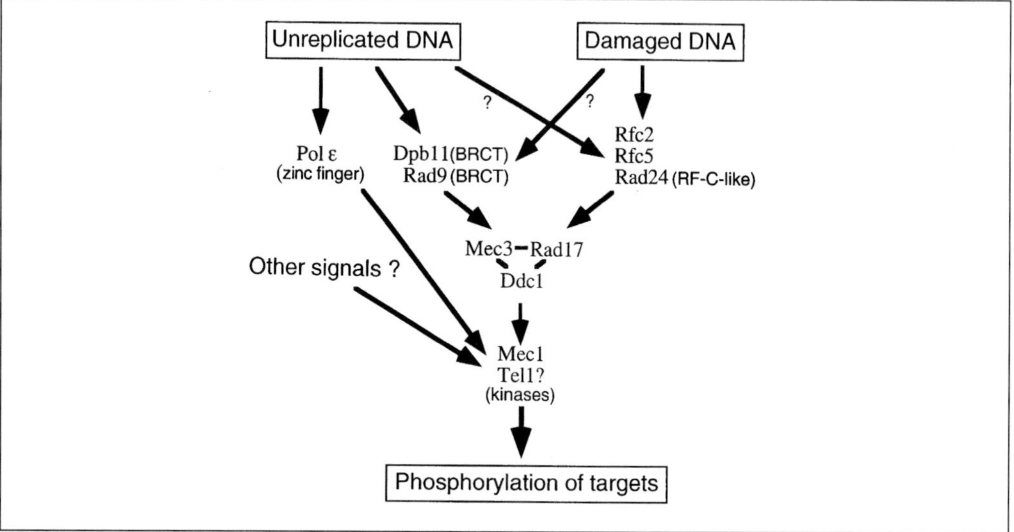

Fig. 1 Budding yeast checkpoints: signals and sensors.

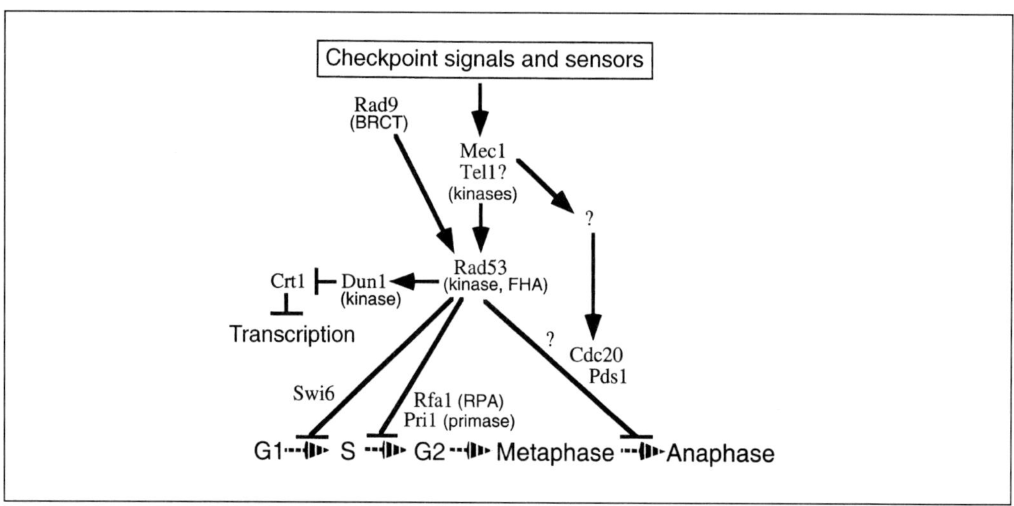

Fig. 2 Budding yeast checkpoints: transducers and targets.

2.1 Genes required for the checkpoint response to DNA damage

In both budding and fission yeast, screens for 'relief of dependence' mutants have identified a set of genes that are required for cell cycle arrest in response to DNA damage and/or unreplicated DNA (4, 12–19). The products of many of these genes seem to be part of the sensing mechanisms that recognize DNA damage or un-

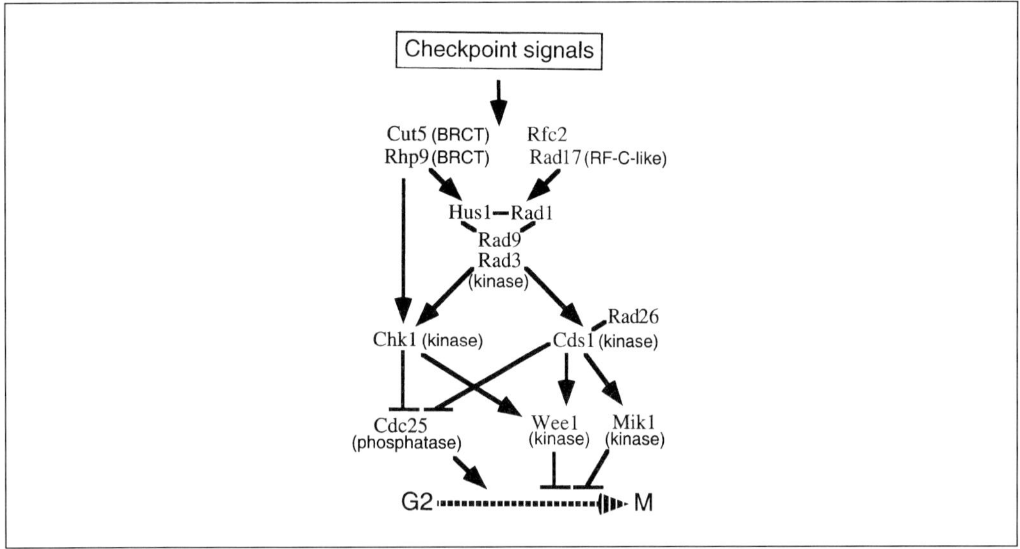

Fig. 3 Fission yeast checkpoint pathway.

replicated DNA. In each yeast, a large phosphatidylinositol 3 (PI-3)-related kinase, Rad3(Sp) or Mec1p(Sc), plays a central role as a sensor (Figs 1, 2, and 3). Each yeast contains a second PI-3-related kinase, Tel1(Sc or Sp), which is required primarily for telomere maintenance, but may play a minor role in the checkpoint reponse as well (20–23). The yeast proteins are related to the human ATM (*ataxia telangiectasia mutated*) and ATR (*ataxia-* and *rad-related*) proteins (2, 24, 25). Human cells having *ATM* mutations show checkpoint defects, while human *ATM* patients suffer from the autosomal recessive disease ataxia telangiectasia, which has a variety of symptoms including a predisposition to cancer (reviewed in Reference 26), establishing the importance of checkpoint response in cancer prevention.

A model for the function of these proteins has been proposed based on the similarity of the conserved PI-3-related kinases involved in checkpoint control to the catalytic subunit of the human DNA-dependent protein kinase (DNA-PK), which is involved in DNA double-strand break (dsb) repair and V(D)J recombination (24; reviewed in Reference 6). Members of this PI-3 kinase-related family were recognized as kinases based on sequence homology to PI-3 lipid kinases, but these related kinases have been shown to phosphorylate proteins rather than lipids (24, 27, 28). *In vitro*, the kinase activity of the DNA-PK catalytic subunit is activated by the binding of its two regulatory subunits (Ku70 and Ku80) to DNA dsbs (reviewed in Reference 29). Thus, the activation of this kinase is dependent both on specific DNA structures and on regulatory co-factors. Although Te11(Sc or Sp), Mec1p(Sc), and Rad3(Sp) are not involved in recombinational repair like DNA-PK, it is possible that their regulation is analogous, that is the activation of these checkpoint kinases depends on the presence of specific DNA structures and on regulatory co-factors (6).

The other sensor genes (*hus1*[+](Sp), *rad1*[+](Sp), *rad9*[+](Sp), *rad17*[+](Sp), and *rad26*[+](Sp); *RAD9(Sc)*, *RAD17(Sc)*, *RAD24(Sc)*, *MEC3(Sc)*, and *DDC1(Sc)*) could encode regulators of Mec1p(Sc) and Rad3(Sp). As shown in Table 2, some of the sensors are found in only one yeast, while others are found in both. The function of these proteins remains to be established, although clues are beginning to emerge from sequence analysis. *RAD24(Sc)/rad17*[+](Sp) is distantly related to subunits of replication factor C (RF-C), a protein that loads proliferating nuclear cell antigen (PCNA) at the primer–template junction during DNA replication (see Chapter 2 for more information about DNA replication). *RAD17(Sc)/rad1*[+](Sp) has similarity to the *Ustilago maydis REC1* gene which has been shown to be an exonuclease *in vitro* (30). The significance of these similarities is discussed in more detail below. In addition, Rad17p(Sc)/Rad1(Sp) is part of a distinct trimeric complex that includes Ddc1p(Sc) and Mec3p(Sc) or Rad9(Sp) and Hus1(Sp) (31, 32). Interestingly, Ddc1p(Sc) and Rad9(Sp) show some sequence similarity, but Mec3p(Sc) and Hus1(Sp) are not obviously related. Thus this complex has two conserved and one variable subunit. In both yeasts, the trimeric complex forms in the absence of checkpoint signals, but is phosphorylated in response to checkpoint signals in a Mec1p(Sc)/Rad3(Sp)-dependent manner (31, 32). This has led to the proposal that the complex functions as a 'sensor' complex that recruits and activates PI-3-related kinase in response to DNA damage. However, so far there is no evidence for a direct interaction between any of these proteins and Mec1p(Sc)/Rad3(Sp). Recently sophisticated computational analysis has detected a possible structural relationship between Rad17p(Sc)/Rad1(Sp), Hus1(Sp), and PCNA (33, 34). Thus it is possible that the trimeric complex, like PCNA, forms a sliding clamp that diffuses along the chromosomes searching for damage. By analogy to the role of RF-C, Rad24p(Sc)/Rad17(Sp) could function to load this trimeric clamp on the DNA. Biochemical studies of the complex will be required to determine the validity of this model. It is also not clear whether a PCNA-like role for Rad17p(Sc)/Rad1(Sp) is compatible with its proposed exonuclease function.

2.2 Possible checkpoint signals

One of the most interesting challenges before the checkpoint field is the identification of primary chromosomal signals that activate the checkpoint response. Any model for such a signal must explain how a wide range of abnormal DNA structures, including DNA adducts, DNA dsbs and, in some cases, unreplicated DNA can all activate a similar checkpoint response. Two general possibilities are currently under consideration. One possibility is that all primary lesions are processed by repair enzymes to some common DNA structure, possibly single-stranded DNA (ssDNA), which is then recognized by checkpoint proteins. A version of this model, the lesion processing model, proposes that one of the roles of the checkpoint proteins is to convert DNA lesions to ssDNA (35). A second, and perhaps not mutually exclusive, possibility is that repair or replication proteins assembled at the site of DNA damage activate the checkpoint response (36). In this section we discuss the evidence in favour of each of these models.

2.2.1 Single-stranded DNA as a potential checkpoint signal

The lesion processing model is based on analysis of the checkpoint-dependent arrest phenotype of the *cdc13(Sc)* temperature-sensitive (ts) mutant. Cdc13p(Sc) is believed to bind to repetitive telomeric DNA and protect it from degradation (37). At the restrictive temperature, *cdc13(Sc)* mutants arrest in G2 with long regions of ssDNA at the telomeres. This arrest is checkpoint dependent, as a *cdc13 rad9(Sc)* double mutant continues to divide several times (38). This suggests that the checkpoint pathway could be sensing the ssDNA at the telomeres. Furthermore, the products of the sensor genes could be involved in the production of ssDNA, as *cdc13 rad24(Sc)* mutants accumulate less ssDNA at the restrictive temperature than *cdc13(Sc)* single mutants (35). Based on this observation, Lydall and Weinert argued that the checkpoint sensors Rad24p(Sc), Rad17p(Sc), and Mec3p(Sc) are required to process DNA lesions, and that the putative exonuclease Rad17p(Sc) could be directly responsible for the degradation (35). However, this model does not account for several observations. First, Rad17p(Sc) is homologous to 3′→5′ exonucleases, while the telomeric DNA in *cdc13(Sc)* is degraded in the 5′ to 3′ direction, which implies that it is being processed by a 5′→3′ exonuclease (35, 38). Second, it has not yet been shown that Rad17p(Sc) is actually a functional exonuclease or that this proposed activity is required for checkpoint activation. In fact, while the *rec1-1* allele of Rec1 (a *Ustilago maydis* homologue of Rad17p(Sc) andRad1(Sp)) is competent as an exonuclease *in vitro*, it is checkpoint defective (39). Finally, the absence of Rad17p(Sc) has no effect on the degradation of an HO-generated DNA dsb, which is known to be processed to ssDNA and to cause a checkpoint-mediated arrest (10).

Although the lesion processing model remains unproven, an independent study suggests that the amount of ssDNA does affect the strength of the checkpoint signal. These studies employ a single dsb generated at the mating-type locus using the HO (*homothallic switching*) endonuclease, in a strain where the break cannot be repaired by homologous recombination. Under these conditions, an unidentified exonuclease degrades the 5′ strand, resulting in the formation of a 3′ single-stranded tail that can be several kilobases in length (40). This structure induces a typical checkpoint response in otherwise wild-type cells (9, 40). In *hdf1(Sc)* mutants (which lack the *S. cerevisiae* Ku70 homologue) exonuclease activity is promoted, a longer single-stranded tail is produced, and the checkpoint-dependent arrest becomes irreversible. This irreversible arrest can be suppressed by mutations in *mre11(Sc)* and *rad50(Sc)* which reduce the rate of degradation of the 5′ end. Thus the strength of the checkpoint signal can apparently be modulated by increasing or decreasing the amount of ssDNA (10). However, this experiment does not establish that ssDNA is the checkpoint signal. To do this it would be necessary to show that no checkpoint signal is generated when the production of ssDNA is completely prevented, an experiment that is not possible for technical reasons. Interestingly this study also showed that a specific mutation in the gene encoding *RFA1(Sc)*, *rfa-t11(Sc)*, also reduced the strength of the checkpoint signal, without affecting the rate of ssDNA production (10). Rfa1p(Sc) is a subunit of replication protein A (RP-A), a complex that binds to ssDNA and is required for replication and recombination. This raises the possibility that the checkpoint signal might be RP-A bound to ssDNA, and that the Rfa-t11p(Sc) protein binds to DNA, but for some reason fails to generate a checkpoint signal.

2.2.2 Repair/replication complexes as checkpoint signals

Another possible source of the checkpoint signal is replication complexes assembled or partially assembled on the chromosome. As recent studies have shown that leading and lagging strand replication enzymes are required for repair of dsbs in *S. cerevisiae* (41), in principle replication complexes could signal the presence of either damaged or unreplicated DNA.

A prediction of this model is that it should be possible to identify genes that are required for repair/replication as well as checkpoint control. Such genes could have been missed in the original screens for checkpoint mutants, which focused on mutant strains that were checkpoint deficient but otherwise viable. Complete deletion of many genes involved in initiation of replication gives a phenotype consistent with a checkpoint defect in sensing unreplicated DNA: cells fail to replicate DNA, but progress into mitosis anyway although they have only a 1C DNA content. In *S. pombe*, such genes include *rfc2$^+$*(Sp), *cut5$^+$/rad4$^+$*(Sp), *cdc18$^+$*(Sp), *hsk1$^+$*(Sp), *orp1$^+$/cdc30$^+$*(Sp), *orp2$^+$*(Sp), *cdc21$^+$*(Sp), *rad11$^+$*(Sp), *cdt1$^+$*(Sp), and *pol1$^+$*(Sp) (polymerase α) (42–49; reviewed in Reference 36). One explanation for these observations is that the corresponding proteins interact directly with the checkpoint machinery, thus allowing cells to detect ongoing replication. However, an alternative possibility is that the requirement for these replication proteins in checkpoint control is indirect. For example, if the checkpoint signal is ssDNA or a protein–DNA complex that forms during DNA replication, in the absence of a replication complex these structures will not form, and thus no checkpoint signal will be generated.

We propose that a replication protein directly involved in checkpoint control might have some of the following properties:

1. In the absence of the protein, cells enter mitosis without replicating their DNA.
2. Alleles that can carry out DNA replication, but are checkpoint deficient, can be identified. These mutants would presumably contain changes that disrupt the surface of the protein that directly contacts known checkpoint proteins.
3. The protein interacts physically with known checkpoint proteins.

2.2.3 Replication factor C and the checkpoint response

Members of the replication factor C (RF-C) complex, which binds to primer–template junctions to load PCNA and the leading strand replication complex, have some of the properties described above. A ts allele of *RFC2*(Sc), *rfc2-1*(Sc), has been identified which enters mitosis in the absence of DNA replication at the restrictive temperature. However, even at the permissive temperature, 30% of the cells fail to arrest when replication is blocked with hydroxyurea (HU; an inhibitor of dNTP synthesis) and enter mitosis like 'classically' checkpoint-defective cells (50). No information on the DNA damage checkpoint response of this mutant is available. A ts allele of *RFC5*(Sc), *rfc5-1*(Sc), also enters mitosis without completing DNA replication at the restrictive temperature (51). PCNA overexpression can suppress the DNA replication defect of *rfc5-1*(Sc), but not the replication checkpoint defect (51, 52). The *rfc5-1*(Sc) allele interacts genetically with known checkpoint genes (52, 53). Finally, a physical interaction between Rfc5 and the checkpoint gene Rad24p(Sc) has been demonstrated (52). As discussed above, Rad24p(Sc) has some sequence similarity to RF-C subunits. These data are consistent with a model in which RF-C, assembled at a primer–template junction, could be a primary signal to the checkpoint pathway.

Although the role of RF-C in the checkpoint response in *S. pombe* has been studied in less detail, the available data are consistent with this model. Deletion of *rfc2*$^+$(Sp) results in cells that fail to replicate DNA but still enter mitosis (44). Deletion of genes required for synthesis of replication primers, such as polymerase α, or deletion of other genes required for origin activation, also results in cells that enter mitosis with unreplicated DNA (42, 47, 48; reviewed in Reference 36). These mutants may fail to activate a checkpoint response because they do not synthesize primers and thus the RF-C–primer–template complex cannot form. In contrast, deletion of PCNA or the replication polymerases δ or ε leads to checkpoint activation and cell cycle arrest (54–56). Perhaps the checkpoint response is still activated in their absence because the RF-C complex is still able to form in cells lacking these polymerases or PCNA (44).

2.2.4 XRCC1-related yeast proteins and the checkpoint response

Another protein that meets some of the criteria for direct involvement in checkpoint control (listed in section 2.2.2) is Cut5/Rad4(Sp), which has some sequence similarity to mammalian XRCC1. *XRCC1* is an essential gene that was cloned by complementation of radiation-sensitive rodent cell lines. It interacts physically with DNA ligase III via a BRCT (*BRCA1 C-terminal*) domain, a motif that is conserved in many proteins implicated in DNA repair including the human breast cancer gene *BRCA1* (57). The interaction with DNA ligase suggests that it may be involved in repair of DNA breaks. In fission yeast, loss of Cut5(Sp) causes a phenotype similar to loss of RF-C; cells fail to complete DNA replication, do not activate the checkpoint, and enter mitosis with unreplicated DNA (45, 58). No *cut5*(Sp) alleles that are viable but checkpoint defective have been identified. However, when *cut5*(Sp)-ts cells are arrested with HU at the permissive temperature and shifted to the non-permissive temperature, cells enter mitosis with unreplicated DNA, indicating that Cut5(Sp) may be required to block mitosis in response to unreplicated DNA (58). There are also genetic interactions between Cut5(Sp) and two known checkpoint genes, *rhp9*$^+$/*crb2*$^+$(Sp) (related to *RAD9*(Sc)) and *chk1*$^+$(Sp), a serine/threonine kinase involved in the transduction of the checkpoint signal (59). Physical interactions between these proteins have been observed in a two-hybrid system, although it is not known whether such direct interactions also take place under physiological conditions (59). In addition, *cut5*$^-$ alleles are sensitive to DNA-damaging agents (45), but no role in the checkpoint response to damaged DNA has been demonstrated, despite the observed interactions with known components of the DNA damage response.

A *cut5*$^+$/*rad4*$^+$(Sp)-related gene, *DPB11*(Sc), has also been identified as a suppressor of mutations in *POL2(Sc)*, which encodes polymerase ε (60). At the restrictive temperature, a *dpb11(Sc)*-ts mutant exhibits defects in DNA replication, but, like a *cut5(Sp)* mutant, it fails to activate the checkpoint and then enters mitosis without completing DNA replication. At the permissive temperature the *dpb11(Sc)*-ts mutant is sensitive to inhibitors of DNA replication and DNA-damaging agents, and is also defective in arrest in response to unreplicated DNA (60, 61). The mutant's response to damage has not been characterized. A Dpb11p(Sc) interacting protein, Drc1p(Sc), which is essential for normal DNA replication and may also be involved in the replication checkpoint, has also been identified (61). The *S. pombe* genome contains a *DRC1(Sc)* homologue, but its function has not been explored (61). It is also not known whether Dpb11p(Sc) interacts with Rad9p(Sc) or any of the other checkpoint sensors in *S. cerevisiae*.

2.2.5 Polymerase ε and the checkpoint response

In the preceding discussion, we have considered the signals generated by unreplicated and damaged DNA together. They are likely to be similar in *S. pombe*, because all the sensors required for the response to unreplicated DNA are also involved in the response to damaged DNA. In contrast, of the 'classical' checkpoint genes in *S. cerevisiae*, only Mec1p(Sc) and Rad53p(Sc) are required for both the response to damage and to unreplicated DNA; mutants in *RAD9(Sc)*, *RAD17(Sc)*, *RAD24(Sc)*, *MEC3(Sc)*, and *DDC1(Sc)* all arrest normally when treated with HU (12, 19, 62, 63). Thus it appears that many of the genes used in *S. pombe* to respond to either DNA damage or replication arrest are used in *S. cerevisiae* exclusively for the response to damage (See Figs 1, 3). On the other hand, proteins that we have suggested may generate the primary signal, RF-C and XRCC1, may well function in both the damage and replication checkpoints in *S. cerevisiae*, as mutations in the corresponding genes cause UV sensitivity (50, 60). Furthermore, although *rad24(Sc)* mutants are not defective in the replication checkpoint, *rad24 rfc5(Sc)* double mutants show more severe replication defects than *rfc5(Sc)* single mutants, implying that there may be some overlap between the damage and replication checkpoints even in *S. cerevisiae*. Finally, it is probably important to remember that most experiments with unreplicated DNA have been done using HU, which blocks dNTP synthesis. Other methods of blocking DNA replication such as polymerase inhibitors and mutants in replication components have been studied much less extensively. Conceivably, the HU signal could be a combination of a unique 'limiting nucleotide signal' and a more general 'partially replicated DNA signal'. The latter signal may resemble the DNA damage signal. The relative contributions of each signal may vary between the two yeasts.

A good candidate for a component of the unique replication signal is the zinc finger at the C-terminus of polymerase ε which is required only for the checkpoint response to HU in *S. cerevisiae*. Two mutant alleles of polymerase ε, *pol2-11* and *pol2-12*, show defects in the checkpoint response to HU at the permissive temperature, although they arrest normally in response to DNA damage (64). Both these mutations truncate polymerase ε, eliminating the conserved zinc finger in the C-terminus of the protein. However unlike *rfc2-1(Sc)* and *rfc5-1(Sc)*, these mutants arrest normally at the non-permissive temperature, although they do not complete DNA replication (64). Thus the checkpoint signal at the non-permissive temperature must be different from the checkpoint signal generated at the permissive temperature in HU. The role of polymerase ε (*cdc20$^+$*(Sp)) in *S. pombe* is not possible to evaluate as the equivalent alleles are not viable, but, like the *S. cerevisiae* mutants, they arrest with unreplicated DNA without entering mitosis (54). So far, direct interactions between polymerase ε and known checkpoint proteins have not been demonstrated. *pol2-11*(Sc) mutants are synthetically lethal with mutations in the *XRCC1*-related gene *DPB11*(Sc) (60); however, as this interaction is also observed with *pol2*(Sc) alleles that are not checkpoint defective, its importance in the checkpoint response is unclear.

2.2.6 Checkpoint signals: conclusions

As the preceding discussion has illustrated, there is evidence suggesting that at least four essential replication proteins, RF-A, RF-C, XRCC1, and polymerase ε, could be primary checkpoint signals. Although the evidence is tantalizing, many experiments clearly remain to be done to establish definitively that any of these proteins act as checkpoint signals. In addition, more replication or repair proteins activating the checkpoint pathway may remain to be identified. The number of different proteins that can act as signals may suggest that there is no single checkpoint signal. Rather, 'classical' checkpoint proteins may have the capacity to be activated by any of several different protein–DNA complexes. Some of these complexes may form in response to specific structures such as ssDNA or primer–template junctions. The type of structures formed will vary depending on the initial insult. Thus we may expect to see that some proteins, such as polymerase ε, are required only for the checkpoint response to certain insults. It is also possible that two or more repair/replication complexes could function redundantly, which means that, if one repair system is inactivated, a checkpoint signal may still be generated by other complexes. This may explain why the checkpoint phenotypes of the proteins we have discussed above are relatively weak. Finally, we should expect each yeast to exhibit some unique features since the two yeasts show many differences in DNA repair systems and cell cycle progression.

2.3 Transducers of the checkpoint response

Once the PI-3-related kinases Rad3(Sp) and Mec1p(Sc) are activated, evidence from both yeasts suggests that they phosphorylate a relatively small number of targets, which in turn transduce this signal to cell cycle regulators. In budding yeast, the serine/threonine kinase Rad53p(Sc) is essential for checkpoint control. This protein becomes phosphorylated in response to both DNA damage and incomplete replication, in a manner that is dependent on the sensor proteins, and on Mec1p(Sc) (22, 65). Overexpression of Rad53p(Sc) can bypass deficiencies in *MEC1*(Sc) and *RAD9*(Sc) (22, 62). Thus, Rad53p(Sc) is believed to act downstream of Mec1p(Sc) in the checkpoint pathway (see Fig. 2).

The activation of Rad53p(Sc) by checkpoint signals may be mediated by Rad9p(Sc), which becomes hyperphosphorylated in response to checkpoint signals. This phosphorylation requires *MEC1(Sc)*, *RAD17(Sc)*, *RAD24(Sc)*, *MEC3(Sc)*, and *DDC1(Sc)* (23, 66, 67). Phosphorylated Rad9p(Sc) interacts with the C-terminal *forkhead homology-associated* (FHA2) domain of Rad53p(Sc) (23, 67). In addition to decreasing the ability of Rad53p(Sc) to interact with Rad9p(Sc), mutation of Rad53p(Sc) FHA2 domain impairs G2–M cell cycle arrest and induction of *RNR3(Sc)* transcription (67). Quantitative analysis reveals that *rad53⁻(Sc)* is less defective in the checkpoint response than *mec1⁻(Sc)*, which argues that Mec1p(Sc) could have other targets (68), for example Pds1p(Sc) (see section 2.4.2) or Dun1p(Sc) (see section 4.1).

In fission yeast, the major targets of the activated PI-3-related kinase Rad3(Sp) are likely to be two serine/threonine kinases, Cds1(Sp) and Chk1(Sp) (see Fig. 3) (69, 70). Cds1(Sp) is structurally related to Rad53p(Sc) and also has an FHA domain (67, 69). A gene related to *chk1⁺(Sp)* has been identified in *S. cerevisiae*; however, it does not play a significant role in the checkpoint response. In fission yeast, both Cds1(Sp) and Chk1(Sp) are phosphorylated in response to checkpoint signals in a Rad3(Sp)-dependent manner (28, 71). Both kinases can also be directly phosphorylated by Rad3(Sp) *in vitro* (28). Like Rad53p(Sc), Cds1(Sp) is believed to be downstream of other checkpoint genes because overexpression of Cds1(Sp) can rescue the HU sensitivity, and partially rescue the UV sensitivity, of mutants in *rad1⁺(Sp)*, *rad3⁺(Sp)*, *rad9⁺(Sp)*, and *rad26⁺(Sp)* (69, 71). Likewise, Chk1(Sp) is believed to be downstream of the Rad3(Sp) and other checkpoint genes because overexpression of Chk1(Sp) can partially rescue the UV sensitivity of mutants in *rad1⁺(Sp)* or *rad3⁺(Sp)*, although this rescue may be non-specific because overexpression of Chk1(Sp) can cause cell cycle delay (28, 70). *chk1(Sp)* mutants have clear defects in the checkpoint response to damaged DNA, although they arrest normally when DNA replication is blocked with HU (70). Curiously, *cds1(Sp)* mutants arrest normally in response to HU, although they lose viability rapidly (69). Several studies suggest that this is because Cds1(Sp) and Chk1(Sp) function redundantly in the response to unreplicated DNA, although the sequences of the two kinases are not particularly similar to one another outside their kinase domains. For example, low-level overexpression of Chk1(Sp) can rescue *cds1Δ(Sp)* in the presence of HU (72). In addition, *chk1Δ cds1Δ(Sp)* double mutants are much more sensitive than either single mutant to unreplicated or damaged DNA (71–74). In fact, the checkpoint defect of the double mutant is indistinguishable from the defects of *rad3(Sp)* mutants, arguing that Cds1(Sp) and Chk1(Sp) are the major targets of Rad3(Sp) (72). Normally Chk1(Sp) is phosphorylated in response to DNA damage, but not HU; however, Chk1(Sp) is phosphorylated in HU in *cds1Δ(Sp)* cells (71, 75), suggesting that, in the absence of Cds1(Sp), Chk1(Sp) is activated. A recent model suggests that the two kinases are playing equivalent but cell cycle-specific roles; that is, Cds1(Sp) is important for responding to damage or replication blocks while the cells are in S phase, but Chk1(Sp) is responsible for transducing these signals during late S–G2 (28). Consistent with this idea, Cds1(Sp) is phosphorylated and activated in response to DNA damage or unreplicated DNA most strongly when the cells are in S phase (71). Chk1(Sp) gets phosphorylated in response to damage during late S and G2, but not M or G1–early S (71). There is also evidence arguing that Chk1(Sp) and Cds1(Sp) induce cell cycle arrest by the same mechanism, despite their lack of sequence similarity (see section 2.4.1). Given that Cds1(Sp) and Chk1(Sp) appear to be used in response to different types of checkpoint signals, it will be interesting to determine what determines which one is activated, particularly because the same sensors are required for the response to all checkpoint signals.

2.4 Targets of the checkpoint pathway

The signal transducers must transmit a checkpoint signal to the cell cycle machinery, which arrests the cell cycle to allow time for repair to take place. The targets of the checkpoint pathway thus depend on the point of cell cycle arrest. Moreover the cell cycle targets in the two yeasts are apparently dissimilar, probably due to the different cell cycle arrangements of the two yeasts (see section 1.1 of Chapter 3, and Reference 76). Finally, although the PI-3-related kinases are required for every checkpoint response, not all the sensors are needed at each cell cycle stage. For this reason we will divide the following discussion according to the point at which cell cycle arrest occurs. Where necessary we will also consider the two yeasts separately. We will begin at the end of the cell cycle, with G2–M, because historically this is the first control point that was analysed, and it is also the one that has been studied most extensively.

2.4.1 G2–M arrest in fission yeast

Fission yeast regulate the normal G2–M transition by modulating the phosphorylation of Cdc2(Sp) tyrosine 15, and this regulation is critical for the checkpoint responses (77–79; reviewed in Reference 76). When Cdc2(Sp) is maintained in its tyrosine-phosphorylated inactive form, the cells are arrested and do not enter mitosis. The balance between the activities of the Cdc2(Sp)-inhibitory kinases, Wee1(Sp) and Mik1, and the Cdc2(Sp)-activating phosphatases, Cdc25(Sp) and Pyp3, determines whether a cell will pass the G2–M boundary (76). Cells in which this balance is altered are checkpoint defective. For example, a strain that constitutively over-produces Cdc25(Sp) lacks the DNA replication checkpoint (80). A *cdc2-3w* strain, which renders Cdc2(Sp) activation independent of Cdc25(Sp), is also defective in this checkpoint (80). A *wee1-50 mik1Δ*(Sp) strain, which has greatly decreased tyrosine kinase activity due to a ts allele of *wee1*$^+$(Sp) and the deletion of the *mik1*$^+$(Sp) gene, is checkpoint defective even at the permissive temperature for *wee1-50*(Sp). At the non-permissive temperature, these cells are inviable (81, 82). A Cdc2F15 mutant, which replaces the tyrosine with a non-phosphorylatable phenylalanine residue, is unable to delay mitosis (77). Thus, regulation of Cdc2(Sp) tyrosine phosphorylation is crucial for checkpoint control in fission yeast.

The kinases Cds1(Sp) and Chk1(Sp) are believed to transduce the checkpoint signal from the sensors to Cdc2(Sp). (Refer to section 2.3 for further information about Chk1(Sp) and Cds1(Sp)). Cds1(Sp) and Chk1(Sp) phosphorylate Cdc25(Sp) on identical sets of serine residues (72, 79). This phosphorylation of Cdc25(Sp) creates phosphoserine motifs which are the consensus-binding sites for members of the 14-3-3 protein family (72, 83–85). Members of the 14-3-3 family promote particular protein–

protein interactions and modulate the activity of many signal transduction pathways in higher eukaryotes (86).

The fission yeast genome contains two members of the evolutionarily conserved 14-3-3 family, Rad24(Sp) and Rad25(Sp), which are 71% identical to each other (87). The *rad24⁻*(Sp) mutant was initially identified as a radiation sensitive mutant, while *rad25⁺*(Sp) was isolated as a suppressor of the *rad24⁻*(Sp) mutant (16, 87). The *rad24⁻*(Sp) and *rad25⁻*(Sp) mutants are defective in delaying their cell cycles following irradiation, and partially defective in their responses to unreplicated DNA (16, 73, 87). A double mutant *rad24⁻ rad25⁻*(Sp) is inviable, so its checkpoint response cannot be evaluated (87).

Fission yeast 14-3-3 proteins (presumably either Rad24(Sp) or Rad25(Sp) *in vivo*, although only Rad24(Sp) has been tested *in vitro*) bind the phosphoserine motifs of Cdc25(Sp) following phosphorylation by Cds1(Sp) and Chk1(Sp) kinase (72, 83). The functional significance of Cdc25(Sp)'s serine phosphorylations is demonstrated by the fact that fission yeast cells expressing a mutant version of Cdc25(Sp) that lacks the phosphorylation sites are checkpoint defective *in vivo* and show reduced binding of 14-3-3p to Cdc25 in an *in vitro* assay (72). Binding to 14-3-3p inhibits Cdc25(Sp) activity by a mechanism that does not seem to involve inhibition of Cdc25(Sp)'s phosphatase activity (83). Instead, a current model suggests it may involve localization of Cdc25(Sp) to the cytoplasm where it no longer has access to its substrate, Cdc2(Sp). Cdc25(Sp) binds Rad24(Sp), which contains a nuclear export sequence, and the nuclear export machinery promotes the export of Cdc25(Sp) (presumably bound to Rad24(Sp)) from the nucleus (88). In the absence of active Cdc25(Sp) in the nucleus, Cdc2(Sp) remains in a tyrosine-phosphorylated, inactive conformation, and the cells remain in G2. In addition to phosphorylating Cdc25(Sp), Chk1(Sp) and Cds1(Sp) can each phosphorylate Wee1(Sp) *in vitro*, but the *in vivo* functional significance of these phosphorylations has not been established (74, 89). It is possible that these kinases may also phosphorylate Mik1(Sp).

In fission yeast, the arrest in response to unreplicated DNA seems to be extremely similar to the arrest in response to damaged DNA. As with the G2 damage checkpoint, inhibitory tyrosine phosphorylation of Cdc2(Sp), the catalytic subunit of CDK, is required (79, 80; reviewed in Reference 76). Despite the similar biochemistry, there must be some differences in the signal generated by unreplicated DNA, as there is a mutant allele of *cdc2*, *cdc2-3w* that is defective in the replication checkpoint but responds normally to DNA damage (15). The *cdc2-3w* allele is believed to make Cdc2(Sp) activation Cdc25(Sp) independent (80), but it is not clear why this should make the mutant specifically defective in the replication checkpoint.

2.4.2 G2–M arrest in budding yeast

G2–M arrest in response to checkpoint signals differs markedly between the two yeasts. This probably reflects the differences in their cell cycles. Fission yeast have a clearly defined and well regulated G2–M transition when cytoplasmic microtubules are replaced by an intranuclear spindle. In contrast, budding yeast assemble a mitotic spindle shortly after Start, and spindle morphogenesis overlaps with chromosome replication. Possibly for this reason, DNA damage appears to block cell cycle progression from metaphase to anaphase in budding yeast, instead of G2 to M as in fission yeast and higher eukaryotes. Indeed, tyrosine phosphorylation of the budding yeast Cdc2(Sp) family member, Cdc28p(Sc), is not critical for cell cycle progression and is not required for the DNA damage checkpoint response (90–92).

The transition from metaphase to anaphase is activated by a conserved complex of proteins, the anaphase-promoting complex (APC; also called the cyclosome), an E3 ubiquitin–protein ligase complex that tags proteins with ubiquitin to mark them as substrates for ubiquitin-dependent proteolysis (see Chapters 3 and 5, or Reference 93, for more detailed information about the APC and cell cycle control). The APC associates with substrate-specific activators which help it recognize the correct substrate proteins; at anaphase the activator is Cdc20p(Sc) which helps the APC target the anaphase inhibitor Pds1p(Sc) for degradation (94–96). Once Pds1p(Sc) is degraded, anaphase begins. It is possible that the DNA damage checkpoint could delay the cell cycle by preventing activation of the APC, or by modifying APC substrates so that they are not recognized. Cdc20p(Sc) and Pds1p(Sc) have both been implicated as possible downstream targets of the DNA damage checkpoint pathway. (Cdc20p(Sc) may also be a target of the spindle assembly checkpoint; see section 6.2.) For example, cells overexpressing Cdc20p(Sc) lose viability when treated with DNA-damaging agents, and no longer arrest the cell cycle in response to damage (92, 97). Likewise, *pds1*(Sc) mutants are sensitive to DNA damage and fail to arrest in response to γ irradiation (95). Pds1p(Sc) is phosphorylated in response to UV or γ irradiation in a manner dependent on Mec1p(Sc) and Rad9p(Sc), but not on Rad24p(Sc), Rad17p(Sc), Mec3p(Sc), Ddc1p(Sc), or Rad53p(Sc) (98). However, it is not known whether this phosphorylation is needed for checkpoint function. So far it is unclear whether there are direct physical connections between the upstream components of the arrest pathway and the APC, Cdc20p(Sc), or Pds1p(Sc); perhaps there are as yet unidentified intermediate proteins between Mec1p(Sc) and its putative targets. This topic awaits further investigation.

Pds1p(Sc) is not required for the response to unreplicated DNA. Indeed, arrest in response to unreplicated DNA differs markedly from the response to damaged DNA in *S. cerevisiae*. Arrest requires the PI-3-related kinase, Mec1p(Sc) and the downstream kinase Rad53p(Sc); however, the sensors Rad9p(Sc), Rad17p(Sc), Rad24p(Sc), and Mec3p(Sc) are not required (12, 62, 63). Instead, the checkpoint response to unreplicated DNA may employ polymerase ε (see section on checkpoint signals), which plays no role in the response to DNA damage. As in the case of DNA damage, Rad53p(Sc) becomes phosphorylated in response to incomplete replication, in a manner that is dependent on Mec1p(Sc) (22, 65); however, there is no information on subsequent steps.

2.5 G1 arrest

In both yeasts, the G1 arrest in response to DNA damage has been studied much less than the G2 arrest. This arrest is not likely to be seen in unperturbed fission yeast cells because most wild-type cells are in G2, and thus damage is most likely to induce arrest at G2–M. This may be an adaptation to the fission yeast's haploid lifestyle: since haploid organisms are much more sensitive to DNA damage during G1 than G2, they spend a minimal amount of time in G1. A recent study suggests fission yeast are able to delay the cell cycle following UV irradiation in G1, but it is not clear what genes are necessary for this arrest (99).

In contrast to fission yeast, budding yeast do have a G1 phase in their normal cell cycle. In the budding yeast DNA damage checkpoint, Sidorova *et al.* have suggested that Rad53p(Sc) may delay the cell cycle in G1 by inhibiting the transcription of cyclins (*CLN1* and *CLN2*) needed to bind to Cdc28p(Sc) and promote S (100). A transcription factor necessary for *CLN* gene transcription (*SWI6(Sc)*; see Chapter 3) is phosphorylated in a Rad53p(Sc)-dependent manner following DNA damage in an *in vitro* assay, consistent with this model (100).

3. S-phase delay

During S phase, a damage-sensitive checkpoint slows the rate of DNA replication in response to DNA damage. (This is distinct from the checkpoint that inhibits mitotic entry when the cells are blocked in S phase by inhibition of replication.) Intra-S checkpoints were known to slow the rate of ongoing S-phase progression in *E. coli* and mammalian cells (reviewed in Reference 101) and have recently been shown to exist in budding yeast (102) and fission yeast (71, 99).

An intra-S checkpoint in yeast was first described by Paulovich and Hartwell, who demonstrated that wild-type budding yeast exposed to sublethal doses of the alkylating agent methyl methanesulfonate (MMS) went through S very slowly (102). In contrast, mutants in *MEC1(Sc)* or *RAD53(Sc)* replicated their DNA at a similar rate in the presence or absence of MMS and were MMS sensitive. This suggested that replication was not being slowed by lesions or repair complexes physically blocking replication forks, but that a surveillance mechanism in wild-type cells was slowing replication and that this response was abrogated in the checkpoint mutants (102). This response is specific to checkpoint mutants, since mutants in numerous nucleotide excision repair or recombination genes are able to delay S phase normally (103).

Further studies have demonstrated that, in addition to *MEC1(Sc)* and *RAD53(Sc)*, some of the genes involved in sensing DNA damage in other parts of the cell cycle are also needed for the intra-S response. Mutants in *RAD9(Sc)*, *RAD17(Sc)*, *RAD24(Sc)*, *MEC3(Sc)*, and *DDC1(Sc)* do not lengthen S phase in the presence of DNA damage as much as wild-type cells, suggesting they are involved in this response (19, 63, 103). However, the *rad9Δ rad17Δ rad24Δ(Sc)* triple mutant has an S-phase length between that of wild-type and *rad53(Sc)* or *mec1-1(Sc)*, suggesting these sensors may play a lesser or qualitatively different role than Mec1p(Sc) and Rad53p(Sc) (103). Mutation of *RFC5(Sc)* also reduces the damage-induced S-phase delay, possibly because the RFC complex may be signalling the presence of damage to the checkpoint machinery (53) (see section 2.2.3).

3.1 Possible targets of the intra-S checkpoint

Mutations in DNA primase (*PRI1*(Sc)), or the large subunit (*RFA1*(Sc)) of the single-stranded binding protein RP-A, are also defective in slowing the rate of S-phase progression in the presence of DNA damage (104, 105). (See Chapter 2 for further information on DNA replication proteins.) These results suggest that primase and RP-A could be targets of the intra-S pathway, since phosphorylation of Rad53p(Sc) is intact in these mutants, indicating that Mec1p(Sc) activation has occurred normally (104, 105). The large subunit of RP-A is phosphorylated in a Mec1p(Sc)-dependent manner, suggesting it could be a checkpoint target (106). Other components of the replication apparatus may be involved as well, but they have not been tested yet.

There is also evidence suggesting that the activation of late origins of replication is inhibited by checkpoint signals in *S. cerevisiae* (101, 102). By monitoring the genomic footprints of pre-replicative complexes at replication origins, and observing the presence of replication intermediates from particular origins, Santocanale and Diffley showed that Mec1p(Sc) and Rad53p(Sc) inhibit the firing of late origins in the presence of HU (107). Using two-dimensional gel analysis of origins in the presence of MMS, Shirahige *et al.* also observed inhibition of the firing of late origins (108). This delay was absent in *rad53*(Sc) mutants, but *mec1-1*(Sc) behaved like wild-type in this assay (108). The molecular mechanism of this effect is unclear. It might involve DNA primase, RP-A or other regulators of origin activation.

Less is known about the intra-S checkpoint in fission yeast, and most of the known checkpoint genes have yet to be tested for defects in this response. To date, mutants in three genes in fission yeast have been shown to be defective in an S-phase monitoring function: *rad3$^+$*(Sp), *cds1$^+$*(Sp), and *rad26$^+$*(Sp) (71, 99, 109). Both *cds1*(Sp) and *rad26*(Sp) mutants are initially able to arrest their cell cycle after DNA damage or inhibition of replication by HU, but the cells lose viability very rapidly even though they are not entering mitosis at the time (i.e. the S–M checkpoint seems intact) (16, 69). It has been suggested that *cds1$^-$*(Sp) and *rad26$^-$*(Sp) are losing viability due to a defect in an S-phase function, referred to as 'recovery', which is needed to survive S in the presence of damage or blocked replication (see section 4.2) (16, 69).

It would be premature to draw a pathway for an intra-S checkpoint or to speculate about what the effectors of such a pathway could be in fission yeast, but some interesting interactions amongst the known actors have been observed. Intriguingly, Cds1(Sp) is phosphorylated and activated by DNA damage during S, but not during G1 or G2 (71). This phosphorylation is dependent on *rad3$^+$*(Sp) (71), and Rad3(Sp) can phosphorylate Cds1(Sp) *in vitro* (28), suggesting Cds1(Sp) may be a downstream target of Rad3(Sp). *cds1$^+$*(Sp) can act as a multi-copy suppressor of a ts *pol1*(Sp) (polymerase α), suggesting Cds1(Sp) could be involved in monitoring DNA replication (69). In addition, *cds1$^+$*(Sp) can act as a multi-copy suppressor of the *rad26-T12*(Sp) mutant allele, and Cds1(Sp) and Rad26(Sp) can be co-immunoprecipitated from yeast lysates, although the interaction appears to be indirect (71).

4. Other targets of the damaged and unreplicated DNA checkpoint pathways

Until now, we have focused on cell cycle control in response to damaged or un-replicated DNA. However, the same signals may be used to activate other processes as well. The role of the checkpoint genes in transcription in *S. cerevisiae* has been particularly well studied. There is also evidence for a role of the *S. pombe* genes in a unique repair process that has been termed 'recovery' or 'tolerance'. Finally, in *S. cerevisiae*, null alleles of *rad53*(Sc) and *mec1*(Sc) are inviable, and recent studies have shed light on the essential function of these genes.

4.1 Transcription

In addition to arresting the cell cycle upon encountering conditions of DNA damage or blocked replication, cells may induce transcription of genes needed to counter the effects of the lesions. In bacteria, for instance, the SOS system is responsible for both inhibiting cell division and inducing transcription of repair and recombination proteins (110). In *S. cerevisiae* over 50 genes are estimated be induced by DNA-damaging agents, including genes involved in excision repair, recombinational repair, and DNA metabolism. These genes have a variety of promoter elements, so there is probably not a shared system of induction for all of them (reviewed in Reference 110). Some of these genes are induced in a checkpoint-dependent manner. For example, transcriptional induction in response to UV of about 15 genes, including the *ribonucleotide reductase* (*RNR*) genes, was shown to be dependent on Rad9p(Sc) (111). It was not determined whether this transcriptional induction involved any of the genes downstream of Rad9p(Sc) in the checkpoint pathway. Also, as mentioned in section 2.5, a transcription factor, SWI6p(Sc), needed to induce transcription of G1 cyclins (*CLN1* and *CLN2*), can be phosphorylated *in vitro* by Rad53p(Sc) immunoprecipitated from cells in which the damage checkpoint has been activated (100). This phosphorylation prevents *CLN* transcription, leading to G1 cell cycle arrest (100). There is no known equivalent to this system in fission yeast.

RAD53(Sc) has long been known to be involved in DNA damage-induced transcription (see Fig. 2). The transcription of four ribonucleotide reductase subunits, the *RNR1–4* genes, is dependent on Rad53p(Sc) (62, 112). Rad53p(Sc) is likely to work upstream of Dun1p(Sc), a kinase that is required for transcriptional induction of *RNRs* but not for cell cycle arrest after DNA damage, since Dun1p(Sc) is not activated in a *rad53*(Sc) mutant (62). A transcription factor, Crt1p(Sc), binds to sequences in the promoter regions of *RNRs* (and the *CRT1* gene itself) and recruits general transcriptional repressors (113). Crt1p(Sc) becomes hyperphosphorylated in a Mec1p(Sc)-, Rad53p(Sc)-, Dun1p(Sc)-dependent fashion following DNA damage, and the hyperphosphorylated Crt1p(Sc) is not able to bind DNA, leading to the loss of repression of the *RNRs* (113). Mutation of the promoter elements can also lead to the loss of repression (113). In summary, the current model suggests that, in the absence of DNA damage, Crt1p(Sc) represses transcription of a set of genes including *RNRs* and itself.

In the presence of DNA damage, activated Rad53p(Sc) signals Dun1p(Sc), which leads to phosphorylation of Crt1p(Sc), and its release from the promoters of inducible genes.

Less is known about transcriptional induction following damage or unreplicated DNA in fission yeast, but a few transcriptionally-induced genes have been identified. In one study, four UV-inducible transcripts were identified, but it is not known whether their induction is dependent on genes in the checkpoint pathway (114). Transcription of both the large and small subunits of ribonucleotide reductase is induced by exposure to HU (115, 116). Transcription of the small subunit is also induced by a UV-mimetic drug that induces DNA damage, and this response requires $rad1^+$(Sp) (117). The nucleoside analogue 5-azacytidine, which may be both interfering with DNA replication and causing DNA damage, also causes inducible transcription of the small subunit (116). This induction requires $rad1^+$(Sp), $rad3^+$(Sp), $rad9^+$(Sp), $rad17^+$(Sp), and $hus1^+$(Sp), but not $chk1^+$(Sp) (116). It is not yet known whether this transcriptional induction in fission yeast requires a gene similar to *CRT1*(Sc).

4.2   Recovery and tolerance

In fission yeast, mutation of the checkpoint sensors and $rad3^+$(Sp) makes cells much more sensitive to HU or UV radiation than mutations of targets such as $cdc2^+$(Sp) or $chk1^+$(Sp). Moreover, cell cycle analysis reveals that sensor mutations undergo a lethal event in S phase in addition to entering M with damaged or unreplicated DNA. To explain these observations it has been proposed that the checkpoint pathway in fission yeast signals a 'recovery' or 'tolerance' pathway in addition to blocking cell cycle progression. Without this function, even if M can be arrested by some other mechanism, exposure to DNA damage or inhibition of replication is still lethal (14, 15). Some evidence suggests that components of the tolerance pathway could include the RecQ-related helicase, Rqh1(Sp) (118), the recombination repair proteins Rhp51(Sp) (related to Rad51p(Sc)) and Rhp54 (related to Rad54p(Sc)), Rad2(Sp) (related to *FEN-1* endonuclease) and Rad18(Sp) (related to structural maintenance of chromosomes (SMC) proteins) (119); however, a direct physical interaction between these proteins and the checkpoint proteins has not been demonstrated.

It is also possible that the sensitivity of checkpoint mutants to UV and HU could reflect loss of other checkpoints, particularly the intra-S checkpoint, which require the same proteins as mitotic arrest (see section 3.1) (71). Finally, as in *S. cerevisiae*, the checkpoint proteins in fission yeast could be required for inducible transcription of enzymes that are essential for efficient repair. Clearly further characterization of the recovery and tolerance function will be necessary to distinguish between these possibilities.

4.3   Essential function of MEC1(Sc) and RAD53(Sc)

Both *MEC1*(Sc) and *RAD53*(Sc) are essential genes in *S. cerevisiae*; however, the lethality of the mutations can be suppressed by increasing expression of *RNR1*(Sc), which encodes a large subunit of ribonucleotide reductase (120). $mec1^-$(Sc) and $rad53^-$(Sc) strains that were reported to be viable but checkpoint deficient have been shown to harbour suppressor mutations in a gene called *SML1*(Sc). Sml1p(Sc) binds to ribonucleotide reductase and may negatively regulate its activity (121). The lethality of *mec1⁻*(Sc) and *rad53⁻*(Sc) strains can also be suppressed by overexpression of Rnr1p(Sc) (120), or by deletion of the G1 cyclins *CLN1* and *CLN2*, possibly because cells delayed in G1 synthesize more Rnr1p(Sc) (122). To explain these observations, it has been proposed that deoxynucleotides are limiting for DNA replication in *S. cerevisiae*, and that because of this Mec1p(Sc) and Rad53p(Sc) are required every cell cycle to control replication progression carefully. Upregulation of *RNR1*(Sc) transcription by any mechanism increases nucleotide pools to an extent where they no longer limit replication, and under these conditions *MEC1*(Sc) and *RAD53*(Sc) are no longer essential (120–122). This effect could be exacerbated if full expression of *RNR1*(Sc) required positive regulation by the Mec1p(Sc)–Rad53p(Sc) pathway. However, it is puzzling that other mutants with defects in the intra-S and replication checkpoint defects are still viable. In addition it is intriguing that the equivalent *S. pombe* genes, *rad3⁺*(Sp) and *cds1⁺*(Sp) are not essential. Perhaps *S. pombe* has higher levels of deoxynucleotides than *S. cerevisiae*. Alternatively, ribonucleotide reductase expression could be less dependent on the checkpoint proteins in *S. pombe*.

5. Checkpoint response to damaged and unreplicated DNA: summary and perspectives

Ten years ago the role of Rad9p(Sc) in G2–M arrest in response to DNA damage was first described. Since then the efforts of many groups have shown Rad9p(Sc) to be one of a set of proteins controlling many aspects of cell cycle progression in response to abnormal DNA structures. As shown in Table 2, human genes related to nearly every yeast checkpoint gene have been identified, strongly suggesting that these genes are used by all eukaryotic cells to prevent cell cycle progression in the presence of abnormal DNA structures. As the preceding discussion has illustrated, in the years since the identification of the first checkpoint system almost every important cell cycle transition, from origin-firing to CDK and APC activation, has been shown to be regulated by a checkpoint pathway. Clearly we are well on the way to understanding these fascinating signal-transduction pathways.

In our opinion, further breakthroughs may depend critically on developing *in vitro* systems for the analysis of the checkpoint response. Such systems could be used to dissect the early steps of the pathway, in particular to learn what DNA structures and/or proteins activate the PI-3-related kinases. Such an approach should also permit a more detailed understanding of the biochemical function of each checkpoint protein. We hope that this very productive period of genetic analysis of the checkpoint pathways will be followed by an equally exciting and fruitful period of biochemical investigation.

6. The spindle checkpoint

During mitosis, the parent cell must correctly distribute a copy of each of its chromosomes to both of its daughter cells (see Chapter 5). In addition to the checkpoints ensuring faithful replication and repair of the chromosomes which we discussed in previous sections, cells have a checkpoint that monitors the segregation of their genetic material during mitosis and cell division. To ensure that their chromosomes are properly segregated during anaphase, cells monitor the attachment of the chromosomes to the mitotic spindle and the integrity of the spindle. Chromosomes are segregated by the attachment of each sister chromatid, at its kinetochore, to the microtubules emanating from each pole of the mitotic spindle. If even a single kinetochore is not connected to the microtubules linking it to one of the poles of the spindle, or if the spindle becomes depolymerized, the cell cycle is arrested before entry into anaphase at the 'spindle assembly checkpoint' (reviewed in References 123, 124). This delay allows time for the detached kinetochore(s) to become attached or for the spindle to be reassembled before mitosis proceeds.

The molecular details of how cells monitor the attachment of their chromosomes to the mitotic spindle and the integrity of the spindle are unclear. Some clues as to what might be happening have been obtained in studies of higher eukaryotes. Some intriguing experiments indicate that the tension on the kinetochore from bipolar attachment to the spindle may be necessary for cells to enter anaphase (reviewed in Reference 125). A chromosome that is attached improperly, to only one of the two poles, is not under tension, whereas a properly attached chromosome is being pulled by both poles. In insect cells, if an improperly attached chromosome is artificially placed under tension by micromanipulation, the cells do not delay entry into mitosis. Insect kinetochores that are not attached to both poles of the mitotic spindle stain brightly with the 3F3/2 antibody (which recognizes a kinetochore-associated phospho-epitope), but the staining fades when the chromosome becomes attached correctly. Microinjection of this antibody causes a delay in the loss of staining and in exit from mitosis. Laser ablation of the kinetochore of the last chromosome not properly attached to the spindle releases cells from the anaphase delay, even though the un-attached chromosome is still present. These experiments suggest that cells may be monitoring the tension on the kinetochores, and that this mechanical force is some-how converted into a chemical change, such as the desphosphorylation of certain epitopes (125). However, the absence of tension is not the only difference between the improperly attached chromosome and the remaining chromosomes; an improperly attached chromosome is also incorrectly positioned in the cell, not aligned on the metaphase plate with the other chromosomes (125). Thus, the abnormal location of the stray chromosome could also be acting as a checkpoint signal. In addition, a recent study using paclitaxel (Taxol), which reduces tension at kinetochores, suggests that merely decreasing the tension without disrupting the kinetochore–microtubule interactions is not sufficient to generate a checkpoint signal (126).

6.1 Sensors and transducers of the spindle checkpoint signal

While it is not clear how the checkpoint signal is generated, work in yeast has led to the identification of a number of proteins that are likely to be involved in this process. The budding yeast genes involved in the spindle surveillance system were initially

Table 3 Spindle assembly checkpoint genes in *S. pombe* and *S. cerevisiae*

Budding yeast gene	Fission yeast gene	Potential human homologue	Sequence motifs
Sensors and/or Transducers			
MPS1 (129)	*mph1*$^+$ (132)		Kinase
MAD1 (170)	*mad1*$^+$ (123)		
MAD2 (138)	*mad2*$^+$ (133)	*hMAD2* (138)	
MAD3 (127)			
BUB1 (134, 135)	*bub1*$^+$ (171)	*hBUB1* (3)	Serine/threonine kinase
BUB2 (128)	*cdc16*$^+$ (172)		
BUB3 (128)	Z99163	*hBUB3* (173)	
YHR115 or			
YNL116 (130)	*dma1*$^+$ (130)		FHA domain, RING-H2 finger domain
Receivers			
CDC20 (92, 94)	*slp1*$^+$ (140)	p55CDC (141)	WD repeat/ubiquitin targeting specificity factor
PDS1 (96, 143)	*cut2*$^+$ (144)		

identified in screens for mutants that lost viability rapidly in the presence of chemicals that disassemble the spindle. '*MAD*' (*mitotic arrest deficient*) and '*BUB*' (*budding uninhibited by benzimidazole*) mutants that failed to delay the cell cycle in response to spindle depolymerizing drugs were identified (127, 128). (Refer to Table 3 for a summary of the genes involved in the spindle surveillance system, including the *MAD* and *BUB* genes.) Although *MAD* and *BUB* mutants do not segregate their chromosomes or complete cytokinesis in the presence of spindle-depolymerizing drugs, they do undergo cell cycle events such as bud emergence and DNA replication which would normally occur in the next cell cycle following cytokinesis. The *MAD* and *BUB* genes are not essential for viability under normal growth conditions (i.e. when mitotic spindle function is not perturbed) (127–129). The *MPS1*(Sc) gene, initially identified for its role in spindle pole body duplication, is also required for the checkpoint monitoring spindle integrity (130). Thus far, components of the spindle checkpoint appear to be quite conserved; *S. pombe* homologues are listed in Table 3 (reviewed in Reference 123). In addition, a fission yeast multicopy suppressor of a *cdc16*(Sp) ts mutant, *dma1*$^+$(Sp) (*defective in mitotic arrest*), has also been shown to be necessary for the spindle assembly checkpoint (131). Curiously, no analysis of mutants in the genes similar to *dma1*$^+$(Sp) in the *S. cerevisiae* genome has been reported to date; therefore, it is not known whether budding yeast has a functional homologue of *dma1*$^+$(Sp). Higher eukaryotic homologues (in *Xenopus*, mouse and human) of several *S. cerevisiae MAD* and *BUB* genes have been identified (see section 7 and Table 3).

Exactly how these proteins are monitoring the status of the kinetochores and spindle, and transmitting the signal to arrest the cell cycle, is not clear. One model is shown in Fig. 4. The Mad and Bub proteins may somehow sense the microtubule–kinetochore interactions, the tension produced by these interactions, or the chromosome alignment. For example, the *MPS1*(Sc) and *BUB1*(Sc) genes encode protein kinases, but the remaining *MAD* and *BUB* genes do not have informative motifs (reviewed in References 123, 124). Mps1p(Sc) has been proposed to act early in the

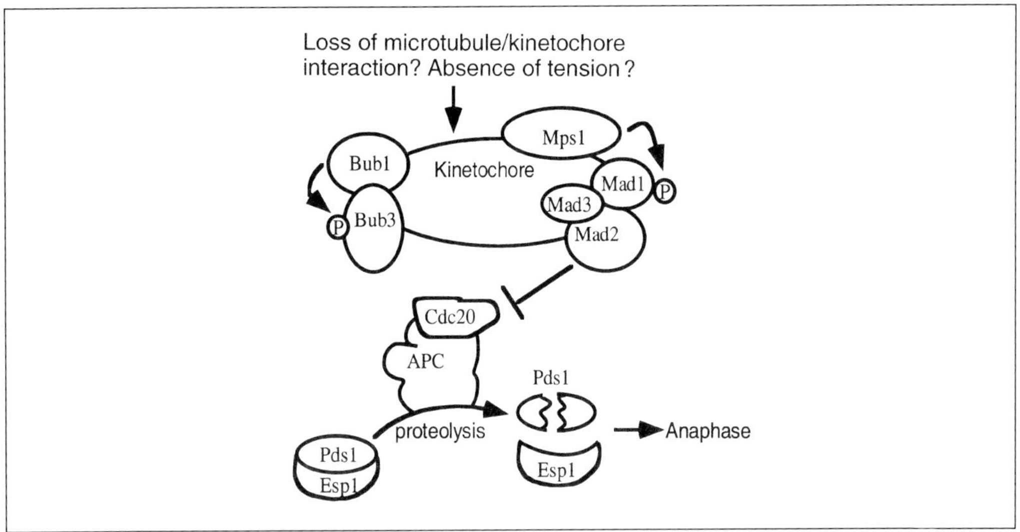

Fig. 4 Budding yeast spindle checkpoint.

checkpoint pathway, because overexpression of Mps1 constitutively activates the spindle checkpoint even in wild-type cells, a response that is dependent on the other *MAD* and *BUB* genes (132). Similarly, overexpression of *mph1*$^+$(Sp) or *mad2*$^+$(Sp) in fission yeast causes a metaphase arrest (133, 134). In budding yeast, Mad1p(Sc) is phosphorylated by Mps1p(Sc) when the checkpoint is activated, in a manner dependent on Bub1p(Sc), Bub3p(Sc), and Mad2p(Sc) (132). Bub1p(Sc) kinase is also a candidate signal transducer. Overexpression of a dominantly acting allele, *Bub1-5(Sc)*, causes a cell cycle delay in wild-type cells in the absence of detectable spindle damage, in a manner dependent on the presence of the other Mad and Bub proteins, Mps1p(Sc), and a functional Bub1p(Sc) kinase domain (135). It is interesting that mitotic delay induced by overexpression of Mps1p(Sc) requires Bub1p(Sc), whereas mitotic delay induced by overexpression of Bub1-5p(Sc) requires Mps1p(Sc). This suggests the functions of the two kinases are interdependent; perhaps they are required at the same step in checkpoint activation (135).

It is possible that some or all the Mad and Bub proteins involved in sensing and transducing the spindle assembly checkpoint may form a complex located on kinetochores (123). Some interactions between these proteins have been identified in budding yeast. For example, Mad2p(Sc) interacts with Mad1p(Sc) (123); Bub1p(Sc) forms a complex with Bub3p(Sc) and is able to phosphorylate it *in vitro* (129). Other interactions amongst the yeast proteins remain to be investigated. Some experiments in higher eukaryotes also support the idea that there may be a complex on kinetochores. The *Xenopus* Mad1 and Mad2, human Mad2, and mouse Bub1 homologues all localize to the kinetochores in the earliest parts of mitosis, from prophase to prometaphase; as chromosomes become aligned on the metaphase plate, these proteins diffuse, but they remain localized on misaligned chromosomes (136–138). This suggests the sensors could be associated with all the chromosomes early in a normal mitosis, then 'mark' the misalignedchromosomes as mitosis progresses.

6.2 Does the spindle checkpoint target the APC?

In wild-type cells, the spindle checkpoint prevents exit from mitosis by arresting the cells at the metaphase to anaphase transition point in mitosis. An important regulator of progression through mitosis is the APC (also called the cyclosome), an E3 ubiquitin–protein ligase complex that targets proteins for ubiquitin-dependent degradation (see Chapter 3 or Reference 93 for more detailed information about the APC and cell cycle control, and see Table 3 for a list of genes involved in the spindle checkpoint). With the assistance of substrate-specific accessory factors, the APC promotes the degradation of proteins at two transitions during mitosis. With the accessory factor Cdc20(Sc) (or its homologue Slp1(Sp) or p55CDC in humans), the APC promotes degradation of anaphase inhibitors such as Pds1(Sc) (or Cut2(Sp)) at the metaphase-to-anaphase transition (94–96, 139–141). During metaphase, sister chromatids are held together in a manner dependent on 'cohesion' proteins such as Scc1(Sc), which disappears from chromosomes at the entry to anaphase (142). (See Chapter 5 for further details about the mechanics of chromosome separation.) The degradation of Pds1(Sc) frees the Pds1(Sc)-associated protein, Esp1(Sc), which is essential for the dissociation of Scc1(Sc) from sister chromatids and their separation (143). Similarly, the fission yeast homologue of Esp1(Sc), Cut1(Sp), binds to Cut2(Sp), and proteolysis of Cut2(Sp) is necessary to release Cut1(Sp) for promotion of sister chromatid separation (144). Later in the cell cycle, the APC also promotes exit from mitosis by targeting the spindle component Ase1p and the B-type cyclins for proteolysis, using a different accessory factor, Hct1/Cdh1(Sc) (94, 145, 146). The mechanism of cell cycle arrest by the spindle assembly checkpoint is likely to be inhibition of the APC at the metaphase-to-anaphase transition, because cells arrested at the checkpoint have stabilized Pds1(Sc), high mitotic cyclin levels, high Cdk activity, and synapsed sister chromatids.

Several recent papers have suggested that there are direct interactions between spindle checkpoint components and the APC or its substrate-specific accessory factor, Cdc20(Sc). For example, human p55CDC, a Cdc20(Sc) family member, is found at the kinetochore (139). Mad2(Sc) and Mad3(Sc) can be co-immunoprecipitated with Cdc20(Sc) in a Mad1(Sc)-dependent manner, although this association does not change with the activation of the checkpoint (92). In fission yeast, Mad2(Sp) physically interacts with the Cdc20(Sc) family member, Slp1(Sp) (140).

In addition to these physical interactions between spindle checkpoint components and Cdc20(Sc) family members, there are additional reasons to suspect that the spindle checkpoint acts via Cdc20(Sc) and the APC. For example, budding yeast cells overexpressing Cdc20(Sc) no longer arrest in mitosis in response to the spindle-depolymerizing drug nocodazole, or in response to overexpression of the upstream kinase Mps1(Sc) (92). Furthermore, overexpression of Cdc20(Sc) cannot induce cells to enter anaphase if they are defective in APC function (147). Dominant mutants of *CDC20(Sc)* that do not respond to checkpoint arrest caused by overexpression of Mps1(Sc) have been identified (92, 147), and some of these mutants abolish binding to Mad2(Sc) and Mad3(Sc) (92). Similarly, mutants of the Cdc20(Sc) family member, Slp1(Sp), which do not arrest the cell cycle when Mad2(Sp) is overexpressed, have been identified, and these mutants fail to bind Mad2(Sp) (140). Thus, the current model (Fig. 4) for the receivers of the spindle assembly checkpoint signal is that Mad2 and Mad3 (possibly in conjunction with other Mad or Bub proteins as well) interact with Cdc20 and some components of the APC, to keep it inactive during early mitosis or when the checkpoint is activated. Presumably release of the Mad proteins is necessary for the APC and Cdc20 to become active again and promote the transition into anaphase, but this is not yet proven. The spindle checkpoint may also have additional targets. Current experiments are certain to further elucidate the roles of the checkpoint components, including Bub2(Sc) whose function is currently unclear.

7. Checkpoint control and human disease

In their initial analysis, Weinert and Hartwell noted that *rad9(Sc)* mutants were viable in the absence of DNA damage, but showed increase rates of chromosome loss (4). This led to the prediction that loss of checkpoint control might be an early step in the evolution of tumour cells. If this is the case, the human proteins related to yeast checkpoint proteins would be predicted to behave as tumour suppressors.

This prediction has been borne out most dramatically in the case of the gene responsible for the severe congenital cancer-prone syndrome ataxia telangiectasia (AT), *ATM* (*a*taxia *t*elangiectasia *m*utated). Affected individuals are prone to cancer and also exhibit severe neurological symptoms. Early studies showed that affected individuals are also extremely sensitive to X-rays, and that cultured AT cells exhibit X-ray sensitivity. This sensitivity correlates with a defect in cell cycle control. When normal cell lines were irradiated, cell cycle progression was blocked as judged by measurement of DNA synthesis. In contrast, irradiated AT cells continue to synthesize DNA, a phenomenon known as radiation resistant DNA synthesis (reviewed in Reference 110). Positional cloning of the *ATM* gene revealed that it encodes a protein with striking structural similarity to the yeast PI-3 related kinases Mec1p(Sc), Rad3(Sp), Tel1p(Sc), and Tel1(Sp) (2, 25). This finding establishes the universal role of this surveillance pathway in the response to DNA damage, as well as confirming the link between genomic instability and carcinogenesis.

Human genes related to many more yeast checkpoint genes including *rad9*$^+$*(Sp)*/ *DDC1(Sc)*, *hus1*$^+$*(Sp)*, and *RAD17(Sc)*/*rad1*$^+$*(Sp)* have been discovered (see Tables 2 and 3). It is expected that more homologues will be found as the Human Genome Project progresses. Intriguingly, hRad1, hHus1, and hRad9 are associated in a complex, like the *S. pombe* and *S. cerevisiae* proteins (148). It will be important to determine whether these genes also function in damage surveillance and checkpoint control in mammalian cells. Kinases related to both Chk1(Sp) and Cds1(Sp)/Rad53p(Sc) have also been identified and shown to play a role in the checkpoint response to DNA damage (83, 149–151). We can expect mammalian cells to have more checkpoint targets than yeast since the normal response to damage in mammalian cells includes growth inhibition, and in some cases apoptosis, as well as cell cycle arrest.

Other human genes with no yeast relatives have also been implicated in a DNA damage surveillance pathway. The tumor suppressor p53 plays a central role in growth control, cell cycle control, and apoptosis, and p53 mutants show numerous defects in the response to DNA damage (152, 153). The breast cancer gene *BRCA1* contains a recently identified domain called 'BRCT', which is also found in *RAD9*(Sc), *DPB11*(Sc), *rhp9*$^+$/*crb2*$^+$(Sp), and *cut5*$^+$/*rad4*$^+$(Sp)(154). However, the human gene is not particularly similar to the yeast genes apart from the BRCT domain, so its significance is still under investigation.

The spindle assembly surveillance system is also evolutionarily conserved. Genes related to *S. cerevisiae MAD* and *BUB* genes have now been discovered in *Xenopus*, mouse, and humans. Many of these genes localize to the mitotic spindle and have been shown to function in the checkpoint response in these systems (see Table 3; reviewed in Reference 123). There is also evidence linking defects in this surveillance system to cancer. Certain diploid human colorectal cancer cell lines treated with microtubule-depolymerizing agents are able to arrest their cell cycle with condensed chromosomes, while aneuploid cell lines pass through mitosis and enter a new round of DNA synthesis, suggesting they are defective in a spindle checkpoint (3). In addition, alterations in two human homologues of yeast spindle checkpoint genes have been observed in cancer cells. In one study, dominant negative mutations in *hBUB1* were identified in two cancer cell lines (3). In another study, a human breast tumour cell line had reduced *hMAD2* expression and failed to arrest in mitosis after nocodazole treatment (138). This tentative connection between checkpoint failure and genomic instability is quite intriguing because karyotypic studies have indicated that the majority of human cancers have lost or gained chromosomes (152). However, more extensive studies will be required to determine whether checkpoint failure plays a major role in the genomic instability of human tumors. Studies of mouse models with altered spindle checkpoint genes, for example, may be useful in addressing this issue.

8. Future prospects

As the preceding discussion illustrates, basic genetic studies of genomic instability in yeasts have led to the discovery of surveillance pathways that are conserved in all eukaryotic cells and play a critical role in preventing cancer in humans. If yeasts can be used to identify causes of cancers, can they also be used in the search for a cure? A number of systematic attempts to use yeast mutants in drug discovery are currently underway. Briefly, the rationale for these studies is that, because tumour cells exhibit genomic instability, they may lack functions such as DNA repair or checkpoint control. Yeast strains with such defects are dramatically more sensitive to some insults than wild-type yeast. Therefore, it may be possible to develop a therapy strategy that exploits the differences in DNA metabolism between normal and tumour cells. One approach to identifying such selective drugs is to search for those that kill analogous yeast mutants without affecting wild-type yeast cells. For example, a drug that selectively killed *bub1*(Sc) mutants might be used clinically to treat colorectal cancers that show high rates of aneuploidy. To identify such selective drugs, panels of isogenic yeast mutants are currently being used to screen libraries of chemical compounds (155). Perhaps these studies are the beginning of the next exciting chapter in the history of checkpoints and surveillance mechanisms.

Acknowledgements

The authors thank Scott Schuyler and members of the Enoch laboratory for comments on this manuscript. Work in the laboratory of T. E. is supported by a grant from the NIH (GM50015).

References

1. Hartwell, L. and Weinert, T. (1989) Checkpoints: controls that ensure the order of cell cycle events. *Science,* **246**, 629.
2. Savitsky, K., Bar-Shira, A., Gilad, S., Rotman, G., Ziv, Y., Vanagaite, L., Tagle, D. A., Smith, S., Uziel, T., Sfez, S., Ashkenazi, M., Pecker, I., Frydman, M., Harnik, R., Patanjali, S. R., Simmons, A., Clines, G. A., Sartiel, A., Gatti, R. A., Chessa, L., Sanal, O., Lavin, M. F., Jaspers, N. G.J., Taylor, A. M. R., Arlett, C. F., Miki, T., Weissman, S. M., Lovett, M., Collins, F. S., and Shiloh, Y. (1995) A single ataxia telangiectasia gene with a product similar to PI-3 kinase. *Science,* **268**, 1749.
3. Cahill, D. P., Lengauer, C., Yu, J., Riggins, G. J., Willson, J. K., Markowitz, S. D., Kinzler, K. W., and Vogelstein, B. (1998) Mutations of mitotic checkpoint genes in human cancers. *Nature,* **392**, 300.
4. Weinert, T. A. and Hartwell, L. H. (1988) The *RAD9* gene controls the cell cycle response to DNA damage in *Saccharomyces cerevisiae. Science,* **241**, 317.
5. Weinert, T. (1998) DNA damage checkpoints update: getting molecular. *Curr. Opin. Genet. De.v,* **8**, 185.
6. Stewart, E. and Enoch, T. (1996) S-phase and DNA damage checkpoints: a tale of two yeasts. *Curr. Opin. Cell Biol.,* **8**, 781.
7. Sipiczki, M. (1989) Taxonomy and Phylogenesis. In *Molecular biology of the fission yeast* (ed A. Nasim, P. Young, and B. F. Johnson), p. 431. Academic Press, San Diego.
8. Longhese, M. P., Foiani, M., Muzi-Falconi, M., Lucchini, G., and Plevani, P. (1998) DNA damage checkpoint in budding yeast. *EMBO J,* **17**, 5525.
9. Sandell, L. L. and Zakian, V. A. (1993) Loss of a yeast telomere: arrest, recovery, and chromosome loss. *Cell,* **75**, 729.
10. Lee, S. E., Moore, J. K., Holmes, A., Umezu, K., Kolodner, R. D., and Haber, J. E. (1998) *Saccharomyces* Ku70, mre11/rad50 and RPA proteins regulate adaptation to G2/M arrest after DNA damage. *Cell,* **94**, 399.
11. Toczyski, D. P., Galgoczy, D. J., and Hartwell, L. H. (1997) CDC5 and CKII control adaptation to the yeast DNA damage checkpoint. *Cell,* **90**, 1097.
12. Weinert, T. A., Kiser, G. L., and Hartwell, L. H. (1994) Mitotic checkpoint genes in budding yeast and the dependence of mitosis on DNA replication and repair. *Genes Dev.,* **8**, 652.

13. Weinert, T. A. and Hartwell, L. H. (1993) Cell cycle arrest of cdc mutants and specificity of the *RAD9* checkpoint. *Genetics*, **134**, 63.
14. Enoch, T., Carr, A. M., and Nurse, P. (1992) Fission yeast genes involved in coupling mitosis to completion of DNA replication. *Genes Dev.*, **6**, 2035.
15. al-Khodairy, F. and Carr, A. M. (1992) DNA repair mutants defining G2 checkpoint pathways in *Schizosaccharomyces pombe*. *EMBO J.*, **11**, 1343.
16. al-Khodairy, F., Fotou, E., Sheldrick, K. S., Griffiths, D. J., Lehmann, A. R., and Carr, A. M. (1994) Identification and characterization of new elements involved in checkpoint and feedback controls in fission yeast. *Mol. Biol. Cell*, **5**, 147.
17. Willson, J., Wilson, S., Warr, N., and Watts, F. Z. (1997) Isolation and characterization of the *Schizosaccharomyces pombe rhp9* gene: a gene required for the DNA damage checkpoint but not the replication checkpoint. *Nucleic Acids Res.*, **25**, 2138.
18. Rowley, R., Subramani, S., and Young, P. G. (1992) Checkpoint controls in *Schizosaccharomyces pombe: rad1*. *EMBO J.*, **11**, 1335.
19. Longhese, M. P., Paciotti, V., Fraschini, R., Zaccarini, R., Plevani, P., and Lucchini, G. (1997) The novel DNA damage checkpoint protein Ddc1p is phosphorylated periodically during the cell cycle and in response to DNA damage in budding yeast. *EMBO J.*, **16**, 5216.
20. Morrow, D. M., Tagle, D. A., Shiloh, Y., Collins, F. S., and Hieter, P. (1995) *TEL1*, an *S. cerevisiae* homolog of the human gene mutated in ataxia telangiectasia, is functionally related to the yeast checkpoint gene *MEC1*. *Cell*, **82**, 831.
21. Naito, T., Matsuura, A., and Ishikawa, F. (1998) Circular chromosome formation in a fission yeast mutant defective in two ATM homologues. *Nature Genet.*, **20**, 203.
22. Sanchez, Y., Desany, B. A., Jones, W. J., Liu, Q., Wang, B., and Elledge, S. J. (1996) Regulation of RAD53 by the ATM-like kinases MEC1 and TEL1 in yeast cell cycle checkpoint pathways. *Science*, **271**, 357.
23. Vialard, J. E., Gilbert, C. S., Green, C. M., and Lowndes, N. F. (1998) The budding yeast Rad9 checkpoint protein is subjected to Mec1/Tel1-dependent hyperphosphorylation and interacts with Rad53 after DNA damage. *EMBO J.*, **17**, 5679.
24. Hartley, K. O., Gell, D., Smith, G. C., Zhang, H., Divecha, N., Connelly, M. A., Admon, A., Lees-Miller, S. P., Anderson, C. W., and Jackson, S. P. (1995) DNA-dependent protein kinase catalytic subunit: a relative of phosphatidylinositol 3-kinase and the ataxia telangiectasia gene product. *Cell*, **82**, 849.
25. Lavin, M., Khanna, K., Beamish, H., Spring, K., Watters, D., and Shiloh, Y. (1995) Relationship of ataxia-telangiectasia protein ATM to phosphoinositide 3-kinase. *TIBS*, **20**, 382.
26. Lehmann, A. R. and Carr, A. M. (1995) The ataxia-telangiectasia gene: a link between checkpoint controls, neurodegeneration and cancer. *Trends Genet.*, **11**, 375.
27. Bentley, N., Holtzman, D. A., Flaggs, G., Keegan, K. S., DeMaggio, A., Ford, J. C., Hoekstra, M., and Carr, A. M. (1996) The *Schizosacharomyces pombe rad3*[+] checkpoint gene. *EMBO J.*, **15**, 6641.
28. Martinho, R. G., Lindsay, H. D., Flaggs, G., DeMaggio, A. J., Hoekstra, M. F., Carr, A. M., and Bentley, N. J. (1998) Analysis of Rad3 and Chk1 protein kinases defines different checkpoint responses. *EMBO J.*, **17**, 7239.
29. Lieber, M. R., Grawunder, U., Wu, X., and Yaneva, M. (1997) Tying loose ends: roles of Ku and DNA-dependent protein kinase in the repair of double-strand breaks. *Curr. Opin. Genet. Dev.*, **7**, 99.
30. Thelen, M. P., Onel, K., and Holloman, W. K. (1994) The *REC1* gene of *Ustilago maydis* involved in the cellular response to DNA damage encodes an exonuclease. *J. Biol. Chem.*, **269**, 747.

31. Kostrub, C. F., Knudsen, K., Subramani, S., and Enoch, T. (1998) Hus1p, a conserved fission yeast checkpoint protein, interacts with Rad1p and is phosphorylated in response to DNA damage. *EMBO J.*, **17**, 2055.

32. Paciotti, V., Lucchini, G., Plevani, P., and Longhese, M. P. (1998) Mec1p is essential for phosphorylation of the yeast DNA damage checkpoint protein Ddc1p, which physically interacts with Mec3p. *EMBO J.*, **17**, 4199.

33. Thelen, M. P., Venclovas, C., and Fidelis, K. (1999) A sliding clamp model for the Rad1 family of cell cycle checkpoint proteins. *Cell*, **96**, 769.

34. Aravind, L., Walker, D. R., and Koonin, E. V. (1999) Conserved domains in DNA repair proteins and evolution of repair systems. *Nucleic Acids Res.*, **27**, 1223.

35. Lydall, D. and Weinert, T. (1995) Yeast checkpoint genes in DNA damage processing: implications for repair and arrest. *Science*, **270**, 1488.

36. Humphrey, T. and Enoch, T. (1995) Keeping mitosis in check. *Curr. Biol.*, **5**, 376.

37. Nugent, C. I., Hughes, T. R., Lue, N. F., and Lundblad, V. (1996) Cdc13p: a single-strand telomeric DNA-binding protein with a dual role in yeast telomere maintenance. *Science*, **274**, 249.

38. Garvik, B., Carson, M., and Hartwell, L. (1995) Single-stranded DNA arising at telomeres in *cdc13* mutants may constitute a specific signal for the *RAD9* checkpoint. *Mol. Cell. Biol.*, **15**, 6128.

39. Onel, K., Koff, A., Bennett, R. L., Unrau, P., and Holloman, W. K. (1996) The *REC1* gene of *Ustilago maydis*, which encodes a $3' \rightarrow 5'$ exonuclease, couples DNA repair and completion of DNA synthesis to a mitotic checkpoint. *Genetics*, **143**, 165.

40. White, C. I. and Haber, J. E. (1990) Intermediates of recombination during mating type switching in *Saccharomyces cerevisiae*. *EMBO J.*, **9**, 663.

41. Holmes, A. M. and Haber, J. E. (1999) Double-strand break repair in yeast requires both leading and lagging strand DNA polymerases. *Cell*, **96**, 415.

42. Bhaumik, D. and Wang, T. S.F. (1998) Mutational effect of fission yeast polα on cell cycle events. *Mol. Biol. Cell*, **9**, 2107.

43. D'Urso, G., Grallert, B., and Nurse, P. (1995) DNA polymerase alpha, a component of the replication initiation complex, is essential for the checkpoint coupling S phase to mitosis in fission yeast. *J. Cell Sci.*, **108**, 3109.

44. Reynolds, N., Fantes, P. A., and MacNeill, S. A. (1999) A key role for replication factor C in DNA replication checkpoint function in fission yeast. *Nucleic Acids Res.*, **27**, 462.

45. Saka, Y. and Yanagida, M. (1993) Fission yeast *cut5*$^+$, required for S phase onset and M phase restraint, is identical to the radiation-damage repair gene *rad4*$^+$. *Cell*, **74**, 383.

46. Kelly, T. J., Martin, G. S., Forsburg, S. L., Stephen, R. J., Russo, A., and Nurse, P. (1993) The fission yeast *cdc18+* gene product couples S phase to START and mitosis. *Cell*, **74**, 371.

47. Grallert, B. and Nurse, P. (1996) The *ORC1* homolog *orp1* in fission yeast plays a key role in regulating onset of S phase. *Genes Dev.*, **10**, 2644.

48. Leatherwood, J., Lopez-Girona, A., and Russell, P. (1996) Interaction of Cdc2 and Cdc18 with a fission yeast ORC2-like protein. *Nature*, **379**, 360.

49. Parker, A. E., Clyne, R. K., Carr, A. M., and Kelly, T. J. (1997) The *Schizosaccharomyces pombe rad11*$^+$ gene encodes the large subunit of replication protein A. *Mol. Cell. Biol.*, **17**, 2381.

50. Noskov, V. N., Araki, H., and Sugino, A. (1998) The *RFC2* gene, encoding the third-largest subunit of the replication factor C complex, is required for an S-phase checkpoint in *Saccharomyces cerevisiae*. *Mol. Cell. Biol.*, **18**, 4914.

51. Sugimoto, K., Shimomura, T., Hashimoto, K., Araki, H., Sugino, A., and Matsumoto, K. (1996) Rfc5, a small subunit of replication factor C complex, couples DNA replication and mitosis in budding yeast. *Proc. Natl. Acad. Sci. U.S.A.*, **93**, 7048.

52. Shimomura, T., Ando, S., Matsumoto, K., and Sugimoto, K. (1998) Functional and physical interaction between Rad24 and Rfc5 in the yeast checkpoint pathways. *Mol. Cell. Biol.*, **18**, 5485.

53. Sugimoto, K., Ando, S., Shimomura, T., and Matsumoto, K. (1997) Rfc5, a replication factor C component, is required for regulation of Rad53 protein kinase in the yeast checkpoint pathway. *Mol. Cell. Biol.*, **17**, 5905.

54. D'Urso, G. and Nurse, P. (1997) *Schizosaccharomyces pombe cdc20$^+$* encodes DNA polymerase epsilon and is required for chromosomal replication but not for the S phase checkpoint. *Proc. Natl. Acad. Sci. U.S.A.*, **94**, 12491.

55. Waseem, N. H., Labib, K., Nurse, P., and Lane, D. P. (1992) Isolation and analysis of the fission yeast gene encoding polymerase delta accessory protein PCNA. *EMBO J.*, **11**, 5111.

56. Francesconi, S., Park, H., and Wang, T. S. (1993) Fission yeast with DNA polymerase delta temperature-sensitive alleles exhibits cell division cycle phenotype. *Nucleic Acids Res.*, **21**, 3821.

57. Taylor, R. M., Wickstead, B., Cronin, S., and Caldecott, K. W. (1998) Role of a BRCT domain in the interaction of DNA ligase III-alpha with the DNA repair protein XRCC1. *Curr. Biol.*, **8**, 877.

58. Saka, Y., Fantes, P., Sutani, T., McInerny, C., Creanor, J., and Yanagida, M. (1994) Fission yeast *cut5$^+$* links nuclear chromatin and M phase regulator in the replication checkpoint control. *EMBO J.*, **13**, 5319.

59. Saka, Y., Esashi, F., Matsusaka, T., Mochida, S., and Yanagida, M. (1997) Damage and replication checkpoint control in fission yeast is ensured by interactions of Crb2, a protein with BRCT motif, with Cut5 and Chk1. *Genes Dev.*, **11**, 3387.

60. Araki, H., Leem, S. H., Phongdara, A., and Sugino, A. (1995) Dpb11, which interacts with DNA polymerase II(ε) in *Saccharomyces cerevisiae*, has a dual role in S-phase progression and at a cell cycle checkpoint. *Proc. Natl. Acad. Sci. U.S.A.*, **92**, 11791.

61. Wang, H. and Elledge, S. J. (1999) DRC1, DNA replication and checkpoint protein 1, functions with DPB11 to control DNA replication and the S-phase checkpoint in *Saccharomyces cerevisiae*. *Proc. Natl. Acad. Sci. U.S.A.*, **96**, 3824.

62. Allen, J. B., Zhou, Z., Siede, W., Friedberg, E. C., and Elledge, S. J. (1994) The SAD1/RAD53 protein kinase controls multiple checkpoints and DNA damage-induced transcription in yeast. *Genes Dev.*, **8**, 2401.

63. Longhese, M. P., Fraschini, R., Plevani, P., and Lucchini, G. (1996) Yeast *pip3/mec3* mutants fail to delay entry into S phase and to slow DNA replication in response to DNA damage, and they define a functional link between Mec3 and DNA primase. *Mol. Cell. Biol.*, **16**, 3235.

64. Navas, T. A., Zhou, Z., and Elledge, S. J. (1995) DNA polymerase epsilon links the DNA replication machinery to the S phase checkpoint. *Cell*, **80**, 29.

65. Sun, Z., Fay, D. S., Marini, F., Foiani, M., and Stern, D. F. (1996) Spk1/Rad53 is regulated by Mec1-dependent protein phosphorylation in DNA replication and damage checkpoint pathways. *Genes Dev.*, **10**, 395.

66. Emili, A. (1998) MEC1-dependent phosphorylation of Rad9p in response to DNA damage. *Mol. Cell*, **2**, 183.

67. Sun, Z., Hsiao, J., Fay, D. S., and Stern, D. F. (1998) Rad53 FHA domain associated with phosphorylated Rad9 in the DNA damage checkpoint. *Science*, **281**, 272.

68. Pati, D., Keller, C., Groudine, M., and Plon, S. E. (1997) Reconstitution of a *MEC1*-independent checkpoint in yeast by expression of a novel human fork head cDNA. *Mol. Cell. Biol.,* **17**, 3037.

69. Murakami, H. and Okayama, H. (1995) A kinase from fission yeast responsible for blocking mitosis in S phase. *Nature,* **374**, 817.

70. Walworth, N., Davey, S., and Beach, D. (1993) Fission yeast chk1 protein kinase links the rad checkpoint pathway to cdc2. *Nature,* **363**, 368.

71. Lindsay, H. D., Griffiths, D. J., Edwards, R. J., Christensen, P. U., Murray, J. M., Osman, F., Walworth, N., and Carr, A. M. (1998) S-phase-specific activation of Cds1 kinase defines a subpathway of the checkpoint response in *Schizosaccharomyces pombe. Genes Dev.,* **12**, 382.

72. Zeng, Y., Chrispell Forbes, K. L., Wu, Z., Moreno, S., Piwnica-Worms, H., and Enoch, T. (1998) Replication checkpoint requires phosphorylation of the phosphatase Cdc25 by Cds1 or Chk1. *Nature,* **395**, 507.

73. Chrispell Forbes, K., Humphrey, T., and Enoch, T. (1998) Suppressors of cdc25p over-expression identify two pathways that influence the G2/M checkpoint in fission yeast. *Genetics,* **150**, 1361.

74. Boddy, M. N., Furnari, B., Mondesert, O., and Russell, P. (1998) Replication checkpoint enforced by kinases Cds1 and Chk1. *Science,* **280**, 909.

75. Walworth, N. C. and Bernards, R. (1996) rad-dependent response of the chk1-encoded protein kinase at the DNA damage checkpoint. *Science,* **271**, 353.

76. Russell, P. (1998) Checkpoints on the road to mitosis. *Trends Biochem. Sci.,* **23**, 399.

77. Gould, K. L. and Nurse, P. (1989) Tyrosine phosphorylation of the fission yeast $cdc2^+$ protein kinase regulates entry into mitosis. *Nature,* **342**, 39.

78. Rhind, N., Furnari, B., and Russell, P. (1997) Cdc2 tyrosine phosphorylation is required for the DNA damage checkpoint in fission yeast. *Genes Dev.,* **11**, 504.

79. Rhind, N. and Russell, P. (1998) Tyrosine phosphorylation of cdc2 is required for the replication checkpoint in *Schizosaccharomyces pombe. Mol. Cell. Biol.,* **18**, 3782.

80. Enoch, T. and Nurse, P. (1990) Mutation of fission yeast cell cycle control genes abolishes dependence of mitosis on DNA replication. *Cell,* **60**, 665.

81. Sheldrick, K. S. and Carr, A. M. (1993) Feedback controls and G2 checkpoints: fission yeast as a model system. *Bioessays,* **15**, 775.

82. Lundgren, K., Walworth, N., Booher, R., Dembski, M., Kirschner, M., and Beach, D. (1991) *mik1* and *wee1* cooperate in the inhibitory tyrosine phosphorylation of *cdc2. Cell,* **64**, 1111.

83. Peng, C. Y., Graves, P. R., Thoma, R. S., Wu, Z., Shaw, A. S., and Piwnica-Worms, H. (1997) Mitotic and G2 checkpoint control: regulation of 14-3-3 protein binding by phosphorylation of Cdc25C on serine-216. *Science,* **277**, 1501.

84. Muslin, A. J., Tanner, J. W., Allen, P. M., and Shaw, A. S. (1996) Interaction of 14-3-3 with signaling proteins is mediated by the recognition of phosphoserine. *Cell,* **84**, 889.

85. Yaffe, M., Rittinger, K., Volinia, S., Caron, P., Aitken, A., Leffers, H., Gamblin, S., Smerdon, S., and Cantley, L. (1997) The structural basis for 14-3-3:phosphopeptide binding specificity. *Cell,* **91**, 961.

86. Wang, W. and Shakes, D. C. (1996) Molecular evolution of the 14-3-3 protein family. *J. Mol. Evol.,* **43**, 384.

87. Ford, J. C., al-Khodairy, F., Fotou, E., Sheldrick, K. S., Griffiths, D. J., and Carr, A. M. (1994) 14-3-3 protein homologs required for the DNA damage checkpoint in fission yeast. *Science,* **265**, 533.

88. Lopez-Girona, A., Furnari, B., Mondesert, O., and Russell, P. (1999) Nuclear localization of Cdc25 is regulated by DNA damage and a 14-3-3 protein. *Nature,* **397**, 172.

89. O'Connell, M. J., Raleigh, J. M., Verkade, H. M., and Nurse, P. (1997) Chk1 is a wee1 kinase in the G2 DNA damage checkpoint inhibiting cdc2 by Y15 phosphorylation. *EMBO J.*, **16**, 545.

90. Sorger, P. K. and Murray, A. W. (1992) S-phase feedback control in budding yeast independent of tyrosine phosphorylation of p34cdc28. *Nature*, **355**, 365.

91. Amon, A., Surana, U., Muroff, I., and Nasmyth, K. (1992) Regulation of p34CDC28 tyrosine phosphorylation is not required for entry into mitosis in *S. cerevisiae*. *Nature*, **355**, 368.

92. Hwang, L. H., Lau, L. F., Smith, D. L., Mistrot, C. A., Hardwick, K. G., Hwang, E. S., Amon, A., and Murray, A. W. (1998) Budding yeast Cdc20: a target of the spindle checkpoint. *Science*, **279**, 1041.

93. Townsley, F. M. and Ruderman, J. V. (1998) Proteolytic ratchets that control progression through mitosis. *Trends Cell Biol.*, **8**, 238.

94. Visintin, R., Prinz, S., and Amon, A. (1997) *CDC20* and *CDH1*: a family of substrate-specific activators of APC-dependent proteolysis. *Science*, **278**, 460.

95. Yamamoto, A., Guacci, V., and Koshland, D. (1996) Pds1p, an inhibitor of anaphase in budding yeast, plays a critical role in the APC and checkpoint pathway(s). *J. Cell Biol.*, **133**, 99.

96. Cohen-Fix, O., Peters, J. M., Kirschner, M. W., and Koshland, D. (1996) Anaphase initiation in *Saccharomyces cerevisiae* is controlled by the APC-dependent degradation of the anaphase inhibitor Pds1p. *Genes Dev.*, **10**, 3081.

97. Lim, H. H. and Surana, U. (1996) Cdc20, a beta-transducin homologue, links *RAD9*-mediated G2/M checkpoint control to mitosis in *Saccharomyces cerevisiae*. *Mol. Gen. Genet.*, **253**, 138.

98. Cohen-Fix, O. and Koshland, D. (1997) The anaphase inhibitor of *Saccharomyces cerevisiae* Pds1p is a target of the DNA damage checkpoint pathway. *Proc. Natl. Acad. Sci. U.S.A.*, **94**, 14361.

99. Rhind, N. and Russell, P. (1998) The *Schizosaccharomyces pombe* S-phase checkpoint differentiates between different types of DNA damage. *Genetics*, **149**, 1729.

100. Sidorova, J. M. and Breeden, L. L. (1997) Rad53-dependent phosphorylation of Swi6 and down-regulation of CLN1 and CLN2 transcription occur in response to DNA damage in *Saccharomyces cerevisiae*. *Genes Dev.*, **11**, 3032.

101. Paulovich, A. G., Toczyski, D. P., and Hartwell, L. H. (1997) When checkpoints fail. *Cell*, **88**, 315.

102. Paulovich, A. G. and Hartwell, L. H. (1995) A checkpoint regulates the rate of progression through S phase in *S. cerevisiae* in response to DNA damage. *Cell*, **82**, 841.

103. Paulovich, A. G., Margulies, R. U., Garvik, B. M., and Hartwell, L. H. (1997) RAD9, RAD17, and RAD24 are required for S phase regulation in *Saccharomyces cerevisiae* in response to DNA damage. *Genetics*, **145**, 45.

104. Longhese, M. P., Neecke, H., Paciotti, V., Lucchini, G., and Plevani, P. (1996) The 70 kDa subunit of replication protein A is required for the G1/S and intra-S DNA damage checkpoints in budding yeast. *Nucleic Acids Res.*, **24**, 3533.

105. Marini, F., Pellicioli, A., Paciotti, V., Lucchini, G., Plevani, P., Stern, D. F., and Foiani, M. (1997) A role for DNA primase in coupling DNA replication to DNA damage response. *EMBO J.*, **16**, 639.

106. Brush, G. S., Morrow, D. M., Hieter, P., and Kelly, T. J. (1996) The ATM homologue MEC1 is required for phosphorylation of replication protein A in yeast. *Proc. Natl. Acad. Sci. U.S.A.*, **93**, 15075.

107. Santocanale, C. and Diffley, J. F. (1998) A Mec1- and Rad53-dependent checkpoint controls late-firing origins of DNA replication. *Nature,* **395**, 615.

108. Shirahige, K., Hori, Y., Shiraishi, K., Yamashita, M., Takahashi, K., Obuse, C., Tsurimoto, T., and Yoshikawa, H. (1998) Regulation of DNA-replication origins during cell-cycle progression. *Nature,* **395**, 618.

109. Uchiyama, M., Galli, I., Griffiths, D. J., and Wang, T. S. (1997) A novel mutant allele of *Schizosaccharomyces pombe rad26* defective in monitoring S-phase progression to prevent premature mitosis. *Mol. Cell. Biol.,* **17**, 3103.

110. Friedberg, E. C., Walker, G., and Siede, W. (1995) *DNA repair and mutagenesis.* ASM Press, Washington, DC.

111. Aboussekhra, A., Vialard, J. E., Morrison, D. E., de la Torre-Ruiz, M. A., Cernakova, L., Fabre, F., and Lowndes, N. F. (1996) A novel role for the budding yeast *RAD9* checkpoint gene in DNA damage-dependent transcription. *EMBO J.,* **15**, 3912.

112. Huang, M. and Elledge, S. J. (1997) Identification of RNR4, encoding a second essential small subunit of ribonucleotide reductase in *Saccharomyces cerevisiae. Mol. Cell. Biol.,* **17**, 6105.

113. Huang, M., Zhou, Z., and Elledge, S. J. (1998) The DNA replication and damage checkpoint pathways induce transcription by inhibition of the Crt1 repressor. *Cell,* **94**, 595.

114. Lee, J. K., Park, E. J., Chung, H. K., Hong, S. H., Joe, C. O., and Park, S. D. (1994) Isolation of UV-inducible transcripts from *Schizosaccharomyces pombe. Biochem. Biophys. Res. Commun.,* **202**, 1113.

115. Fernandez Sarabia, M. J., McInerny, C., Harris, P., Gordon, C., and Fantes, P. (1993) The cell cycle genes *cdc22*⁺ and *suc22*⁺ of the fission yeast *Schizosaccharomyces pombe* encode the large and small subunits of ribonucleotide reductase. *Mol. Gen. Genet.,* **238**, 241.

116. Taylor, E. M., McFarlane, R. J., and Price, C. (1996) 5-Azacytidine treatment of the fission yeast leads to cytotoxicity and cell cycle arrest. *Mol. Gen. Genet.,* **253**, 128.

117. Harris, P., Kersey, P. J., McInerny, C. J., and Fantes, P. A. (1996) Cell cycle, DNA damage and heat shock regulate *suc22*⁺ expression in fission yeast. *Mol. Gen. Genet.,* **252**, 284.

118. Stewart, E., Chapman, C. R., Al-Khodairy, F., Carr, A. M., and Enoch, T. (1997) *rqh1*⁺, a fission yeast gene related to the Bloom's and Werner's syndrome genes, is required for reversible S phase arrest. *EMBO J.,* **16**, 2682.

119. Murray, J. M., Lindsay, H. D., Munday, C. A., and Carr, A. M. (1997) Role of *Schizosaccharomyces pombe* RecQ homolog, recombination, and checkpoint genes in UV damage tolerance. *Mol. Cell. Biol.,* **17**, 6868.

120. Desany, B. A., Alcasabas, A. A., Bachant, J. B., and Elledge, S. J. (1998) Recovery from DNA replicational stress is the essential function of the S-phase checkpoint pathway. *Genes Dev.,* **12**, 2956.

121. Zhao, X., Muller, E. G., and Rothstein, R. (1998) A suppressor of two essential checkpoint genes identifies a novel protein that negatively affects dNTP pools. *Mol. Cell,* **2**, 329.

122. Vallen, E. A. and Cross, F. R. (1999) Interaction between the MEC1-dependent DNA synthesis checkpoint and G1 cyclin function in *Saccharomyces cerevisiae. Genetics,* **151**, 459.

123. Hardwick, K. G. (1998) The spindle checkpoint. *TIG,* **14**, 1.

124. Straight, A. (1997) Checkpoint proteins and kinetochores. *Curr. Biol.,* **7**, R613.

125. Nicklas, R. B. (1997) How cells get the right chromosomes. *Science,* **275**, 632.

126. Waters, J. C., Chen, R. H., Murray, A. W., and Salmon, E. D. (1998) Localization of Mad2 to kinetochores depends on microtubule attachment, not tension. *J. Cell Biol.,* **141**, 1181.

127. Li, R., and Murray, A. (1991) Feedback control of mitosis in budding yeast. *Cell,* **66**, 519.

128. Hoyt, M. A., Totis, L., and Roberts, B. T. (1991) *S. cerevisiae* genes required for cell cycle arrest in response to loss of microtubule function. *Cell*, **66**, 507.

129. Roberts, R. T., Farr, K. A., and Hoyt, M. A. (1994) The *Saccharomyces cerevisiae* checkpoint mutant *BUB1* encodes a novel protein kinase. *Mol. Cell. Biol.*, **14**, 8282.

130. Weiss, E. and Winey, M. (1996) The *Saccharomyces cerevisiae* spindle pole body duplication gene MPS1 is part of a mitotic checkpoint. *J. Cell Biol.*, **132**, 111.

131. Murone, M. and Simanis, V. (1996) The fission yeast *dma1* gene is a component of the spindle assembly checkpoint, required to prevent septum formation and premature exit from mitosis if spindle function is compromised. *EMBO J.*, **15**, 6605.

132. Hardwick, K. G., Weiss, E., Luca, F. C., Winey, M., and Murray, A. W. (1996) Activation of the budding yeast spindle assembly checkpoint without mitotic spindle disruption. *Science*, **273**, 953.

133. He, X., Jones, M. H., Winey, M., and Sazer, S. (1998) Mph1, a member of the Mps1-like family of dual specificity protein kinases, is required for the spindle checkpoint in *Schizosaccharomyces pombe. J. Cell Sci.*, **111**, 1635.

134. He, X., Patterson, T. E., and Sazer, S. (1997) The *Schizosaccharomyces pombe* spindle checkpoint protein mad2p blocks anaphase and genetically interacts with the anaphase-promoting complex. *Proc. Natl. Acad. Sci. U.S.A.*, **94**, 7965.

135. Farr, K. A. and Hoyt, M. A. (1998) Bub1p kinase activates the *Saccharomyces cerevisiae* spindle assembly checkpoint. *Mol. Cell. Biol.*, **18**, 2738.

136. Chen, R. H., Waters, J. C., Salmon, E. D., and Murray, A. W. (1996) Association of spindle assembly checkpoint component XMAD2 with unattached kinetochores. *Science*, **274**, 242.

137. Taylor, S. S. and McKeon, F. (1997) Kinetochore localization of murine Bub1 is required for normal mitotic timing and checkpoint response to spindle damage. *Cell*, **89**, 727.

138. Li, Y. and Benezra, R. (1996) Identification of a human mitotic checkpoint gene: *hsMAD2*. *Science*, **274**, 246.

139. Kallio, M., Weinstein, J., Daum, J. R., Burke, D. J., and Gorbsky, G. J. (1998) Mammalian p55CDC mediates association of the spindle checkpoint protein Mad2 with the cyclosome/anaphase-promoting complex, and is involved in regulating anaphase onset and late mitotic events. *J. Cell Biol.*, **141**, 1393.

140. Kim, S., Lin, D., Matsumoto, S., Kitazono, A., and Matsumoto, T. (1998) Fission yeast Slp1: an effector of the Mad2-dependent spindle checkpoint. *Science*, **279**, 1045.

141. Weinstein, J., Jacobsen, F. W., Hsu-Chen, J., Wu, T., and Baum, L. G. (1994) A novel mammalian protein, p55CDC, present in dividing cells is associated with protein kinase activity and has homology to the *Saccharomyces cerevisiae* cell division cycle proteins Cdc20 and Cdc4. *Mol. Cell. Biol.*, **14**, 3350.

142. Michaelis, C., Ciosk, R., and Nasmyth, K. (1997) Cohesins: chromosomal proteins that prevent premature separation of sister chromatids. *Cell*, **91**, 35.

143. Ciosk, R., Zachariae, W., Michaelis, C., Shevchenko, A., Mann, M., and Nasmyth, K. (1998) An ESP1/PDS1 complex regulates loss of sister chromatid cohesion at the metaphase to anaphase transition in yeast. *Cell*, **93**, 1067.

144. Kumada, K., Nakamura, T., Nagao, K., Funabiki, H., Nakagawa, T., and Yanagida, M. (1998) Cut1 is loaded onto the spindle by binding to Cut2 and promotes anaphase spindle movement upon Cut2 proteolysis. *Curr. Biol.*, **8**, 633.

145. Juang, Y. L., Huang, J., Peters, J. M., McLaughlin, M. E., Tai, C. Y., and Pellman, D. (1997) APC-mediated proteolysis of Ase1 and the morphogenesis of the mitotic spindle. *Science*, **275**, 1311.

146. Schwab, M., Lutum, A. S., and Seufert, W. (1997) Yeast Hct1 is a regulator of Clb2 cyclin proteolysis. *Cell,* **90**, 683.

147. Schott, E. J. and Hoyt, M. A. (1998) Dominant alleles of *Saccharomyces cerevisiae CDC20* reveal its role in promoting anaphase. *Genetics,* **148**, 599.

148. Volkmer, E. and Karnitz, L. M. (1999) Human homologs of *Schizosaccharomyces pombe rad1, hus1,* and *rad9* form a DNA damage-responsive protein complex. *J. Biol. Chem.,* **274**, 567.

149. Matsuoka, S., Huang, M., and Elledge, S. J. (1998) Linkage of ATM to cell cycle regulation by the Chk2 protein kinase. *Science,* **282**, 1893.

150. Sanchez, Y., Wong, C., Thoma, R. S., Richman, R., Wu, Z., Piwnica-Worms, H., and Elledge, S. J. (1997) Conservation of the Chk1 checkpoint pathway in mammals: linkage of DNA damage to Cdk regulation through Cdc25. *Science,* **277**, 1497.

151. Blasina, A., de Weyer, I. V., Laus, M. C., Luyten, W., Parker, A. E., and McGowan, C. H. (1999) A human homologue of the checkpoint kinase Cds1 directly inhibits Cdc25 phosphatase. *Curr. Biol.,* **9**, 1.

152. Lengauer, C., Kinzler, K. W., and Vogelstein, B. (1998) Genetic instabilities in human cancers. *Nature,* **396**, 643.

153. Zhang, H., Tombline, G., and Weber, B. L. (1998) BRCA1, BRCA2, and DNA damage response: collision or collusion? *Cell,* **92**, 433.

154. Bork, P., Hofmann, K., Bucher, P., Neuwald, A. F., Altschul, S. F., and Koonin, E. V. (1997) A superfamily of conserved domains in DNA damage-responsive cell cycle checkpoint proteins. *FASEB J.,* **11**, 68.

155. Hartwell, L. H., Szankasi, P., Roberts, C. J., Murray, A. W., and Friend, S. H. (1997) Integrating genetic approaches into the discovery of anticancer drugs. *Science,* **278**, 1064.

156. Woollard, A. and Nurse, P. (1995) G1 regulation and checkpoints operating around START in fission yeast. *Bioessays,* **17**, 481.

157. Stern, B. and Nurse, P. (1996) A quantitative model for the cdc2 control of S phase and mitosis in fission yeast. *Trends Genet.,* **12**, 345.

158. Muhua, L., Adames, N. R., Murphy, M. D., Shields, C. R., and Cooper, J. A. (1998) A cytokinesis checkpoint requiring the yeast homologue of an APC-binding protein. *Nature,* **393**, 487.

159. McMillan, J. N., Sia, R. A.L., and Lew, D. J. (1998) A morphogenesis checkpoint monitors the actin cytoskeleton in yeast. *J. Cell Biol.,* **142**, 1487.

160. Mendenhall, M. D. and Hodge, A. E. (1998) Regulation of Cdc28 cyclin-dependent protein kinase activity during the cell cycle of the yeast *Saccharomyces cerevisiae. Microbiol. Mol. Biol. Rev.,* **62**, 1191.

161. Parker, A. E., Van de Weyer, I., Laus, M. C., Oostveen, I., Yon, J., Verhasselt, P., and Luyten, W. H. (1998) A human homologue of the *Schizosaccharomyces pombe rad1*[+] checkpoint gene encodes an exonuclease. *J. Biol. Chem.,* **273**, 18332.

162. Freire, R., Murguia, J. R., Tarsounas, M., Lowndes, N. F., Moens, P. B., and Jackson, S. P. (1998) Human and mouse homologs of *Schizosaccharomyces pombe rad1*[+] and *Saccharomyces cerevisiae RAD17*: linkage to checkpoint control and mammalian meiosis. *Genes Dev.,* **12**, 2560.

163. Griffiths, D. J.F., Garbet, N. C., McCready, S., Lehmann, A. R., and Carr, A. M. (1995) Fission yeast *rad17*: a homologue of budding yeast *RAD24* that shares regions of sequence similarity with DNA polymerase accessory proteins. *EMBO J.,* **14**, 5812.

164. Parker, A. E., Van de Weyer, I., Laus, M. C., Verhasselt, P., and Luyten, W. H. (1998) Identification of a human homologue of the *Schizosaccharomyces pombe rad17*[+] checkpoint gene. *J. Biol. Chem.,* **273**, 18340.

165. Lieberman, H. B., Hopkins, K. M., Nass, M., Demetrick, D., and Davey, S. (1996) A human homolog of the *Schizosaccharomyces pombe rad9+* checkpoint control gene. *Proc. Natl. Acad. Sci. U.S.A.*, **93**, 13890.

166. Kostrub, C. F., Al-Khodairy, F., Ghazizadeh, H., Carr, A. M., and Enoch, T. (1997) Molecular analysis of *hus1+*, a fission yeast gene required for S-M and DNA damage check points. *Mol. Gen. Genet.*, **254**, 389.

167. Kato, R. and Ogawa, H. (1994) An essential gene, *ESR1*, is required for mitotic cell growth, DNA repair and meiotic recombination in *Saccharomyces cerevisiae*. *Nucleic Acids Res.*, **22**, 3104.

168. Greenwell, P. W., Kronmal, S. L., Porter, S. E., Gassenhuber, J., Obermaier, B., and Petes, T. D. (1995) *TEL1*, a gene involved in controlling telomere length in *S. cerevisiae*, is homologous to the human ataxia telangiectasia gene. *Cell*, **82**, 823.

169. Roberts, R. L., Mosch, H. U., and Fink, G. R. (1997) 14-3-3 proteins are essential for RAS/MAPK cascade signaling during pseudohyphal development in *S. cerevisiae*. *Cell*, **89**, 1055.

170. Hardwick, K. G. and Murray, A. W. (1995) Mad1p, a phosphoprotein component of the spindle assembly checkpoint in budding yeast. *J. Cell Biol.*, **131**, 709.

171. Bernard, P., Hardwick, K., and Javerzat, J. P. (1998) Fission yeast bub1 is a mitotic centromere protein essential for the spindle checkpoint and the preservation of correct ploidy through mitosis. *J. Cell Biol.*, **143**, 1775.

172. Fankhauser, C., Marks, J., Reymond, A., and Simanis, V. (1993) The *Schizosaccharomyces pombe cdc16* gene is required both for maintenance of p34cdc2 kinase activity and regulation of septum formation: a link between mitosis and cytokinesis? *EMBO J.*, **12**, 2697.

173. Taylor, S. S., Ha, E., and McKeon, F. (1998) The human homologue of Bub3 is required for kinetochore localization of Bub1 and a Mad3/Bub1-related protein kinase. *J. Cell Biol.*, **142**, 1.

5 | Mechanisms of nuclear division

HIROHISA MASUDA and YASUSHI HIRAOKA

1. Introduction

Mitotic nuclear division is the process that ensures proper segregation of duplicated copies of the genome at cell division. The orderly segregation of genetic information depends on its reorganization into condensed chromosomes, and on the assembly and functioning of the mitotic spindle. The mitotic spindle is the machinery that segregates sister chromatids into the two daughter nuclei at mitosis. These events are precisely regulated to produce genetically identical daughter cells.

In contrast, meiotic nuclear division is the process that generates genetic diversity in the offspring. Recombination of genetic information from cells with opposite mating types occurs at meiotic prophase. Meiosis encompasses two successive nuclear divisions: the first division involves cohesion of sister chromatids and disjunction of homologous chromosomes, while the second division disjoins sister chromatids in a process similar to mitotic division.

In this chapter we first describe mechanisms of mitotic nuclear division with an emphasis on the mechanistic aspect of the mitotic spindle and chromosomes; next, we describe mechanisms of meiotic nuclear division with a particular emphasis on chromosome organization specific to meiosis.

2. Mitosis

2.1 Overview of mitotic nuclear division

The mitotic spindle is organized from nuclear microtubules that are nucleated at the spindle pole body (SPB), the microtubule organizing centre embedded into the nuclear envelope (1–5). The mitotic events that occur in the yeast nucleus seem similar to those found in higher eukaryotes, although the nuclear envelope in yeast cells does not break down at mitosis but persists through the cell cycle (Fig. 1). Two sets of microtubules nucleated by the duplicated, opposing SPBs interdigitate in the spindle midzone to form a bipolar spindle. Duplicated and condensed chromosomes attach via their centromere/kinetochore regions to microtubules emanating from each pole

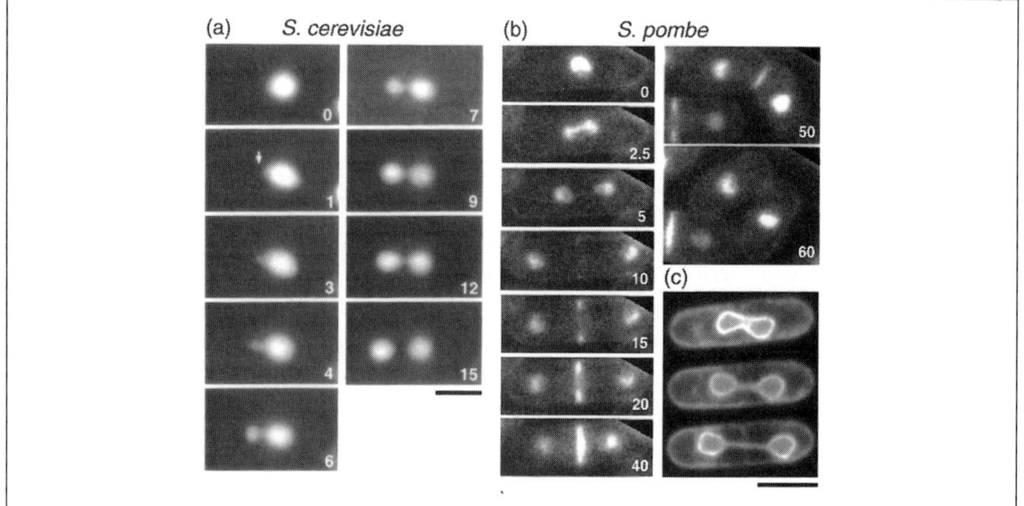

Fig. 1 Live observations of mitotic nuclear division. (A) Nuclear division of a *S. cerevisiae* cell. Cells expressing a green fluorescent protein (GFP)-tagged version of nuclcoplasmin, a nuclear protein, were observed at 30°C on a fluorescence microscope stage. The arrow indicates the position of the bud neck. (B) Nuclear and cell division of a *S. pombe* cell stained with Hoechst 33342, a DNA dye. In addition to chromosomes, the septum is also stained with the dye. (C) Nuclear division of *S. pombe* cells expressing a GFP-tagged NADPH–cytochrome P450 reductase. The nuclear envelope, the cell membrane, and endoplasmic reticulum are visible in the cell (D. Q. Ding and Y. Hiraoka, unpublished observations). Numbers at the right corner of panels A and B represent time in minutes. Bar, 5 μm.

of the mitotic spindle. Sister chromatids disjoin at the onset of anaphase, and are separated from each other by their movement to the spindle poles (anaphase A) and by elongation of the spindle that they are attached to (anaphase B). Microtubule-based motor proteins mediate both formation and elongation of the spindle. As the spindle elongates, the nucleus elongates and acquires a dumbbell shape with separating chromosomes at both ends. The spindle then elongates to cell ends and disassembles, finally the nucleus is divided into two and segregates in the daughter cells.

Some differences in nuclear division do exist between *Schizosaccharomyces pombe* and *Saccharomyces cerevisiae*. *S. pombe* cells divide by 'fission' at the cell centre, whereas *S. cerevisiae* cells divide by 'budding'; therefore the nucleus has to move from the mother cell through a narrow bud neck into the daughter cell. In *S. cerevisiae*, positioning of the nucleus to the bud neck and orientation of the spindle along the bud–mother axis is important for distributing chromosomes equally into the mother and daughter cells. In addition, in *S. pombe* activation of cyclin-dependent kinase 1 (CDK1) (Cdc2) at the end of G2 phase induces chromosome condensation and spindle formation, both being the landmark events of the onset of M phase (see Chapter 3). In contrast, the onset of M phase is not clear in *S. cerevisiae*. At present it is not known precisely when spindle formation and chromosome condensation occur relative to DNA replication and the onset of anaphase, nor how they are regulated during the cell cycle. The spindle is formed at a point between late S phase and early G2 phase.

Chromosome condensation occurs at a point between the end of S phase and the onset of anaphase, and may be regulated independently of spindle formation. Therefore we refer to the period between the end of S phase and the onset of anaphase as the 'G2–M phase'.

Here the nuclear division machinery, whose function and assembly lead to nuclear division, consists of nuclear and cytoplasmic microtubules, the SPB, microtubule-based motor proteins and the chromosomes. We will first review the microtubule organization and the behaviour of the nucleus during the mitotic cell cycle and then describe roles for components of the machinery in nuclear division.

2.2 Microtubule organization and nuclear behaviour during the mitotic cell cycle

Figure 2 illustrates microtubule organization and the behaviour of the nucleus during the yeast cell cycle. Immunofluorescence microscopy (6–8), live observation of spindles and the nucleus using digitally enhanced and video-enhanced differential interference contrast microscopy (9), and live observation of the spindle, the SPB, or cytoplasmic microtubules using green fluorescent protein (GFP)-tagging techniques (10–13) have revealed the dynamic property of spindle and cytoplasmic microtubules and the behaviour of the nucleus during the cell cycle.

In *S. cerevisiae*, nuclear and cytoplasmic microtubules are nucleated exclusively

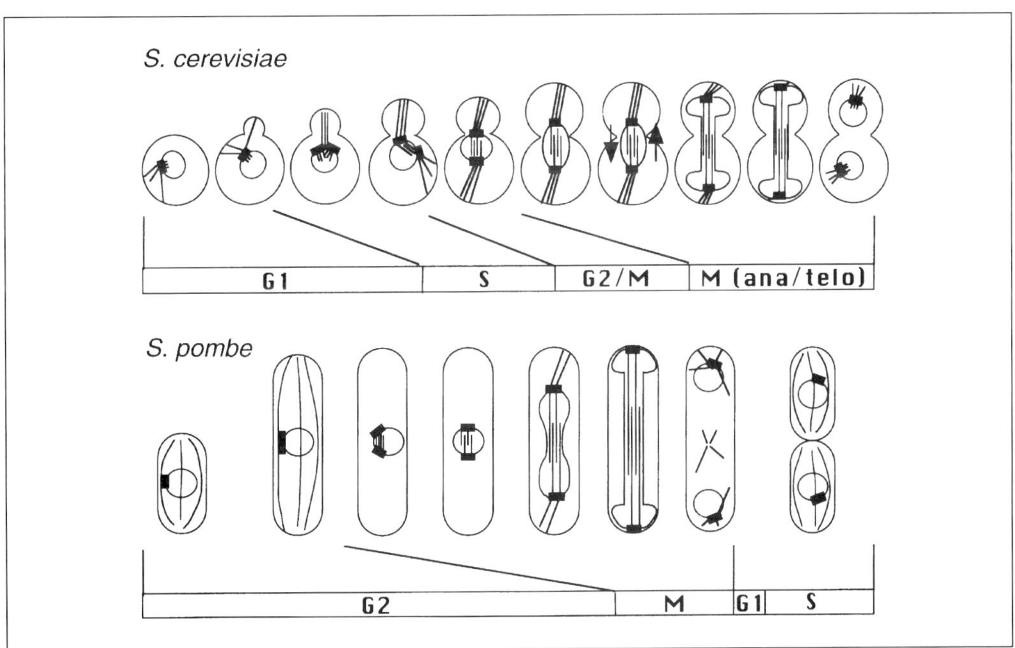

Fig. 2 Microtubule organization and behaviour of the nucleus during the mitotic cell cycle. The solid box represents the SPB. See text for detail.

from the SPB throughout the cell cycle. Cytoplasmic microtubules are also called astral microtubules. In unbudded G1 cells, astral microtubules are nucleated from the single SPB, and push against the cell cortex to propel the nucleus around the cell interior. Initiation of DNA replication (S phase), bud emergence, and SPB duplication take place almost simultaneously. Once a bud has formed, astral microtubules penetrate into the bud, the nucleus moves toward the bud neck, and the spindle is assembled as a short bar on the side of the nucleus. In G2–M cells, the spindle lies across the nucleus, and remains relatively constant in size (about 1.5–2.0 μm in length). The spindle observed before the onset of anaphase is called preanaphase spindle. At the onset of anaphase, the spindle aligns itself along the mother–bud axis and pushes the nucleus into the bud neck. The spindle elongates as the nucleus rapidly extends into the bud. The elongated nucleus then oscillates over 1–2 μm within the neck. The spindle remains at a constant length (4–5 μm) during this oscillation. Spindle elongation then resumes; the spindle elongates to its maximal length (11–12 μm). At telophase, the spindle disassembles, and the divided nuclei move back to a central position in the mother and daughter cells.

In *S. pombe*, microtubule organization and the ability of the SPB to nucleate cytoplasmic and nuclear microtubules depend on the position of the cell in the cell cycle. After cytokinesis the cell starts to grow longitudinally at the older of the two ends, and then switches to grow at both ends. During the cell growth, the nucleus continues to reside at the cell centre. Recently divided small cells are already in G2 phase. In G2 cells, several cytoplasmic microtubules run along the long axis of the cell; some of them are found to run close to the SPB. These interphase cytoplasmic microtubules are not static, but dynamically growing and shrinking. No nuclear microtubules are formed during interphase. Interphase chromatin resembles a hemisphere with two protrusions into the nucleolus. At mitosis, the chromatin region becomes more compact, and the nucleolar protrusions disappear. Cytoplasmic microtubules disassemble, and the short mitotic spindle is formed on the side of the nucleus. The spindle elongates to lie across the nucleus, and then remains at a constant length (2.5 μm). At the onset of anaphase, the spindle starts to elongate, and separating chromosomes show a U shape; chromosomes are further separated as the spindle elongates. Astral microtubules associated with the SPB are observed at metaphase through anaphase. After the spindle elongates to the cell ends (12–15 μm) and then disassembles, the nuclei move back toward the cell centre. This nuclear movement is accompanied by the appearance of SPB-associated microtubules. Cytoplasmic microtubules are assembled at the cell centre where septation and cytokinesis will take place.

2.3   Structure and duplication of the SPB

The SPB duplicates during interphase and nucleates microtubules required for spindle formation. It is the functional equivalent of the centrosome in animal cells, although it differs from the centrosome in structure. In *S. cerevisiae* the SPB is a disc-shaped, multilayered structure (95–100 nm in height and 80–200 nm in diameter in diploid cells) (14) embedded in the nuclear envelope throughout the cell cycle, and

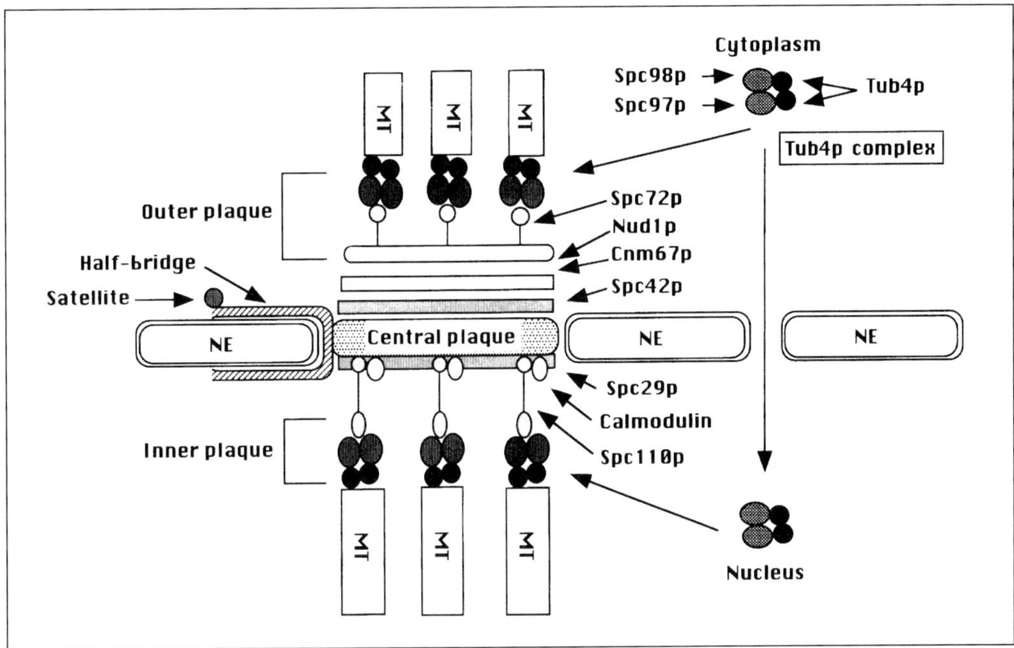

Fig. 3 Structure of the *S. cerevisiae* SPB. Spc42p, Cnm67p, Nud1p, and Spc72p are cytoplasmic SPB components. Spc29p, Spc110p, and calmodulin are nuclear SPB components. Tub4p, Spc98p, and Spc97p are component of both the inner and outer plaques, and form a complex. The Tub4p complex binds to Spc72p for nucleating astral microtubules (MTs). The Tub4p complex is also imported from the cytoplasm into the nucleus through the nuclear pore of the nuclear envelope (NE), and binds to Spc110p for assembling nuclear microtubules. The half-bridge is attached to the central plaque. In late G1, the satellite, the precursor of the nascent SPB, is formed at the cytoplasmic side of the half-bridge.

has nuclear and cytoplasmic faces to form nuclear and cytoplasmic microtubules, respectively (2).

The structure of the SPB has been well characterized in *S. cerevisiae* by electron microscopy. Three-layer structures were originally proposed: the central plaque (the darkly staining disc embedded in the nuclear envelope), and the inner and outer plaques (lightly staining discs layered over each flat surface of the central plaque). Nuclear and cytoplasmic microtubules have their proximal ends associated with the inner and outer plaques respectively. Development of the procedure for isolating a large quantity of the SPBs made it possible to identify the components and to study the structure in more details (15). A minimum of six layers is recognized in frozen hydrated SPBs and thin sections (14). Analysis of the SPB by matrix-assisted laser desorption/ionization (MALDI) mass spectrometry has identified 23 proteins localized to the SPB (16). Among those proteins, Spc42p, Cnm67p, Nud1p, and Spc72p are cytoplasmic SPB components, whereas Spc29p and Spc110p (a calmodulin-binding protein) are nuclear SPB components (17) (Fig. 3). Spc98p, Spc97p, and Tub4p (γ-tubulin) are localized to both the outer and inner plaques, and form a γ-tubulin complex required for microtubule nucleation (18, 19) (see next section).

Duplication of the SPB is temporally integrated with cell cycle events. In *S. cerevisiae*, the SPB duplicates at the G1 to S transition. In this accordance the putative promoter region of the *SPC110* gene shares a consensus sequence with DNA synthesis genes (20). Before duplication, the SPB has the structure called half-bridge on one side of the central plaque. Late in G1 the half-bridge has a spherical structure called the satellite attached to its cytoplasmic side, which probably is the precursor of the nascent SPB (Fig. 3). SPB components, Cnm67p, Nud1p, Spc42p, and Spc29p, are localized to the satellite (17). Mating pheromone arrests the cell at an unbudded G1 state with an unduplicated SPB with the satellite on the half-bridge. After duplication the two SPBs are connected by fused half-bridges (now called the bridge). As the bud enlarges, the duplicated SPBs are separated and interact via nuclear microtubules to form a spindle. The half-bridge and the bridge have an ability to nucleate cytoplasmic microtubules. Early in the cell cycle before spindle formation, cytoplasmic microtubules are nucleated not only by the outer plaque, but also by the half-bridge and the bridge (2).

Defects in SPB duplication generally arrest cells at G2–M, resulting in the formation of a monopolar spindle with a high CDK1 kinase activity (reviewed in Reference 21). One class of mutations including *cdc31* and *kar1* has defects in early steps of SPB duplication. The other class of mutations including *mps2* and *ndc1* allows SPB duplication to proceed, but the newly formed aberrant SPB is not inserted into the nuclear envelope. Thus, mutants in these genes have defects in late steps of SPB duplication, and also display a monopolar spindle. The G2–M arrest shown by the mutant cells requires the activation of the spindle checkpoint that induces mitotic arrest in response to spindle assembly defects (see Chapter 4). Interestingly, the strains carrying *mps1* mutations that disrupt an early step of SPB duplication do not arrest cell division, but proceed through monopolar mitosis and cytokinesis. Mps1p is a protein kinase that is essential for SPB duplication and, in addition, performs a non-essential function in the spindle checkpoint (21). Overexpression of Mps1p activates the spindle checkpoint and causes mitotic arrest, suggesting that Mps1p is involved in a mechanism that couples completion of SPB duplication with the onset of anaphase (see Chapter 4).

In *S. pombe*, the SPB is not a multilayered structure, but observed as an oblate ellipsoid (90 nm thick and about 180 nm in diameter). It resides on the cytoplasmic surface of the nuclear envelope during most of interphase as does the centrosome (22). The osmiophilic material is found near the inner surface of the nuclear envelope located beneath the interphase SPB, which is the site of nuclear microtubule nucleation during mitosis. As the cell enters mitosis, the nuclear envelope invaginates beneath the SPB and forms an opening into which the SPB settles. After settling, nuclear microtubules are nucleated from the osmiophilic material. During anaphase, the SPBs are extruded back into the cytoplasm.

The *S. pombe* SPB has not been isolated yet, but a few proteins homologous to the *S. cerevisiae* SPB components have been identified including gamma-tubulin (23), Alp4 and Alp6 (Spc97 and Spc98 homologues; T. Toda, personal communication), Pcp1 (a Spc110p homologue; M. Flory and T. Davis, personal communication) and calmodulin (24). In addition, Cut11, Cut12, and Sad1 seem to have important roles in the function of the *S. pombe* SPB. Cut11 appears to be required for anchoring the SPB in the nuclear envelope during mitosis (25). It localizes to the nuclear pore complexes and the mitotic SPBs, and has limited sequence similarity and functional homology to *S. cerevisiae* Ndc1p. Cut12 is required for microtubule formation at the SPB (26; see next section). Sad1 seems to be an SPB component integrated into the nuclear envelope (27). Deletion of the *sad1*$^+$ gene, or a temperature-sensitive mutation of *sad1*, causes a defect in bipolar spindle formation, but not in microtubule formation.

The *S. pombe* SPB duplicates at late G2 (4). Early G2 cells have a single SPB with an appendage that projects from one edge of the ellipsoid. The function of the appendage probably is similar to that of the half bridge and bridge in *S. cerevisiae*. The mechanism of SPB duplication in *S. pombe* at the molecular level remains unknown. So far, Dph1 was found to be possibly involved in SPB duplication (28).

2.4 Mechanism of microtubule nucleation at the SPB

γ-Tubulin, the third class of tubulin, has been shown to function in microtubule formation at the microtubule-organizing centres, including the *Aspergillus nidulans* SPB, and the vertebrate centrosomes. In centrosomes, γ-tubulin is located on the pericentriolar material, the centrosomal substructure that nucleates microtubules. Microtubule nucleating activity of the centrosome is activated at the onset of mitosis, and the activation seems to be required for promoting spindle assembly. It is not clear how the activity is regulated during the cell cycle.

In *S. cerevisiae*, as mentioned above, the SPB assembles nuclear and cytoplasmic microtubules throughout the cell cycle. Unbudded cells containing a single SPB have 16 nuclear microtubules, enough to attach to each of the 16 chromosomes (29). This suggests that the nuclear microtubules may interact with chromosomes throughout the cell cycle. In contrast, 20 to 30 nuclear microtubules per SPB are observed in cells forming a short spindle (5), suggesting that microtubule-nucleating activity of the nuclear face of the SPB is increased for spindle formation.

The *TUB4* gene encoding γ-tubulin is essential for cell viability, and Tub4p is localized to the SPB as mentioned in the previous section (30; reviewed in Reference 31). Expression of *TUB4* is regulated in a cell cycle-dependent manner. In temperature-sensitive *tub4* mutant cells the SPB duplicates, but spindle formation or elongation is defective, owing to defects in microtubule orgnization at the SPB. Tub4p functions in microtubule nucleation by forming a complex with Spc98p and Spc97p at the SPB. Spc98p, Spc97p, and Tub4p form a complex in the cytoplasm. Nuclear import of the complex is mediated by the nuclear localization sequence of Spc98p (19). The complex imported into the nucleus binds to the nuclear side of the SPB through an interaction of Spc98p and Spc97p with Spc110p. The Tub4p complex binds to the cytoplasmic side of the SPB via Spc72p for astral microtubule formation (18). It is not clear whether binding of the Tub4p complex to Spc110p or Spc72p is sufficient for microtubule nucleation at the SPB. It is also unclear how the number of microtubules assembled is determined. About 1000 copies of Spc110p seem to be present on the

SPB (16), whereas only 20 to 30 spindle microtubules are nucleated at the SPB (5).

In *S. pombe*, microtubule nucleation at the SPB is also regulated during the cell cycle. No nuclear microtubules are observed during interphase. At the onset of mitosis, the SPB settles into an opening of the nuclear envelope and assembles more than 20 nuclear microtubules. Microtubule formation and spindle assembly are induced by the CDK1–cyclin B (Cdc2/Cdc13) complex at the onset of mitosis. Cdc2/Cdc13 localize to the SPB at the early stages of mitosis (32). Cut12, a SPB component, may be a regulator or substrate of Cdc2/Cdc13 (26). A *cut12* loss of function mutant is defective in spindle formation, and nucleates spindle microtubules from only one of the two duplicated SPBs. A gain of function mutant bypasses the requirement for Cdc25, a normally essential tyrosine phosphatase that removes an inhibitory tyrosine from Cdc2.

The γ-tubulin gene ($gtb^+/tug1^+$) is essential for viability (23, 33); gene-disrupted cells show defects in either spindle formation or chromosome disjunction (23). On the other hand, γ-tubulin localizes to the SPB throughout the cell cycle (22, 23, 34). Electron microscope immunocytometry shows that in G2 cells γ-tubulin localizes to the osmiophilic material that lies near the inner surface of the nuclear envelope, immediately adjacent to the SPB (22). Thus, the presence of γ-tubulin is not sufficient for microtubule nucleation.

In vitro, interphase γ-tubulin complexes are activated downstream of CDK1–cyclin B by incubation in mitotic extracts prepared from unfertilized *Xenopus* eggs (34, 35). The SPB activator has been isolated from *Xenopus* egg mitotic extracts and identified as the large subunit of ribonucleotide reductase (the *S. pombe* Cdc22) (S. Takada and H. Masuda, unpublished observation), which is known to act at S phase for supplying substrates of DNA synthesis. This result suggests that controlling microtubule nucleation at the SPB may be involved in a mechanism that couples S phase to M phase.

2.5 Roles of cytoplasmic microtubules and motor proteins in nuclear positioning

In *S. cerevisiae*, the positioning of the nucleus to the bud neck and the orientation of the mitotic spindle along the mother–bud axis before initation of anaphase are crucial for properly distribution of the two nuclei between the mother and the daughter cell. Astral microtubules nucleated from the SPB are required for these events (36, 37). Interestingly, *cnm67* mutant cells that lack a detectable outer plaque of the SPB still show cytoplasmic microtubules attached to the half-bridge throughout the cell cycle, and are able to perform correct nuclear migration and spindle orientation at low frequency (38).

Disruption of the functions of microtubule-based motor proteins revealed that the motor proteins are involved in nuclear positioning and spindle orientation via interaction with astral microtubules (reviewed in Reference 39). Among the motor proteins, cytoplasmic dynein (Dhc1p/Dyn1p) is a putative minus end-directed motor that moves on microtubules toward the minus end, and localizes on astral microtubules.

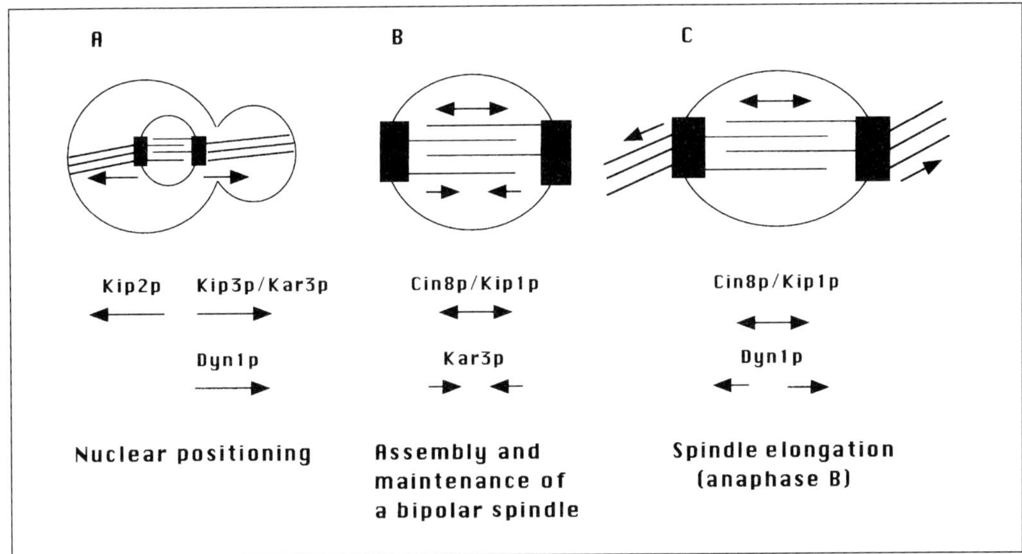

Fig. 4 Role of motor proteins in nuclear positioning, formation, and maintenance of a bipolar spindle, and spindle elongation in *S. cerevisiae*. (A) At pre-anaphase stages, Kip3p and Kar3p act synergistically to locate the nucleus towards the bud neck. Kip2p is antagonistic to their action. At anaphase, Dyn1p is required for nuclear positioning. (B) Formation and maintenance of a bipolar spindle require a balance of forces that act antagonistically. Cin8p and Kip1p provide the force that connects antiparallel microtubules nucleated from the two spindle poles and elongates the spindle, whereas Kar3p provides the force that shrinks the spindle by depolymerizing microtubules at the spindle pole. (C) Spindle elongation is mediated by forces that act synergistically. Cin8p and Kip1p push the spindle poles via antiparallel microtubules in the nucleus, whereas Dyn1p pulls the spindle poles from the cytoplasm via astral microtubules.

The other type of motor proteins is the family of kinesin-related proteins (Kar3p, Cin8p, Kip1p, Kip2p, Kip3p, and Smy1p). Kinesin-related proteins generally act as plus end-directed motors; Cin8p and Kip1p belong to this class. Kar3p, which belongs to the other class of kinesin-related protein, is a minus end-directed motor that localizes to the SPB and induces depolymerization of microtubules.

A combination of motor proteins that act either synergistically or antagonistically to one another is required for nuclear positioning at different stages of the cell cycle (Fig. 4A) (reviewed in References 39, 40). At pre-anaphase, Kip3p is required for the movement of the nucleus to the bud neck and the alignment of the spindle along the mother–daughter axis. Kar3p and Kip2p act synergistically and antagonistically to the movement respectively (41, 42). At anaphase, cytoplasmic dynein (Dyn1p/ Dhc1p) and components of a dynein-activating protein complex dynactin (Jnm1p, Act5p, and Nip100p) are required for translocating the nucleus through the neck. Kip2p is antagonistic to the action.

The functions of motor proteins may be regulated for nuclear positioning during the cell cycle. Indeed, Clb5p-dependent Cdc28p kinase activity is required for proper nuclear positioning (43; see Chapter 3). Clb5p is a B-type cyclin required for efficient DNA replication. Homozygous diploid cells carrying a *cdc28*ts allele and *clb5* deletion are defective in spindle positioning, leading to migration of the undivided nucleus into the bud. In the mutant cells, long curving astral microtubules are found oriented towards the mother cell cortex. In contrast, mutant cells such as *kip3*, *dhc1*, *act5*, and *jnm1* in which the nuclei frequently fail to migrate to the bud neck have long curving astral microtubules directing specially towards the bud. Disruption of the functions of motor proteins and their cell cycle-dependent modification may influence the dynamic property of microtubules at the SPB and the cell cortex, resulting in nuclear positioning defects.

Interaction of astral microtubules and the cell cortex also is crucial to nuclear positioning (reviewed in Reference 40). The interaction depends on the presence of filamentous actin, Kar9p, Bni1p, and Bub6p on the attachment site in the cortex of the bud. Nuclear positioning is defective if attachment of astral microtubules to the bud tip is impaired by mutations in genes encoding those proteins.

In *S. pombe*, nuclear positioning seems to be dependent on cytoplasmic microtubules and SPB-mediated nuclear migration (8). For instance, nuclear movement towards the cell equator after mitosis is accompanied by the appearance of SPB-associated microtubules (8). The presence of microtubule-destabilizing chemicals, or mutations in tubulin or microtubule-binding protein genes, can displace the nuclei from the centre of the cells and often impair the control of cell shape (44–46). So far, eight kinesin-related motor proteins and one cytoplasmic dynein have been found in *S. pombe*. However, involvement of the motor proteins in the process has not yet been demonstrated.

2.6 Roles of motor proteins in spindle assembly

In either yeast, formation and maintenance of the bipolar spindle require the functions of motor proteins that act antagonistically. Cytoplasmic microtubules are not required for those processes (6, 37, 46).

In *S. cerevisiae*, kinesin-related, plus end-directed motor proteins, Cin8p and Kip1p, are required for bipolar spindle formation (47–49). *kip1* and *cin8* double mutant cells arrest in mitosis with large buds and duplicated, but unseparated, SPBs. When the function of Kip1p and Cin8p is eliminated after spindle formation, the bipolar spindle collapses with previously separated poles being drawn together (49). Deletion of the *KAR3* gene partially suppresses the phenotype. As mentioned in the previous section, Kar3p belongs to a class of kinesin-related, minus end-directed motor proteins. It localizes at the SPB, and seems to depolymerize spindle microtubules so that it may provide the pulling force that decreases the spindle length (50). In contrast, Kip1p and Cin8p are thought to provide the pushing force that elongates the spindle. These results suggest that formation and maintenance of the bipolar spindle need a balance of forces counteracting each other (Fig. 4B). In addition to kinesin-related proteins, Stu1p, a non-motor protein localizing to the spindle microtubules, is required for spindle assembly (51). Clearly, other participating factors remain to be isolated.

In *S. pombe*, a kinesin-related protein Cut7 is required for bipolar spindle formation (52, 53). Cut7 belongs to a class of plus end-directed motor proteins. In *cut7* mutant cells, microtubules are nucleated at the SPB during mitosis, but cannot interdigitate with those nucleated from the other SPB. Disruption of the $pkl1^+$ gene encoding another kinesin-related protein suppresses mutations in Cut7 (54). Overexpression of Pkl1 causes a similar phenotype to that seen in cut7 mutants. These phenotypes can be explained if Cut7 and Pkl1 provide opposing forces in the spindle.

2.7 Chromosome condensation and sister chromatid cohesion

Chromosome condensation is essential for proper chromosome segregation, as it helps to eliminate the tangles between sister chromatids that arise during DNA replication. In *S. cerevisiae*, 16 individual condensed chromosomes are not cytologically distinguishable. A fluorescence *in situ* hybridization (FISH) analysis indicates that chromosome condensation occurs during mitosis, but the extent is less than that observed for human chromosomes (55). These results suggest that not all aspects of chromosome condensation are conserved between *S. cerevisiae* and higher eukaryotes. In *S. pombe*, three individual condensed chromosomes can be observed in mutant cells arrested in mitosis (44, 45). Chromosome condensation at mitosis has been confirmed by FISH studies (56).

Topoisomerase II, but not topoisomerase I, is required for chromosome condensation and chromosome segregation. Whereas topoisomerase I mutants of *S. pombe* and *S. cerevisiae* are viable, mutations in topoisomerase II are lethal as the cell enters mitosis (57–59). The decatenating activity of topoisomerase II is required to form proper folds during chromosome condensation, and to remove the interlocks between sister chromatids before the onset of anaphase.

A class of the SMC family proteins has essential functions for chromosome condensation. The SMC family shares a common head (nucleotide binding motif)–rod (two coiled coil regions with hinge)–tail (DA box motif) configuration (60; reviewed in Reference 61). Two *Xenopus* SMC family members, XCAP-C and XCAP-E, and three other proteins were identified as components of a protein complex, condensin, that promotes chromosome condensation in *Xenopus* egg mitotic extracts. The 13S condensin complex introduces positive supercoils into plasmid DNA in the presence of ATP and topoisomerase I (61). The yeast SMC proteins, *S. pombe* Cut3 and Cut14, and *S. cerevisiae* Smc2p, are required for chromosome condensation (56, 62). Cut3 and Cut14 form a complex that displays DNA renaturation activity *in vitro* (63), but a larger complex similar to the 13S condensin may be required for promoting chromosome condensation.

S. pombe Cut15, a homologue of importin α that mediates translocation of proteins containing a nuclear localization signal (NLS) through the nuclear pore complex, seems to be involved in chromosome condensation (64; see Chapter 9). Cut15 may be involved in import of condensin or other M phase-specific factors required for chromosome condensation. Clearly further work is required to identify its role in chromosome condensation.

Another class of the SMC family seems to function to establish and maintain

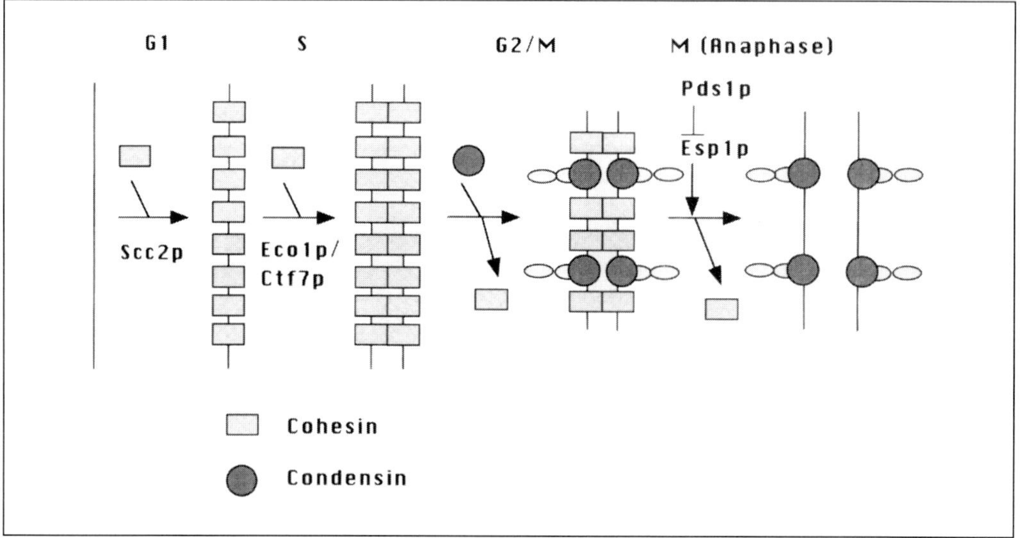

Fig. 5 Sister chromatid cohesion and chromosome condensation in *S. cerevisiae*. At late G1, the cohesin complex binds to chromatin in the presence of Scc2p. At S phase, sister chromatid cohesion is established in the presence of Eco1p/Ctf7p. At G2–M, the condensin complex promotes chromosome condensation. The anaphase inhibitor Pds1p suppresses Esp1p function. At anaphase, Esp1p dissociates the cohesin complex, and sister chromatid separation occurs. See text for further details.

cohesion between sister chromatids. Holding sister chromatids together until the onset of anaphase is required for distributing chromosomes equally into the two daughter cells. In *S. cerevisiae*, cohesion between sister chromatids depends on a multi-subunit complex, called cohesin, which contains at least four subunits: Scc1p/Mcd1p, Scc3p, Smc1p, and Smc3p (reviewed in Reference 65). The cohesin complex associates with chromosomes from late G1 until the onset of anaphase (Fig. 5). Scc2p is required for association of the cohesin complex with chromosomes. Eco1p/Ctf7p is necessary for the establishment of cohesion during DNA replication. At the onset of anaphase, Scc1p/Mcd1p dissociates from chromosomes by proteolysis stimulated by Esp1p, resulting in sister chromatid separation (65). In *S. pombe*, Rad21, Mis4 (66), and Rec11 seem to be homologues of Scc1p, Scc2p, and Scc3p respectively. A cohesion complex similar to that of *S. cerevisiae* may exist and function in *S. pombe*.

Interestingly, Scc1p/Mcd1p is also involved in chromosome condensation (67). The cohesin complex probably functions for both sister chromatid cohesion and chromosome condensation (Fig. 5). Thus, cohesion and condensation activities are overlapping in time. In contrast, the *Xenopus* cohesin complex binds to interphase chromosomes and dissociates from them at the onset of mitosis when the condesin complex binds to them (61). This difference may reflect the observations that, in vertebrate cells, a major chromosome reorganization occurs at the onset of mitosis; in *S. cerevisiae* such a chromosome reorganization is not observed, and less compacted chromosomes are formed at mitosis (61).

2.8 Interaction between kinetochores and microtubules

Chromosomes attach spindle microtubules via the kinetochore region. Kinetochores are composed of centromeric DNA and associated proteins. The chromatin structure of the centromere is highly differentiated (see Chapter 7). In higher eukaryotes, centromeres consist of several mega-base pairs (Mbp) of DNA containing repetitive sequences. In contrast, functional centromere in *S. cerevisiae* is very small in size, 125 bp, with no repetitive sequences. This chromatin domain has a specialized nucleosome structure that is protected against nuclease digestion. The budding yeast centromere consists of three elements, called CDEI, CDEII, and CDEIII. CDEII and CDEIII are essential for centromere function. Structure of *S. pombe* centromeric DNA resembles that of higher eukaryotes. The three centromeres in *S. pombe* are much larger than those in *S. cerevisiae*, have repetitive DNA sequences, and vary in size. Each centromere contains an inner region of about 15 kbp and outer repeat sequences of 20–100 kbp. The inner region has a specialized chromatin structure, which seems to be essential for centromere function. Mis6, which binds to the inner region, is required for maintaining the inner centromere structure; in *mis6* mutant cells chromosomes are frequently missegregated, and the specialized chromatin in the inner centromere is disrupted (68).

In higher eukaryotes, the kinetochore is seen as a trilaminar disc-shaped structure under an electron microscope, and a few or dozen of microtubules terminate at the kinetochore region. In *S. cerevisiae*, each centromere seems to be associated with a single microtubule. No kinetochore-like structure, however, is observed at the ends of the putative kinetochore microtubules by electron microscopy (5). In *S. pombe*, each of the three centromeres is in contact with two to four microtubules. Some of the microtubules are shown to end on fibrillar material, a potential kinetochore structure (4). The kinetochore structure of *S. pombe* may be similar to but smaller than that of higher eukaryotes. The small size of *S. cerevisiae* centromeric DNA and electron microscopy observations suggest that the kinetochore structure is much more simpler than those of *S. pombe* and higher eukaryotes.

A protein complex that binds to the CDEIII region (CBF3 complex) is essential for the *S. cerevisiae* kinetochore function. The CBF3 complex isolated by affinity chromatography consists of p23 (Skp1p), p58 (Ctf13p), p64 (Cep3p), and p110 (Ndc10p) (69, 70). Temperature-sensitive mutations in any of these four genes increase the rate of chromosome loss (reviewed in Reference 71). *In vivo* kinetochore activity in the mutant cells correlates with *in vitro* CDEIII-binding activity of the CBF3 complex in yeast extracts (72, 73). No mammalian or *S. pombe* homologues of the components have been reported, suggesting that the CBF3 complex may be unique to the *S. cerevisiae* centromere funtion.

The binding of centromeric DNA to microtubules can be reconstituted *in vitro* using yeast cell extracts (74). The binding is more efficient in extracts prepared from mitotically arrested cells than from G1-arrested cells. Centromeric DNA-attached beads that are incubated with the crude CBF3 complex attach to and move on microtubules toward the minus end (75). The binding of the CBF3 complex to CDEIII is essential, but not sufficient for linking DNA to microtubules (76, 77). Additional cellular factor(s) must interact with the CBF3 complex to form active microtubule-binding complexes. Kinetochores reconstituted on centromeric DNA attached to beads bind preferentially to GTP forms of the microtubule lattice and to the plus end of the lattice where a small cap of GTP–tubulin exists (78). This result is consistent with the observations in mammalian cells that kinetochores stay at the plus ends of microtubules during most of mitosis.

Microtubule-binding activity of the kinetochore complex seems to be regulated by a balance between the activities of Ipl1p protein kinase and Glc7p type 1 protein phosphatase (79–81). *ipl1* mutant cells show missegregation of chromosomes at high frequency (82); they separate sister chromatids but are defective for chromosome segregation (79). Ipl1p phosphorylates a CBF3 component, p110/Ndc10p, *in vitro*. In contrast, mutants in Glc7p arrest at G2–M with a high CDK1 kinase activity. Ndc10p is hyperphosphorylated in *glc7* mutant cell extracts. Kinetochores assembled in the *glc7* mutant cell extracts exhibit lowered microtubule-binding activity. These results clearly suggest that Ipl1p and Glc7p regulate microtubule-binding activity of kinetochores via Ndc10p phosphorylation.

The spindle checkpoint induces mitotic arrest in cells with spindle defects (see Chapter 4). In mammalian cells, some of the proteins involved in the spindle check-point are shown to localize to the kinetochore. In *S. cerevisiae*, mitotic arrest of *glc7* mutants is abolished if cells carry additional mutations in the spindle checkpoint proteins (80, 81). This is consistent with the current view that defects in kinetochore–microtubule interactions activate the spindle checkpoint for the mitotic arrest.

The molecular mechanisms responsible for the organization of the *S. pombe* kinetochore and the interaction of kinetochores with microtubules remain unknown at present. Considering the differences in sequence of centromeric DNA between *S. pombe* and *S. cerevisiae* as mentioned above, the *S. pombe* kinetochore may be organized by fundamentally different mechanisms. In contrast, microtubule-binding activity of kinetochores may be regulated by a similar mechanism. *S. pombe* cells deleted for the type 1 protein phosphatases Dis2p and Sds21p show a phenotype similar to *glc7* mutants (83). The budding yeast spindle checkpoint proteins that seem to monitor the interaction between kinetochores and microtubules have homologues in fission yeast and higher eukaryotes (see Chapter 4).

2.9 Molecular mechanism responsible for the onset of anaphase

Anaphase is defined as the stage in mitotic division when sister chromatids are separated. The onset of anaphase requires destruction of anaphase inhibitors, Pds1p in *S. cerevisiae* and Cut2p in *S. pombe*, which suppress premature sister chromatid separation (reviewed in References 84, 85). What signal initiates anaphase is not yet known.

In *S. cerevisiae*, Pds1p forms a complex with Esp1p, and inhibits function of Esp1p. Destruction of Pds1p activates the function of Esp1p, which in turn dissociates a cohesin subunit Scc1p from chromosomes (Fig. 5). In *S. pombe*, Cut2p forms a complex with Cut1p, and localizes to the pre-anaphase spindle, but not to chromosomes. Cut1p is activated when Cut2p is degraded. Cut1p remains associated with the elongating spindle at anaphase. Without Cut1p or Cut2p function, sister chromatid separation is blocked. The functional similarity between Cut2p and Pds1p, and the partial sequence similarity of Cut1p to Esp1p, suggests that Cut1p–Cut2p complex may be required for regulating dissociation of *S. pombe* cohesins. However, this complex may also be required for other processes, as it localizes to the mitotic spindle, and because Cut2p, but not Pds1p, is an essential protein.

A ubiquitin–protein ligase complex called the anaphase-promoting complex or cyclosome (APC/C) is required for the polyubiquitination that mediates the degradation of anaphase inhibitors (reviewed in Reference 85). Polyubiquitination requires a nine-residue motif called the destruction box present on the target proteins. Polyubiquitinated proteins are degraded by the 26S proteasome. *S. cerevisiae* and *S. pombe* mutants in the 26S proteasome are blocked at metaphase. The activity of the APC/C may be regulated by a balance of activities between cAMP-dependent protein kinase A and type 1 protein phosphatase (86).

The APC/C also ubiquitinates mitotic cyclins and spindle proteins. Substrate-specific activators of the APC/C function have been identified in *S. cerevisiae*. Cdc20p is required for degradation of Pds1p and the mitotic cyclin Clb3p, whereas Cdh1p/Hct1p are required for degradation of the mitotic cyclin Clb2p and the spindle protein Ase1 (reviewed in Reference 85). Degradation of mitotic cyclins results in CDK1 inactivation, which is required for the exit from mitosis. The spindle checkpoint seems to arrest cells before the onset of anaphase by inhibiting the activity of Cdc20p (see Chapter 4 for details).

2.10 Dynamics of chromosome separation

Chromosome segregation at anaphase after separation of sister chromatids is achieved by two kinds of movement: chromosome movement towards the spindle poles (anaphase A) and separation of the spindle poles by elongation of the spindle (anaphase B). In *S. pombe* and *S. cerevisiae*, anaphase B is a major contributor to chromosome separation, since the spindles elongate from 2–3 to 10–15 μm during anaphase. Do yeast cells have anaphase A movement?

In *S. cerevisiae*, FISH and live observations of chromosome regions close to the centromere using probes located 9–24 kb from centromeric DNA show anaphase A-like movement of the labelled spots (87, 88). Centromere dynamics, however, remain to be determined using centromeric probes. In *S. pombe*, visualization of centromeres by FISH and live observation of centromere dynamics show the presence of anaphase A (13, 89). As shown with isolated mammalian chromosomes, anaphase A movement may be mediated by microtubule-based motors that are located on the kinetochore and move toward the spindle pole while attaching to the ends of depolymerizing microtubules.

Spindle elongation is mediated by the activity of microtubule-based motor proteins. Although electron microscopic observations show that the length of interdigitating microtubules emanating from the SPBs increases during anaphase B (4, 5), polymerization of tubulin itself is not sufficient for spindle elongation. *S. pombe* spindles in permeabilized cells are able to elongate in the presence of ATP and exogenous tubulin (90). Exogenous tubulin is incorporated into the plus ends of interdigitating microtubules in the elongating spindle, but, in the absence of ATP, tubulin incorporation does not induce spindle elongation. This suggests that full length of the microtubule network is obtained by sliding of antiparallel microtubules and simultaneous incorporation of tubulin into the plus ends of the microtubules.

In *S. cerevisiae*, spindle elongation is mediated by the combined actions of kinesin-related proteins, Kip1p and Cin8p, and of a cytoplasmic dynein, Dyn1p (91) (see Fig. 4). As described above, Kip1p and Cin8p provide the pushing force required for formation and maintenance of the pre-anaphase spindle. At anaphase, they provide the pushing force for spindle elongation, although deletion of Kip1p and Cin8p function does not block, but delays, spindle elongation. Further loss of Dyn1p function blocks spindle elongation completely. Dyn1p localizes on astral microtubules, but not on spindle microtubules (12). These results suggest that Dyn1p increases the pole-to-pole distance by pulling the spindle poles toward the cell ends via astral microtubules. Bik1p and Ase1p, microtubule-associated proteins, are not required for spindle assembly, but are essential for anaphase spindle elongation (92). Ase1p is localized to the midzone of the anaphase spindle (92), and degraded by the APC/C when cells exit from mitosis (93). Expression of non-degradable Ase1p delays spindle disassembly, suggesting that the APC/C regulates spindle disassembly.

In *S. pombe*, motor proteins responsible for the spindle elongation at anaphase have not been determined. So far, Cut7p is an interesting candidate responsible for the pushing force. Although astral microtubules that apparently associate laterally with the cytoplasmic face of the SPBs are found during anaphase (7, 94), disruption of cytoplasmic dynein Dhc1p in *S. pombe* does not delay spindle elongation (95). This suggests either that the pulling force does not contribute to spindle elongation or that there is another motor protein responsible for the pulling force.

3. Meiosis

3.1 Overview of meiotic nuclear division

Meiosis is the special type of cell division that produces haploid progeny from diploid parental cells. In this process, a single round of DNA replication is followed by two rounds of chromosome segregation: first, segregation of homologous chromosomes (reductional division) at meiosis I, and then segregation of sister chromatids (equational division) at meiosis II (see Fig. 6). Proper disjunction of chromosomes in both meiosis I and II is ensured by formation of a bipolar spindle and force balance within a bipolar spindle as in mitosis. A unique aspect of the chromosome behaviour at meiosis is the pairing interaction between homologous chromosomes whereas

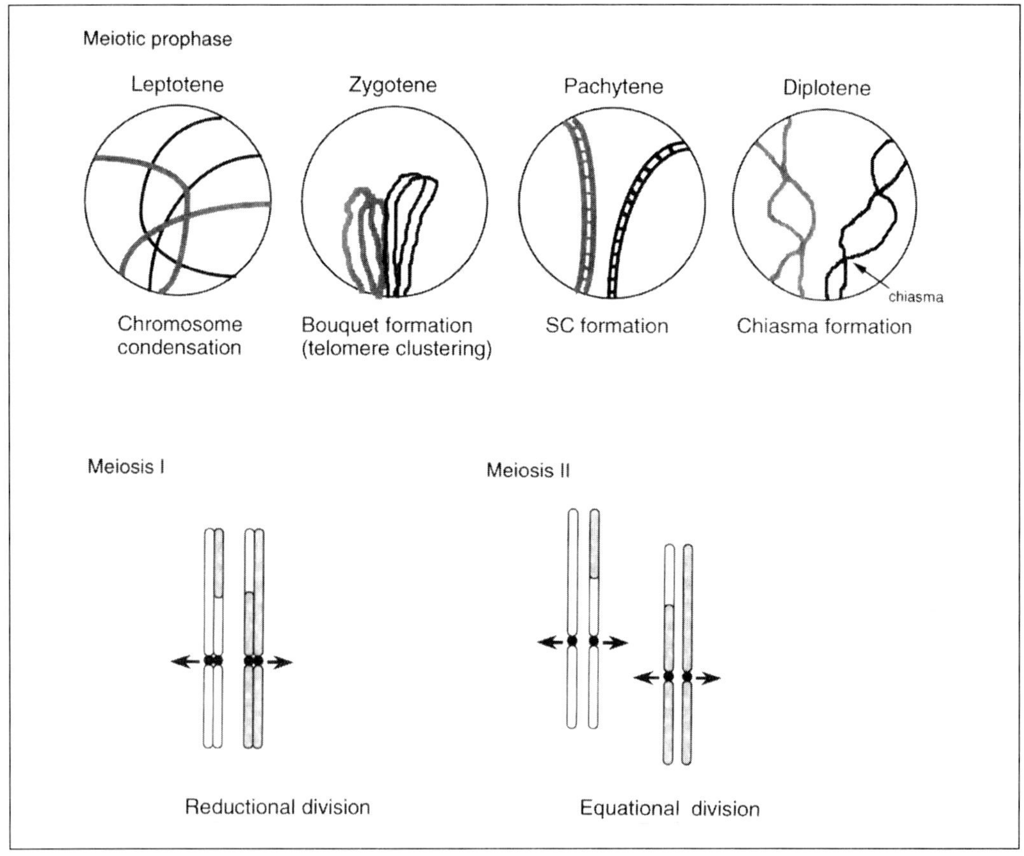

Fig. 6 Meiotic landmarks. Top panel: The behaviour of two pairs of homologous chromosomes in a single nucleus undergoing meiosis. Meiotic prophase is subdivided into the stages leptotene, zygotene, pachytene, and diplotene. In leptotene, chromosomes start condensation; in zygotene, homologous chromosomes pair with each other; in pachytene, a structure called the synaptonemal complex (SC) is formed between homologous chromosomes; and in diplotene, homologous chromosomes repel each other and are held together only at the sites of crossing-over, or chiasmata. Bottom panel: Meiotic prophase is followed by segregation of homologous chromosomes at meiosis I (reductional division) and segregation of sister chromatids at meiosis II (equational division). The arrows indicate the movement of the homologous chromosomes or the sister chromatids at meiosis I and II respectively. Occurrence of a crossing-over event is denoted by the exchange of genetic information (symbolized as grey versus white portion) between homologous chromosomes.

they behave independently of each other in mitosis. In meiosis I, connection between homologous chromosomes generated by homologous recombination achieves the force balance between the spindle poles whereas this role is carried out in meiosis II by the cohesion between sister chromatids. Thus, pairing and recombination of homologous chromosomes in meiotic prophase are necessary for faithful segregation of chromosomes during meiotic divisions. Meiotic prophase is subdivided into several stages (leptotene, zygotene, pachytene, and diplotene) as defined in animals and plants based on the cytological appearance of chromosomes (Fig. 6).

3.2 Behaviour of chromosomes and microtubules in meiosis

In *S. pombe* meiotic prophase, the nucleus elongates and oscillates between the ends of the cell (Fig. 7B); during this process, microtubules significantly change their organization (Fig. 7A). In mitotic interphase, arrays of cytoplasmic microtubules

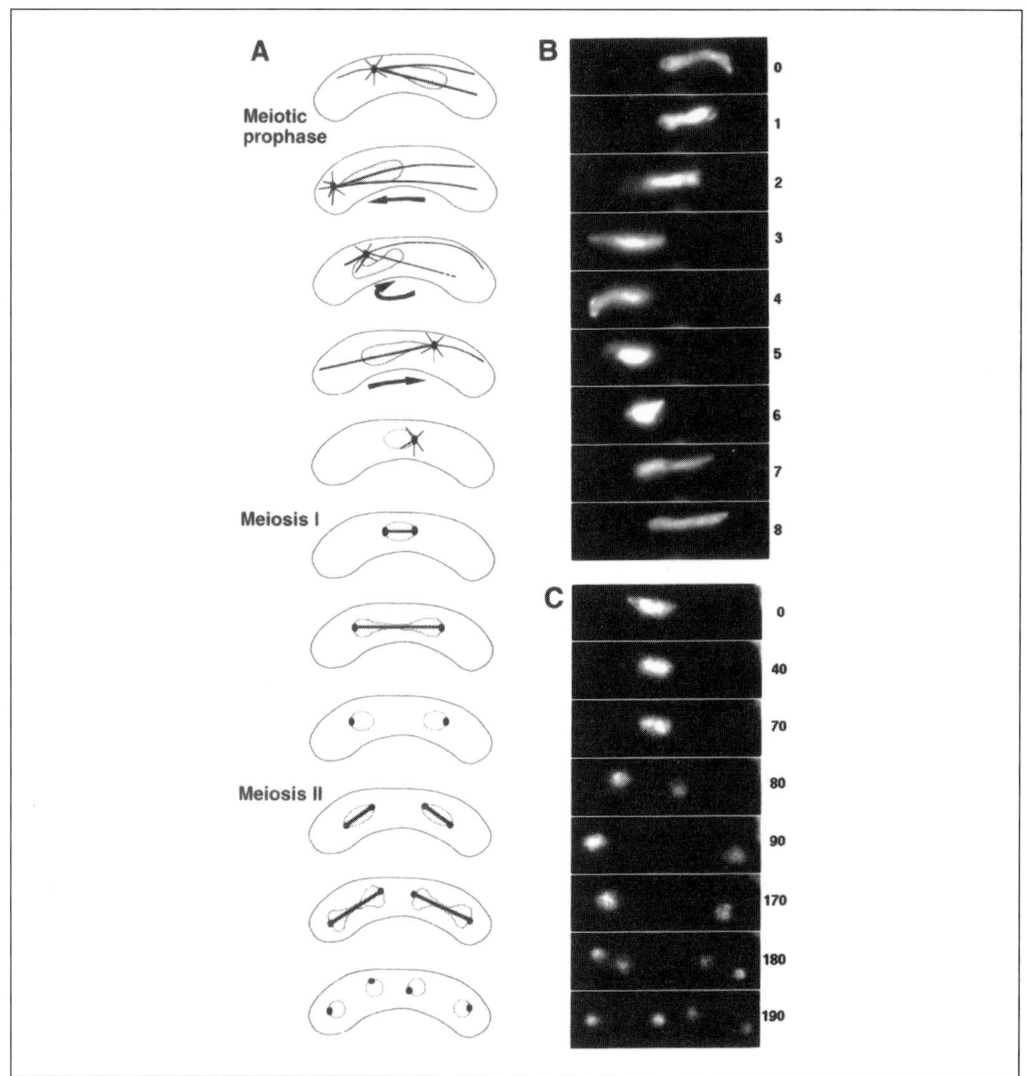

Fig. 7 Meiotic prophase and meiotic divisions in *S. pombe*. (A) A cartoon of nuclear behaviour during meiosis in the fission yeast *S. pombe*. A nucleus is sketched in grey and microtubules are represented by solid black lines emanating from the SPB. *S. pombe* meiotic prophase is characterized by an oscillating nucleus: the nucleus starts oscillatory movement immediately after nuclear fusion, and continues the movement for 2–3 h until the first meiotic division (109). Oscillatory movements are mediated by continuous reorganization of microtubules (97). (B, C) Live observation of nuclear behaviour during meiosis. Nuclei were stained with a DNA-specific fluorescent dye, Hoechst 33342, and observed in a living cell undergoing meiosis. Values on the right represent time in minutes.

extend along the length of the cell, but astral microtubules do not radiate from the SPB (7). When *S. pombe* cells are induced to undergo meiosis, the SPB acquires microtubule-nucleating activity, and cytoplasmic microtubules originate exclusively from the SPB (27, 96, 97). Also in *S. cerevisiae*, astral microtubules change their organization during meiotic prophase (98). However, in *S. cerevisiae*, meiotic prophase nuclei do not show such a striking elongation nor movement as observed in *S. pombe* (Fig. 8).

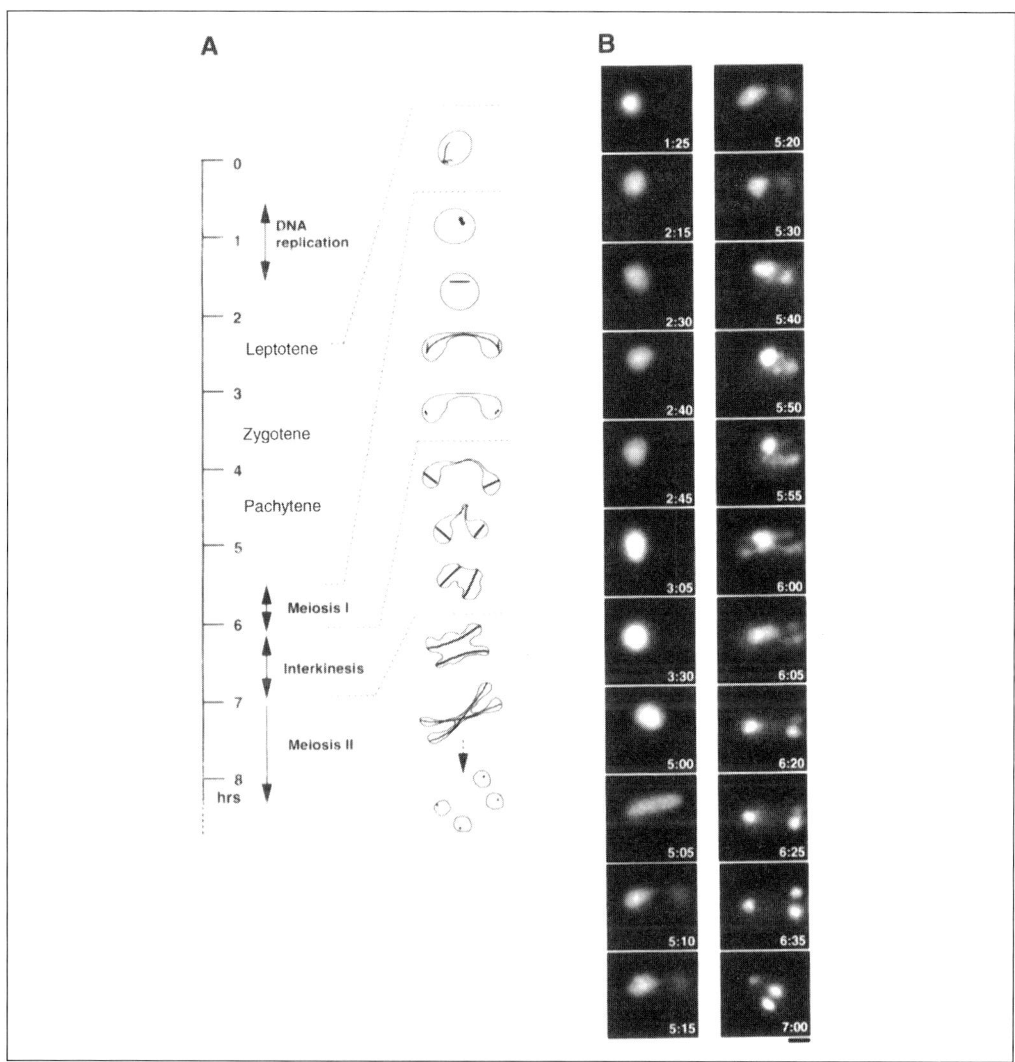

Fig. 8 Meiotic prophase and meiotic divisions in *S. cerevisiae*. (A) Nuclear behaviour landmarks during meiosis in the budding yeast *S. cerevisiae*. Solid black lines represent microtubules within an area surrounded by the nuclear membrane; the cell shape is not shown. Two rounds of chromosome segragation take place in a single undivided nucleus (101). (B) Live observation of nuclear behaviour during meiosis. Nuclei were stained with GFP-conjugated nucleoplasmin, and observed in a living meiotic cell (98). Values in each frame represent time in hours:minutes after induction of meiosis. Bar, 2 μm.

In *S. pombe*, the oscillatory movement of the meiotic prophase nucleus is driven by the reorganization of astral microtubules originating from the SPB with the cytoplasmic dynein as a microtubule motor protein (95, 97). The dynein heavy chain is localized predominantly at the SPB and astral microtubules. When the nucleus approaches the cell end, the dynein heavy chain is also found on a spot at which astral microtubules reach the cell cortex (95). It is likely that nuclear movements are mediated by the interaction of astral microtubules with the cell cortex generating a pushing force at the SPB and/or a pulling force at the cell cortex.

In *S. cerevisiae*, similar patterns of interactions between astral microtubules and the cell cortex are found in other examples of microtubule-mediated nuclear migration: the positioning of the nucleus to the bud neck (11, 36, 37) and the fusion of haploid nuclei during karyogamy (99). Involvement of astral microtubules in meiotic prophase chromosomal events has not yet been demonstrated in *S. cerevisiae*. However, *S. cerevisiae* mutants in the *KAR3* gene, which encodes a kinesin-related microtubule motor protein, are blocked at meiotic prophase with a monopolar spindle and an incompletely formed synaptonemal complex, suggesting a meiotic role of microtubules in synapsing homologous chromosomes (100).

Meiotic prophase is followed by two consecutive rounds of chromosome segregation; at each of these, the SPB duplicates and a spindle is formed (Figs 7A and 8A). During *S. cerevisiae* mitosis, nuclear division takes place asymmetrically; that is, a portion of the nucleus gradually enters into the bud through the narrow channel of the bud neck (see Fig. 1). During meiosis, on the other hand, the nuclear mass is divided into two equal masses symmetrically within a single cell (98) (Fig. 8B). Furthermore, chromosome segregation proceeds in a single, undivided nucleus; the nuclear membrane divides only at the end of meiosis (101) (see Fig. 8A). This differs from *S. pombe*, where the nuclear membrane divides to form separate nuclei after each round of chromosome segregation (102) (see Fig. 7A).

3.3 Nuclear positioning of telomeres and centromeres in meiosis

Nuclear architecture changes upon entry into meiosis, both in *S. cerevisiae* and *S. pombe*. In mitotic interphase, the *S. pombe* three chromosomes are organized in a polarized configuration as illustrated in Fig. 9A; centromeres are clustered at a single site near the SPB located at the opposite side to the nucleolus, whereas telomeres are located at the side of the nucleolus (reviewed in Reference 103).

A similar polarized orientation of chromosomes is also observed in *S. cerevisiae*: centromeres are clustered near the SPB at the region opposite to the nucleolus (104, 105) (also see Fig. 9C); five to eight clusters of telomeres are distributed from the SPB side to the nucleolus side (98, 106) (see Fig. 9C). It should be pointed out that in *S. cerevisiae* the arrangement of chromosomes within the nucleus may be spatially restricted because of the small size of chromosomes. As shown in Fig. 10, sizes of chromosome arms (the portion of chromosome from the centromere to either end of the chromosome) ranges from 90 kbp to 1.1 Mbp in *S. cerevisiae* (107), whereas it ranges

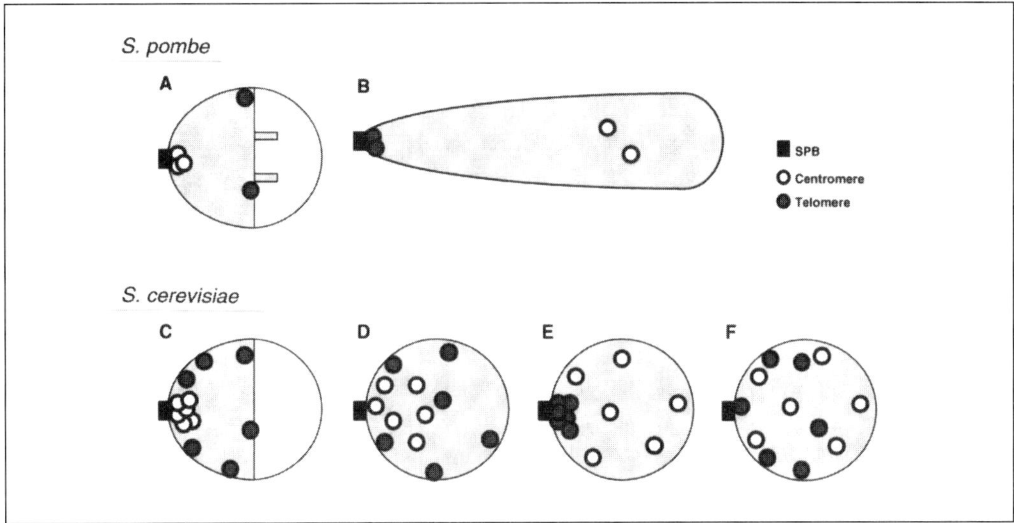

Fig. 9 Nuclear positions of centromeres and telomeres in yeasts: a summary. Cartooned is the nuclear architecture of a single nucleus of *S. pombe* (panels A, B) or *S. cerevisiae* (panels C–F). In panels A and C, the open hemisphere represents the nucleolus. (A) *S. pombe* mitotic interphase. (B) *S. pombe* meiotic prophase. (C) *S. cerevisiae* mitotic interphase. (D–F) *S. cerevisiae* meiotic prophase.

from 1.5 to 3.7 Mbp in *S. pombe* (108). Thus, when centromeres are clustered at the SPB, the positions of telomeres are spatially constrained according to the length of chromosome arms (see the legend to Fig. 10). This may provide a mechanistic explanation for the observed nuclear architecture differences between *S. cerevisiae* and *S. pombe*.

As the cell enters meiotic prophase, cytological observation in both yeasts showed that the centromeres and the telomeres undergo a repositioning within the nucleus. In *S. pombe* meiotic prophase, telomeres are clustered at the SPB and centromeres are separated from the SPB, in contrast with the mitotic chromosomes (109) (compare Fig. 9B with Fig. 9A). After meiotic prophase, chromosomes resume the mitotic configuration in which centromeres are again associated with the SPB, and persist through meiotic chromosome segregation.

Also in *S. cerevisiae*, it was recently demonstrated that telomeres transiently form a single cluster near the SPB for a limited period during meiotic prophase (110) (see Fig. 9E). In contrast to this ephemeral event of telomere clustering, *S. cerevisiae* has demonstrated a dramatic repositioning of centromeres upon entering meiotic prophase: interphase centromeres are clustered at the SPB, but become scattered within the nucleus in meiotic prophase (98, 105) (Fig. 9C–E). Centromere scattering may relieve spatial constraint, thereby facilitating the homologous chromosome pairing.

3.4 Homologous chromosome pairing in meiotic prophase

Presynaptic pairing, or alignment, of homologous chromosomes is prerequisite for physical synapsis and recombination. How homologous chromosomes find each other

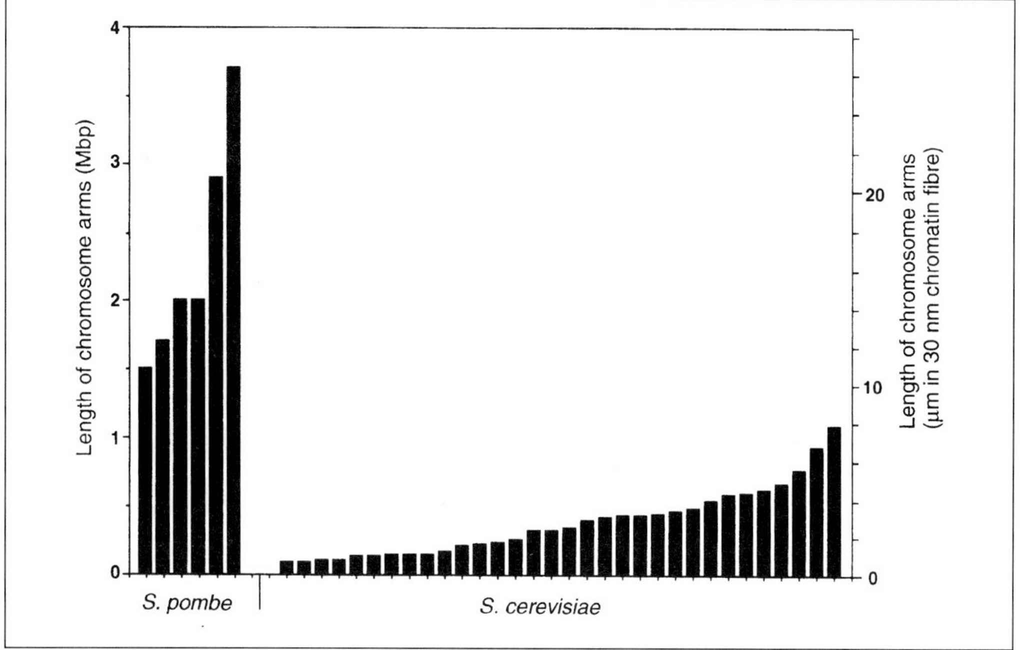

Fig. 10 Distribution of chromosomal arm lengths in budding and fission yeasts. Lengths of the chromosomal arms in *S. pombe* (total of six) and *S. cerevisiae* (total of 32). Left axis represents the length of chromsosome arms as DNA in Mbp; right axis represents the length of chromsosome arms as the 30-nm chromatin fiber in micrometres (μm). The 30-nm chromatin fibre has a DNA compaction ratio of 40–50; this means that 1 Mbp of DNA, which has the physical length of 340 μm, is packaged into about 8 μm of the 30-nm chromatin fibre. The 30-nm chromatin fibre is an extended form of chromatin, and thus the length of chromosome arms should be shorter when the chromatin is more condensed than the 30-nm fibre. Note that the size of a meiotic prophase nucleus in *S. pombe* is about 7–8 μm; thus the shortest chromosome arm in a form of the 30-nm fibre can reach the entire length of the nucleus. On the other hand, in *S. cerevisiae* most of the chromosome arms in a form of the 30-nm fibre are far smaller than its nuclear diameter of about 2 μm. Thus, when centromeres are clustered at the SPB, telomeres are distributed from the SPB side to the nucleolus side according to the length of chromosome arms (see Fig. 9D).

remains largely unknown. Strategies for homologous chromosome pairing seem to be diverse among species (reviewed in Reference 111). The choice of strategy may be related to mechanistic features, such as the size and the number of chromosomes.

In *S. pombe*, several lines of evidence have supported the idea that telomere clustering at the SPB and subsequent telomere-led nuclear movement are necessary for the proper pairing of homologous chromosomes. In several mutants defective in telomere clustering at the SPB (*taz1*, *lot2*, and *kms1*), recombination frequency is markedly reduced, indicating improper pairing of homologous chromosomes (112–114; reviewed in Reference 115). In addition, a mutant in the dynein heavy chain (designated *dhc1*), a microtubule motor protein specifically required for nuclear movements at meiotic prophase, also displays reduced frequency of meiotic recombination and improper positioning of homologous loci (95). Thus, *S. pombe* appears to rely primarily on chance contact between telomere-aligned homologous chromosomes, this chance being increased by the telomere-led chromosome movement as postulated previously (97, 103, 109).

S. cerevisiae has provided a different view of homologous chromosome pairing. In this organism, it has been suggested that pairing of homologous loci occurs in conjunction with recombination events during meiotic interphase and results in the co-localization of homologous chromosomes to conjoined areas during chromosome condensation in meiotic prophase (116, 117). On the other hand, analysis of homologous pairing by *in situ* hybridization in mutants of *S. cerevisiae* has shown that homologous pairing is independent of meiotic recombination and synaptonemal complex formation (118). Several observations suggest that, in *S. cerevisiae* also, telomeres are involved in meiotic events as in *S. pombe*: the telomeres of *S. cerevisiae* form a cluster in meiotic prophase (110), and genetic analysis shows that telomeres play a role in homologous chromosome recognition (119). One possible candidate in mediating such a telomere-promoted, homologous chromosome recognition is Tam1p/Ndj1p, a meiosis-specific telomere-binding protein. Mutations in *TAM1/ NDJ1* affect chromosome synapsis (120, 121). Furthermore, in cells disrupted for the *TAM1/NDJ1* gene, telomeres do not cluster, suggesting this meiosis-specific telomere-binding protein is involved in formation of the telomere bouquet in meiotic prophase (H. Scherthan, personal communication). A homologous gene for *TAM1/NDJ1* has not yet been found in *S. pombe*.

3.5 Homologous chromosome synapsis in meiotic prophase

In higher eukaryotes, the so-called synaptonemal complex (SC) is formed along the entire length of paired homologous chromosomes in the pachytene stage (Fig. 11). In diplotene, the SC disappears, and homologous chromosomes are connected only at chiasmata which are formed after recombination intermediates are resolved (see Fig. 6 for a chiasma). Formation of chiasmata is necessary for proper disjunction of homologous chromosomes at meiosis I; at least one chiasma for each pair of chromosomes ensures the force balance within a bipolar spindle.

Likewise, in *S. cerevisiae*, condensed homologous chromosomes held by a SC can be seen in spread preparation at the pachytene stage of meiotic prophase (122). In *S. cerevisiae*, proteins involved in SC formation have been isolated (reviewed in Reference 123): Zip1p is a component of the central region of the mature SC. In the absence of Zip1p, axial elements are formed but not synapsed; the unsynapsed axial elements are connected at a few sites, called 'axial associations'. Zip2p, another SC component, localizes to the axial associations; Zip2p is required for the initiation of synapsis, as axial associations are believed to be initiation sites for synapsis. Hop1p and Red1p are components of the axial and lateral elements, localizing to the mature SC and also to the unsynapsed axial elements in a *zip1* mutant; localization of Hop1p is dependent on Red1p.

In *S. pombe*, a typical tripartite form of the SC is not formed although linear elements resembling axial element of the SC are formed (124). In addition, *S. pombe* does not show obvious condensation of pachytene chromosomes (109, 125, 126). Lack

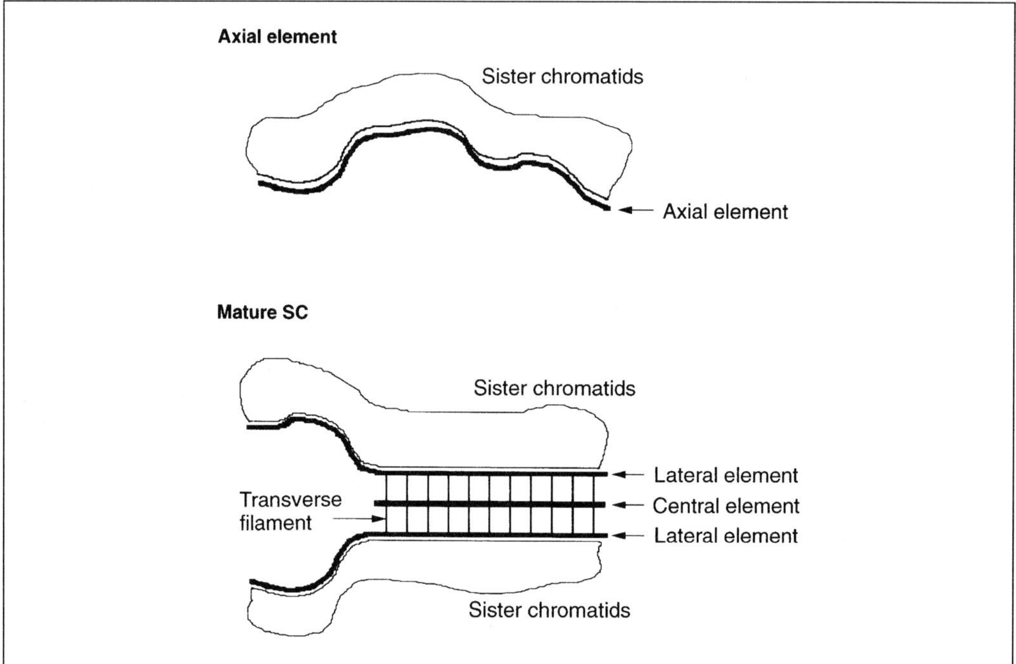

Fig. 11 Formation of the synaptonemal complex (SC). An axial element is formed on each pair of sister chromatids, and axial elements of homologous chromosomes become closely connected to form a tripartite structure of the SC. In the mature SC, axial elements are referred to as lateral elements.

of the SC, together with lack of chromosome condensation, results in a flexible chromosome structure, and the highly mobile meiotic prophase chromosomes may be reflective of this unusual condition. *S. pombe* may thus have evolved a strategy of telomere-led nuclear movement for homologous chromosome searching due to— perhaps to compensate for—the lack of a stable SC structure.

3.6 Sister chromatid cohesion during meiotic divisions

A major difference in chromosomal behaviour between mitosis and meiosis is that sister chromatids are held together during segregation of homologous chromosomes in meiosis I. Cohesion of sister chromatids and their stepwise separation are necessary for proper meiotic chromosome behaviours.

The exact underlying molecular mechanism remains to be elucidated, but cohesion at centromeres seems to exert a crucial role. This stems from studies in *Drosophila*. According to studies in *Drosophila*, cohesion along sister chromatid arms is released at anaphase of meiosis I whereas cohesion near centromeres is maintained until anaphase of meiosis II; MEI-S332 protein is involved in sister chromatid cohesion at the centromere; mutations in the *MEI-S332* gene lead to precocious separation of sister chromatids in anaphase of meiosis I, resulting in chromosome loss and missegregation in meiosis II (127).

Studies in *S. cerevisiae* have shown that mature SC is not required for meiotic sister chromatid cohesion, but axial elements may play a role in keeping sister chromatids together. In the *zip1* mutant, axial elements are formed but not synapsed, nevertheless precocious separation of sister chromatids is infrequent (128). In the *red1* mutant, which lacks axial elements, chromosomes undergo crossing-over but missegregate (129). This clearly points to a role of axial elements in sister chromatid cohesion.

The role of axial elements for sister chromatid cohesion is also inferred in *S. pombe*. There, although no typical SC is observed, linear elements resembling the axial element of the SC are formed (124). *S. pombe* Rec8 is a meiosis-specific protein involoved in sister chromatid cohesion; it is related to the mitotic cohesin Rad21 (Scc1p in *S. cerevisiae*) (130). A *rec8* mutant of *S. pombe* fails to produce linear elements and undergoes precocious separation of sister chromatids at high frequency (131), suggesting that axial elements are important in proper sister chromatid cohesion also in *S. pombe*. Consistantly, the Rec8 protein localizes along the chromosomes during meiotic prophase, and largely disappears from the chromosome arms during meiosis I but remains near centromeres until anaphase of meiosis II (132). The behaviour of the Rec8 protein is reminiscent of that of the MEI-S332 protein, but relationship between these proteins is not known.

4. Conclusion

A key requirement to a proper nuclear division is the coordination of nuclear and cytoskeletal events. These are basically independent, but coordinated, or interdependent, with one another under genetic control. Many aspects of molecular machines and mechanisms underlying nuclear division are shared between *S. cerevisiae* and *S. pombe*, as well as with higher eukaryotes. Differences between *S. cerevisiae* and *S. pombe* may come from the different modes of cell division (i.e. budding versus fission), thereby generating different cell geometry during nuclear division. In *S. cerevisiae*, mitotic nuclear division takes place asymmetrically owing to the asymmetrical mother–bud geometry; thus mechanisms to determine the site of budding and to position the dividing nucleus to the bud neck are required for proper nuclear division. In *S. pombe*, a nucleus divides symmetrically at the centre of the cell, and a septum is formed after nuclear division; thus mechanisms are required to form the septum at the right place at the right time.

Meiotic nuclear division may look like a variation of mitotic nuclear division and partly shares molecular machinery with mitosis. However, as outlined above, faithful chromosome segregation in meiosis relies largely on chromosomal events in meiotic prophase, and many of the meiotic prophase chromosomal events use meiosis-specific machinery. Despite a large body of cytological observation of meiotic prophase chromosomal events accumulated in animals and plants, the underlying molecular mechanisms remain largely unknown. Yeasts have provided a unique experimental system to elucidate molecular mechanisms for meiosis. Studies in *S. pombe* have provided insights into the role of meiotic telomeres because their telomere cluster is formed stably throughout meiotic prophase, whereas it is an ephemeral structure in most organisms. On the other hand, *S. cerevisiae* forms the synaptonemal complex similar to that in higher eukaryotes, and thus has provided an excellent model system for studying molecular basis of synapsis and recombination of homologous chromosomes. Yet strategies for meiosis seem to be diverse among species; molecular mechanisms for meiotic events remain to be elucidated for further understanding of essentials of meiosis.

Acknowledgements

The authors apologize to colleagues whose work could not be cited owing to space limitations. They thank Hubert Renauld for critical reading and English editing of the manuscript. They also thank Tokuko Haraguchi, Kentaro Nabeshima, Akira Nabetani, Ayumu Yamamoto, and Yukiko Yamashita for comments on the manuscript, and Trisha Davis, Mark Flory, Harry Scherthan, and Takashi Toda for communicating results prior to publication.

References

1. McCully, E. K. and Robinow, C. F. (1971) Mitosis in the fission yeast *Schizosaccharomyces pombe*: a comparative study with light and electron microscopy. *J. Cell Sci.*, **9**, 475.
2. Byers, B. and Goetsch, L. (1975) Behaviour of spindles and spindle plaques in the cell cycle and conjugation of *Saccharomyces cerevisiae*. *J. Bacteriol.*, **124**, 511.
3. Tanaka, K. and Kanbe. T. (1986) Mitosis in the fission yeast *Schizosaccharomyces pombe* as revealed by freeze–substitution electron microscopy. *J. Cell Sci.*, **80**, 253.
4. Ding, R., McDonald, K. L., and McIntosh, J. R. (1993) Three-dimensional reconstruction and analysis of mitotic spindles from the yeast *Schizosaccharomyces pombe*. *J. Cell Biol.*, **120**, 141.
5. Winey, M., Mamay, C. L., O'Toole, E. T., Mastronarde, D. N., Giddings, T. H., Jr., McDonald, K. L., and McIntosh, J. R. (1995) Three-dimentional ultrastructural anaysis of the *Saccharomyces cerevisiae* mitotic spindle. *J. Cell Biol.*, **129**, 1601.
6. Jacobs, C. W., Adams, A. E. M., Szaniszlo, P. J., and Pringle, J. R. (1988) Functions of microtubules in the *Saccharomyces cerevisiae* cell cycle. *J. Cell Biol.*, **107**, 1409.
7. Hagan, I. M. and Hyams, J. S. (1988) The use of cell division cycle mutants to investigate the control of microtubule distribution in the fission yeast *Schizosaccharomyces pombe*. *J. Cell Sci.*, **89**, 343.
8. Hagan, I. and Yanagida, M. (1997) Evidence of a cell cycle specific, spindle pole body mediated, nuclear positioning in fission yeast. *J. Cell Sci.*, **110**, 1851.
9. Yeh, E., Skibbens, R., Cheng, J., Salmon, E., and Bloom, K. (1995). Spindle dynamics and cell cycle regulation of dynein in the budding yeast *Saccharomyces cerevisiae*. *J. Cell Biol.*, **130**, 687.
10. Kahana, J. A., Schnapp, B. J., and Silver, P. A. (1995) Kinetics of spindle pole body separation in budding yeast. *Proc. Natl. Acad. Sci. U.S.A.*, **92**, 9707.
11. Carminati, J. L. and Stearns, T. (1997) Microtubules orient the mitotic spindle in yeast through dynein-dependent interactions with the cell cortex. *J. Cell Biol.*, **138**, 629.
12. Shaw, S. L., Yeh, F., Maddox, P., Salmon, E. D., and Bloom, K. (1997) Astral microtutule dynamics in yeast: a microtubule-based searching mechanism for spindle orientation and nuclear migration into the bud. *J. Cell Biol.*, **139**, 985.

13. Nabeshima, K., Nakagawa, T., Straight, A. F., Murray, A., Chikashige, Y., Yamashita, Y. M., Hiraoka, Y., and Yanagida, M. (1998) Dynamics of centromeres during metaphase-anaphase transition in fission yeast: dis1 is implicated in force balance in metaphase bipolar spindle. *Mol. Biol. Cell*, **9**, 3211.

14. Bullitt, E., Rout, M. P., Kilmartin, J. V., and Akey, C. W. (1997) The yeast spindle pole body is assembled around a central crystal of Spc42p. *Cell*, **89**, 1077.

15. Rout, M. P. and Kilmartin, J. V. (1990) Components of the yeast spindle and spindle pole body. *J. Cell Biol.*, **111**, 1913.

16. Wigge, P. A., Jensen, O. N., Holmes, S., Soues, S., Mann, M., and Kilmartin, J. V. (1998) Analysis of the *Saccharomyces* spindle pole by matrix-assisted laser desorption/ionization (MALDI) mass spectrometry. *J. Cell Biol.*, **141**, 967.

17. Adams, I. R. and Kilmartin, J. V. (1999) Localization of core spindle pole body (SPB) components during SPB duplication in *Saccharomyces cerevisiae*. *J. Cell Biol.*, **145**, 809.

18. Knop, M. and Schiebel, E. (1998) Receptors determine the cellular localization of a γ-tubulin complex and thereby the site of microtubule formation. *EMBO J.*, **17**, 3952.

19. Pereira, G., Knop, M., and Schiebel, E. (1998) Spc98p directs the yeast γ-tubulin complex into the nucleus and is subject to cell cycle-dependent phosphorylation on the nuclear side of the spindle pole body. *Mol. Biol. Cell*, **9**, 775.

20. Kilmartin, J. V., Dyos, S. L., Kershaw, D., and Finch, J. T. (1993) A spacer protein in the *Saccharomyces cerevisiae* spindle pole body whose transcription is cell-cycle regulated. *J. Cell Biol.*, **123**, 1175.

21. Weiss, E. and Winey, M. (1996) The *Saccharomyces cerevisiae* spindle pole body duplication gene MPS1 is part of a mitotic checkpoint. *J. Cell Biol.*, **132**, 111.

22. Ding, R., West, R. R., Morphew, D. M., Oakley, B. R., and McIntosh, J. R. (1997) The spindle pole body of *Schizosaccharomyces pombe* enters and leaves the nuclear envelope as the cell cycle proceeds. *Mol. Biol. Cell*, **8**, 1461.

23. Horio, T., Uzawa, S., Jung, M. K., Oakley, B. R., Tanaka, K., and Yanagida, M. (1991) The fission yeast γ-tubulin is essential for mitosis and is localized at microtubule organizing centers. *J. Cell Sci.*, **99**, 693.

24. Moser, M. J., Flory, M. R., and Davis, T. N. (1997) Calmodulin localizes to the spindle pole body of *Schizosaccharomyces pombe* and performs an essential function in chromosome segregation. *J. Cell Sci.*, **110**, 1805.

25. West, R. R., Vaisberg, E. V., Ding, R., Nurse, P., and McIntosh, J. R. (1998) *cut11*[+]: a gene required for cell cycle-dependent spindle pole body anchoring in the nuclear envelope and bipolar spindle formation in *Schizosaccharomyces pombe*. *Mol. Biol. Cell*, **9**, 2839.

26. Bridge, A. J., Morphew, M., Bartlett, R., and Hagan, I. M. (1998) The fission yeast SPB component Cut12 links bipolar spindle formation to mitotic control. *Genes Dev.*, **12**, 927.

27. Hagan, I. and Yanagida, M. (1995) The product of the spindle formation gene *sad1*[+] associates with the fission yeast spindle pole body and is essential for viability. *J. Cell Biol.*, **129**, 1033.

28. He, X., Jones, M. H., Winey, M., and Sazer, S. (1998) mph1, a member of the Mps1-like family of dual specificity protein kinase, is required for the spindle checkpoint in *S. pombe*. *J. Cell Sci.*, **111**, 1635.

29. O'Toole, E. T., Winey, M., and McIntosh, J. R. (1999) High-voltage electron tomography of spindle pole bodies and early mitotic spindles in the yeast *Saccharomyces cerevisiae*. *Mol. Biol. Cell*, **10**, 2017.

30. Sobel, S.G. and Snyder, M. (1995) A highly divergent γ-tubulin gene is essential for cell growth and proper microtubule organization in *Saccharomyces cerevisiae*. *J. Cell Biol.*, **131**, 1775.

31. Pereira, G. and Schiebel, E. (1997) Centrosome-microtubule nucleation. *J. Cell Sci.*, **110**, 295.

32. Alfa, C. E., Ducommun, B., Beach, D., and Hyams, J. S. (1990) Distinct nuclear and spindle pole body population of cyclin–cdc2 in fission yeast. *Nature*, **347**, 680.

33. Stearns, T., Evans, L., and Kirschner, M. (1991) γ-tubulin is a highly conserved component of the centrosome. *Cell*, **65**, 825.

34. Masuda, H., Sevik. M., and Cande, W. Z. (1992) *In vitro* microtubule nucleating activity of spindle pole bodies in fission yeast *Schizosaccharomyces pombe*: cell cycle-dependent activation in *Xenopus* cell-free extracts. *J. Cell Biol.*, **117**, 1055.

35. Masuda, H. and Shibata, T. (1996) Role of gamma-tubulin in mitosis specific microtubule nucleation from the *Schizosaccharomyces pombe* spindle pole body. *J. Cell Sci.*, **109**, 165.

36. Palmer, R. E., Sullivan, D. S., Huffaker, T. H., and Koshland, D. (1992) Role of astral microtubules in spindle orientation and migration in the budding yeast, *Saccharomyces cerevisiae*. *J. Cell Biol.*, **119**, 583.

37. Sullivan, D. S. and Huffaker, T. C. (1992) Astral microtubules are not required for anaphase B in *Saccharomyces cerevisiae*. *J. Cell Biol.*, **119**, 379.

38. Brachat, A., Kilmartin, J. V., Wach, A., and Philippsen, P. (1998) *Saccharomyces cerevisiae* cells with defective spindle pole body outer plaques accomplish nuclear migration via half-bridge organized microtubules. *Mol. Biol. Cell*, **9**, 977.

39. Stearns, T. (1997) Motoring to the finish: kinesin and dynein work together to orient the yeast mitotic spindle. *J. Cell Biol.*, **138**, 957.

40. Heil-Chapdelaine, R. A., Adames, N. R., and Cooper, J. A. (1999) Formin' the connection between microtubules and the cell cortex. *J. Cell Biol.*, **144**, 809.

41. Huyett, A., Kahana, J., Silver, P., Zeng, X., and Saunders, W. (1998) The Kar3p and Kip2p motors function antagonistically at the spindle poles to influence cytoplasmic microtubule numbers. *J. Cell Sci.*, **111**, 295.

42. Miller, R. K., Heller, K. K., Frisen, L., Wallack, D. L., Loayza, D., Gammie, A. E., and Rose, M. D. (1998) The kinesin-related proteins, Kip2p and Kip3p, function differently in nuclear migration in yeast. *Mol. Biol. Cell*, **9**, 2051.

43. Segal, M., Clarke, D., and Reed, S. I. (1998) Clb5-associated kinase activity is required early in the spindle pathway for correct preanaphase nuclear positioning in *Saccharomyces cerevisiae*. *J. Cell Biol.*, **143**, 135.

44. Umesono, K., Toda, T., Hayashi, S., and Yanagida, M. (1983) Cell division cycle genes *nda2* and *nda3* of the fission yeast *Schizosaccharomyces pombe* control microtubular organization and sensitivity to anti-mitotic benzimidazole compounds. *J. Mol. Biol.*, **168**, 271.

45. Hiraoka, Y., Toda, T. and Yanagida, M. (1984) The *NDA3* gene of fission yeast encodes β-tubulin: a cold-sensitive *nda3* mutation reversibly blocks spindle formation and chromosome movement in mitosis. *Cell*, **39**, 349.

46. Beinhauer, J. D., Hagan, I. M., Hegemann, J. H., and Fleig, U. (1997) Mal3, the fission yeast homologue of the human APC-interacting protein EB-1, is required for microtubule integrity and the maintenance of cell form. *J. Cell Biol.*, **139**, 717.

47. Hoyt, M. A., He, L., Loo, K. K., and Saunders, W. S. (1992) Two *Saccharomyces cerevisiae* kinesin-related gene-products required for mitotic spindle assembly. *J. Cell Biol.*, **118**, 109.

48. Roof, D. M., Meluh, P. B., and Rose, M. D. (1992) Kinesin-related proteins required for assembly of the mitotic spindle. *J. Cell Biol.*, **118**, 95.

49. Saunders, W. S. and Hoyt, M. A. (1992) Kinesin-related proteins required for structural integrity of the mitotic spindle. *Cell*, **70**, 451.

50. Saunders, W., Hornack, D., Lengyel, V., and Deng, C. (1997) The *Saccharomyces cerevisiae* kinesin-related motor Kar3p acts at preanaphase spindle poles to limit the number and length of cytoplasmic microtubules. *J. Cell Biol.*, **137**, 417.

51. Pasqualone, D. and Huffaker, T. C. (1994). *STU1*, a suppressor of a beta tubulin mutation, encodes a novel and essential component of the yeast mitotic spindle. *J. Cell Biol.*, **127**, 1973.

52. Hagan, I. and Yanagida, M. (1990) Novel potential mitotic motor protein encoded by the fission yeast *cut7*$^+$ gene. *Nature*, **347**, 563.

53. Hagan, I. and Yanagida, M. (1992) Kinesin-related cut7 protein associates with mitotic and meiotic spindles in fission yeast. *Nature*, **356**, 74.

54. Pidoux, A, L., LeDizet, M., and Cande, W. Z. (1996) Fission yeast pkl1 is a kinesin-related protein involved in mitotic spindle function. *Mol. Biol. Cell*, **7**, 1639.

55. Guacci, V., Hogan, E., and Koshland, D. (1994) Chromosome condensation and sister chromatid pairing in budding yeast. *J. Cell Biol.*, **125**, 517.

56. Saka, Y., Sutani, T., Yamashita, Y., Saitoh, S., Takeuchi, M., Nakaseko, Y., and Yanagida, M. (1994) Fission yeast cut3 and cut14, members of a ubiquitous protein family, are required for chromosome condensation and segregation in mitosis. *EMBO J.*, **13**, 4938.

57. Uemura, T. and Yanagida, M. (1984) Isolation of type I and II topoisomerase mutants from fission yeast: single and double mutants show different phenotypes in cell growth and chromatin organization. *EMBO J.*, **3**, 1737.

58. Holm, C., Goto, T., Wang, J. C., and Botstein, D. (1985) DNA topoisomerase II is required at the time of mitosis in yeast. *Cell*, **41**, 553.

59. Uemura, T., Ohkura, H., Adachi, Y., Morino, K., Shiozaki, K., and Yanagida, M. (1987) DNA topoisomerase II is required for condensation and separation of mitotic chromosomes in *S. pombe*. *Cell*, **50**, 917.

60. Strunnikov, A. V., Larionov, V. L., and Koshland, D. (1993) *SMC1*: an essential yeast gene encoding a putative head–rod–tail protein is required for nuclear division and defines a new ubiquitous protein family. *J. Cell Biol.*, **123**, 1635.

61. Hirano, T. (1999) SMC-mediated chromosome mechanics: a conserved scheme from bacteria to vertebrate? *Genes Dev.*, **13**, 11.

62. Strunnikov, A. V., Hogan, E., and Koshland, D. (1995) *SMC2*, a *Saccharomyces cerevisiae* gene essential for chromosome segregation and condensation, defines a subgroup within the SMC-family. *Genes Dev.*, **9**, 587.

63. Sutani, T. and Yanagida, M. (1997) DNA renaturation activity of the SMC complex implicated in chromosome condensation. *Nature*, **388**, 798.

64. Matsusaka, T., Imamoto, N., Yoneda, Y., and Yanagida, M. (1998) Mutations in fission yeast Cut15, an importin a homolog, lead to mitotic progression without chromosome condensation. *Curr. Biol.*, **8**, 1031.

65. Uhlmann, F., Lottspeich, F., and Nasmyth, K. (1999) Sister-chromatid separation at anaphase onset is promoted by cleavage of the cohesin subunit Scc1. *Nature*, **400**, 37.

66. Furuya, K., Takahashi, K., and Yanagida, M.(1998) Faithful anaphase is ensured by Mis4, a sister chromatid cohesion protein molecule required in S phase and not destroyed in the G1 phase. *Genes Dev.*, **12**, 3408.

67. Guacci, V., Koshland, D., and Strunnikov, A. (1997) A direct link between sister chromatid cohesion and chromosome condensation revealed through the analysis of MCD1 in *S. cerevisiae*. *Cell*, **91**, 47.

68. Saitoh, S., Takahashi, K., and Yanagida, M. (1997) Mis6, a fission yeast inner centromere protein, acts during G1/S and forms specialized chromatin required for equal segregation. *Cell*, 90, 131–143.

69. Lechner, J. and Carbon, J. (1991) A 240kd multisubunit protein complex, CBF3, is a major component of the budding yeast centromere. *Cell*, **64**, 717.

70. Stemmann, O. and Lechner, J. (1996) The *Saccharomyces cerevisiae* kinetochore contains a cyclin–CDK complexing homologue, as identified by *in vitro* reconstitution. *EMBO J.*, **15**, 3611.

71. Connelly, C. and Hieter, P. (1996) Budding yeast *SKP1* encodes an evolutionary conserved kinetochore protein required for cell cycle progression. *Cell*, **86**, 275.

72. Sorger, P. K., Doheny, K. F., Hieter, P., Kopski, K. M., Huffaker, T. C., and Hyman, A. A. (1995) Two genes required for the binding of an essential *Saccharomyces cerevisiae* kineto-chore complex to DNA. *Proc. Natl. Acad. Sci. U.S.A.*, **92**, 12026.

73. Kaplan, K. B., Hyman, A. A., and Sorger, P. K. (1997) Regulating the yeast kinetochore by ubiquitin-dependent degradation and Skp1p mediated phosphorylation. *Cell*, **91**, 491.

74. Kingsbury, J. and Koshland, D. (1991) Centromere-dependent binding of yeast mini-chromosomes to microtubules *in vitro*. *Cell*, **66**, 483.

75. Hyman, A. A., Middleton, K., Centola, M., Mitchison, T. J., and Carbon, J. (1992) Micro-tubule motor activity of a yeast centromere-binding protein complex. *Nature*, **359**, 536.

76. Kingsbury, J. and Koshland, D. (1993) Centromere function on minichromosomes isolated from budding yeast. *Mol. Biol. Cell*, **4**, 859.

77. Sorger, P. K., Severin, F. F., and Hyman, A. A. (1994) Factors required for the binding of reassembled yeast kinetochores to microtubules *in vitro*. *J. Cell Biol.*, **127**, 995.

78. Severin, F. F., Sorger, P. K., and Hyman, A. A. (1997) Kinetochores distinguish GTP from GDP forms of the microtubule lattice. *Nature*, **388**, 888.

79. Biggins, S., Severin, F. F., Bhalla, N., Sassoon, A. A., and Murray, A. W. (1999) The con-served protein kinase Ipl1 regulates microtubule binding to kinetochores in budding yeast. *Genes Dev.*, **13**, 532.

80. Bloecher, A. and Tatchell, K. (1999) Defects in *Saccharomyces cerevisiae* protein phosphatase type 1 activate the spindle/kinetochore checkpoint. *Genes Dev.*, **13**, 517.

81. Sassoon, I., Severin, F. F., Andrews, P. D., Taba, M. R., Kaplan, K. B., Ashford, A. J., Stark, M. J. R., Sorger, P. K., and Hyman, A. A. (1999) Regulation of *Saccharomyces cerevisiae* kinetochores by the type 1 phosphatase Glc7p. *Genes Dev.*, **13**, 545.

82. Francisco, L., Wang, W., and Chan, C. S. (1994) Type 1 protein phosphatase acts in opposition to Ipl1 protein kinase in regulating yeast chromosome segregation. *Mol. Cell. Biol.*. **14**, 4731.

83. Ishii, K., Kumada, K., Toda, T., and Yanagida, M. (1996) Requirement for PP1 phosphatase and 20S cyclosome/APC for the onset of anaphase is lessened by the dosage increase of a novel gene $sds23^+$. *EMBO J.*, **15**, 6629.

84. Yanagida, M. (1998) Fission yeast *cut* mutations revisited: control of anaphase. *Trends Cell Biol.*, **8**, 144.

85. Zachariae, W. and Nasmyth, K. (1999) Whose end is destruction: cell division and the anaphase-promoting complex. *Genes Dev.*, **13**, 2039.

86. Yamada, H., Kumada, K., and Yanagida, M. (1997) Distinct subunit functions and cell cycle regulated phosphorylation of 20S APC/cyclosome required for anaphase in fission yeast. *J. Cell Sci.*, **110**, 1793.

87. Guacci, V., Hogan, E., and Koshland, D. (1997) Centromere position in budding yeast: evidence for anaphase A. *Mol. Biol. Cell*, **8**, 957.

88. Straight, A. F., Marshall, W. F., Sedat, J. W., and Murray, A. W. (1997) Mitosis in living budding yeast: anaphase A but no metaphase plate. *Science*, **277**, 574.

89. Funabiki, H., Hagan, I., Uzawa, S., and Yanagida, M. (1993) Cell cycle dependent specific positioning and clustering of centromeres and telomeres in fission yeast. *J. Cell Biol.*, **121**, 961.

90. Masuda, H., Hirano, T., Yanagida, M., and Cande, W. Z. (1990) *In vitro* reactivation of spindle elongation in fission yeast *nuc2* mutant cells. *J. Cell Biol.*, **110**, 417.

91. Saunders, W. S., Koshland, D., Eshel, D., Gibbons, I. R., and Hoyt, M. A. (1995) *Saccharomyces cerevisiae* kinesin- and dynein-related proteins required for anaphase chromosome segregation. *J. Cell Biol.*, **128**, 617.

92. Pellman, D., Bagget, M., Tu, Y. H., and Fink, G. R. (1995) Two microtubule associated proteins required for anaphase spindle movement in *Saccharomyces cerevisiae. J. Cell Biol.*, **130**, 1373 (erratum *J. Cell Biol.*, **131,** 561).

93. Juang, Y. L., Huang, J., Peters, J. M., McLaughlin, M. E.,Tai, C. Y. and Pellman, D. (1997) APC-mediated proteolysis of Ase1 and the morphogenesis of the mitotic spindle. *Science*, **275**, 1311.

94. Hagan, I. M. and Hyams, J. S. (1996) Forces acting on the fission yeast anaphase spindle. *Cell Motil. Cytoskel.*, **34**, 69.

95. Yamamoto, A., West, R. R., MaIntosh, J. R., and Hiraoka, Y. (1999) A cytoplasmic dynein heavy chain is required for oscillatory nuclear movement of meiotic prophase and efficient meiotic recombination in fission yeast. *J. Cell Biol.*, **145**, 1233.

96. Svoboda, A., Bahler, J., and Kohli, J. (1995) Microtubule-driven nuclear movements and linear elements as meiosis-specific characteristics of the fission yeasts *Schizosaccharomyces versatilis* and *Schizosaccharomyces pombe. Chromosoma*, **104**, 203.

97. Ding, D. Q., Chikashige, Y., Haraguchi, T., and Hiraoka, Y. (1998) Oscillatory nuclear movement in fission yeast meiotic prophase is driven by astral microtubules, as revealed by continuous observation of chromosomes and microtubules in living cells. *J. Cell Sci.*, **111**, 701.

98. Hayashi, A., Ogawa, H., Kohno, K., Gasser, S., and Hiraoka, Y. (1998) Meiotic behaviors of chromosomes and microtubules in budding yeast: relocalization of centromeres and telomeres during meiotic prophase. *Genes Cells*, **3**, 587.

99. Maddox, P., Chin, E., Mallavarapu, A., Yeh, E., Salmon, E. D., and Bloom, K. (1999) Microtubule dynamics from mating through the first zygotic division in the budding yeast *Saccharomyces cerevisiae. J. Cell Biol.*, **144**, 977.

100. Bascom-Slack, C. A. and Dawson, D. S. (1997) The yeast motor protein, Kar3p, is essential for meiosis I. *J. Cell Biol.*, **139**, 459.

101. Moens, P. B. and Rapport, E. (1971) Spindles, spindle plaques, and meiosis in the yeast *Saccharomyces cerevisiae. J. Cell Biol.*, **50**, 344.

102. Hirata, A. and Tanaka, K. (1982) Nuclear behavior during conjugation and meiosis in the fission yeast *Schizosaccharomyces pombe. J. Gen. Appl. Microbiol.*, **28**, 263.

103. Chikashige, Y., Ding, D.-Q., Imai, Y., Yamamoto, M., Haraguchi, T., and Hiraoka, Y. (1997) Meiotic nuclear reorganization: switching the position of centromeres and telomeres in fission yeast *Schizosaccharomyces pombe. EMBO J.*, **16**, 193.

104. Yang, C. H., Lambie, E. J., Hardin, J., Craft, J., and Snyder, M. (1989) Higher order structure is present in the yeast nucleus: autoantibody probes demonstrate that the nucleolus lies opposite the spindle pole body. *Chromosoma*, **98**, 123.

105. Jin, Q., Trelles-Sticken, E., Scherthan, H., and Loidl, J. (1998) Yeast nuclei display prominent centromere clustering that is reduced in nondividing cells and in meiotic prophase. *J. Cell Biol.*, **41**, 21.

106. Gotta, M., Laroche, T., Formenton, A., Maillet, L., Scherthan, H., and Gasser, S. M. (1996) The clustering of telomeres and colocalization with Rap1, Sir3 and Sir4 proteins in wild-type *Saccharomyces cerevisiae*. *J. Cell Biol.*, **134**, 1349.

107. Goffeau, A. *et al.* (1997) The yeast genome directory. *Nature* (suppl.), **387**, 1.

108. Fan, J. B., Chikashige, Y., Smith, C. L., Niwa, O., Yanagida, M., and Cantor, C. R. (1989) Construction of a Not I restriction map of the fission yeast *Schizosaccharomyces pombe* genome. *Nucleic Acids Res.*, **17**, 2801.

109. Chikashige, Y., Ding, D. Q., Funabiki, H., Haraguchi, T., Mashiko, S., Yanagida, M., and Hiraoka, Y. (1994) Telomere-led premeiotic chromosome movement in fission yeast. *Science*, **264**, 270.

110. Trelles-Sticken, E., Loidl, J., and Scherthan, H. (1999) Bouquet formation in budding yeast: initiation of recombination is not required for meiotic telomere clustering. *J. Cell Sci.*, **112**, 651.

111. Haber, J. E. (1998) Searching for a partner. *Science*, **279**, 823.

112. Shimanuki, M., Miki, F., Ding, D. Q., Chikashige, Y., Hiraoka, Y., Horio, T., and Niwa, O. (1997) A novel fission yeast gene, $kms1^+$, is required for the formation of meiotic prophase-specific nuclear architecture. *Mol. Gen. Genet.*, **254**, 238.

113. Cooper, J. P., Watanabe, Y., and Nurse, P. (1998) Fission yeast Taz1 protein is required for meiotic telomere clustering and recombination. *Nature*, **23**, 828.

114. Nimmo, E. R., Pidoux, A. L., Perry, P. E., and Allshire, R. C. (1998) Defective meiosis in telomere-silencing mutants of *Schizosaccharomyces pombe*. *Nature*, **23**, 825.

115. Hiraoka, Y. (1998) Meiotic telomeres: a matchmaker for homologous chromosomes. *Genes Cells*, **3**, 405.

116. Weiner, B. M. and Kleckner, N. (1994) Chromosome pairing via multiple interstitial interactions before and during meiosis in yeast. *Cell*, **77**, 977.

117. Kleckner, N. (1996) Meiosis: how could it work? *Proc. Natl. Acad. Sci. U.S.A.*, **6**, 8167.

118. Loidl, J., Klein, F., and Scherthan, H. (1994) Homologous pairing is reduced but not abolished in asynaptic mutants of yeast. *J. Cell Biol.*, **125**, 1191.

119. Rockmill, B. and Roeder, G. R. (1998) Telomere-mediated chromosome pairing during meiosis in budding yeast. *Genes Dev.*, **12**, 2574.

120. Chua, P. R. and Roeder, G. S. (1997) Tam1, a telomere-associated meiotic protein, functions in chromosome synapsis and crossover interference. *Genes Dev.*, **11**, 1786.

121. Conrad, M. N., Dominguez, A. M., and Dresser, M. E. (1997) Ndj1, a meiotic telomere protein required for normal chromosome synapsis and segregation in yeast. *Science*, **276**, 1252.

122. Goetsch, L. and Byers, B. (1982) Meiotic cytology of *Saccharomyces cerevisiae* in protoplast lysates. *Mol. Gen. Genet.*, **187**, 54.

123. Roeder, G. S. (1997) Meiotic chromosomes: it takes two to tango. *Genes Dev.*, **11**, 2600.

124. Bahler, J., Wyler, T., Loidl, J., and Kohli, J. (1993) Unusual nuclear structures in meiotic prophase of fission yeast: a cytological analysis. *J. Cell Biol.*, **121**, 241.

125. Scherthan, H., Bahler, J., and Kohli, J. (1994) Dynamics of chromosome organization and pairing during meiotic prophase in fission yeast. *J. Cell Biol.*, **127**, 273.

126. Hartsuiker, E., Bahler, J., and Kohli, J. (1998) The role of topoisomerase II in meiotic chromosome condensation and segregation in *Schizosaccharomyces pombe*. *Mol. Biol. Cell*, **9**, 2739.

127. Miyazaki, W. Y. and Orr-Weaver, T. L. (1994) Sister-chromatid cohesion in mitosis and meiosis. *Annu. Rev. Genet.*, **28**, 167.

128. Sym, M. and Roeder, G. S. (1995) Zip1-induced changes in synaptonemal complex structure and polycomplex assembly. *J Cell Biol.*, **128**, 455.
129. Rockmill, B. and Roeder, G. S. (1990) Meiosis in asynaptic yeast. *Genetics*, **126**, 563.
130. Parisi, S., McKay, M. J., Molnar, M., Thompson, M. A., van der Spek, P. J., van Drunen-Schoenmaker, E., Kanaar, R., Lehmann, E., Hoeijmakers, J. H., and Kohli, J. (1999) Rec8p, a meiotic recombination and sister chromatid cohesion phosphoprotein of the rad21p family conserved from fission yeast to humans. *Mol. Cell. Biol.*, **19**, 3515.
131. Molnar, M., Bahler, J., Sipiczki, M., and Kohli, J. (1995) The *rec8* gene of *Schizosaccharomyces pombe* is involved in linear element formation, chromosome pairing and sister-chromatid cohesion during meiosis. *Genetics*, **141**, 61.
132. Watanabe, Y. and Nurse, P. (1999) Meiosis specific cohesin Rec8p is required for reductional chromosome segregation at meiosis. *Nature*, **400**, 461.

6 | RNA polymerase II transcriptional machinery

MICHAEL HAMPSEY

1. Introduction

Transcription initiation by RNA polymerase II (RNAP II) requires interaction between *cis*-acting promoter elements and *trans*-acting factors. The eukaryotic promoter consists of core elements that include the TATA box and other DNA sequences that define transcription start sites, and regulatory elements that either enhance or repress transcription in a gene-specific manner. The core promoter is the site of assembly of the transcription preinitiation complex (PIC), which includes RNAP II and the general transcription factors (GTFs) TATA-binding protein (TBP), TFIIB, TFIIE, TFIIF, and TFIIH. Regulatory elements bind gene-specific factors to affect the rate of transcription by interacting, either directly or indirectly, with components of the general transcriptional machinery. Biochemical characterization of the requirements for activation identified another class of transcription factors, designated coactivators or mediators, that are required for activation by many, but not all, activators. Accordingly, coactivators are neither gene-specific factors nor GTFs and include the SRB–MED, TFIID, and SAGA complexes, as well as other factors.

In this chapter there is emphasis on the combined roles of biochemistry and genetics in defining the GTF and the coactivator requirements for transcription, as well as a section on general transcriptional repressors. As eukaryotic transcription does not occur on naked DNA templates, but rather in the context of chromatin, a section on chromatin and its effect on transcription is also included. This information is presented primarily from the perspective of the yeast *Saccharomyces cerevisiae*. However, this subject cannot be considered separately from the RNAP II transcriptional machinery of higher eukaryotic organisms, where studies of human, rat, and fly systems have often led the way. Therefore, an attempt has been made to integrate information from the yeast and metazoan systems wherever appropriate. The chapter is organized into the following categories:

- Promoter structure: core and regulatory elements
- RNAP II
- GTFs: TBP, TFIIB, TFIIE, TFIIF, and TFIIH

- Coactivators: SRB–MED, TFIID, HAT, and SWI–SNF complexes
- General repressors: Ydr1p–Bur6p, Mot1p, and HDA complexes
- Mechanisms of transcriptional activation

The chapter finishes with a brief discussion of the directions this field is likely to take in the immediate future.

2. Promoter elements

Eukaryotic promoters consist of two functionally distinct elements: the core promoter and regulatory elements (Fig. 1). The core promoter is the site of PIC assembly and typically includes a TATA box and an initiator sequence (Inr) encompassing the transcription start site. Regulatory elements are gene-specific sequences located upstream

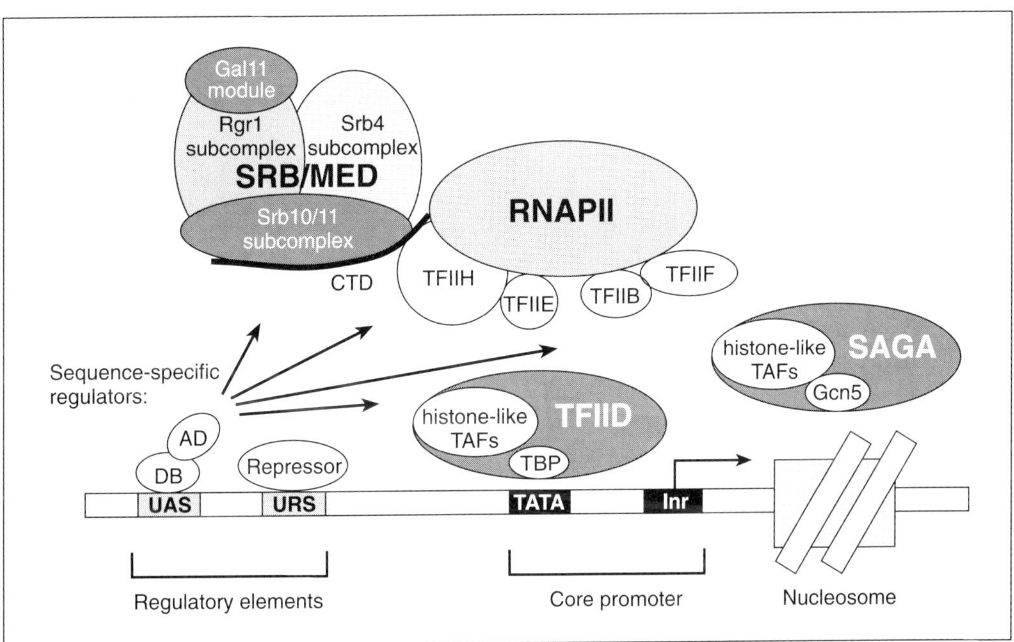

Fig. 1 The RNAP II transcription initiation machinery. A typical RNAP II promoter includes core elements (TATA box and Inr region), which function to position RNAP II at the initiation site, and regulatory elements, typically located 100–1500 bp upstream of the core promoter, which regulate the rate of transcription. Regulatory regions include enhancer (UAS) and repressor (URS) elements that bind sequence-specific transcription factors either to stimulate or to repress transcription. The transcriptional machinery shown here includes: (1) RNAP II and the GTFs TBP, TFIIB, TFIIE, TFIIF, and TFIIH; (2) the SRB–MED complex; (3) TFIID, comprised of TBP and TBP-associated factors (TAFs); and (4) the SAGA histone acetyltransferase complex, comprised of Gcn5, the same histone-like TAFs as TFIID, and other subunits. The SRB–MED complex is composed of functionally distinct subcomplexes and associates with RNAP II through the C-terminal repeat domain (CTD) of Rpb1 to form a holoenzyme complex. Sequence-specific activators are bipartite, including a DNA-binding domain (DB) and an activation domain (AD). Activators interact with GTFs, SRB–MED, TFIID, SAGA, or other coactivators to facilitate assembly of the preinitiation complex. Repressors impair transcription by several distinct mechanisms, some of which involve interaction with cofactors, including the Sin3–Rpd3 or Tup1–Ssn6 complex (not shown).

of the core promoter and include upstream activation sequences (UASs) and upstream repressor sequences (URSs). UAS and URS elements function as binding sites for activators and repressors of transcription. Accordingly, core elements define the site of initiation by RNAP II, whereas regulatory elements control the rate of transcription in response to physiological or developmental cues.

The TATA box dictates either the initiation site or the window within which initiation can occur. Mutational analyses defined TATAAA as the consensus TATA sequence in *S. cerevisiae* (1). Derivatives of this sequence also confer TATA function, albeit with diminished activity. Many yeast promoters contain more than one functional TATA box, each directing initiation from distinct Inr sites (2). Yeast promoters lacking canonical TATA boxes have also been identified; regardless, transcription from 'TATA-less' promoters remains TBP dependent (3). Thus, a TATA-less promoter is unlikely to be a fundamentally distinct promoter, but is instead one that binds TBP with relatively weak affinity.

Inr elements are less well defined than the TATA box. In organisms other than *S. cerevisiae*, including *Schizosaccharomyces pombe*, initiation typically occurs approximately 30 bp from TATA, suggesting that the Inr is defined simply by spacing from TATA. However, in *S. cerevisiae* initiation occurs within a window located 40–120 bp from TATA. The TATA element defines the window within which initiation can occur, but specific sequences within the window define the Inr element (2, and references therein). Preferred Inr sequences have been identified, but no clearly defined consensus sequence has emerged (4). In metazoan organisms, Inr elements were initially defined at TATA-less promoters and were implicated in transcriptional control by nucleating PIC assembly in a TATA-independent manner (reviewed in Reference 5). Metazoan Inr-binding proteins have also been identified. It is not clear whether yeast and metazoan Inr elements are functional counterparts; specific Inr-binding proteins have not been identified in yeast.

UAS and URS elements are typically located 100–1,500 bp upstream of the core promoter. Once associated with their cognate UAS elements, transcriptional activators stimulate the rate of transcription either by direct contact with GTFs or indirectly through coactivators. Conversely, repressors bind URS elements and impair transcription, in some cases by mechanisms analogous to those of activators. For example, repressors have been identified that bind URS elements and interact directly with transcriptional corepressors such as the Sin3p–Rpd3p or Tup1p–Ssn6p complexes (6, 7). Other classes of repressors have also been described, including those that interfere with activator function and others that interfere directly with PIC assembly (reviewed in References 8–10).

3. RNA polymerase II

There are three functionally distinct RNA polymerases in eukaryotic cells. RNAP I is dedicated to transcription of ribosomal DNA (rDNA) genes (excluding 5S rRNA). RNAP II transcribes protein-encoding genes and is therefore responsible for all mRNA synthesis, as well as synthesis of some small nuclear RNAs (snRNAs). RNAP

III transcribes genes encoding tRNAs, 5S rRNA, and some snRNAs. RNAP II and its machinery are the focus of this chapter.

3.1 RNAP II structure

RNAP II of *S. cerevisiae* comprises 12 subunits encoded by the *RPB1—RPB12* genes (11). Although once thought to lack counterparts of Rpb4p and Rpb9p, *S. pombe* RNAP II is also composed of 12 subunits, homologous to Rpb1p–Rpb12p (12, and references therein). These subunits are highly conserved among other eukaryotic RNAP IIs; indeed, six subunits of human RNAP II can functionally replace their counterparts in yeast (13). The two largest subunits, Rpb1p ($\sim$200 kDa) and Rpb2p ($\sim$150 kDa), are homologues of the β' and β subunits respectively of bacterial RNA polymerase. Rpb3p and Rpb11p form a heterodimer that comprises the counterpart of the α subunit of bacterial RNA polymerase (14). None of the RNAP II subunits appears to be closely related to the bacterial σ-subunit family, although structural and functional similarities between σ and certain GTFs have been noted (reviewed in Reference 15).

RNAP I and III include homologues of Rpb1p, Rpb2p, Rpb3p, and Rpb11p. Five other subunits (Rpb5p, Rpb6p, Rpb8p, Rpb10p, and Rpb12p) are common to all three RNA polymerases. Only Rpb4p, Rpb7p, and Rpb9p are unique to RNAP II; Rpb4p and Rpb7p form a subcomplex that can be dissociated from RNAP II. Thus, RNAPs are assembled from common as well as class-specific subunits. Ten of the yeast genes encoding RNAP II subunits are essential for cell viability. Only the *RPB4* and *RPB9* genes are dispensable, although deletion of either gene confers conditional growth phenotypes. The structure of RNAP II from *S. cerevisiae* was recently solved to 5 Å resolution by X-ray crystallography (16) and the structure of an elongating RNAP II–DNA–RNA ternary complex has been solved by electron crystallography (17).

3.2 The C-terminal repeat domain

A unique feature of RNAP II is the presence of tandem repeats of a heptapeptide sequence at the C-terminus of the largest subunit (Rpb1p) of RNAP II (reviewed in Reference 18). The C-terminal repeat domain (CTD) has the consensus sequence Tyr-Ser-Pro-Thr-Ser-Pro-Ser and is highly conserved among eukaryotic organisms. Although the CTD is a ubiquitous feature of RNAP II, the repeat length varies. *S. cerevisiae* Rpb1p includes 26 to 27 repeats, *C. elegans* 34 repeats, *Drosophila* 43 repeats, and human 52 repeats, suggesting that repeat length increases with increasing genome complexity.

There are two forms of RNAP II *in vivo*, designated RNAP IIO, which is extensively phosphorylated at the CTD, and RNAP IIA, which is unphosphorylated (reviewed in Reference 19). RNAP IIA preferentially enters the PIC, whereas RNAP IIO is found in the elongating complex. Conversion of RNAP IIA to RNAP IIO occurs concomitant with or shortly after the transition from initiation to elongation. These results implicate the CTD in conversion of RNAP II from a form involved in promoter recognition to an elongation-competent form.

Recruitment of RNAP IIA to the PIC and subsequent conversion to RNAP IIO implies the existence of a CTD kinase. Several CTD kinases have been identified, including TFIIH, the Srb10p–Srb11p kinase–cyclin pair, and the CTD kinase I complex (CTDK-I) (reviewed in Reference 15). TFIIH and Srb10p are indistinguishable in their ability to phosphorylate the CTD. However, TFIIH is a positive effector of transcription, whereas Srb10p is a negative effector. This paradox was recently explained (20). Srb10p is uniquely capable of phosphorylating the CTD prior to PIC formation, thereby blocking entry of RNAP II into the PIC. Conversely, TFIIH phosphorylates the CTD only after PIC assembly, allowing transition from the abortive to elongation competent form of the enzyme. Thus, the timing of CTD phosphorylation affects transcriptional control.

The necessity to recycle RNAP IIO to RNAP IIA for recruitment also implies the existence of a CTD phosphatase. A protein required for dephosphorylation of the CTD has been identified in human cells (21). Its yeast counterpart is encoded by the essential *FCP1* gene (22, 23). CTD phosphatase activity is regulated by TFIIB and TFIIF (24). The RAP74 subunit of TFIIF stimulates CTD phosphatase activity, whereas TFIIB inhibits the stimulatory activity of TFIIF. These results suggest that the CTD phosphatase, TFIIF and TFIIB interact to regulate RNAP II recycling.

Additional functions have been attributed to the CTD. Its presence in RNAP II, but absence from RNAP I and RNAP III, implies a role unique to RNAP II transcription. Genetic and physical interactions between the CTD and the SRB–MED complex (see section 6.1) defined a function for the CTD in transcriptional activation. The CTD has also been implicated in 5′-cap addition, pre-mRNA splicing, and 3′-end formation (25–28). Accordingly, the CTD appears to function as a platform for the recruitment and assembly of factors involved both in regulation of RNAP II transcription and pre-mRNA processing (reviewed in Reference 29; see also Chapter 8, section 7.2).

3.3 RNAP II holoenzyme complexes

Although the CTD is essential for cell growth, all but 8 to 10 repeats can be deleted from Rpb1p without loss of viability. However, strains with a minimum number of CTD repeats exhibit a cold-sensitive growth defect. This phenotype was exploited to isolate extragenic suppressors of CTD truncations (30). These suppressor genes are designated *SRBs* (suppressor of *RNA polymerase B*). Antibodies directed against SRB proteins identified an RNAP II holoenzyme complex that includes SRB proteins and a subset of the GTFs (31). Unlike core RNAP II, holoenzyme responds to transcriptional activators *in vitro*. RNAP II holoenzyme was independently discovered based on the association of RNAP II with a 'mediator' complex (MED), required for transcriptional activation *in vitro* (32). The two RNAP II holoenzyme complexes are similar though not identical. Both include subcomplexes of SRB and MED proteins, as well as a subcomplex that includes Gal11p, Rgr1p, Sin4p, and Med3p (see section 6.1).

A second, distinct, form of RNAP II holoenzyme has also been discovered in *S. cerevisiae* (33). This complex was identified by affinity purification using RNAP II immobilized through the CTD and consequently lacks most of the SRB and MED proteins. Components of this holoenzyme include TFIIB, TFIIF, TFIIS, the Gal11p–Rgr1p–Sin4p–Med3p subcomplex, and a novel subcomplex that includes Paf1p, Cdc73p, Ccr4p, and Hpr1p (34). This form of the holoenzyme is less abundant than the SRB–MED-containing holoenzyme and appears to play a more specific role in regulation, possibly affecting gene expression downstream of the protein kinase C signal transduction pathway (34).

RNAP II holoenzyme complexes have also been identified in mammalian cells (reviewed in References 35–37). Like the yeast holoenzymes, but distinct from core RNAP II, these complexes are able to respond to transcriptional activators *in vitro*. In addition to the GTFs, an intriguing array of proteins has been found in these various preparations, including DNA repair proteins (38), splicing and polyadenylation factors (28), the breast cancer tumour suppressor BRCA1 (39), and mammalian homologues of SRBs (38, 40). Recently, mammalian SRB–MED complexes have been described that are free of RNAP II (reviewed in Reference 37). Like yeast SRB–MED, these complexes mediate transcriptional activation and repression. Moreover, one complex, denoted SMCC–TRAP, mediates activation in a TAF-independent manner *in vitro* (41). This result emphasizes the functional relationship between SMCC–TRAP and SRB–MED, and underscores the extent to which the core transcriptional machinery is conserved among eukaryotic organisms.

4. General transcription factors

A breakthrough in understanding the mechanism of transcription initiation by RNAP II followed the discovery by Roeder and colleagues that RNAP II selectively and accurately initiates transcription from template DNA when supplemented with a crude cell extract (42). This activity provided an assay for the purification of the GTFs, defined as factors required for accurate, basal-level transcription initiation *in vitro* (43). Similar assays were developed to identify GTFs from other organisms, including rats, flies, and yeast. The GTFs are remarkably conserved in structure and function among these phylogenetically diverse organisms. Accordingly, the experimental advantages of one organism can be exploited to learn about transcription in other systems.

Five GTFs have been identified, including TBP, TFIIB, TFIIE, TFIIF, and TFIIH (Table 1). TFIIA, although initially defined as a GTF, is dispensable for accurate initiation but stimulates the rate of transcription in a TBP-dependent manner. Accordingly, TFIIA is a coactivator of transcription rather than a GTF (reviewed in Reference 15). Order of addition experiments demonstrated that TBP binds DNA to nucleate PIC assembly, followed sequentially by TFIIB, RNAP II/TFIIF, TFIIE, and TFIIH (reviewed in References 15, 44, 45). Although these experiments describe PIC assembly *in vitro*, the presence of GTFs in RNAP II holoenzyme complexes implies that some GTFs are recruited to the promoter in association with RNAP II *in vivo*.

Table 1 *S. cerevisiae* GTFs

Factor	Mass (kDa)	Gene	Characteristics
TBP	27	*SPT15*	Binds TATA box; nucleates PIC assembly; recruits TFIIB
TFIIB	38	*SUA7*	Stabilizes TATA–TBP interaction; recruits RNAPII; affects start site selection
TFIIF	82	*TFG1/SSU71*	Facilitates RNAPII-promoter targeting; functional interaction with TFIIB; stimulates elongation
	47	*TFG2*	σ factor homology; destabilizes nonspecific RNA pol II–DNA interactions
	27	*TFG3/SWP29/TAF30*	Common subunit of TFIID, TFIIF, and the SWI–SNF complex; only GTF subunit not essential for cell viability
TFIIE	66	*TFA1*	Recruits TFIIH; stimulates TFIIH catalytic activities; functions in promoter melting and clearance
	43	*TFA2*	
TFIIH	95	*SSL2/RAD25*	Functions in promoter melting and clearance; ATP-dependent DNA helicase (3′–5′); DNA-dependent ATPase. ATPase/helicase required for both transcription and nucleotide excision repair (NER)
	85	*RAD3*	ATP-dependent DNA helicase (5′–3′); DNA-dependent ATPase; ATPase/helicase required for NER, but not transcription
	73	*TFB1*	Required for NER
	59	*TFB2*	Required for NER
	50	*SSL1*	Required for NER
	47, 45	*CCL1*	TFIIK subcomplex with Kin28
	37	*TFB4*	
	32	*TFB3*	Links core–TFIIH with TFIIK
	33	*KIN28*	TFIIK subcomplex with Ccl1

4.1   TBP

TBP is a universal transcription factor, required for initiation by RNAP I, RNAP II, and RNAP III (reviewed in Reference 46). The function of TBP is to nucleate PIC formation by binding the TATA box and subsequently recruiting TFIIB to the promoter. TBP was initially identified as a subunit of TFIID, a 750 KDa complex comprising TBP and TAFs. Whereas TBP is sufficient for basal transcription, TFIID is required for the response to transcriptional activators in metazoan *in vitro* systems.

S. cerevisiae TBP is a monomer of 27 KDa encoded by the essential *SPT15* gene (47, 48). The *S. pombe* gene encoding TBP was cloned from a cDNA library by complementation of a *S. cerevisiae spt15* mutant (49). The two proteins exhibit limited sequence similarity at their N-termini, but their core domains, which encompass the C-terminal two-thirds of the proteins, are 93% identical. Furthermore, the human and *S. cerevisiae* proteins are functionally interchangeable *in vitro*, underscoring the conserved nature of TBP.

Crystal structures have been reported for the core domain of TBP from *S. cerevisiae* (50), the plant *Arabidopsis thaliana* (51), and the archaebacterium *Pyrococcus woesei* (52). When bound to DNA, TPB forms a molecular 'saddle' that sits astride the TATA box, bending the DNA 80° toward the major groove (Fig. 2). This structure appears to provide the appropriate topology for TFIIB association, which binds DNA on both

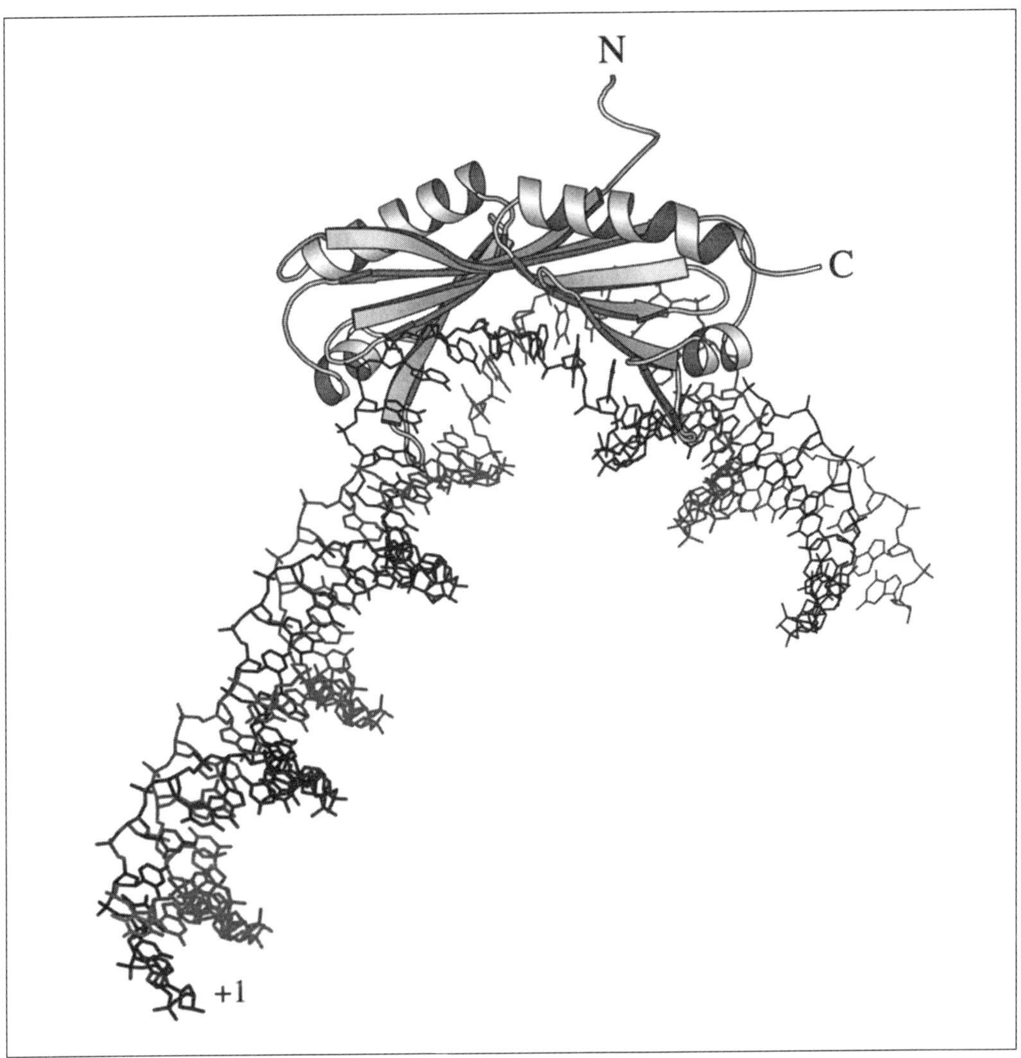

Fig. 2 Three-dimensional structure of TBP complexed with core promoter DNA. The molecular 'saddle' of TBP is depicted by a ribbon drawing, and the DNA template is shown as a stick figure with the transcription start site labelled +1. TBP–DNA association induces an 80° bend in the DNA template. Figure courtesy of Stephen Burley and reproduced with permission from *Nature*, 1993, **365,** 520 (Reference 51). © Macmillan Magazine Limited.

sides of the TATA box (53). The convex seat of the saddle is then potentially available for interaction with other factors (reviewed in Reference 54).

TBP is a direct target of activators *in vitro*. Tethering TBP to a promoter by a heterologous DNA-binding domain bypasses the need for an activator, suggesting that activators function, at least in part, to recruit TBP to the promoter (55–57). Consistent with these results, acidic activators enhance the kinetics of TBP recruitment *in vivo* (58). Moreover, activation defective yeast TBP mutants have been described that are without effect on basal transcription (59–61). Interestingly, human TBP has been reported to dimerize in solution, yet binds to DNA as a monomer, raising the possibility that certain activators might regulate TBP dimer dissociation (62).

4.2 TFIIB

S. cerevisiae TFIIB is a 38-kDa monomer encoded by the essential *SUA7* gene (63). *SUA7* was first identified as a consequence of a downstream shift in transcription start-site selection associated with *sua7* mutations (63, 64). Thus, TFIIB functions, at least in part, to define initiation sites. The same genetic selection that uncovered *sua7* also identified mutations in *RPB1* (*sua8* alleles) (65). The *sua7* and *sua8* mutations confer similar effects on start-site selection, and *sua7 sua8* double mutants are inviable. This information suggests that TFIIB interacts with the Rpb1p subunit of RNAP II to define start sites (65). This premise is consistent with biochemical results demonstrating that pairwise replacement of *S. cerevisiae* TFIIB and RNAP II with their counterparts from *S. pombe* is sufficient to confer the start site pattern characteristic of *S. pombe* in an *in vitro* transcription system (66).

TFIIB includes a zinc-binding domain near the N-terminus (nTFIIB), two imperfect repeats encompassing the C-terminal two-thirds of the molecule (cTFIIB), and a phylogenetically conserved region located between these two domains (63, 67). cTFIIB interacts with TBP, whereas nTFIIB interacts with RNAP II and TFIIF (reviewed in Reference 15). A solution structure for cTFIIB and a crystal structure for a TATA–TBP–cTFIIB ternary complex have been solved (53, 68). cTFIIB consists of two similar domains, each comprising five α helices. The metal-binding motif of nTFIIB forms a zinc ribbon (69). No tertiary structural information is available for the highly conserved domain linking the zinc ribbon with cTFIIB.

TFIIB has also been implicated in the mechanism of transcriptional activation. As for TBP, several activators directly bind TFIIB (70, 71). Moreover, nTFIIB and cTFIIB form an intramolecular interaction that is disrupted by specific activators (72). The physiological significance of this interaction was recently demonstrated by a *sua7* mutation that impairs transcription in an activator-specific manner (73). These results define a role for TFIIB in activation and suggest that certain activators induce a conformational change in TFIIB as part of their mechanism of transcriptional stimulation.

4.3 TFIIF

TFIIF was initially identified based on co-purification with RNAP II (74) and functions as both an initiation and elongation factor. Distinct biochemical properties have been assigned to TFIIF:

- high-affinity binding to RNAP II
- suppression of non-specific RNAP II–DNA binding
- stabilization of the PIC

- start-site recognition
- stimulation of elongation

These characteristics are reminiscent of the roles of σ factors in bacterial transcription. Indeed, both subunits of TFIIF exhibit limited sequence similarities to σ^{70} (reviewed in Reference 15).

TFIIF from *S. cerevisiae* comprises three subunits encoded by the *TFG1*, *TFG2*, and *TFG3* genes (75), and is functionally interchangeable with its counterpart from *S. pombe* (66). The *ssu71* allele of *TFG1* was identified as a suppressor of a *sua7* mutation, thereby implicating TFIIF in the mechanism of start-site selection (76). The Tfg1p and Tfg2p proteins are structurally similar to the RAP30 and RAP74 subunits of human TFIIF. Interestingly, Tfg3p is not unique to TFIIF, but is also a subunit of TFIID (Taf30p), the SWI-SNF chromatin remodelling complex (Swp29p) and the NuA3 histone acetyltransferase complex. As such, Tfg3p establishes a connection between the general and regulatory components of the RNAP II transcriptional machinery (75). Despite its multiple functions, Tfg3p is the only GTF that is dispensable for cell viability.

4.4 TFIIE

TFIIE enters the PIC subsequent to RNAP II-TFIIF, but before TFIIH. The function of TFIIE is unclear, but is linked to TFIIH. This was demonstrated by factor swap experiments where *S. cerevisiae* TFIIE could functionally replace *S. pombe* TFIIE in a reconstituted transcription system only when exchanged as a TFIIE–TFIIH pair (66). Functions attributed to TFIIE include recruitment of TFIIH to the PIC, stimulation of TFIIH-dependent phosphorylation of the RNAP II CTD, and stimulation of TFIIH-dependent ATP hydrolysis (reviewed in Reference 15). Two-dimensional crystallography of TFIIE–RNAP II suggests that TFIIE participates in a conformational switch at the RNAP II active centre upon RNAP II–DNA interaction (77).

S. cerevisiae TFIIE is encoded by the essential *TFA1* and *TFA2* genes (78). Mutational analysis of Tfa1p defined two functionally distinct domains (79, 80). One domain includes a zinc-binding motif that is structurally similar to domains within TFIIB, the Rpb9p subunit of RNAP II, and the elongation factor TFIIS. However, no specific function has been assigned to these structures. The requirement for TFIIE is promoter specific (80, 81). This observation might be related to the promoter-specific requirement for TFIIE and TFIIH in metazoan *in vitro* transcription systems, where promoter topology defines the TFIIE–TFIIH requirement (82–84).

4.5 TFIIH

TFIIH is the only GTF with known enzymatic activities. These include:

- DNA-dependent ATPase
- ATP-dependent DNA helicase (5′→3′ and 3′→5′)
- CTD kinase

Since TFIIE, TFIIH, and ATP are dispensable for initiation from supercoiled promoter DNA and from a premelted template, open complex formation appears to be catalysed by the ATP-dependent DNA helicases. The CTD kinase facilitates the transition from initiation to elongation.

TFIIH is composed of nine essential subunits and can be separated into a core–TFIIH complex and TFIIK (85). TFIIK comprises the Kin28p protein kinase and the Ccl1p cyclin. Kin28p is a member of the Cdc2 family of cyclin-dependent protein kinases and is homologous to CDK7, the catalytic subunit of human TFIIH; Ccl1p is homologous to cyclin H. The human CDK7–cyclin H complex includes a third subunit, MAT1 (86). This heterotrimeric complex comprises the CDK-activating kinase (CAK), which functions in cell cycle control (reviewed in Reference 87). Therefore, human TFIIH also affects cell cycle progression. However, this structural arrangement is different in *S. cerevisiae*. The MAT1 homologue, Tfb3p, is a component of core–TFIIH, not TFIIK (88). Moreover, CAK is monomeric in yeast (Cak1p) and is not a component of TFIIH. Accordingly, yeast appears to have evolved two distinct protein kinases, Kin28p and Cak1p, to carry out the functions attributed to human CDK7.

TFIIH also functions in nucleotide excision repair (NER) (reviewed in Reference 89). This role was uncovered by characterization of the p89 subunit of human TFIIH (yeast Ssl2p/Rad25p homologue), which turned out to be the DNA excision repair protein ERCC3. ERCC3 complements the DNA repair deficiency associated with the *XPB* gene defect in patients with xeroderma pigmentosum. Additional subunits of TFIIH were subsequently identified as components of the NER complex. A direct role for TFIIH in NER was established by the ability of purified TFIIH to rescue the repair deficiency of mammalian and yeast TFIIH mutants (90). NER requires core–TFIIH, but is independent of TFIIK (91). The discovery that TFIIH plays a dual role in transcription and DNA repair accounts nicely for earlier observations demonstrating that transcriptionally active genes undergo preferential NER, and underscores the complex nature of TFIIH.

5. Chromatin

Transcription in eukaryotic cells does not occur on naked DNA templates, but instead on DNA in the form of chromatin. As might be expected, chromatin has a generally repressive effect on transcription. Regulatory mechanisms have evolved both to maintain and to overcome chromatin-mediated repression. In this section a brief overview of chromatin structure is presented and the strategies used by the cell to regulate transcription in the context of chromatin are summarized. Examples of transcriptional coactivators that function to relieve chromatin-mediated repression, or general repressors that function through chromatin, are reviewed in subsequent sections.

5.1 Chromatin structure

Chromatin is composed by mass of approximately 50% DNA and 50% protein. Histones are the principal protein components of chromatin and comprise five types, designated H1, H2A, H2B, H3, and H4. The fundamental structural unit of chromatin is the nucleosome. Each nucleosome includes a histone octamer, organized as a $(H3)_2(H4)_2$ tetramer and two H2A–H2B dimers around which 146 base pairs of DNA is spooled as 1.65 turns of left-handed helical DNA (92). This structure is known as the nucleosome 'core' particle and was initially defined by micrococcal nuclease digestion of chromatin (reviewed in Reference 93). 'Linker' DNA of variable length connects adjacent nucleosome core particles, which recur approximately every 200 base pairs. Histone H1 binds linker DNA adjacent to the core particles and facilitates assembly of repeating nucleosome cores into higher-order chromatin structures.

Nucleosome core particles have been crystallized and a high-resolution X-ray structure has been solved (92). This structure confirms the general arrangement of DNA wrapped around the histone octamer and defines specific histone–histone and histone–DNA interactions. Both sets of interactions depend upon 'histone-fold' domains, defined as three α helices connected by two loops, present in all four histones. A key feature of the nucleosome structure is that the N-terminal regions of the four histones pass over and between the gyres of the superhelical DNA such that they are exposed as unstructured 'tails' on the surface of the nucleosome. The histone tails of neighbouring nucleosome particles interact and contribute to the assembly of nucleosomes into chromatin fibres.

5.2 Chromatin and transcription

The presence of nucleosomes on a DNA template prevents transcription initiation *in vitro* (reviewed in Reference 93). A generally repressive role for nucleosomes was demonstrated *in vivo* by nucleosome depletion experiments in *S. cerevisiae*. The gene encoding histone H4 was placed under control of a *GAL* promoter such that addition of glucose led to nucleosome loss and commensurate activation of specific genes (94). This effect occurred independent of UAS elements and implied that nucleosome loss or displacement is a prerequisite for activation of at least certain genes. Moreover, these results confirmed the generally repressive role of chromatin in gene expression.

How does the transcriptional machinery contend with the repressive effects of chromatin to activate gene expression? One clue came from the observation in the early 1960s that histones are sometime heavily acetylated and that acetylation correlates with transcriptionally active chromatin (95). This observation suggested either that histone acetylation facilitates transcriptional activation or that activation promotes acetylation. The distinction between these two possibilities was resolved by the isolation and characterization of a nuclear histone acetyltransferase (HAT) from *Tetrahymena*, which turned out to be homologous to the product of the *GCN5* gene from *S. cerevisiae* (96). This was a key discovery because *GCN5* had been found in two separate genetic selections for transcriptional coactivators. Gcn5p was subsequently shown to catalyse acetylation of specific lysine residues in the tails of histones H3 and H4 (97). Furthermore, this activity is required for transcriptional activation (98, 99). Lysine acetylation neutralizes its positive charge, which presumably relieves nucleosome condensation, thereby enhancing access to DNA by the transcriptional machinery.

If histone acetylation facilitates activation, then histone deacetylation might be expected to cause repression. Indeed, a histone deacetylase (HDA) was purified from human cells and found to be homologous to the product of the yeast *RPD3* gene (100), which had been identified in a genetic selection for transcriptional repressors (101). Rpd3p exhibits specific histone deacetylase catalytic activity and this activity is required for transcriptional repression (102, 103).

Since the histone tails lie outside of the core particle, histone acetylation is unlikely to disrupt nucleosome structure. Yet transcriptional activators and components of the core transcriptional machinery, including RNA polymerase II, are unable to bind nucleosomal DNA. This implies the existence of factors that displace nucleosomes and that these factors would facilitate transcriptional activation. Such chromatin-remodelling complexes have been identified in yeast and other eukaryotic organisms and include the SWI–SNF complex, which catalyses nucleosome displacement in an ATP-dependent manner (reviewed in Reference 104). The components of the SWI–SNF complex were first identified genetically based on mutations that impaired activation of specific genes (reviewed in Reference 105). Moreover, a role for the SWI–SNF complex in chromatin remodeling was suggested by suppressors of *swi* and *snf* mutations, which encode either histones or histone-related proteins (106).

These results demonstrate that two distinct classes of transcriptional coactivators can remodel chromatin. One catalyses acetylation of histone tails and presumably relieves nucleosome condensation, whereas the other catalyses displacement of the nucleosome core in an ATP-dependent manner. Recently, the temporal relationship between histone acetylation and ATP-dependent chromatin remodelling was investigated at the *HO* gene in *S. cerevisiae*. Chromatin immunoprecipitation studies demonstrated that the Swi5p transcriptional activator recruits the SWI–SNF complex, which in turn recruits the SAGA histone acetyltransferase complex (107, 108). These two factors facilitate binding of the Swi4p/Swi6p complex, which promotes PIC assembly. Thus, chromatin remodeling can occur in separate and successive steps to facilitate gene activation.

6. Transcriptional coactivators

Coactivators are defined as transcription factors that mediate activation. They are distinct from GTFs in that they are dispensable for basal-level transcription *in vitro* and distinct from activators in that most do not bind DNA directly. Several functionally distinct coactivator complexes have been described, including the SRB–MED complex that associates with RNAP II, the TAF subunits of TFIID, SAGA and other histone acetyltransferase complexes, and the SWI–SNF chromatin remodelling complex (reviewed in Reference 15). An overview of these coactivators is presented here.

6.1 The SRB–MED complex

Evidence for the existence of a transcriptional mediator came from the ability of one activator to squelch activation by another *in vitro*, presumably by sequestering an intermediary molecule that mediates the interaction between activator and core machinery (reviewed in Reference 109). This coactivator, or 'mediator', was subsequently purified from yeast based on its requirement for activation by RNAP II *in vitro*. Three functions have been attributed to mediator: (1) stimulation of basal transcription; (2) response to transcriptional activators; and (3) stimulation of the CTD kinase activity of TFIIH.

Mediator was independently discovered based on its genetic interaction with the CTD of RNAP II (reviewed in Reference 110). Nine different *SRB* genes were identified as suppressors of the growth defects associated with truncations of the CTD heptapeptide repeat domain. The SRB proteins exist as a multisubunit complex that is tightly associated with RNAP II through the CTD. Furthermore, Srb2p, Srb4p, Srb5p, Srb6p, and Srb7p were identified as subunits of the mediator complex. Accordingly, this complex is now designated 'SRB–MED'. The Srb8p, Srb9p, Srb10p, and Srb11p proteins are found in one SRB–MED preparation (111, 112), but not in another (32, 113). The Srb8p–Srb11p subunits form a subcomplex that functions in repression rather than activation. This offers an explanation for the structural distinction between the two complexes: Mediator was purified based on an assay for transcriptional activation, whereas the SRB complex was purified independent of a functional requirement. Association of SRB–MED with RNAP II and its ability to respond to activators *in vitro* defined the RNAP II 'holoenzyme' (see section 3.3). A conditional *srb4* mutant rapidly ceases mRNA synthesis at the non-permissive temperature, thereby defining a general requirement for the RNAP II holoenzyme *in vivo* (114).

A subset of the genes encoding SRB–MED subunits was identified in a genetic selection for factors involved in glucose repression (reviewed in Reference 115). These include *SSN2/SRB9, SSN3/SRB10, SSN4/SIN4, SSN5/SRB8,* and *SSN7/ROX3.* The *SIN4, ROX3,* and *RGR1* genes were also identified in independent screens for transcriptional repressors (116–119). The Gal11p subunit of SRB–MED, although initially uncovered based on its requirement for full activation of galactose-inducible genes, has also been implicated in transcriptional repression (120). Thus, the SRB–MED complex has both positive and negative effects on transcription and might therefore be more appropriately termed a transcriptional 'co-factor', rather than a 'coactivator'.

The third group of SRB–MED components was not identified genetically, but as purified subunits of the complex. These are the MED proteins and include Med1p, Med2p, Med4p, Med6p, Med7p, Med8p, Med9p, and Med10p (121–125). Deletion of the nonessential *MED1* gene affected both activation and repression *in vivo*, a result reminiscent of the effects of *srb10* and *srb11* mutations (123). A conditional *med6* mutation diminished activated, but not uninduced, transcription (122). Furthermore, *med6* was suppressed by a dominant *SRB4* allele, pointing to a functional interaction between the MED and SRB proteins (126).

The functional differences among subunits of the SRB–MED complex suggested that specific subunits might exist within structurally distinct subcomplexes (Fig. 1, Table 2). Indeed, urea treatment dissociated SRB–MED into two subcomplexes, one

Table 2 *S. cerevisiae* SRB/MED subunits

Factor	Mass (kDa)	Gene	Subcomplex
Srb2	23	*SRB2*	Srb4
Srb4	78	*SRB4*	Srb4
Srb5	34	*SRB5*	Srb4
Srb6	14	*SRB6*	Srb4
Srb7	16	*SRB7*	Rgr1
Srb8	167	*SRB8/SSN5/ARE2*	Srb10
Srb9	160	*SRB9/SSN2/UME2*	Srb10/11
Srb10	63	*SRB10/SSN3/UME5/ARE1*	Srb10/11
Srb11	36	*SRB11/SSN8/UME3*	Srb10/11
Gal11	38	*GAL11/SPT13/SDS4/RAR3*	Gal11
Sin4	111	*SIN4/SSN4/TSF3*	Gal11
Rgr1	123	*RGR1*	Rgr1
Rox3	25	*ROX3/SSN7*	Srb4
Med1	64	*MED1*	Rgr1
Med2	48	*MED2*	Rgr1
Med3/Pgd1/Hrs1	47	*MED3/PGD1/HRS1*	Gal11
Med4	32	*MED4*	Rgr1
Med6	33	*MED6/MTR32*	Srb4
Med7	32	*MED7*	Rgr1
Med8	25	*MED8*	Rgr1
Med9	18	*MED9/NUT2*	Rgr1
Med10	17	*MED10/CSE2*	Rgr1
Med11	15	*MED11*	Rgr1

(Srb4p subcomplex) including all of the SRB proteins along with Med6p and Rox3p, the other (Rgr1p subcomplex) including all other MED proteins, Rgr1p, and a distinct module composed of Gal11p, Sin4p, and Med3p. Genetic and biochemical dissection of functions associated with these subcomplexes defined a role for the Gal11p module of the Rgr1p subcomplex in activation but not basal transcription, whereas the Srb4p subcomplex functions in basal and activated transcription (122, 127, 128). Thus, the Rgr1p subcomplex appears to be required for recruitment of the RNAP II holoenzyme to the promoter, whereas the Srb4p subcomplex affects RNAP II activity (127). The Srb8p, Srb9p, Srb10p, and Srb11p proteins comprise a third SRB–MED subcomplex (Srb10p subcomplex), functioning as a negative regulator of transcription by phosphorylating the CTD before PIC formation (20). Whether this subcomplex is associated with SRB–MED *in vitro* presumably depends upon the conditions used to isolate the complex.

6.2   TFIID

Although TBP is sufficient for promoter recognition and subsequent assembly of other factors into a functional PIC, transcriptional activation in metazoan systems was observed only when the PIC was assembled with the multisubunit TFIID complex. This observation suggested that TAFs might be requisite mediators of transcriptional

Table 3 *S. cerevisiae* TAFs

Factor	Mass (kDa)	Gene	Complex
Taf150	155	*TSM1*	TFIID
Taf145/130	121	*TAF130*	TFIID
Taf90	89	*TAF90*	TFIID and SAGA
Taf67	67	*TAF67*	TFIID
Taf61/68	61	*TAF61*	TFIID and SAGA
Taf60	58	*TAF60*	TFIID and SAGA
Taf47	47	*TAF47*	TFIID
Taf40	41	TAF40	TFIID
Taf30	27	*TAF30/TFG3/SWP29*	TFIID, TFIIF, SWI/SNF, NuA3
Taf25	23	*TAF25*	TFIID and SAGA
Taf19	19	*TAF19/FUN81*	TFIID
Taf17	17	*TAF17*	TFIID and SAGA

activation (reviewed in Reference 129). Consistent with this possibility, some TAFs directly bind activator proteins, whereas other TAFs directly bind GTFs or promoter DNA (130). Thus, TAFs were proposed to function in transcriptional activation by relaying information from activators to the core transcriptional machinery.

The *S. cerevisiae* counterpart of metazoan TFIID was identified by immunopurification of yeast TBP (131) and by affinity purification with TBP as the ligand (132). In both cases, TFIID functioned as a coactivator, allowing transcriptional activation by RNAP II *in vitro*. Yeast TFIID comprises at least 12 subunits (Table 3). With the exceptions of Taf47p and Taf30p (Tfg3/Swp29), all subunits are homologous to metazoan TAFs (reviewed in Reference 15). Conversely, all metazoan TAFs have a homologue in yeast, with the exception of TAF110, which is required for activation by the glutamine-rich activator Sp1; interestingly, glutamine-rich activators do not function in yeast, perhaps due to the absence of a TAF110 homologue (133).

All the *TAF* genes, except *TFG3*, are essential for yeast cell viability. This was assumed to reflect their requisite role as coactivators of transcription. However, depletion or inactivation of Taf145p (Taf130p), Taf90p, Taf68p, Taf60p, Taf47p, and Taf19p did not compromise transcriptional activation *in vivo*, although depletion of Taf145p and Taf19p diminished expression from promoters containing weak TATA boxes (134, 135). Whole-genome, microarray analysis confirmed that Taf145p is not required for transcription of all genes (136). Thus, in contrast to the TFIID requirement for activation in metazoan *in vitro* transcription systems, TAFs were suggested to be dispensable for activation in yeast.

What then is the essential function of TAFs? One possibility is that TAFs are in fact essential coactivators, but only for transcription of a subset of genes. This possibility is supported by the failure of a *taf90^{ts}* mutant to progress through the G2–M phase of the cell cycle at the restrictive temperature (137). Furthermore, a *taf145^{ts}* mutation blocks transcription of G1–S cyclin genes (138). Indeed, a role for TAFs in regulation of cell cycle-specific genes was suggested earlier by the identification of mammalian TAF250 as the product of the *CCG1* gene, which is required for passage through the

G1 phase of the cell cycle (139). Surprisingly, the promoter element that renders a gene dependent upon Taf145p is not the UAS, but rather the core element (140). Moreover, core dependence is determined by the sequences flanking, but not including, the TATA box. Thus, Taf145p appears to position TFIID at TATA-less promoters by making specific contacts with the core promoter. This conclusion is consistent with the earlier observation that Taf145p is important for transcription from promoters lacking consensus TATA sequences (135) and with binding of *Drosophila* TFIID to Inr and downstream promoter elements (141).

Gene expression in the absence of Taf145p is especially significant because Taf145p is thought to be the scaffold for TFIID assembly. Presumably, depletion of Taf145p results in disassembly of TFIID. It is therefore striking that more recent analysis of TAF requirements revealed that Taf61p (Taf68p), Taf60p, and Taf17p, unlike Taf145p, are broadly, though not universally, required for transcription. Interestingly, these three TAFs are structurally related to histones H2B (Taf61p), H3 (Taf17p), and H4 (Taf60p). Furthermore, genetic and structural studies argue for the existence of a histone octamer-like structure within TFIID (142, 143).

The general requirement for the histone-like TAFs for transcription might not reflect a TFIID requirement (reviewed in Reference 144). These TAFs, along with Taf90p and Taf25p, were recently identified as subunits of the SAGA coactivator complex (see section 6.3), which is composed of ADA, SPT, and TAF polypeptides, but does not include Taf145p (145, 146) (reviewed in Reference 147). However, elimination of Spt20p/Ada5p, which is critical for SAGA function, does not cause a general defect in transcription (142). Given the redundant nature of coactivators, perhaps both TFIID and SAGA must be eliminated in order to see general defects in transcription.

Nonetheless, the role of TAFs remains controversial. Recently, mutations in *TAF40*, whose product is a component of TFIID, but not SAGA, were reported to cause a general defect in transcription (148). Another recent report suggests that TAFs might be more than transcriptional coactivators: the *ptr6* gene of S. *pombe* encodes a homolog of S. *cerevisiae* Taf67p and human TAF55, yet functions in nucleocytoplasmic transport of mRNA (149). Clearly, much remains to be learned regarding TAFs.

6.3   Histone acetyltransferases

A fundamental question in the field of gene regulation is how the transcriptional machinery contends with the chromatin structure of eukaryotic DNA to activate gene expression. Histone acetylation is a hallmark of transcriptionally active chromatin, suggesting that acetylation weakens histone–DNA interactions, thereby relieving the repressive effects of chromatin (reviewed in Reference 150). A convergence of genetics and biochemistry provided insight into the role of histone acetylation in controlling gene expression (reviewed in Reference 151). A nuclear histone acetyltransferase was purified from *Tetrahymena* and identified as a homologue of the yeast Gcn5p protein (96). As mentioned in section 5.2, this was a key discovery because *gcn5* had been identified in two separate genetic selections for transcriptional adaptors – molecules suggested to bridge the interaction between activators and the core transcriptional apparatus (152–154). Gcn5 was subsequently shown to catalyse acetylation of specific lysine residues in the histone tails (97). Even though Gcn5p turns out not to be an adapter molecule *per se*, its identity provided the first direct link between histone acetylation and gene activation.

Genetic evidence suggested that Gcn5p did not act alone to mediate transcriptional activation. In addition to *gcn5*, the genetic selection for transcriptional co-factors uncovered four other genes, *ada1–ada3* and *ada5* (reviewed in Reference 151). Mutations in each of these genes confer similar pleiotropic phenotypes. Furthermore, Ada1p, Ada2p, Ada3p, and Gcn5p physically interact with one another, suggesting the existence of an ADA–Gcn5 complex.

An independent selection for genes encoding transcription factors was based on suppression of initiation defects associated with Ty or solo δ elements within the *HIS4* and *LYS2* promoters (reviewed in Reference 155). These genes, designated *SPT*, fall into either of two classes, one encoding histones or proteins that affect chromatin, the other encoding TBP or factors that affect TBP function. The TBP class includes the *SPT3*, *SPT7*, *SPT8*, *SPT15* (TBP), and *SPT20* genes. The SPT and ADA stories merged when *SPT20* and *ADA5* were found to be identical (156, 157). Furthermore, Spt20p/Ada5p physically associates with Spt3p, Spt7p, and Spt8p (158). This suggested that Gcn5p functions within a large complex that includes SPT and ADA proteins.

The existence of a Gcn5 HAT complex was also suggested by the ability of recombinant Gcn5p to acetylate free histones, but not histones assembled into nucleosomes (97, 159). Perhaps other components of the putative complex were required for recognition of nucleosomes. Consistent with this notion, HAT activity and interaction with Ada2p are both required for Gcn5p function *in vivo* (99). The results of a collaboration among several laboratories identified four distinct nucleosomal HAT complexes (160). Two of these complexes, with apparent molecular masses of 1.8 and 0.8 MDa, included Gcn5p and Ada2p. Other laboratories have independently identified similar ADA-containing complexes (161, 162). The genetic relationship among the *GCN5*, *ADA*, and *SPT* genes provided a clue to the identity of the other components. Specifically, the 1.8-MDa complex copurified with Gcn5p, Ada2p, Spt3p, Spt7p, and Spt20p/Ada5p, and the integrity of the complex was dependent upon intact *GCN5*, *ADA2*, *ADA3*, *SPT7*, and *SPT20/ADA5* genes. This complex has been named SAGA (*SPT–ADA–Gcn5p–acetyltransferase*) and links nucleosomal histone acetylation with the transcriptional activation associated with ADA and SPT proteins. Remarkably, SAGA also includes histone-like TAFs (see section 6.2), suggesting that the effects of mutations in these *TAF* genes might be manifest, at least in part, through SAGA.

The three other nucleosomal HAT complexes have now been defined. The 0.8-MDa Gcn5p complex has been designated ADA. The ADA complex is structurally related to SAGA, but also includes a novel subunit, Ahc1p, which is required for the integrity of the ADA complex but is not found in SAGA (163). Thus, ADA is not simply a subcomplex of SAGA, but is structurally and functionally distinct. The other HAT complexes do not include Gcn5p as their catalytic subunit. The NuA3 complex

(0.4–0.5 MDa) contains the Sas3p histone acetyltransferase, as well as Taf30p, which is not found in SAGA (164). Accordingly, Taf30p has now been found as a component of four complexes: ADA, TFIID, TFIIF, and SWI–SNF. Interestingly, NuA3 binds Spt16p, which is a component of the yeast CP (Cdc 68p / Spt16p-Pob3p) complex and homologous to a subunit of human FACT (*facilitates chromatin transcription*), a complex that facilitates transcription elongation through chromatin templates (165). The NuA4 complex (1.3 MDa) contains Esa1p, an essential protein required for cell cycle progression, as its histone acetyltransferase subunit (166). Like SAGA, NuA4 loses its ability to respond to acidic activators *in vitro* in the absence of its catalytic subunit. This result is consistent with the ability of transcriptional activators to direct SAGA and NuA4 to nucleosomes (167). Thus, SAGA and NuA4, but not ADA and NuA3, catalyse histone acetylation in response to transcriptional activators.

6. 4   The SWI–SNF chromatin remodelling complex

Chromatin remodelling complexes distinct from HAT complexes have been described. These include SWI–SNF, RSC, NURF (*nucleosome remodeling factor*), CHRAC (*chromatin accessibility complex*), and ACF (*ATP-utilizing chromatin assembly and remodeling factor*), all of which use the energy of ATP hydrolysis to alter chromatin structure (reviewed in Reference 104). The composition and function of the yeast SWI–SNF complex was unravelled by linking disparate genetic systems (reviewed in Reference 105). Mutations in two classes of genes, designated *SWI* and *SNF*, were found to impair transcription of genes required for mating type switching (*HO*) and sucrose fermentation (*SUC2*). The functional relationship between the *SWI* and *SNF* genes was revealed when *SWI2* was found to be identical to *SNF2*. The connection to chromatin was made when the *SIN* and *SSN* genes, identified as suppressors of *swi* and *snf* mutations respectively, were found to encode either histones or factors that affect chromatin structure.

The SWI–SNF complex has been purified from yeast as a 2-MDa 11-subunit complex (104, 168, 169). Swi2/Snf2 is the most well characterized component and, as a DNA-dependent ATPase, is the only subunit with known enzymatic activity. As noted above, the Swp29p subunit is common to SWI–SNF, TFIID, TFIIF, and the NuA3 HAT complex. The SWI–SNF complex has also been reported to be a component of the RNA pol II holoenzyme (112). This could endow the holoenzyme with the ability to promote PIC assembly by disrupting nucleosomal DNA.

Based on homology to components of the SWI–SNF complex, a second chromatin remodelling complex, RSC (*remodels the structure of chromatin*), was isolated from yeast (170). Like SWI–SNF, RSC is a DNA-dependent ATPase whose activity is stimulated by both free and nucleosomal DNA. RSC is a 15-subunit complex that includes several SWI–SNF-related polypeptides and two actin-related subunits, Arp7p and Arp9p, that are shared between SWI–SNF and RSC (171). RSC is approximately 10-fold more abundant than SWI–SNF, present at several thousand molecules per cell (170). Mutations in components of the SWI–SNF and RSC confer unique phenotypes, implying that SWI–SNF and RSC are functionally distinct (172, 173). Chromatin remodelling by SWI–SNF and RSC involves unwrapping of nucleosomal DNA and, in the case of RSC, this leads to transfer of the entire histone octamer from the nucleosome to naked DNA (174). Interestingly, this mechanism is distinct from that of the *Drosophila* NURF and CHRAC ATP-dependent chromatin remodelling complexes, which catalyse sliding of the histone octamer along the DNA (175, 176).

7. General repressors

General transcriptional repressors are comparable to transcriptional coactivators, albeit with opposite effects. In contrast to the GTFs, most of which were initially identified biochemically, many of the general repressors were uncovered initially in genetic selections for mutations that enhanced transcription, either in the absence of a gene-specific transcriptional activator or its cognate UAS promoter element.

Two classes of general repressors have been identified (reviewed in Reference 10). One class operates through the core promoter and includes Ydr1p–Bur6p, Mot1p, the Ccr4p–NOT complex, and the Srb10p–Srb11p subcomplex of SRB–MED. The second class is functionally related to chromatin and includes histones, histone-related proteins, and histone deacetylases. Another general repressor, Ssn6p–Tup1p, has been reported to target both chromatin and components of the core transcriptional machinery. Both classes of general repressors are reviewed here, focusing on Ydr1p–Bur6p and Mot1p as examples of repressors that act through TBP, and histone deacetylases as repressors that act through chromatin.

7.1 Ydr1p–Bur6p

A complex defined as Dr1–Drap1 was isolated from human cells based on its ability to repress transcription by RNAP II in a reconstituted system independent of activators (177, 178). The Dr1 subunit participates directly in repression, whereas Drap1 is a regulatory subunit that enhances the activity of Dr1. Dr1 and Drap1 from a heterodimer through histone fold motifs present in each subunit.

A counterpart to Dr1–Drap1, denoted Ydr1p–Bur6p (also known as NC2), has been identified in *S. cerevisiae* (179, 180). The Drap1 homologue was identified in two different genetic selections. In one case, mutants were identified that increased transcription of the *SUC2* gene in the absence of its UAS element (*suc2ΔUAS*). One suppressor, *bur6*, not only enhanced *suc2ΔUAS* expression, but conferred multiple, pleiotropic phenotypes, consistent with the premise that Ydr1p–Bur6p plays a general role in transcription (181). In the other case, mutations in the genes encoding both Ydr1p and Bur6p were isolated as suppressors of a mutation in *SRB4*, which encodes a component of the SRB–MED coactivator complex (180). These results confirm that Ydr1p–Bur6p is a general repressor and underscore the interdependence of coactivators and general repressors in the regulation of gene expression.

Human Dr1 binds TBP and blocks PIC formation. Yeast Ydr1p–Bur6p appears to function analogously. Overexpression of Ydr1p diminishes mRNA accumulation and confers a slow growth defect that is suppressed by overexpression of TBP (179). Furthermore, TBP mutations have been identified that suppress *suc2ΔUAS*, some of which are defective for interaction with Ydr1p–Bur6p (182). These mutations are clustered in a previously undefined domain of TBP adjacent to the TFIIB-binding domain. Thus, Ydr1p–Bur6p appears to repress transcription by preventing association of TFIIB with TBP.

7.2 Mot1p

Mot1p was identified biochemically as a factor (termed ADI (*ATP-d*ependent *i*nhibitor)) from yeast nuclear extracts that inhibited TBP binding to DNA in an ATP-dependent manner (183). ADI-mediated TBP displacement was not promoter specific and could be counteracted by TFIIA. Mot1p was also identified genetically in a screen for mutants with enhanced basal expression of several unrelated genes (184). Consistent with a functional relationship among Mot1p, TBP, and TFIIA, over-expression of either TBP or TFIIA suppressed the growth defect associated with a dominant negative *MOT1* allele. Consistent with its ability to repress transcription, Mot1p (Bur3p) was also uncovered in the screen for suppressors of the *suc2ΔUAS* mutation described in section 7.1 (185). Mot1p is a member of the Swi2/Snf2 family of ATPases, but in contrast to Swi2/Snf2, its activity is not stimulated by DNA. Yeast Mot1p exists in a complex with TBP that is distinct from the TFIID complex (131). Similarly, a human homologue of Mot1p was identified as the TAF$_{II}$170 subunit of B-TFIID, an alternative form of TFIID that includes TBP (186).

MOT1 was also identified in a genetic screen for factors that functionally interact with Spt3p. Specifically, a *mot1* allele conferred synthetic lethality in combination with an *spt3* disruption (187). Surprisingly, *mot1* and *spt3* mutations cause similar phenotypes, including suppression of the *his4-912δ* allele and diminished levels of certain other transcripts. Furthermore, mutation in the *TOA1* gene, encoding the larger subunit of TFIIA, resulted in lethality in combination with either *mot1* or *spt3*, and *TOA1* or *TOA2* overexpression suppressed *spt3* phenotypes (187). A *mot1* mutation also markedly diminished transcription from TATA-less elements in the *HIS3* and *HIS4* promoters (188). These results led to the proposal that Mot1p, Spt3p, and TFIIA regulate TBP–DNA interactions: Mot1p by displacing TBP from non-functional TATA boxes, and Spt3p and TFIIA by enhancing TBP association with functional TATA boxes (187). Mot1p can also function as either an activator or repressor *in vitro*, depending upon the relative concentrations of Mot1p, TBP, and DNA. The model that emerges from these studies is that Mot1p is not a *bona fide* transcriptional repressor, but rather a factor that regulates the distribution of TBP between promoter and non-promoter sites (189). Despite its ability to repress transcription under certain conditions, Mot1p might be more appropriately considered to be a coactivator of relatively weak promoters, including TATA-less promoters, rather than a general repressor.

7.3 Histone deacetylases

The positive role of histone acetylation in transcription and the long-standing correlation between histone acetylation and transcriptionally active chromatin implies the existence of histone deacetylases that would negatively regulate transcription. Reminiscent of the HAT story (96), a histone deacetylase was purified from human cells and found to be homologous to the yeast Rpd3p protein (100). This discovery was insightful because the *RPD3* gene was identified in a genetic selection for transcriptional repressors (101). Rpd3p forms a complex with Sin3p and is recruited to promoter DNA by sequence-specific repressor proteins, resulting in a highly localized effect on chromatin structure (6, 103, 190). Rpd3p activity is directed to specific acetylated lysine residues in the histone tails (102). Accordingly, the Sin3p–Rpd3p repressor complex is comparable to the SAGA and NuA4 coactivator complexes, albeit with the opposite effects on histone acetylation and transcription.

Sin3p–Rpd3p exists as a 2-MDa complex in *S. cerevisiae (191)*. Other multisubunit histone deacetylase complexes have also been identified in yeast (192). One complex includes Rpd3p, whereas the other includes the Rpd3p homologue, Hda1p (193). Three other yeast genes, designated *HOS1, HOS2,* and *HOS3,* were identified based on sequence similarity to *RPD3* and *HDA1* (193). Mutations in all of these genes increase acetylation of histone H3 and H4 tails, implying that each affects histone acetylation *in vivo.*

Large complexes containing histone deacetylases and a Sin3p homologue have also been identified in mammalian cells (reviewed in Reference 194). HDA complexes mediate repression by the MAD–MAX repressor and by unliganded nuclear hormone receptors. These effects suggest a simple model to account for the genetic switch from repression to activation observed at genes targeted by these factors: MAD–MAX or unliganded receptor complexes bind DNA and recruit the HDA complex, resulting in histone deacetylation and repression. The switch to activation occurs when MYC–MAX replaces MAD–MAX, or upon hormone binding by the cognate receptor, resulting in replacement of the HDA complex with a HAT complex and subsequent histone acetylation. In essence, the model invokes targeted histone acetylation/deacetylation as a toggle between transcriptional activation/repression.

Histone deacetylases also affect silencing, defined as position-dependent, gene-independent repression (see Chapter 7). Silent DNA is packaged into hypoacetylated nucleosomes that exhibit a pattern of histone acetylation reminiscent of metazoan heterochromatin (195). Three classes of silent loci have been described in *S. cerevisiae,* including the *HM* silent mating-type loci, telomeres, and the rDNA locus (reviewed in Reference 196). In contrast to its effects on repression, deletion of *RPD3* enhances silencing at all three loci (197, 198). A homologue of Rpd3p was recently identified in *S. pombe,* encoded by the *hda1⁺* gene (199). Similar to *rpd3* deletion mutants of *S. cerevisiae, hda1⁻* deletion mutants are viable and enhance silencing at the silent mating-type loci, telomeres, and centromeres. It has not been reported whether the Hda1 histone deacetylase functions as a component of a larger complex, perhaps comparable to the Rpd3–Sin3p complex of *S. cerevisiae,* although a Sin3 homologue has been identified in *S. pombe* (200).

8. Mechanisms of transcriptional activation

Transcriptional control of gene expression is a complex and highly regulated process. Activator proteins can stimulate transcription at any of several steps during mRNA synthesis. These include displacement of transcriptional repressors, including histones, from DNA; recruitment of GTFs and RNAP II to the promoter; induction of conformational changes in the PIC; covalent modifications of PIC components; and stimulation of promoter clearance and chain elongation. This subject is reviewed elsewhere (44).

'Activation by recruitment' has been postulated to be the principal mechanism by which activators stimulate transcription (201). Accordingly, activators bind promoter DNA in a sequence-specific manner and stimulate PIC assembly, either by direct interaction with components of the core machinery or indirectly through coactivators (see Fig. 1). One model for stimulation of PIC assembly is activator-mediated recruitment of TBP. Strong support for this model is provided by tethering experiments, where the DNA-binding domain of either Gal4p or bacterial LexA is fused to TBP (55–57). These fusion proteins bypass the activator requirement for transcription in yeast. However, when altered forms of TBP were analysed in the bypass experiment, two classes of mutants were identified (202). The class defective for promoter binding was rescued by tethering, whereas the other class was not. This led to the proposal that activation by recruitment is at least a two-step process, with one step occurring subsequent to TBP recruitment (reviewed in Reference 203).

Other experiments also support the two-step model for activation. In these cases, tethering of either the Gal11p or Srb2p components of the SRB–MED complex was sufficient for activation (204, 205). As expected for the two-step recruitment model, activation in these experiments is dependent upon a functional TATA box (206). Strong support for the validity of these experiments is provided by fusions of activation domains to components of either TFIID or SRB–MED. None of these hybrid proteins stimulated transcription (207). Direct contact between activators and Gal11p (204), Srb4p (208), and Med6p (122) lends further support to the physiological relevance of these results.

The two-step recruitment model is also supported by template immobilization experiments where PIC assembly occurred by two independent steps, one involving TFIID recruitment independent of RNAP II, the other involving RNAP II recruitment in an SRB- and CTD-dependent manner (209). A similar conclusion was reached by analysing the mechanism of synergistic activation of the human interferon-β gene by multiple activators (210). The yeast experiments suggest the involvement of at least one additional step subsequent to TFIID and RNAP II recruitment for transcription. This implies that formation of the PIC by activator-mediated recruitment is not sufficient for transcription, but that steps subsequent to PIC formation can also be targeted by activators.

9. Summary and perspectives

The convergence of genetics, biochemistry, and structural biology has now defined the components and fundamental mechanisms involved in transcription by RNAP II. The core transcriptional machinery, including RNAP II, GTFs, and coactivator complexes, comprises about 100 polypeptides. Transcriptional activators, which bind to upstream promoter elements in a sequence-specific manner, target the core transcriptional machinery to stimulate PIC assembly. In total, an estimated 250 different activators are responsible for controlling expression of the approximately 6200 structural genes in *S. cerevisiae*. 'Activation by recruitment' appears to be a principal mechanism by which activators work. Accordingly, activators directly contact components of the core machinery, either facilitating assembly or stabilization of the PIC.

Although the framework for understanding eukaryotic transcription is now in place, many important questions remain unanswered. Activator bypass experiments support recruitment as the predominant mechanism for transcriptional activation in yeast. But which factors are targeted? Physical interactions between activators and many different components of the core machinery have been described *in vitro*, but which of these interactions are relevant *in vivo*? Also, to what extent can steps subsequent to PIC assembly—including promoter isomerization, transcription initiation, promoter clearance and elongation—be rate-limiting for transcription?

It was especially startling to learn that TAF subunits of TFIID are not generally required for activation. Yet more recent results demonstrate differential TAF requirements for activation. Whereas Taf145p is required for transcription of only about 16% of yeast genes, Taf17 affects approximately 67% of genes (136). What accounts for these differences? The discovery that the histone-like TAFs are components of both TFIID and SAGA might explain this difference, yet the Gcn5p component of SAGA affects expression of only about 5% of genes (136). Furthermore, Taf145p dependence is specified by core promoter elements rather than the UAS (140). This discovery suggests that the core promoter and the general factors play a more significant role in activation than is commonly assumed. The roles of the core promoter and general factors in activation need to be further investigated and will surely yield unexpected results.

Ultimately the spectrum of genes affected by specific components of the core transcriptional machinery must be defined. Yeast will remain a seminal organism in these studies, offering the combined power of classical and molecular genetics methods, facilitated by the complete sequence of the *S. cerevisiae* genome. Moreover, microarray analysis offers an unprecedented opportunity to define the effects of single gene mutations on the expression of virtually every gene in the yeast genome. Results from the past decade have defined the framework for understanding transcription. The stage is now set for unravelling how the transcriptional machinery receives and responds to the myriad physiological and developmental signals encountered by a cell.

Acknowledgements

I am grateful to Danny Reinberg and the members of my laboratory for many fruitful discussions, and to Stephen Burley for generously providing Fig. 2. Research in the author's laboratory is supported by NIH grant GM-39484.

References

1. Singer, V. L., Wobbe, C. R., and Struhl, K. (1990) A wide variety of DNA sequences can functionally replace a yeast TATA element for transcriptional activation. *Genes Dev.*, **4**, 636.
2. Li, W. Z. and Sherman, F. (1991) Two types of TATA elements for the *CYC1* gene of the yeast *Saccharomyces cerevisiae. Mol. Cell. Biol.*, **11**, 666.
3. Iyer, V. and Struhl, K. (1995) Mechanism of differential utilization of the his3 T-R and T-C TATA elements. *Mol. Cell. Biol.*, **15**, 7059.
4. Furter-Graves, E. M. and Hall, B. D. (1990) DNA sequence elements required for transcription initiation of the *Schizosaccharomyces pombe ADH* gene in *Saccharomyces cerevisiae. Mol. Gen. Genet.*, **223**, 407.
5. Weis, L. and Reinberg, D. (1992) Transcription by RNA polymerase II: initiator-directed formation of transcription-competent complexes. *FASEB J.*, **6**, 3300.
6. Kadosh, D. and Struhl, K. (1997) Repression by Ume6 involves recruitment of a complex containing Sin3 corepressor and Rpd3 histone deacetylase to target promoters. *Cell*, **89**, 365.
7. Keleher, C. A., Redd, M. J., Schultz, J., Carlson, M., and Johnson, A. D. (1992) Ssn6–Tup1 is a general repressor of transcription in yeast. *Cell*, **68**, 709.
8. Johnson, A. D. (1995) The price of repression. *Cell*, **81**, 655.
9. Hanna-Rose, W. and Hansen, U. (1996) Active repression mechanisms of eukaryotic transcription repressors. *Trends Genet.*, **12**, 229.
10. Maldonado, E., Hampsey, M., and Reinberg, D. (1999) Repression: targeting the heart of the matter. *Cell*, **99**, 455.
11. Woychik, N. A. and Young, R. A. (1994) Exploring RNA polymerase II structure and function. In *Transcription: mechanisms and regulation* (ed R. C. Conaway and J. W. Conaway), p. 227. Raven Press, New York.
12. Sakurai, H., Mitsuzawa, H., Kimura, M., and Ishihama, A. (1999) The rpb4 subunit of fission yeast *Schizosaccharomyces pombe* RNA polymerase II is essential for cell viability and similar in structure to the corresponding subunits of higher eukaryotes. *Mol. Cell. Biol.*, **19**, 7511.
13. McKune, K., Moore, P. A., Hull, M. W., and Woychik, N. A. (1995) Six human RNA polymerase subunits functionally substitute for their yeast counterparts. *Mol. Cell. Biol.*, **15**, 6895.
14. Kimura, M., Ishiguro, A., and Ishihama, A. (1997) RNA polymerase II subunits 2, 3, and 11 form a core subassembly with DNA binding activity. *J. Biol. Chem.*, **272**, 25851.
15. Hampsey, M. (1998) Molecular genetics of the RNA polymerase II general transcriptional machinery. *Microbiol. Mol. Biol. Rev.*, **62**, 465.
16. Fu, J., Gnatt, A. L., Bushnell, D. A., Jensen, G. J., Thompson, N. E., Burgess, R. R., David, P. R., and Kornberg, R. D. (1999) Yeast RNA polymerase II at 5 Å resolution. *Cell*, **98**, 799.
17. Poglitsch, C. L., Meredith, G. D., Gnatt, A. L., Jensen, G. J., Chang, W. H., Fu, J., and Kornberg, R. D. (1999) Electron crystal structure of an RNA polymerase II transcription elongation complex. *Cell*, **98**, 791.
18. Young, R. A. (1991) RNA polymerase II. *Annu. Rev. Biochem.*, **60**, 689.
19. Dahmus, M. E. (1996) Reversible phosphorylation of the C-terminal domain of RNA polymerase II. *J. Biol. Chem.*, **271**, 19009.
20. Hengartner, C. J., Myer, V. E., Liao, S. M., Wilson, C. J., Koh, S. S., and Young, R. A. (1998) Temporal regulation of RNA polymerase II by Srb10 and Kin28 cyclin-dependent kinases. *Mol. Cell*, **2**, 43.

21. Chambers, R. S. and Dahmus, M. E. (1994) Purification and characterization of a phosphatase from HeLa cells which dephosphorylates the C-terminal domain of RNA polymerase II. *J. Biol. Chem.*, **269**, 26243.

22. Archambault, J., Chambers, R. S., Kobor, M. S., Ho, Y., Cartier, M., Bolotin, D., Andrews, B., Kane, C. M., and Greenblatt, J. (1997) An essential component of a C-terminal domain phosphatase that interacts with transcription factor IIF in *Saccharomyces cerevisiae*. *Proc. Natl. Acad. Sci. U.S.A.*, **94**, 14300.

23. Kobor, M. S., Archambault, J., Lester, W., Holstege, F. C., Gileadi, O., Jansma, D. B., Jennings, E. G., Kouyoumdjian, F., Davidson, A. R., Young, R. A., and Greenblatt, J. (1999) An unusual eukaryotic protein phosphatase required for transcription by RNA polymerase II and CTD dephosphorylation in S. cerevisiae. *Mol. Cell*, **4**, 55.

24. Chambers, R. S., Wang, B. Q., Burton, Z. F., and Dahmus, M. E. (1995) The activity of COOH-terminal domain phosphatase is regulated by a docking site on RNA polymerase II and by the general transcription factors IIF and IIB. *J. Biol. Chem.*, **270**, 14962.

25. Cho, E. J., Takagi, T., Moore, C. R., and Buratowski, S. (1997) mRNA capping enzyme is recruited to the transcription complex by phosphorylation of the RNA polymerase II carboxy-terminal domain. *Genes Dev.*, **11**, 3319.

26. McCracken, S., Fong, N., Rosonina, E., Yankulov, K., Brothers, G., Siderovski, D., Hessel, A., Foster, S., Program, A. E., Shuman, S., and Bentley, D. L. (1997) 5′-Capping enzymes are targeted to pre-mRNA by binding to the phosphorylated carboxy-terminal domain of RNA polymerase II. *Genes Dev.*, **11**, 3306.

27. Yue, Z., Maldonado, E., Pillutla, R., Cho, H., Reinberg, D., and Shatkin, A. J. (1997) Mammalian capping enzyme complements mutant *Saccharomyces cerevisiae* lacking mRNA guanylyltransferase and selectively binds the elongating form of RNA polymerase II. *Proc. Natl. Acad. Sci. U.S.A.*, **94**, 12898.

28. McCracken, S., Fong, N., Yankulov, K., Ballantyne, S., Pan, G. H., Greenblatt, J., Patterson, S. D., Wickens, M., and Bentley, D. L. (1997) The C-terminal domain of RNA polymerase II couples mRNA processing to transcription. *Nature*, **385**, 357.

29. Steinmetz, E. J. (1997) Pre-mRNA processing and the CTD of RNA polymerase II: the tail that wags the dog? *Cell*, **89**, 491.

30. Nonet, M. L. and Young, R. A. (1989) Intragenic and extragenic suppressors of mutations in the heptapeptide repeat domain of *Saccharomyces cerevisiae* RNA polymerase II. *Genetics*, **123**, 715.

31. Koleske, A. J. and Young, R. A. (1994) An RNA polymerase II holoenzyme responsive to activators. *Nature*, **368**, 466.

32. Kim, Y.-J., Bjorklund, S., Li, Y., Sayre, M. H., and Kornberg, R. D. (1994) A multiprotein mediator of transcriptional activation and its interaction with the C-terminal repeat domain of RNA polymerase II. *Cell*, **77**, 599.

33. Shi, X., Chang, M., Wolf, A. J., Chang, C. H., Frazer-Abel, A. A., Wade, P. A., Burton, Z. F., and Jaehning, J. A. (1997) Cdc73p and Paf1p are found in a novel RNA polymerase II-containing complex distinct from the Srbp-containing holoenzyme. *Mol. Cell. Biol.*, **17**, 1160.

34. Chang, M., French-Cornay, D., Fan, H. Y., Klein, H., Denis, C. L., and Jaehning, J. A. (1999) A complex containing RNA polymerase II, Paf1p, Cdc73p, Hpr1p, and Ccr4p plays a role in protein kinase C signaling. *Mol. Cell. Biol.*, **19**, 1056.

35. Greenblatt, J. (1997) RNA polymerase II holoenzyme and transcriptional regulation. *Curr. Opin. Cell. Biol.*, **9**, 310.

36. Parvin, J. D. and Young, R. A. (1998) Regulatory targets in the RNA polymerase II holoenzyme. *Curr. Opin. Genet. Dev.*, **8**, 565.

37. Hampsey, M. and Reinberg, D. (1999) RNA polymerase II as a control panel for multiple coactivator complexes. *Curr. Opin. Genet. Dev.*, **9**, 132.
38. Maldonado, E., Shiekhattar, R., Sheldon, M., Cho, H., Drapkin, R., Rickert, P., Lees, E., Anderson, C. W., Linn, S., and Reinberg, D. (1996) A human RNA polymerase II complex associated with SRB and DNA-repair proteins. *Nature*, **381**, 86.
39. Scully, R., Anderson, S. F., Chao, D. M., Wei, W., Ye, L., Young, R. A., Livingston, D. M., and Parvin, J. D. (1997) BRCA1 is a component of the RNA polymerase II holoenzyme. *Proc. Natl. Acad. Sci. U.S.A.*, **94**, 5605.
40. Chao, D. M., Gadbois, E. L., Murray, P. J., Anderson, S. F., Sonu, M. S., Parvin, J. D., and Young, R. A. (1996) A mammalian SRB protein associated with an RNA polymerase II holoenzyme. *Nature*, **380**, 82.
41. Ito, M., Yuan, C. X., Malik, S., Gu, W., Fondell, J. D., Yamamura, S., Fu, Z. Y., Zhang, X., Qin, J., and Roeder, R. G. (1999) Identity between TRAP and SMCC complexes indicates novel pathways for the function of nuclear receptors and diverse mammalian activators. *Mol. Cell*, **3**, 361.
42. Weil, P. A., Luse, D. S., Segall, J., and Roeder, R. G. (1979) Selective and accurate initiation of transcription at the Ad2 major late promotor in a soluble system dependent on purified RNA polymerase II and DNA. *Cell*, **18**, 469.
43. Matsui, T., Segall, J., Weil, P. A., and Roeder, R. G. (1980) Multiple factors required for accurate initiation of transcription by purified RNA polymerase II. *J. Biol. Chem.*, **255**, 11992.
44. Orphanides, G., LaGrange, T., and Reinberg, D. (1996) The general initiation factors of RNA polymerase II. *Genes Dev.*, **10**, 2657.
45. Roeder, R. G. (1996) The role of general initiation factors in transcription by RNA polymerase II. *Trends Biochem. Sci.*. **21**, 327.
46. Hernandez, N. (1993) TBP, a universal eukaryotic transcription factor? *Genes Dev.*, **7**, 1291.
47. Eisenmann, D. M., Dollard, C., and Winston, F. (1989) *SPT15*, the gene encoding the yeast TATA binding factor TFIID, is required for normal transcription initiation *in vivo*. *Cell*, **58**, 1183.
48. Hahn, S., Buratowski, S., Sharp, P. A., and Guarente, L. (1989) Isolation of the gene encoding the yeast TATA binding protein TFIID: a gene identical to the *SPT15* suppressor of Ty element insertions. *Cell*, **58**, 1173.
49. Fikes, J. D., Becker, D. M., Winston, F., and Guarente, L. (1990) Striking conservation of TFIID in *Schizosaccharomyces pombe* and *Saccharomyces cerevisiae*. *Nature*, **346**, 291.
50. Kim, Y., Geiger, J. H., Hahn, S., and Sigler, P. B. (1993) Crystal structure of a yeast TBP/TATA-box complex. *Nature*, **365**, 512.
51. Kim, J. L., Nikolov, D. B., and Burley, S. K. (1993) Co-crystal structure of TBP recognizing the minor groove of a TATA element. *Nature*, **365**, 520.
52. Kosa, P. F., Ghosh, G., DeDecker, B. S., and Sigler, P. B. (1997) The 2.1-Å crystal structure of an archaeal preinitiation complex: TATA-box-binding protein/transcription factor (II)B core/TATA-box. *Proc. Natl. Acad. Sci. U.S.A.*, **94**, 6042.
53. Nikolov, D. B., Chen, H., Halay, E. D., Usheva, A. A., Hisatake, K., Lee, D. K., Roeder, R. G., and Burley, S. K. (1995) Crystal structure of a TFIIB–TBP–TATA-element ternary complex. *Nature*, **377**, 119.
54. Nikolov, D. B. and Burley, S. K. (1994) 2.1 Å resolution refined structure of a TATA box-binding protein (TBP). *Nature Struct. Biol.*, **1**, 621.
55. Chatterjee, S. and Struhl, K. (1995) Connecting a promoter-bound protein to TBP bypasses the need for a transcriptional activation domain. *Nature*, **374**, 820.

56. Klages, N. and Strubin, M. (1995) Stimulation of RNA polymerase II transcription initiation by recruitment of TBP *in vivo*. *Nature*, **374**, 822.

57. Xiao, H., Friesen, J. D., and Lis, J. T. (1995) Recruiting TATA-binding protein to a promoter: transcriptional activation without an upstream activator. *Mol. Cell. Biol.*, **15**, 5757.

58. Klein, C. and Struhl, K. (1994) Increased recruitment of TATA-binding protein to the promoter by transcriptional activation domains *in vivo*. *Science*, **266**, 280.

59. Strubin, M. and Struhl, K. (1992) Yeast and human TFIID with altered DNA-binding specificity for TATA elements. *Cell*, **68**, 721.

60. Reddy, P. and Hahn, S. (1991) Dominant negative mutations in yeast TFIID define a bipartite DNA-binding region. *Cell*, **65**, 349.

61. Arndt, K. M., Ricupero-Hovasse, S., and Winston, F. (1995) TBP mutants defective in activated transcription *in vivo*. *EMBO J.*, **14**, 1490.

62. Jackson-Fisher, A. J., Chitikila, C., Mitra, M., and Pugh, B. F. (1999) A role for TBP dimerization in preventing unregulated gene expression. *Mol. Cell*, **3**, 717.

63. Pinto, I., Ware, D. E., and Hampsey, M. (1992) The yeast *SUA7* gene encodes a homolog of human transcription factor TFIIB and is required for normal start site selection *in vivo*. *Cell*, **68**, 977.

64. Pinto, I., Wu, W.-H., Na, J. G., and Hampsey, M. (1994) Characterization of *sua7* mutations defines a domain of TFIIB involved in transcription start site selection in yeast. *J. Biol. Chem.*, **269**, 30569.

65. Berroteran, R. W., Ware, D. E., and Hampsey, M. (1994) The sua8 suppressors of *Saccharomyces cerevisiae* encode replacements of conserved residues within the largest subunit of RNA polymerase II and affect transcription start site selection similarly to *sua7* (TFIIB) mutations. *Mol. Cell. Biol.*, **14**, 226.

66. Li, Y., Flanagan, P. M., Tschochner, H., and Kornberg, R. D. (1994) RNA polymerase II initiation factor interactions and transcription start site selection. *Science*, **263**, 805.

67. Ha, I., Lane, W. S., and Reinberg, D. (1991) Cloning of a human gene encoding the general transcription initiation factor IIB. *Nature*, **352**, 689.

68. Bagby, S., Kim, S. J., Maldonado, E., Tong, K. I., Reinberg, D., and Ikura, M. (1995) Solution structure of the C-terminal core domain of human TFIIB: similarity to cyclin A and interaction with TATA-binding protein. *Cell*, **82**, 857.

69. Zhu, W. L., Zeng, Q. D., Colangelo, C. M., Lewis, L. M., Summers, M. F., and Scott, R. A. (1996) The N-terminal domain of TFIIB from *Pyrococcus furiosus* forms a zinc ribbon. *Nature Struct. Biol.*, **3**, 122.

70. Lin, Y. S., Ha, I., Maldonado, E., Reinberg, D., and Green, M. R. (1991) Binding of general transcription factor TFIIB to an acidic activating region. *Nature*, **353**, 569.

71. Colgan, J., Ashali, H., and Manley, J. L. (1995) A direct interaction between a glutamine-rich activator and the N terminus of TFIIB can mediate transcriptional activation *in vivo*. *Mol. Cell. Biol.*, **15**, 2311.

72. Roberts, S. G.E. and Green, M. R. (1994) Activator-induced conformational change in general transcription factor TFIIB. *Nature*, **371**, 717.

73. Wu, W. H. and Hampsey, M. (1999) An activation-specific role for transcription factor TFIIB *in vivo*. *Proc. Natl. Acac. Sci. U.S.A.*, **96**, 2764.

74. Burton, Z. F., Killeen, M., Sopta, M., Ortolan, L. G., and Greenblatt, J. (1988) RAP30/74: a general initiation factor that binds to RNA polymerase II. *Mol. Cell. Biol.*, **8**, 1602.

75. Henry, N. L., Campbell, A. M., Feaver, W. J., Poon, D., Weil, P. A., and Kornberg, R. D. (1994) TFIIF–TAF–RNA polymerase II connection. *Genes Dev.*, **8**, 2868.

76. Sun, Z. W. and Hampsey, M. (1995) Identification of the gene (*SSU71/TFG1*) encoding the largest subunit of transcription factor TFIIF as a suppressor of a TFIIB mutation in *Saccharomyces cerevisiae. Proc. Natl. Acad. Sci. U.S.A.*, **92**, 3127.

77. Leuther, K. K., Bushnell, D. A., and Kornberg, R. D. (1996) Two-dimensional crystallography of TFIIB– and IIE–RNA polymerase II complexes: implications for start site selection and initiation complex formation. *Cell*, **85**, 773.

78. Feaver, W. J., Henry, N. L., Bushnell, D. A., Sayre, M. H., Brickner, J. H., Gileadi, O., and Kornberg, R. D. (1994) Yeast TFIIE—cloning, expression, and homology to vertebrate proteins. *J. Biol. Chem.*, **269**, 27549.

79. Kuldell, N. H. and Buratowski, S. (1997) Genetic analysis of the large subunit of yeast transcription factor IIE reveals two regions with distinct functions. *Mol. Cell. Biol.*, **17**, 5288.

80. Tijerina, P. and Sayre, M. H. (1998) A debilitating mutation in transcription factor IIE with differential effects on gene expression in yeast. *J. Biol. Chem.*, **273**, 1107.

81. Sakurai, H., Ohishi, T., and Fukasawa, T. (1997) Promoter structure-dependent functioning of the general transcription factor IIE in *Saccharomyces cerevisiae. J. Biol. Chem.*, **272**, 15936.

82. Holstege, F. C.P., Tantin, D., Carey, M., Vandervliet, P. C., and Timmers, H. T.M. (1995) The requirement for the basal transcription factor IIE is determined by the helical stability of promoter DNA. *EMBO J.*, **14**, 810.

83. Parvin, J. D. and Sharp, P. A. (1993) DNA topology and a minimal set of basal factors for transcription by RNA polymerase II. *Cell*, **73**, 533.

84. Goodrich, J. A. and Tjian, R. (1994) Transcription factors IIE and IIH and ATP hydrolysis direct promoter clearance by RNA polymerase II. *Cell*, **77**, 145.

85. Feaver, W. J., Svejstrup, J. Q., Henry, N. L., and Kornberg, R. D. (1994) Relationship of CDK-activating kinase and RNA polymerase II CTD kinase TFIIH/TFIIK. *Cell*, **79**, 1103.

86. Adamczewski, J. P., Rossignol, M., Tassan, J. P., Nigg, E. A., Moncollin, V., and Egly, J. M. (1996) MAT1, cdk7 and cyclin H form a kinase complex which is UV light-sensitive upon association with TFIIH. *EMBO J.*, **15**, 1877.

87. Fisher, R. P. and Morgan, D. O. (1996) CAK in TFIIH: Crucial connection or confounding coincidence? *Biochem. Biophys. Acta*, **1288**, O7.

88. Feaver, W. J., Henry, N. L., Wang, Z., Wu, X., Svejstrup, J. Q., Bushnell, D. A., Friedberg, E. C., and Kornberg, R. D. (1997) Genes for Tfb2, Tfb3, and Tfb4 subunits of yeast transcription/repair factor IIH. Homology to human cyclin-dependent kinase activating kinase and IIH subunits. *J. Biol. Chem.*, **272**, 19319.

89. Svejstrup, J. Q., Vichi, P., and Egly, J. M. (1996) The multiple roles of transcription/repair factor TFIIH. *Trends Biochem. Sci.*, **21**, 346.

90. van Vuuren, A. J., Vermeulen, W., Ma, L., Weeda, G., Appeldoorn, E., Jaspers, N. G., van der Eb, A. J., Bootsma, D., Hoeijmakers, J. H., Humbert, S., *et al.* (1994) Correction of xeroderma pigmentosum repair defect by basal transcription factor BTF2 (TFIIH). *EMBO J.*, **13**, 1645.

91. Sung, P., Guzder, S. N., Prakash, L., and Prakash, S. (1996) Reconstitution of TFIIH and requirement of its DNA helicase subunits, Rad3 and Rad25, in the incision step of nucleotide excision repair. *J. Biol. Chem.*, **271**, 10821.

92. Luger, K., Mader, A. W., Richmond, R. K., Sargent, D. F., and Richmond, T. J. (1997) Crystal structure of the nucleosome core particle at 2.8 Å resolution. *Nature*, **389**, 251.

93. Kornberg, R. D. and Lorch, Y. (1999) Twenty-five years of the nucleosome, fundamental particle of the eukaryote chromosome. *Cell*, **98**, 285.

94. Han, M. and Grunstein, M. (1988) Nucleosome loss activates yeast downstream promoters *in vivo*. *Cell*, **55**, 1137.

95. Allfrey, V. G., Faulkner, R., and Mirsky, A. E. (1964) Acetylation and methylation of histones and their possible role in regulation of RNA synthesis. *Proc. Natl. Acad. Sci. U.S.A.*, **51**, 786.

96. Brownell, J. E., Zhou, J., Ranalli, T., Kobayashi, R., Edmondson, D. G., Roth, S. Y., and Allis, C. D. (1996) Tetrahymena histone acetyltransferase A: a homolog to yeast Gcn5p linking histone acetylation to gene activation. *Cell*, **84**, 843.

97. Kuo, M. H., Brownell, J. E., Sobel, R. E., Ranalli, T. A., Cook, R. G., Edmondson, D. G., Roth, S. Y., and Allis, C. D. (1996) Transcription-linked acetylation by Gcn5p of histones H3 and H4 at specific lysines. *Nature*, **383**, 269.

98. Kuo, M. H., Zhou, J., Jambeck, P., Churchill, M. E., and Allis, C. D. (1998) Histone acetyl-transferase activity of yeast Gcn5p is required for the activation of target genes *in vivo*. *Genes Dev.*, **12**, 627.

99. Candau, R., Zhou, J. X., Allis, C. D., and Berger, S. L. (1997) Histone acetyltransferase activity and interaction with ADA2 are critical for GCN5 function *in vivo*. *EMBO J.*, **16**, 555.

100. Taunton, J., Hassig, C. A., and Schreiber, S. L. (1996) A mammalian histone deacetylase related to the yeast transcriptional regulator Rpd3p. *Science*, **272**, 408.

101. Vidal, M. and Gaber, R. F. (1991) RPD3 encodes a second factor required to achieve maximum positive and negative transcriptional states in *Saccharomyces cerevisiae*. *Mol. Cell. Biol.*, **11**, 6317.

102. Rundlett, S. E., Carmen, A. A., Suka, N., Turner, B. M., and Grunstein, M. (1998) Transcriptional repression by UME6 involves deacetylation of lysine 5 of histone H4 by RPD3. *Nature*, **392**, 831.

103. Kadosh, D. and Struhl, K. (1998) Histone deacetylase activity of Rpd3 is important for transcriptional repression *in vivo*. *Genes Dev.*, **12**, 797.

104. Burns, L. G. and Peterson, C. L. (1997) Protein complexes for remodeling chromatin. *Biochem. Biophys. Acta*, **1350**, 159.

105. Winston, F. and Carlson, M. (1992) Yeast SNF/SWI transcriptional activators and the SPT/SIN chromatin connection. *Trends Genet.*, **8**, 387.

106. Hirschhorn, J. N., Brown, S. A., Clark, C. D., and Winston, F. (1992) Evidence that SNF2/SWI2 and SNF5 activate transcription in yeast by altering chromatin structure. *Genes Dev.*, **6**, 2288.

107. Cosma, M. P., Tanaka, T., and Nasmyth, K. (1999) Ordered recruitment of transcription and chromatin remodeling factors to a cell cycle- and developmentally regulated promoter. *Cell*, **97**, 299.

108. Krebs, J. E., Kuo, M. H., Allis, C. D., and Peterson, C. L. (1999) Cell cycle-regulated histone acetylation required for expression of the yeast *HO* gene. *Genes Dev.*, **13**, 1412.

109. Bjorklund, S. and Kim, Y. J. (1996) Mediator of transcriptional regulation. *Trends Biochem. Sci.*, **21**, 335.

110. Myer, V. E. and Young, R. A. (1998) RNA polymerase II holoenzymes and subcomplexes. *J. Biol. Chem.*, **273**, 27757.

111. Liao, S. M., Zhang, J. H., Jeffrey, D. A., Koleske, A. J., Thompson, C. M., Chao, D. M., Viljoen, M., Vanvuuren, H. J. J., and Young, R. A. (1995) A kinase–cyclin pair in the RNA polymerase II holoenzyme. *Nature*, **374**, 193.

112. Wilson, C. J., Chao, D. M., Imbalzano, A. N., Schnitzler, G. R., Kingston, R. E., and Young, R. A. (1996) RNA polymerase II holoenzyme contains SWI/SNF regulators involved in chromatin remodeling. *Cell*, **84**, 235.

113. Myers, L. C., Gustafsson, C. M., Bushnell, D. A., Lui, M., Erdjument-Bromage, H., Tempst, P., and Kornberg, R. D. (1998) The Med proteins of yeast and their function through the RNA polymerase II carboxy-terminal domain. *Genes Dev.*, **12**, 45.

114. Thompson, C. M. and Young, R. A. (1995) General requirement for RNA polymerase II holoenzymes *in vivo. Proc. Natl. Acad. Sci. U.S.A.*, **92**, 4587.

115. Carlson, M. (1997) Genetics of transcriptional regulation in yeast: connections to the RNA polymerase CTD. *Annu. Rev. Cell Dev. Biol.*, **13**, 1.

116. Rosenblum-Vos, L. S., Rhodes, L., Evangelista, C. C., Jr., Boayke, K. A., and Zitomer, R. S. (1991) The *ROX3* gene encodes an essential nuclear protein involved in *CYC7* gene expression in *Saccharomyces cerevisiae. Mol. Cell. Biol.*, **11**, 5639.

117. Jiang, Y. W., Dohrmann, P. R., and Stillman, D. J. (1995) Genetic and physical interactions between yeast RGR1 and SIN4 in chromatin organization and transcriptional regulation. *Genetics*, **140**, 47.

118. Li, Y., Bjorklund, S., Jiang, Y. W., Kim, Y. J., Lane, W. S., Stillman, D. J., and Kornberg, R. D. (1995) Yeast global transcriptional regulators Sin4 and Rgr1 are components of mediator complex RNA polymerase II holoenzyme. *Proc. Natl. Acad. Sci. U.S.A.*, **92**, 10864.

119. Gustafsson, C. M., Myers, L. C., Li, Y., Redd, M. J., Lui, M., Erdjument-Bromage, H., Tempst, P., and Kornberg, R. D. (1997) Identification of Rox3 as a component of mediator and RNA polymerase II holoenzyme. *J. Biol. Chem.*, **272**, 48.

120. Fassler, J. S. and Winston, F. (1989) The *Saccharomyces cerevisiae SPT13/GAL11* gene has both positive and negative regulatory roles in transcription. *Mol. Cell. Biol.*, **9**, 5602.

121. Myers, L. C., Gustafsson, C. M., Bushnell, D. A., Lui, M., Erdjument-Bromage, H., Tempst, P., and Kornberg, R. D. (1998) The Med proteins of yeast and their function through the RNA polymerase II carboxy-terminal domain. *Genes Dev.*, **12**, 45.

122. Lee, Y. C., Min, S., Gim, B. S., and Kim, Y.-J. (1997) A transcriptional mediator protein that is required for activation of many RNA polymerase promoters and is conserved from yeast to humans. *Mol. Cell. Biol.*, **17**, 4622.

123. Balciunas, D., Galman, C., Ronne, H., and Bjorklund, S. (1999) The Med1 subunit of the yeast mediator complex is involved in both transcriptional activation and repression. *Proc. Natl. Acad. Sci. U.S.A.*, **96**, 376.

124. Tabtiang, R. K. and Herskowitz, I. (1998) Nuclear proteins Nut1p and Nut2p cooperate to negatively regulate a Swi4p-dependent *lacZ* reporter gene in *Saccharomyces cerevisiae. Mol. Cell. Biol.*, **18**, 4707.

125. Gustafsson, C. M., Myers, L. C., Beve, J., Spahr, H., Lui, M., Erdjument-Bromage, H., Tempst, P., and Kornberg, R. D. (1998) Identification of new mediator subunits in the RNA polymerase II holoenzyme from *Saccharomyces cerevisiae. J. Biol. Chem.*, **273**, 30851.

126. Lee, Y. C. and Kim, Y. J. (1998) Requirement for a functional interaction between mediator components Med6 and Srb4 in RNA polymerase II transcription. *Mol. Cell. Biol.*, **18**, 5364.

127. Lee, Y. C., Park, J. M., Min, S., Han, S. J., and Kim, Y.-J. (1999) An activator binding module of yeast RNA polymerase II holoenzyme. *Mol. Cell. Biol.*, **19**, 2967.

128. Myers, L. C., Gustafsson, C. M., Hayashibara, K. C., Brown, P. O., and Kornberg, R. D. (1999) Mediator protein mutations that selectively abolish activated transcription. *Proc. Natl. Acad. Sci. U.S.A.*, **96**, 67.

129. Pugh, B. F. and Tjian, R. (1992) Diverse transcriptional functions of the multisubunit eukaryotic TFIID complex. *J. Biol. Chem.*, **267**, 679.

130. Chen, J. L., Attardi, L. D., Verrijzer, C. P., Yokomori, K., and Tjian, R. (1994) Assembly of recombinant TFIID reveals differential coactivator requirements for distinct transcriptional activators. *Cell*, **79**, 93.

131. Poon, D., Campbell, A. M., Bai, Y., and Weil, P. A. (1994) Yeast Taf170 is encoded by *MOT1* and exists in a TATA box-binding protein (TBP)–TBP-associated factor complex distinct from transcription factor IID. *J. Biol. Chem.*, **269**, 23135.

132. Reese, J. C., Apone, L., Walker, S. S., Griffin, L. A., and Green, M. R. (1994) Yeast TAF(II)s in a multisubunit complex required for activated transcription. *Nature*, **371**, 523.

133. Moqtaderi, Z., Yale, J. D., Struhl, K., and Buratowski, S. (1996) Yeast homologues of higher eukaryotic TFIID subunits. *Proc. Natl. Acad. Sci. U.S.A.*, **93**, 14654.

134. Walker, S. S., Reese, J. C., Apone, L. M., and Green, M. R. (1996) Transcription activation in cells lacking TAF(II)s. *Nature*, **383**, 185.

135. Moqtaderi, Z., Bai, Y., Poon, D., Weil, P. A., and Struhl, K. (1996) TBP-associated factors are not generally required for transcriptional activation in yeast. *Nature*, **383**, 188.

136. Holstege, F. C., Jennings, E. G., Wyrick, J. J., Lee, T. I., Hengartner, C. J., Green, M. R., Golub, T. R., Lander, E. S., and Young, R. A. (1998) Dissecting the regulatory circuitry of a eukaryotic genome. *Cell*, **95**, 717.

137. Apone, L. M., Virbasius, C. M.A., Reese, J. C., and Green, M. R. (1996) Yeast TAF(II)90 is required for cell-cycle progression through G(2)/M but not for general transcription activation. *Genes Dev.*, **10**, 2368.

138. Walker, S. S., Shen, W.-C., and Green, M. R. (1997) Yeast TAFII145 is required for transcription of G1/S cyclin genes and is regulated by the cellular growth rate. *Cell*, **90**, 607.

139. Ruppert, S., Wang, E. H., and Tjian, R. (1993) Cloning and expression of human TAFII250: a TBP-associated factor implicated in cell-cycle regulation. *Nature*, **362**, 175.

140. Shen, W.-C., Walker, S. S., and Green, M. R. (1997) Yeast $TAF_{II}145$ functions as a core promoter-selectivity factor, not a general co-activator. *Cell*, **90**, 615.

141. Burke, T. W. and Kadonaga, J. T. (1997) The downstream core promoter element, DPE, is conserved from *Drosophila* to humans and is recognized by $TAF_{II}60$ of *Drosophila*. *Genes Dev.*, **11**, 3020.

142. Michel, B., Komarnitsky, P., and Buratowski, S. (1998) Histone-like TAFs are essential for transcription *in vivo*. *Mol. Cell*, **2**, 663.

143. Xie, X., Kokubo, T., Cohen, S. L., Mirza, U. A., Hoffmann, A., Chait, B. T., Roeder, R. G., Nakatani, Y., and Burley, S. K. (1996) Structural similarity between TAFs and the heterotetrameric core of the histone octamer. *Nature*, **380**, 316.

144. Hahn, S. (1998) The role of TAFs in RNA polymerase II transcription. *Cell*, **95**, 579.

145. Grant, P. A., Schieltz, D., Pray-Grant, M. G., Steger, D. J., Reese, J. C., Yates, J. R., III, and Workman, J. L. (1998) A subset of TAF(II)s are integral components of the SAGA complex required for nucleosome acetylation and transcriptional stimulation. *Cell*, **94**, 45.

146. Ogryzko, V. V., Kotani, T., Zhang, X., Schlitz, R. L., Howard, T., Yang, X. J., Howard, B. H., Qin, J., and Nakatani, Y. (1998) Histone-like TAFs within the PCAF histone acetylase complex. *Cell*, **94**, 35.

147. Struhl, K. and Moqtaderi, Z. (1998) The TAFs in the HAT. *Cell*, **94**, 1.

148. Komarnitsky, P. B., Michel, B., and Buratowski, S. (1999) TFIID-specific yeast TAF40 is essential for the majority of RNA polymerase II-mediated transcription *in vivo*. *Genes Dev.*, **13**, 2484.

149. Shibuya, T., Tsuneyoshi, S., Azad, A. K., Urushiyama, S., Ohshima, Y., and Tani, T. (1999) Characterization of the *ptr6*(+) gene in fission yeast: a possible involvement of a transcriptional coactivator TAF in nucleocytoplasmic transport of mRNA. *Genetics*, **152**, 869.

150. Brownell, J. E. and Allis, C. D. (1996) Special HATs for special occasions: linking histone acetylation to chromatin assembly and gene activation. *Curr. Opin. Genet. Dev.*, **6**, 176.

151. Hampsey, M. (1997) A SAGA of histone acetylation and gene expression. *Trends Genet.*, **13**, 427.

152. Berger, S. L., Pina, B., Silverman, N., Marcus, G. A., Agapite, J., Regier, J. L., Triezenberg, S. J., and Guarente, L. (1992) Genetic isolation of *ADA2*: a potential transcriptional adaptor required for function of certain acidic activation domains. *Cell*, **70**, 251.

153. Georgakopoulos, T. and Thireos, G. (1992) Two distinct yeast transcriptional activators require the function of the GCN5 protein to promote normal levels of transcription. *EMBO J.*, **11**, 4145.

154. Marcus, G. A., Silverman, N., Berger, S. L., Horiuchi, J., and Guarente, L. (1994) Functional similarity and physical association between GCN5 and ADA2: putative transcriptional adaptors. *EMBO J.*, **13**, 4807.

155. Winston, F. (1992) Analysis of *SPT* genes: a genetic approach toward analysis of TFIID, histones and other transcription factors of yeast. In *Transcriptional regulation* (ed S. L. McKnight and K. R. Yamamoto), p. 1271. Cold Spring Harbor Laboratory Press, Cold Spring Harbor, NY.

156. Marcus, G. A., Horiuchi, J., Silverman, N., and Guarente, L. (1996) ADA5/SPT20 links the *ADA* and *SPT* genes, which are involved in yeast transcription. *Mol. Cell. Biol.*, **16**, 3197.

157. Roberts, S. M. and Winston, F. (1996) *SPT20/ADA5* encodes a novel protein functionally related to the TATA-binding protein and important for transcription in *Saccharomyces cerevisiae*. *Mol. Cell. Biol.*, **16**, 3206.

158. Roberts, S. M. and Winston, F. (1997) Essential functional interactions of SAGA, a *Saccharomyces cerevisiae* complex of Spt, Ada, and Gcnr proteins, with the Snf/Swi and Srb/mediator complexes. *Genetics*, **147**, 451.

159. Yang, X. J., Ogryzko, V. V., Nishikawa, J., Howard, B. H., and Nakatani, Y. (1996) A p300/CBP-associated factor that competes with the adenoviral oncoprotein E1A. *Nature*, **382**, 319.

160. Grant, P. A., Duggan, L., Cote, J., Roberts, S. M., Brownell, J. E., Candau, R., Ohba, R., Owen-Hughes, T., Allis, C. D., Winston, F., Berger, S. L., and Workman, J. L. (1997) Yeast Gcn5 functions in two multisubunit complexes to acetylate nucleosomal histones: characterization of an Ada complex and the SAGA (Spt/Ada) complex. *Genes Dev.*, **11**, 1640.

161. Horiuchi, J., Silverman, N., Pina, B., Marcus, G. A., and Guarente, L. (1997) ADA1, a novel component of the ADA/GCN5 complex, has broader effects than GCN5, ADA2, or ADA3. *Mol. Cell. Biol.*, **17**, 3220.

162. Saleh, A., Lang, V., Cook, R., and Brandl, C. J. (1997) Identification of native complexes containing the yeast coactivator/repressor proteins NGG1/ADA3 and ADA2. *J. Biol. Chem.*, **272**, 5571.

163. Eberharter, A., Sterner, D. E., Schieltz, D., Hassan, A., Yates, J. R., III, Berger, S. L., and Workman, J. L. (1999) The ADA complex is a distinct histone acetyltransferase complex in *Saccharomyces cerevisiae*. *Mol. Cell. Biol.*, **19**, 6621.

164. John, S., Grant, P. A., Tafrov, S. T., Sternglanz, R., and Workman, J. L. (1999) The something about silencing protein, SAS3, is the catalytic subunit of NuA3. *Genes Dev.*, **14**, 1196.

165. Orphanides, G., Wu, W. H., Lane, W. S., Hampsey, M., and Reinberg, D. (1999) The chromatin-specific transcription elongation factor FACT comprises human SPT16 and SSRP1 proteins. *Nature*, **400**, 284.

166. Allard, S., Utley, R. T., Savard, J., Clarke, A., Grant, P., Brandl, C. J., Pillus, L., Workman,

J. L., and Cote, J. (1999) NuA4, an essential transcription adaptor/histone H4 acetyltransferase complex containing Esa1p and the ATM-related cofactor Tra1p. *EMBO J.*, **18**, 5108.

167. Utley, R. T., Ikeda, K., Grant, P. A., Cote, J., Steger, D. J., Eberharter, A., John, S., and Workman, J. L. (1998) Transcriptional activators direct histone acetyltransferase complexes to nucleosomes. *Nature*, **394**, 498.

168. Cairns, B. R., Kim, Y. J., Sayre, M. H., Laurent, B. C., and Kornberg, R. D. (1994) A multisubunit complex containing the SWI1/ADR6, SWI2/SNF2, SWI3, SNF5, and SNF6 gene products isolated from yeast. *Proc. Natl. Acad. Sci. U.S.A.*, **91**, 1950.

169. Cote, J., Quinn, J., Workman, J. L., and Peterson, C. L. (1994) Stimulation of GAL4 derivative binding to nucleosomal DNA by the yeast SWI/SNF complex. *Science*, **265**, 53.

170. Cairns, B. R., Lorch, Y., Li, Y., Zhang, M. C., Lacomis, L., Erdjumentbromage, H., Tempst, P., Du, J., Laurent, B., and Kornberg, R. D. (1996) RSC, an essential, abundant chromatin-remodeling complex. *Cell*, **87**, 1249.

171. Cairns, B. R., Erdjument-Bromage, H., Tempst, P., Winston, F., and Kornberg, R. D. (1998) Two actin-related proteins are shared functional components of the chromatin-remodeling complexes RSC and SWI/SNF. *Mol. Cell*, **2**, 639.

172. Cao, Y., Cairns, B. R., Kornberg, R. D., and Laurent, B. C. (1997) Sfh1p, a component of a novel chromatin-remodeling complex, is required for cell cycle progression. *Mol. Cell. Biol.*, **17**, 3323.

173. Du, J., Nasir, I., Benton, B. K., Kladde, M. P., and Laurent, B. C. (1998) Sth1p, a *Saccharomyces cerevisiae* Snf2p/Swi2p homolog, is an essential ATPase in RSC and differs from Snf/Swi in its interactions with histones and chromatin-associated proteins. *Genetics*, **150**, 987.

174. Lorch, Y., Zhang, M., and Kornberg, R. D. (1999) Histone octamer transfer by a chromatin-remodeling complex. *Cell*, **96**, 389.

175. Hamiche, A., Sandaltzopoulos, R., Gdula, D. A., and Wu, C. (1999) ATP-dependent histone octamer sliding mediated by the chromatin remodeling complex NURF. *Cell*, **97**, 833.

176. Langst, G., Bonte, E. J., Corona, D. F., and Becker, P. B. (1999) Nucleosome movement by CHRAC and ISWI without disruption or trans-displacement of the histone octamer. *Cell*, **97**, 843.

177. Inostroza, J. A., Mermelstein, F. H., Ha, I., Lane, W. S., and Reinberg, D. (1992) Dr1, a TATA-binding protein-associated phosphoprotein and inhibitor of class II gene transcription. *Cell*, **70**, 477.

178. Mermelstein, F., Yeung, K., Cao, J., Inostroza, J. A., Erdjument-Bromage, H., Eagelson, K., Landsman, D., Levitt, P., Tempst, P., and Reinberg, D. (1996) Requirement of a corepressor for Dr1-mediated repression of transcription. *Genes Dev.*, **10**, 1033.

179. Kim, S., Na, J. G., Hampsey, M., and Reinberg, D. (1997) The Dr1/DRAP1 heterodimer is a global repressor of transcription *in vivo*. *Proc. Natl. Acad. Sci. U.S.A.*, **94**, 820.

180. Gadbois, E. L., Chao, D. M., Reese, J. C., Green, M. R., and Young, R. A. (1997) Functional antagonism between RNA polymerase II holoenzyme and global negative regulator NC2 *in vivo*. *Proc. Natl. Acad. Sci. U.S.A.*, **94**, 3145.

181. Prelich, G. (1997) *Saccharomyces cerevisiae* BUR6 encodes a DRAP1/NC2alpha homolog that has both positive and negative roles in transcription *in vivo*. *Mol. Cell. Biol.*, **17**, 2057.

182. Cang, Y., Auble, D. T., and Prelich, G. (1999) A new regulatory domain on the TATA-binidng protein. *EMBO J.*, **18,** 6662.

183. Auble, D. T. and Hahn, S. (1993) An ATP-dependent inhibitor of TBP binding to DNA. *Genes Dev.*, **7**, 844.

184. Auble, D. T., Hansen, K. E., Mueller, C. G.F., Lane, W. S., Thorner, J., and Hahn, S. (1994) Mot1, a global repressor of RNA polymerase II transcription, inhibits TBP binding to DNA by an ATP-dependent mechanism. *Genes Dev.*, **8**, 1920.

185. Prelich, G. and Winston, F. (1993) Mutations that suppress the deletion of an upstream activating sequence in yeast: involvement of a protein kinase and histone H3 in repressing transcription *in vivo*. *Genetics*, **135**, 665.

186. van der Knaap, J. A., Borst, J. W., van der Vliet, P. C., Gentz, R., and Timmers, H. T.M. (1997) Cloning of the cDNA for the TATA-binding protein-associated factor$_{II}$170 subunit of transcription factor B-TFIID reveals homology to global transcription regulators in yeast and *Drosophila*. *Proc. Natl. Acad. Sci. U.S.A.*, **94**, 11827.

187. Madison, J. M. and Winston, F. (1997) Evidence that Spt3 functionally interacts with Mot1, TFIIA, and TATA-binding protein to confer promoter-specific transcriptional control in *Saccharomyces cerevisiae*. *Mol. Cell. Biol.*, **17**, 287.

188. Collart, M. A. (1996) The *NOT*, *SPT3*, and *MOT1* genes functionally interact to regulate transcription at core promoters. *Mol. Cell. Biol.*, **16**, 6668.

189. Muldrow, T. A., Campbell, A. M., Weil, P. A., and Auble, D. T. (1999) MOT1 can activate basal transcription *in vitro* by regulating the distribution of TATA binding protein between promoter and nonpromoter sites. *Mol. Cell. Biol.*, **19**, 2835.

190. Kadosh, D. and Struhl, K. (1998) Targeted recruitment of the Sin3–Rpd3 histone deacetylase complex generates a highly localized domain of repressed chromatin *in vivo*. *Mol. Cell. Biol.*, **18**, 5121.

191. Kasten, M. M., Dorland, S., and Stillman, D. J. (1997) A large complex containing the yeast Sin3p and Rpd3 transcriptional regulators. *Mol. Cell. Biol.*, **17**, 4852.

192. Carmen, A. A., Rundlett, S. E., and Grunstein, M. (1996) HDA1 and HDA3 are components of a yeast histone deacetylase (HDA) complex. *J. Biol. Chem.*, **271**, 15837.

193. Rundlett, S. E., Carmen, A. A., Kobayashi, R., Bavykin, S., Turner, B. M., and Grunstein, M. (1996) HDA1 and RPD3 are members of distinct yeast histone deacetylase complexes that regulate silencing and transcription. *Proc. Natl. Acad. Sci. U.S.A.*, **93**, 14503.

194. Pazin, M. J. and Kadonaga, J. T. (1997) What's up and down with histone deacetylation and transcription? *Cell*, **89**, 325.

195. Braunstein, M., Sobel, R. E., Allis, C. D., Turner, B. M., and Broach, J. R. (1996) Efficient transcriptional silencing in *Saccharomyces cerevisiae* requires a heterochromatin histone acetylation pattern. *Mol. Cell. Biol.*, **16**, 4349.

196. Sherman, J. M. and Pillus, L. (1997) An uncertain silence. *Trends Genet.*, **13**, 308.

197. Smith, J. S., Caputo, E., and Boeke, J. D. (1999) A genetic screen for ribosomal DNA silencing defects identifies multiple DNA replication and chromatin-modulating factors. *Mol. Cell. Biol.*, **19**, 3184.

198. Sun, Z.-W. and Hampsey, M. (1999) A general requirement for the Sin3–Rpd3 histone deacetylase complex in regulating silencing in *Saccharomyces cerevisiae*. *Genetics*, **152**, 921.

199. Olsson, T. G., Ekwall, K., Allshire, R. C., Sunnerhagen, P., Partridge, J. F., and Richardson, W. A. (1998) Genetic characterisation of *hda1*$^+$, a putative fission yeast histone deacetylase gene. *Nucleic Acids Res.*, **26**, 3247.

200. Dang, V. D., Benedik, M. J., Ekwall, K., Choi, J., Allshire, R. C., and Levin, H. L. (1999) A new member of the Sin3 family of corepressors is essential for cell viability and required for retroelement propagation in fission yeast. *Mol. Cell. Biol.*, **19**, 2351.

201. Ptashne, M. and Gann, A. (1997) Transcriptional activation by recruitment. *Nature*, **386**, 569.

202. Stargell, L. A. and Struhl, K. (1996) A new class of activation-defective TATA-binding protein mutants: evidence for two steps of transcriptional activation *in vivo*. *Mol. Cell. Biol.*, **16**, 4456.

203. Stargell, L. A. and Struhl, K. (1996) Mechanisms of transcriptional activation *in vivo*: two steps forward. *Trends Genet.*, **12**, 311.

204. Barberis, A., Pearlberg, J., Simkovich, N., Farrell, S., Reinagel, P., Bamdad, C., Sigal, G., and Ptashne, M. (1995) Contact with a component of the polymerase II holoenzyme suffices for gene activation. *Cell*, **81**, 359.

205. Farrell, S., Simkovich, N., Wu, Y., Barberis, A., and Ptashne, M. (1996) Gene activation by recruitment of the RNA polymerase II holoenzyme. *Genes Dev.*, **10**, 2359.

206. Gonzalez-Couto, E., Klages, N., and Strubin, M. (1997) Synergistic and promoter-selective activation of transcription by recruitment of transcription factors TFIID and TFIIB. *Proc. Natl. Acad. Sci. U.S.A.*, **94**, 8036.

207. Keaveney, M. and Struhl, K. (1998) Activator-mediated recruitment of the RNA polymerase II machinery is the predominant mechanism for transcriptional activation in yeast. *Mol. Cell*, **1**, 917.

208. Koh, S. S., Ansari, A. Z., Ptashne, M., and Young, R. A. (1998) An activator target in the RNA polymerase II holoenzyme. *Mol. Cell*, **1**, 895.

209. Ranish, J. A., Yudkovsky, N., and Hahn, S. (1999) Intermediates in formation and activity of the RNA polymerase II preinitiation complex: holoenzyme recruitment and a post-recruitment role for the TATA box and TFIIB. *Genes Dev.*, **13**, 49.

210. Kim, T. K., Kim, T. H., and Maniatis, T. (1998) Efficient recruitment of TFIIB and CBP–RNA polymerase II holoenzyme by an interferon-beta enhanceosome *in vitro*. *Proc. Natl. Acad. Sci. U.S.A.*, **95**, 12191.

7 | The structure of yeast centromeres and telomeres and the role of silent heterochromatin

ALISON L. PIDOUX and ROBIN C. ALLSHIRE

1. Introduction

Genome stability and accurate chromosome segregation are of paramount importance to the cell and to the organism. The consequences of mistakes in mitosis and meiosis are severe indeed. In unicellular eukaryotes, aneuploidy is often deleterious or even fatal. For complex eukaryotes, chromosome gain or loss events are a leading cause of genetic diseases, miscarriage, and birth defects, and contribute to the progression of cancer. Chromosome rearrangements and fusions resulting from free DNA ends (e.g. those lacking telomeres) may have profoundly deleterious effects. To ensure their structural integrity and successful transmission, chromosomes must have origin(s) of replication, a single functional centromere and, if linear, telomeres.

The centromere was originally defined cytologically on mammalian chromosomes as the site of the primary constriction. Together with its associated proteinaceous kinetochore, this region of DNA is responsible for chromosome attachment to spindle microtubules and chromosome segregation at cell division (1). Centromeres perform very precise functions; during the S phase of the cell cycle, chromosomes are replicated, but the two chromatids must remain associated at their centromere until mitosis. At the onset of mitosis, sister kinetochores must specifically bind to microtubules emanating from the two opposite poles of the spindle. When all chromosomes are correctly attached and oriented on the metaphase plate, the sister chromatids part company and move to the spindle poles in anaphase, led by their kinetochores. In meiosis the situation is more complex. In the first (reductive) meiotic division, the sister kinetochores must act in unison, attaching to the microtubules from one pole to ensure segregation of homologous chromosomes; in the second (distributive) meiotic division, sister kinetochores act independently to segregate the two chromatids to opposite poles. The kinetochore must contain the machinery necessary to perform these complex tasks and allow the input of regulatory mechanisms. The kinetochore is the site monitored by the spindle checkpoint to ensure attachment of the complete chromosome complement to the spindle before allowing anaphase to proceed (2, 3).

Telomeres are DNA–protein complexes at the ends of chromosomes which protect them from degradation and fusion with other DNA (4). Telomeres also solve the 'end-replication' problem by having their own ribonucleoprotein polymerase, which uses an RNA template and a specialized reverse transcriptase subunit to synthesize repeats *de novo* to the ends of chromosomes, preventing their gradual erosion. The number of repeats added to chromosome ends is monitored by telomere repeat-binding proteins which serve to regulate telomere length.

The structure of centromeres and telomeres in the budding and fission yeasts has been the subject of intensive investigation for many years (5–14). This review aims to give a broad overview of what is known about the DNA and protein components of centromeres and telomeres, and to highlight general unifying features that seem to be important in their specification and structure.

The genomes of budding yeast and fission yeast are similar in size, approximately 14 Mb, but they are organized very differently: *Saccharomyces cerevisiae* has 16 chromosomes, whereas *Schizosaccharomyces pombe* has only three. The genetic tractability of both organisms and the resource of their genome databases makes them ideal model systems in which to study the biology of centromeres and telomeres. The small size and simplicity of budding yeast centromeres and the existence of *in vitro* assays for centromere function make this an excellent system in which to probe the details of a minimal centromere. In contrast, fission yeast centromeres are large and share many features with those of multicellular eukaryotes, making it a tractable system in which to analyse complex centromeres. Investigation of centromere and telomere structure, and specification in these yeasts, provides insights into chromosome biology in all eukaryotes.

The phenomenon of reversible transcriptional silencing is very important for our understanding of centromere and telomere function (for reviews see References 8, 11, 14–19). A well documented example is position effect variegation (PEV) of the white eye colour gene in *Drosophila*. Normal expression of the *white* gene at a euchromatic locus gives red eye pigmentation; however, when the gene is brought into proximity to centromeric heterochromatin through a chromosome rearrangement, it is subject to transcriptional repression, giving white eye colour. Heterochromatin involves a greater compaction of DNA and the incorporation of specific proteins to form a tighter structure which restricts the access of transcription and recombination machinery. In fact, the fly eyes are mottled, indicating an epigenetically modulated system in which stable but reversible states of expression or repression can be propagated for many generations. These interchangeable states do not differ in primary DNA sequence but in particular heritable chromatin structure.

Centromeric and telomeric regions of chromosomes are packaged in a distinct chromatin environment which differs from that of the rest of the genome. As in other organisms, the telomeres of budding yeast and fission yeast, and the centromeres of fission yeast, form transcriptionally silent domains which appear to be heterochromatic (8, 11, 13). Silencing at yeast centromeres and telomeres has much in common with PEV. Even budding yeast centromeres, which are not heterochromatic and do not display silencing, have a chromatin structure that differs from the normal nucleosomal array (20). Transcriptional silencing is presumably not the primary function of heterochromatin since few genes lie within normal centromeres, and telomeres tend to be many kilobases from endogenous genes. Rather, it is the byproduct of the tight packaging. Perhaps one role of heterochromatin is to make regions recombinationally 'silent'? Whatever its function, the phenomenon of silencing has been very useful experimentally in exploring centromere and telomere biology. Studies of mating-type silencing in both yeasts have defined many features of silencing and given a jump-start to the understanding of silencing at centromeres and telomeres, both in conceptual terms and in the identification of common factors required for silencing at different loci (for reviews see References 15, 21).

2. DNA structure of *S. pombe* centromeres

The 13.8-Mb genome of *S. pombe* is organised into three chromosomes, of 5.7, 4.6, and 3.5 Mb (22, 23). Their centromeres are large and complex, ranging in size from 35 to 110 kilobases (6, 7). A combination of chromosome breakage, genomic mapping, construction of artificial minichromosomes, and direct sequencing has contributed to the elucidation of the structure of fission yeast centromeres (24–27) (Fig. 1). Each centromere is composed of a central non-repetitive sequence of 4–7 kb surrounded by at least 15 kb of inverted repeat structures (24–28). Immediately adjacent to the central core are innermost repeat sequences (*imr*), surrounded by outer repeats (*otr*; K/L elements, also called *dg/dh* elements). The elements found in these repeats are organized differently in the three different centromeres (Fig. 1), and there is considerable variation in exact centromere structure between different laboratory and

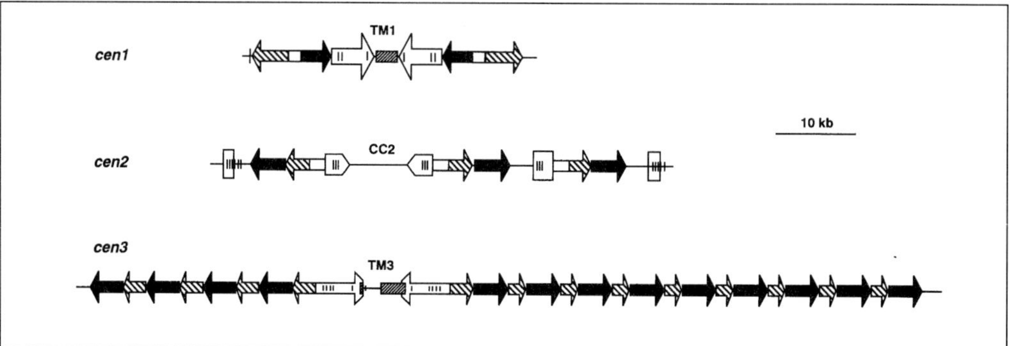

Fig. 1 Organization of *S. pombe* centromeres. Repetitive outer repeats (*otr*) with *dg/dh* elements (including K-type repeats) are shown as filled and cross-hatched arrows. The innermost repeats (*imr*) are represented by larger open arrows and contain tRNA genes (vertical lines). The central domains of *cen1* and *cen3* share homology within the TM element (cross-hatched), while the central domain of *cen2* (CC2) is unique. Centromeres 1, 2, and 3 occupy approximately 35, 60, and 110 kb respectively. Bar, 10 kb.

wild-type isolates of *S. pombe* (29). Even the repeat orientation with respect to the central cores varies, suggesting that it is the inverted repeat organization rather than a particular sequence that is important.

The construction of minichromosomes incorporating various parts of the centromere has narrowed down the regions required for minimal mitotic centromere function to a portion of the central core and at least one copy of K (30; but see below). Fission yeast centromeres naturally contain no PolII-encoded genes, but there are multiple tRNA genes in the *imr* sequences of each centromere and adjacent to each centromere (31, 32); it is not known whether any of these genes are transcriptionally active, but some do correspond to strong DNaseI hypersensitive sites (27). Recombination is suppressed across *S. pombe* centromeres (33). Low levels of recombination presumably exist to prevent loss or gain of centromere function through accidental rearrangements.

Whilst the outer repeat sequences are packaged into regularly spaced nucleosomes indistinguishable from bulk chromatin, the central core regions of fission yeast chromosomes are packaged as irregular chromatin lacking regularly spaced nucleosomes (27, 34). It is possible that this phenomenon reflects the association of the central region with centromere-specific proteins (see below).

It appears that, unlike their budding yeast counterparts, fission yeast centromeres are not specified by some special small *cis*-acting sequence. Instead they consist of vast transcriptionally silent domains that are thought to be the manifestation of kinetochore assembly (35, 36), and are subject to epigenetic regulation (37–39). The studies described below serve to illustrate these important features of fission yeast centromeres.

The centromeres of many multicellular eukaryotes are heterochromatic in nature, i.e. the DNA is packaged into transcriptionally silent chromatin (reviewed in References 16–19). Transcriptional silencing within and close to *Drosophila* centromeres is well documented (16, 18). Silencing has even been documented in a prokaryotic system: partition of the P1 plasmid in *Escherichia coli* requires the function of the ParB protein which binds the ParS centromere site and silences genes flanking it (40). Likewise, when a marker gene is placed at a variety of sites within fission yeast centromeres it is subject to transcriptional silencing (8, 35, 36). Silencing of introduced genes is thought to reflect the packaging of DNA by centromere-specific proteins, which has the effect of preventing access of transcriptional machinery; this idea is supported by the fact that mutations in *swi6$^+$*, a gene that encodes a centromere-associated protein, alleviate silencing (36). Placing the *ade6$^+$* gene at a site in *otr1* (*otr::ade6*) in a strain that is deleted for the endogenous *ade6$^+$* gene gives colonies that are ade– and red on plates low in adenine (owing to the build up of a biosynthetic intermediate); if the gene is placed at a euchromatic site, the colonies are white and ade$^+$ because the gene is expressed (36) (see Fig. 2). Similarly, the *ura4$^+$* gene is transcriptionally repressed when it is placed within the centromere; colonies are phenotypically ura– and resistant to the drug 5-fluoro-orotic acid (FOA). The *ura4$^+$* gene has been placed at 13 different sites within and around centromere I (36) (Fig. 3). The stability and magnitude of silencing imposed at the different sites is not equivalent;

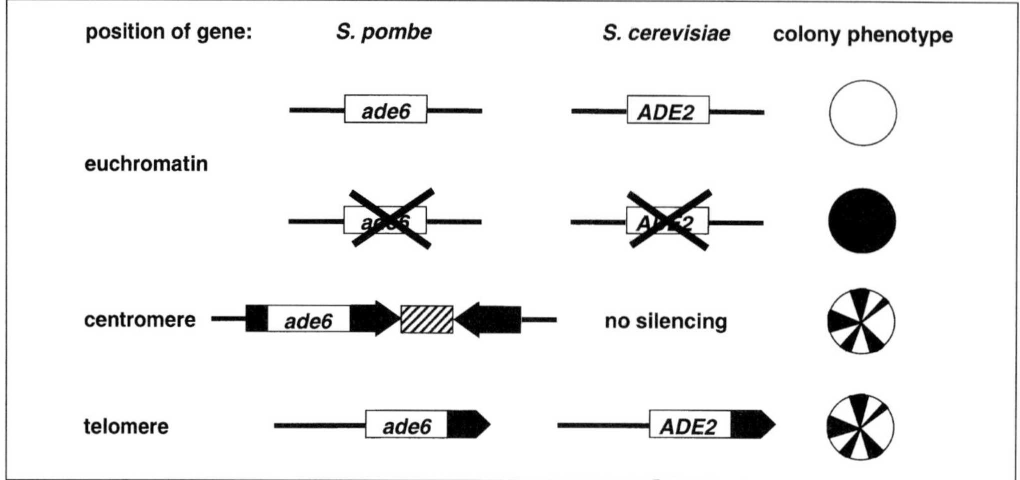

Fig. 2 Transcriptional silencing in *S. pombe* and *S. cerevisiae*. Normal expression of *ade6*[+] (*S. pombe*) or *ADE2* (*S. cerevisiae*) from euchromatic loci produces white colonies. Non-functional alleles of *ade6*[+] or *ADE2* cause the build-up of a red pigment on low-adenine plates, so red colonies form. If the genes are placed within *S. pombe* centromeres or adjacent to telomeres in either yeast, they are subject to reversible transcriptional silencing: colonies are sectored red and white. This reflects the propagation of the repressed (red) or expressed (white) states for many generations and occasional switches from one state to the other. The phenomenon of variegated expression is a feature of heterochromatic regions in many species. *S. cerevisiae* centromeres are not heterochromatic and do not display transcriptional silencing.

silencing is very strong at most positions in the *imr* and *otr* repeats, but considerably less within the central core. At the sites occupied by the tRNA genes in *imr*, silencing seems to be more relaxed and the *ura4*[+] gene is expressed. This correlates with the existence of Dnase-hypersensitive sites at these positions (27). Immediately outside the centromere, the *ura4*[+] gene does not experience silencing. The formation of red–white sectored colonies, and the ability to form colonies on both uracil-lacking media and FOA, indicates that transcriptional silencing in *S. pombe* centromeres is subject to variegation (i.e. states are heritable but reversible). A number of studies have contributed to the idea that fission yeast centromeres can be subject to epigenetic regulation (37–39). The term 'epigenetic' describes phenomena which are not encoded by DNA sequence but are nevertheless propagated through generations (19). When plasmids bearing portions of centromere III were introduced into *S. pombe* it was observed that they sometimes formed functional centromeres and sometimes did not (37). Very careful controls were performed which demonstrated that the difference was not due to DNA sequence; for instance plasmids extracted from 'stable' cells could give rise to both functional and non-functional centromeres when re-introduced into fission yeast, and *vice versa*. The conclusion from this study was that stochastic events due to cell environment/plasmid interaction were the explanation for the formation of functional or non-functional centromeres.

Minimal centromere function can be specified by the central core and a 2.1-kb fragment of the K element, which has been dubbed a 'centromere enhancer' (30). Further

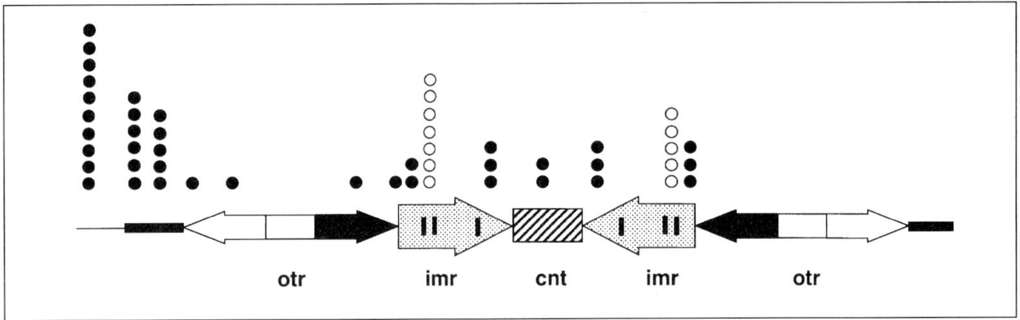

Fig. 3 Transcriptional silencing in *S. pombe* centromeres. The *ura4⁺* gene has been inserted at several sites within and adjacent to centromere I. The levels of expression determined by Northern blotting are indicated by the height of the column of dots above each site. At sites immediately outside the centromere (extreme left) the *ura4⁺* gene is highly expressed. Within the outer (*otr*) and inner repeats (*imr*), the *ura4⁺* gene is subject to transcriptional repression. Silencing is less strong at the central core regions. Sites occupied by tRNA genes (vertical lines within *imr*) are associated with more relaxed silencing, indicated by the open dots. At a distant euchromatic site, represented by the thin line on the left, the *ura4⁺* gene is fully expressed.

investigation of the requirements for centromere function indicated that the 2.1-kb enhancer was in fact dispensable (39). Minichromosomes lacking this region are extremely unstable, but if they are propagated (under selection) for many generations centromere function is gained, in the absence of DNA rearrangement. So, two events appear to be occuring: immediate activation of centromere function upon transformation of naked DNA requires the 2.1-kb enhancer in combination with the central core. However, initially non-functional centromeres (bearing the central core and the non-enhancer part of the K element) are able, if given time in the cellular environment through propagation on selection, to become 'activated' through an epigenetic effect which presumably involves the association of proteins with the centromere and formation of a chromatin configuration that is recognized by the cell as a kinetochore. These observations support a model for centromere function in *S. pombe* in which there is no specific sequence requirement (as there is in budding yeast) but that certain (functionally redundant) elements in combination contribute to a chromatin architecture that specifies centromere activation and function. A certain density of *cis*-acting sites might be required to get efficient nucleation of kinetochore assembly, which would imply that sequence and context are important.

A recent study investigating the role of histone acetylation in centromere function has also provided strong evidence for their epigenetic nature (38). The centromeres (and other heterochromatic regions) of higher eukaryotic chromosomes are known to be less modified by acetylation of histones H3 and H4 than is bulk chromatin (41). The acetylation status of *S. pombe* centromeres was investigated using a technique known as chromatin immunoprecipitation (ChIP). In this procedure, formaldehyde is used to crosslink chromatin *in vivo*; DNA is then fragmented by sonication, and subjected to immunoprecipitation with specific antibodies. Polymerase chain reaction (PCR) or Southern blotting is used to analyse which DNA sequences are specifically associated with the protein of interest. This technique allows *in vivo* composition of chromatin to be analysed (42). It was thus demonstrated that *S. pombe* centromeres are also underacetylated compared with euchromatic non-centromeric sites (38). The functional significance of this hypoacetylation was demonstrated using the drug trichostatin A (TSA) which inhibits histone deacetylase enzymes (effectively the core nucleosomal histones H3 and H4 become more highly acetylated). The fission yeast strains used in these experiments had the *ade6$^+$* gene inserted at the *otr* region (see above); wild-type cells formed red colonies as expected. However, upon treatment with TSA, white colonies were formed, reflecting the derepression of the *ade6$^+$* gene within the centromere. The white/expressing phenotype was propagated along with the hyperacetylated state through many generations in the absence of TSA. This state could 'switch' to the red/repressed state at a low frequency. Furthermore the white/ expressing lineages displayed many features of poorly functioning centromeres (e.g. lagging chromosomes), while the switched-back red derivative colonies had normal chromosome segregation. It was shown genetically that it was the status of the centromere that determined phenotype in *cis*- rather than a *trans*-acting mutation in another gene (38). This important study demonstrated that the function of fission yeast centromeres can be epigenetically regulated: centromeres with the same DNA structure can adopt different functional states dependent on their associated proteins and/or the modification status of these proteins. This supports the notion that context, rather than sequence alone, is critical in centromere specification.

3. Proteins associated with *S. pombe* centromeres or implicated in kinetochore function

Several proteins have been demonstrated to be associated with or have a role at fission yeast centromeres (see Table 1) and a variety of approaches are being used to identify kinetochore components. Transcriptional silencing at the centromere is thought to reflect the formation of a kinetochore complex involving the packaging of centromeric DNA into heterochromatin. A large number of mutants are known which affect silencing/switching at the mating type loci (43–46), and some of these affect centromeric silencing. Mutations in the *clr4$^+$*, *rik1$^+$* and *swi6$^+$* genes derepress silencing at *otr/imr* sites in the flanking regions of the centromere, but silencing at the central core is hardly affected (36). This provides evidence for structural domains within the centromere. *swi6*, *clr4*, and *rik1* mutants display increased rates of mini-chromosome loss, and cytological analyses show defects in chromosome segregation (e.g. lagging chromosomes on anaphase spindles) (36, 47), strongly suggestive of defects in centromere function. Swi6p is a so-called chromo-domain protein (48) and represents a fission yeast homologue of proteins found in heterochromatin such as HP1 in *Drosophila* and mammals (49). Immunolocalization of Swi6 protein gives a pattern of three to five spots on the periphery of the interphase nucleus (47). Fluorescence *in situ* hybridization (FISH) indicates that these spots coincide with the telomeres, the mating type locus, and the centromeres, which are clustered at the spindle pole body during interphase (50, 47). The function of both *rik1$^+$* and *clr4$^+$* is

Table 1 Putative kinetochore components in *S. pombe*

Name	Reference	Characteristics
Swi6*	48	Chromodomain protein; not essential
	47	Mutants derepress silencing in centromere-flanking regions; localized to centromeres, telomeres, mating-type loci
Mis6*	65	Essential; mutants display uneven chromosome segregation; associated specifically with inner centromere region
Chp1	55	Chromodomain protein; not essential
	R.C.A. *et al.* (unpublished)	Similar phenotypes to *swi6* mutants; immuno-localisation pattern consistent with location at centromeres, telomeres and mating-type loci
Clr4	52	Chromo- and SET domain protein; not essential
	51	Required for Swi6 localizsation
Rik1	51	Required for Swi6 localization
Abp1/Cbp1	59	*ars* binding protein; not essential
	60	Binds central core and K-type repeat sequences *in vitro*; homologous to human CENP-B
Cbh1	61	*ars* binding protein; essential; binds K-type repeats sequences *in vitro*; homologous to human CENP-B
Bub1*	66	Required for spindle checkpoint; localized to centromeres by immuno-localization/FISH

*These proteins have been demonstrated to reside at the centromere by ChIP, or by immuno-localization in combination with FISH.

required for the localization of Swi6 to the centromere (51) and it is likely that these proteins associate with centromeric regions. In support of this, the sequence of *clr4*$^+$ reveals that it too is a chromo-domain protein (52) and that it also possesses a SET domain—a 130 amino acid region of homology between three chromosomal proteins that act as modulators of gene activity: Su(var)3-9, Enhancer of zeste and Trithorax (53). Interestingly, mammalian Su(var)3-9 is found at centromeric regions and appears to co-sedimant with HP1 (m31) through sucrose gradients (54).

Chp1 is another chromo-domain protein identified in the *S. pombe* genome sequence database (55; B. Borgstrom and R. C. Allshire, unpublished observations). Disruption of *chp1*$^+$ causes lagging chromosomes and a high rate of chromosome loss. It also strongly alleviates centromeric silencing (B. Borgstrom, J. Partridge, and R. C.Allshire, unpublished observations). Various tagged versions of Chp1p have been localized to discrete foci on the nuclear periphery (one to eight foci per cell), suggesting that Chp1p is associated with centromeres and telomeres. Chp1 has been shown to be located at the outer flanking repeats by ChIP (J. Partridge *et al.*, unpublished results). It has been reported that Chp1 localization does not appear to be dependent on Rik1, Clr4, or Swi6 (55). But other investigations have suggested that Chp1 localization is Rik1 and Clr4 dependent, but that Chp1 is not required for localization of Swi6, and *vice versa* (J. Partridge, B. Borgstrom and R. C. Allshire, unpublished observations).

The phenomenon of transcriptional silencing has been exploited in efforts to identify potential kinetochore components. Twelve *csp* (centromeric suppressor of position effects) mutants have been isolated which alleviate silencing of *otr::ade6* and *imr::ura4*, and display chromosome missegregation phenotypes (56); it is expected that at least some of the *csp* genes will encode additional kinetochore components. Mutants that enhance silencing in the centromere central core have turned out to encode components of the 19S regulatory cap of the 26S proteasome (57). It appears that these mutants have a defect in sister chromatid separation independent of their ability to progress through anaphase and cytokinesis. This suggests that a centromere-associated protein involved in silencing contributes to sister chromatid cohesion and may be regulated by ubiquitin-mediated proteolysis (58).

Two proteins that were initially identified as *ars* (autonomously replicating sequence)-binding proteins have been shown to bind certain regions of *S. pombe* centromeres. Abp1/Cbp1 binds to central core sequences in band-shift assays (59, 60); its deletion or overexpression increases the rate of minichromosome loss. Another protein, Cbh1, binds to sites within the K-type repeats (61), although whether it is required for centromere function is not yet known. Abp1/Cbp1 and Cbh1 have a high degree of homology to each other and to the mammalian protein CENP-B, which localizes to kinetochores and binds to a 17-bp consensus sequence in alphoid satellite DNA which is found at human centromeres (1). Despite this localization, CENP-B may not play an essential role at kinetochores, since a mouse disruption of CENP-B has no centromere-associated phenotypes (see for example Reference 62). Therefore, the functional relevance of these *ars*-binding proteins is not yet clear; it is possible that the chromosome loss phenotypes observed are a consequence of replication defects. Band-shift experiments using crude *S. pombe* extracts and restriction fragments of centromere sequences have defined several regions capable of binding proteins *in vitro* (30, 39). Some of the band-shift patterns are altered when extracts are made from *abp1Δ* strains (39). However, the patterns of protein interactions appear to be complicated and the proteins responsible have not yet been identified.

A minichromosome loss screen resulted in the isolation of 12 temperature-sensitive *minichromosome instability* (*mis*) mutants with high rates of chromosome loss (63). Whilst some of these affect DNA replication, and Mis4 is a sister chromatid cohesion molecule (64), two of them, *mis6* and *mis12*, have phenotypes suggestive of a role at kinetochores (i.e. abnormal mitotic chromosome segregation) (63,65). *mis6*⁺ is an essential gene whose product is associated with the inner centromere region (*imr* and *cnt*, but not *otr*) as demonstrated by ChIP (65). At their restrictive temperature *mis6* mutants display uneven segregation of DNA in mitosis. For this phenotype to be manifested, the cells must pass through G1 and S phases of the cell cycle at the restrictive temperature. Mis6 therefore 'acts' at or before the onset of S phase, although it is unclear how this affects the centromere in molecular terms. Mis6 may need to be modified or interact with a protein synthesized only during G1–S in order to be incorporated into centromeric chromatin (65). Alternatively, Mis6 may need to act on the centromere before replication. The authors propose that Mis6 is required to give kinetochores the property of biorientation, suggesting that it may act as a connector providing rigidity so that sister kinetochores stay together but face away from each other and interact with microtubules from opposite poles (65). In support of a critical role at the centromere, loss of *mis6*⁺ function affects centromeric chromatin structure. As mentioned above, the central region (*cnt* and *imr*) has an unusual chromatin structure in wild-type cells, lacking the usual nucleosome ladder and producing a smear upon miccrococcal nuclease digestion. In *mis6-302* at the restrictive temperature, and also to some extent at the permissive temperature, the smear pattern is lost, and is replaced by a nucleosomal ladder (65). This is further evidence that chromatin structure is important for centromere specification and an indicator of proper functionality.

The spindle checkpoint ensures that all chromosomes are properly attached to the spindle before the cell commences anaphase (2, 3); for a detailed discussion see Chapter 4. *bub1* is essential for the spindle checkpoint response when cells are challenged by spindle damage (microtubule-disrupting drugs) or incompetent kinetochores (e.g. *swi6* mutants; *bub1* is synthetically lethal with *swi6*) (66). Bub1p is also required for accurate chromosome segregation in normal cells; *bub1Δ* has lagging chromosomes. Superb cytological analysis has demonstrated that Bub1 is recruited to fission yeast kinetochores at the onset of mitosis and in response to spindle damage. Twin spots of Bub1 are seen flanking the centromeric FISH signal in *nda3*-blocked cells (66). Localization of other checkpoint components to the kinetochore has not yet been demonstrated.

The picture emerging from studies of fission yeast centromeres is that they are packaged into transcriptionally silent heterochromatin involving underacetylated histones (8, 38). The centromere central core seems to be special: it lacks regularly spaced nucleosomes (34, 27), has weaker silencing than the flanking regions (36), and is specifically associated with Mis6 (65). When *swi6*, *clr4*, or *rik1* is defective, alleviation of flanking region silencing only is observed (36), suggesting that these proteins associate with the outer regions but not the core. ChIP analysis indicates that Swi6 and Chp1 are associated with the flanking region but not the core (J. Partridge, R. C. Allshire, unpublished results). *rik1*[+] and *clr4*[+] functions are required for Swi6 and Chp1 localization to the centromere (51). The CENP-B homologues, Cbh1 and Cbp1/Abp1, may also bind and have a centromeric role (59–61). Current observations thus support the speculative model for kinetochore structure presented in Fig. 4. The orientation and organization of repeat elements varies between the three centromeres but they all have a common structure of inverted repeats surrounded a central core; perhaps the two sides of the centromere flanks interact, forming a hairpin structure which puts the central core into a configuration promoting interaction with spindle microtubules (6, 8, 38). Three-dimensional reconstruction of serial electron micrographs of *S. pombe* spindles has identified structures likely to be kinetochores that are each associated with two to four microtubules (67).

These observations of transcriptional silencing together with the size and relative complexity of fission yeast centromeres, and their association with multiple microtubules, argues that overall they are structurally and functionally similar to the centromeres of higher eukaryotes and may provide a useful genetically tractable model for them.

4. *S. cerevisiae* **centromeres**

Budding yeast has the smallest centromeres known. The mitotic and meiotic functions of the centromere are encompassed within only 125 bp of DNA (5, 10, and

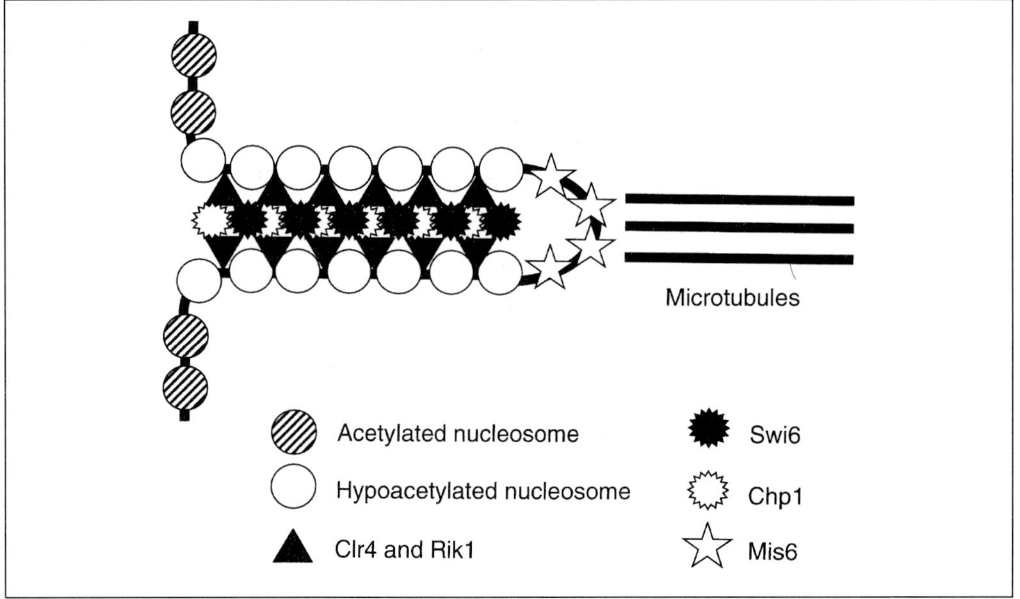

Fig. 4 Speculative model for kinetochore structure in *S. pombe*. In this model, the association of centromere-specific proteins mediates the formation of a hairpin structure in which flanking repeat regions of the centromere interact with one another. This puts the central core region into a favourable configuration for interaction with spindle microtubules. The centromere is known to be associated with nucleosomes containing hypoacetylated histones H3 and H4, and this may influence the binding of kinetochore components. Clr4 and Rik1 are required for the association of the chromodomain proteins Swi6 and, possibly, Chp1, which contribute to the transcriptionally silenced heterochromatic state of the centromere. Mis6 is located at the important central core region of the centromere. These and other proteins (including Cbh1 and Cbp1/Abp1) contribute to the formation of a particular centromere–kinetochore architecture that is required for function. It should be stressed that this is a speculative model and the details of protein association and interactions are not yet known. Based on models proposed in References 6, 8, 38.

references therein), although a larger region of approximately 200 bp is protected in footprinting assays (20). The centromere consists of three centromere determining elements: CDEI, CDEII, and CDEIII (5, 10). CDEI is an 8-bp imperfect palindrome that binds the Cbf1 protein (see below). It is important but not essential for centromere function: its deletion or mutation results in a 10–30-fold increase in chromosome loss. The 80-bp CDEII is essential for centromere function, but it is not highly conserved; it is highly A-T rich, but there are very few invariant bases among all 16 chromosomes. Mutations and small deletions are tolerated within CDEII. Both CDEI and CDEII appear to play important roles in CEN function in meiosis. CDEIII is a 25-bp imperfect palindrome that is absolutely required for centromere function, and can alone bind some of the components of the kinetochore (the CBF3 complex, see below). CDEIII is functionally asymmetrical: mutations in the two halves are not phenotypically identical, and inversion of CDEIII with respect to CDEI/II makes a non-functional centromere.

5. Kinetochore proteins in *S. cerevisiae*

While the tiny centromeres of budding yeast do not conform to the picture of silent heterochromatin observed in other species, they do have a unique chromatin structure. A 200-bp nuclease-resistant region covering CDEI, II, and III is surrounded by ordered nucleosomal arrays (20). Serial electron microscopic reconstructions indicate that budding yeast kinetochores interact with a single microtubule (68). The analysis of the budding yeast centromere–kinetochore complex is at a very advanced stage (5): many components have been identified, kinetochore structure and assembly is beginning to be understood, and something is known of the regulation of kinetochore–microtubule binding. These insights have stemmed from the application of genetic and biochemical approaches that have converged in recent years to remarkable effect.

Several genetic screens assaying the stability of non-essential minichromosomes have been performed, producing a large number of mutants (5, and references therein): *ctf* (chromosome transmission fidelity), *cse* (chromosome segregation), *cin* (chromosome instability), *cep* (centromere protein), *chl* (chromosome loss), *mif* (mitotic fidelity), and *ipl* (increase-in-ploidy) mutants. However, mutants with defects in many processes required for chromosome maintenance, such as DNA replication or spindle integrity, would be garnered in this type of screen. In order to pinpoint the genes likely to encode kinetochore components, additional secondary screens were devised and these have been very successful. In the transcriptional read-through assay (69), a construct is used in which CEN DNA has been placed between the galactose-inducible *GAL1* promoter and the β-galactosidase gene; promoter inducibility makes the centromere conditional. If a kinetochore complex is assembled at the CEN DNA it will block transcription; therefore wild-type strains give white colonies on X-gal plates, but those with defects in kinetochore components allow transcriptional readthrough and form blue colonies. The dicentric stabilization assay utilizes a minichromosome bearing two functional centromere sequences (69). Normally this would be pulled apart on the mitotic spindle and be rapidly lost. In cells defective for kinetochore assembly, the likelihood of forming two functional kinetochores is greatly reduced, and the dicentric minichromosome is therefore correctly segregated in a higher proportion of cells and thus stabilized.

On the biochemical front, two *in vitro* assays have been extremely beneficial to the molecular dissection of kinetochore structure and function. The first assay measures microtubule binding of *in vivo* assembled kinetochores; lysates of cells containing CEN plasmids are incubated with bovine brain microtubules and binding is assessed by microtubule-pelleting and Southern blotting (70, 71). This technique allows kinetochore–microtubule-binding properties of wild-type and mutant strains to be compared. The second assay involves the *in vitro* reconstitution of yeast proteins on to CEN DNA coated on to fluorescent latex beads and subsequent binding to microtubules, which is assessed by microscopy (72, 73). In addition, chromatography, band-shift assays, and formaldehyde crosslinking techniques (ChIP) have been instrumental in the identification of kinetochore components and understanding of kinetochore assembly *in vitro* and *in vivo*.

Table 2 Kinetochore proteins in *S. cerevisiae*

Protein	Gene Names	Reference	Properties
Cbf1p	*CBF1*	5	CDEI-binding protein
	CPF1	75	Not essential
	CEP1	5	Bends CDEI DNA
Ndc10p	*NDC10*	83	Component of CBF3 (CDEIII-binding) complex
	CTF14	69	Essential
	CBF2	82	Forms extended complex
Cep3p	*CEP3*	85	Component of CBF3 complex
	CBF3b	86	Essential
Ctf13p	*CTF13*	69	Component of CBF3 complex
			Essential
Skp1p	*SKP1*	87	Essential
		81	Required for association of Ctf13p in CBF3
Cse4p	*CSE4*	89	Essential, histone variant
		90	Homologous to human CENP-A
		91	CEN-associated by ChIP
		93	
Mif2p	*MIF2*	99	Essential
		97	Limited homology to human CENP-C
		98	CEN-associated by ChIP
		78	
Ctf19p	*CTF19*	100	Non-essential
			CEN-associated by ChIP
Kar3p	*KAR3*	106	Kinesin-related protein
			Non-essential; mitotic defects
			Co-purifies with CBF3
			Possible kinetochore component

The variety of genetic and biochemical approaches has led to a complex nomenclature for kinetochore components. In some cases the genes have several different names (see Table 2) but for simplicity we use only one here.

Cbf1p was identified biochemically as an activity that binds wild-type CDEI but not CDEI with debilitating mutations (74). Purification involving DNA-affinity chromatography allowed protein sequencing and subsequent cloning of the cognate gene (5, and references therein). Cbf1p has a helix-loop-helix motif in common with other DNA-binding proteins. Disruption of *CBF1* is not lethal, but causes a 10-fold increase in chromosome loss and methionine auxotrophy. The defect in methionine biosynthesis is due to a separable role of Cbf1p as a transcription factor at the *MET16* gene (and also at other genes). The degree of chromosome loss caused by *CBF1* deletion echoes that caused by mutation of CDEI. Cbf1p binds CDEI as a homodimer giving a 12–15-bp DNase I footprint (75, 76) and has been shown to bend CDEI DNA (77). Although Cbf1p is able to bind CDEI in isolation *in vitro*, ChIP analysis indicates that Cbf1p–CEN interaction is CDEIII dependent *in vivo* (78).

The CBF3 complex was isolated by affinity chromatography using CDEIII DNA (79, 80). The complex has a DNaseI footprint of 56 bp that is asymmetrical about the 25-bp CDEIII consensus site (the effect of mutations in this seemingly palindromic site is also asymmetrical). The four major components of CBF3, proteins of 110, 64, 58, and 23 kDa, have each been independently identified by genetic means. The formation of a CDEIII-binding complex has been reconstituted from recombinant components (81). The importance of CDEIII in centromere specification is emphasized by the fact that both it and the components of CBF3 are essential.

The protein sequence of the 110-kDa component of CBF3 was used to clone its gene, named *CBF2* (82). It is identical to *NDC10*, identified in a visual chromosome segregation screen (83), and to *CTF14*, identified in a chromosome transmission screen and subsequently focused on because of its behaviour in the transcriptional read-through and dicentric stabilization assays (84, 69). ChIP shows that Ndc10p is associated with centromeres *in vivo* (78). The 64-kDa protein was cloned from its protein sequence (86) and shown to be the same as *CEP3*, which was identified in a genetic screen for mutants that produced synthetically acentric minichromosomes (85). *CTF13* was identified in the same way as *CTF14* and was shown to encode the 58-kDa subunit upon protein sequencing (84, 69, 86). *SKP1* was identified as a high-copy suppressor of *ctf13-30* (87). Antibody supershift experiments indicate that it is a component of CBF3.

The analysis of the CBF3 complex has advanced beyond the simple identification of its components to a situation where we are beginning to understand some of the requirements for assembly of the complex and details of its interaction with CDEIII DNA. Using recombinant proteins expressed in insect cells it has been possible to reconstitute CBF3 complex that is active for CDEIII binding (81). Experiments in this system have led to the proposal that the availability of CBF3 complexes competent for kinetochore assembly is controlled through phosphorylation-dependent activation and ubiquitin-dependent destruction coupled via Skp1p.

DNA–protein crosslinking and binding interference studies have been used to probe the details of CBF3 interaction with CDEIII *in vitro* using recombinant CBF3 components (88). UV-crosslinkable Bromodeoxyuridine or Bromodeoxycytosine was incorporated at various positions within DNA fragments representing the 89-bp region to the 'right' of CDEII in CEN3. These experiments indicate that Ndc10p, Cep3p, and Ctf13p are in direct contact with bases in the major groove of CDEIII DNA and form an asymmetrical 'core' complex. An extended complex can also be formed which has additional DNA-bound Ndc10p and spans 80 bp. Some of the contacts made by CBF3 are sequence specific, and some are at bases known to be important for function. However, the fact that there are many sequence non-specific contacts implies that protein binding does not require a strict consensus sequence, and indeed none exists. This implies an analogy to centromeres of other species.

CSE4 has been identified in a number of genetic screens (89–91), including one in which it was isolated as a suppressor of a histone H4 mutant (90). Cse4p itself is a histone H3 variant and shows sequence and structural similarity to the mammalian centromere protein CENP-A (92). Both a histone H4 (*hhf1-20*) and *cse4* mutants have altered centromeric chromatin as assayed by a nuclease sensitivity assay (93). Cse4p has been demonstrated to reside at the budding yeast centromere by ChIP experiments. It is proposed that Cse4p together with histone H4 forms part of a single specialized centromeric nucleosome, the geometry of which might form the basis for interactions between other components of the centromere–kinetochore complex (93). The finding that this histone variant plays an important role at budding yeast centromeres is exciting in view of the localization of CENP-A to the inner kinetochore plate in mammals (92, 94). The expression of CENP-A is restricted to late S phase (95), a period when centromeres are replicated. This points to a temporal mechanism which acts to ensure its specific localization pattern; ectopic expression of CENP-A (from the early S-phase histone H3 promoter) is deleterious and leads to its incorporation over entire chromosomes and not just at centromeres (95).

Another budding yeast centromere-associated protein, Mif2p, shows limited homology to a mammalian centromere protein, CENP-C (96–98). The homology lies in the C-terminus of Mif2p in a region that is highly conserved in mammalian CENP-C proteins; the functional significance of the homology is supported by the fact that the mutant sites in two alleles of *mif2* lie in the C-terminal region (98). *MIF2* was originally identified through a screen for genes that caused chromosome loss when overexpressed (99). *MIF2* is essential and *mif2* mutants have defects in chromosome segregation and spindle integrity, and show genetic interactions with *cis*- and *trans*-acting centromere components (98). The immunolocalization pattern with anti-Mif2p antibodies is consistent with its localization to centromeres (78). Mif2p interaction with CEN DNA *in vivo* has been confirmed by ChIP experiments. Analysis of segregation incompetent CEN derivatives integrated at the *URA3* locus indicates that intact CDEIII alone is necessary and sufficient for this interaction (78).

Models for possible kinetochore structure have evolved from the studies described above (e.g. 78, 88, 93; see Fig. 5). In such models, a single specialized Cse4p-containing nucleosome exists at the centromere. The facts that CBF3 can bind isolated CDEIII *in vivo* and that Cbf1p binding to CDEI is CDEIII dependent *in vivo* suggest that CBF3 (and Mif2p) may assemble at CDEIII first, followed by binding of Cbf1p and other factors such as motor proteins. Multiple Ndc10p subunits may bind to form an extended kinetochore complex that encompasses approximately 200 bp DNA. Whatever the precise structure of the budding yeast kinetochore, it is clear that a unique DNA and protein configuration is required for function.

A recent addition to the collection of proteins that associate at the budding yeast centromere is Ctf19p (100). Isolated in a chromosome transmission screen, *ctf19* is positive in the transcriptional read-through and dicentric stabilization assays, and is supersensitive to overexpression of Ctf13p. Unusually for a centromere protein, *CTF19* is not essential, but shows genetic interactions with kinetochore components and checkpoint mutants. Centromeric association of Ctf19p has been confirmed by ChIP. Ctf19 mutants are defective in microtubule binding of *in vivo* assembled centromeres. The Ctf19p localization pattern to the nuclear face of the spindle pole body is consistent with a centromeric location, but the pattern is not disrupted by nocodazole, which scatters centromeres. The authors suggest that Ctf19p might form a link between kinetochore and the mitotic spindle, but whether this is the case will require further investigation.

The spindle checkpoint monitors kinetochore attachment to the mitotic spindle (2,

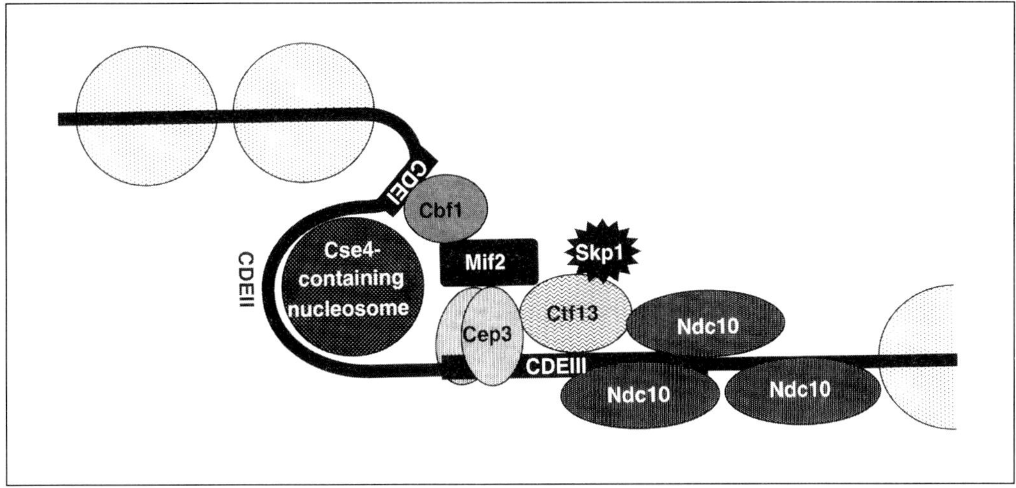

Fig. 5 Model for centromere–kinetochore structure in *S. cerevisiae*. The CEN DNA consisting of CDEI, II, and III is wrapped around a specialized Cse4p-containing nucleosome (dark circle), and surrounded by ordered nuclosome arrays (light circles). Cbf1p (black oval) binds CDEI, and is known to bend the CDEI DNA. The CBF3 complex, consisting of Cep3p, Ctf13p, Ndc10p, and Skp1p, binds to CDEIII. Cep3p, Ctf13p, and Ndc10p all make contacts with DNA, and additional Ndc10p molecules form an extended complex. Ctf13p is phosphorylated in a Skp1p-dependent manner. The location of Mif2p is based on known genetic and biochemical interactions. Model adapted from References 78, 88, 93. See text for details.

3; see Chapters 4 and 5). Many genes encoding kinetochore components (e.g. *CTF13*) display genetic interactions with checkpoint mutants (101). It is highly likely that some of the Mad and Bub proteins act at the budding yeast kinetochores, although actual association has not yet been demonstrated.

Several recent papers have investigated the regulation of microtubule–kinetochore binding. A combination of genetic and biochemical experiments including *in vitro* microtubule-binding assays supports the notion that kinetochore–microtubule binding is regulated by a balance between a protein kinase and a type I protein phosphatase (102–105). In this scenario, Glc7-dependent dephosphorylation of the Ndc10p component of CBF3 would favour binding, while Ipl1-dependent phosphorylation would inhibit microtubule–kinetochore association.

There are a number of proteins that may have a role at kinetochores but their association has not thus far been conclusively demonstrated. These include the kinesin-related protein Kar3p, which co-purifies with the CBF3 complex (72) and might play a role in microtubule–kinetochore interaction and chromosome movement (106). Cbf5 is another minor component that co-purifies with the CBF3 complex; it appears to have nucleolar functions and its possible role at kinetochores remains unclear (107). Additional kinetochore components are expected to be identified through biochemical purification and through existing mutant collections and new secondary genetic screens.

6. Comparison of *S. pombe* and *S. cerevisiae* centromeres and kinetochores

The centromeres of budding yeast and fission yeast appear to be fundamentally different. Centromere function in budding yeast is specified by 125 bp DNA. Fission yeast centromeres appear to have more in common with those of multicellular eukaryotes, being large and complex and involving repetitive DNA, albeit of a different type. Like the centromeric regions of *Drosophila*, for example, they are packaged into transcriptionally silent heterochromatin that can be subject to epigenetic modulation (19). Budding yeast kinetochores appear to bind a single microtubule, in contrast with fission yeast and higher eukaryotic centromeres which are associated with multiple MTs (67, 68).

However, parallels can be drawn between centromeres of these two yeasts. In both, the centromeres are associated with a specialized type of chromatin which differs from that of the rest of the genome; in fission yeast this is heterochromatic, in budding yeast ordered nucleosomal arrays surround a nuclease-resistant region encompassing a specialized Cse4p-containing nucleosome (93). All 16 centromeres in *S. cerevisiae* conform to the CDEI/CDEII/CDEIII structure; however, there is a fair degree of sequence variation, particularly in the AT-rich CDEII element (5). The centromeres of *S. pombe* vary in the unique sequence of their central core and in the organization and number of repeats, although they all follow the same pattern of inverted repeat arrays surrounding a central core.

No *bona fide* kinetochore components have yet been found in common between budding and fission yeasts. There is only one chromo-domain protein in *S. cerevisiae*, but it is of a different type to Swi6, Clr4, and Chp1, having an ATPase/helicase domain, and is involved in chromatin remodelling related to transcription (108). No homologues for CBF3 components have been reported. A *SKP1* homologue exists in human, but its possible role in centromere function is not known (87). However, the fact that *CSE4* and *MIF2* have mammalian homologues, CENP-A and CENP-C respectively, suggests that similar proteins are likely to exist at centromeres in fission yeast. This conservation is perhaps surprising given the apparent lack of sequence and organizational similarity between the centromeres of budding yeast and humans. Presumably some kind of functional conservation exists which we do not yet understand.

Budding yeast centromeres are sufficient to ensure incredibly accurate chromosome segregation. Why, then, do other organisms have such large, complex centromeres? The small size of budding yeast centromeres is not simply a reflection of the small size of the chromosomes they carry, since the largest *S. cerevisiae* chromosome is of a similar size to the smallest *S. pombe* one, but has a centromere 800 times smaller. Perhaps budding yeast centromeres represent a pared-down model involving efficient use of DNA, rather than the large complex ones being 'better'. The 125 bp of DNA and its associated proteins is able to specify microtubule attachment and checkpoint regulation, functions that it shares with the large complex centromeres of other eukaryotes. The specific sequence requirements for larger centromeres are not known in detail. While current thinking is that a specialized chromatin state determines centromere function, there is still the possibility that defined *cis*-acting sequences do exist and complex centromeres consist of multiples of point centromeres. One view of centromere function in *S. pombe* is that specific sequences embedded within repeat elements nucleate a complex that can spread along adjacent nucleosomes if in a favourable state. This heterochromatic structure would provide the correct environment for kinetochore assembly.

7. *S. pombe* telomeres and associated proteins

Whilst the analysis of *S. pombe* telomeres is not at such an advanced stage as that of budding yeast, their structure is known and several factors that affect telomere function have been identified. The sequence of telomeric repeats in fission yeast, T1-2AC(A)G1-8 (109), looks rather different from the vertebrate sequence TTAGGG and even other non-vertebrate species (4). However, a simplification of this sequence has recently been proposed (110) which discards the minor variations in sequence and quotes bases found in the majority of repeats: 5′-GGTTACA-3′, which is not so different from the vertebrate consensus, especially when viewed in a repeat context. The related sequence CGGTTA can be added to primers by an *S. pombe* telomerase activity *in vitro* (111); however, the templating RNA subunit of *S. pombe* telomerase has yet to be identified. The average length of the repeat tracts on wild-type telomeres is 300 bp. Internal to the repeats on chromosomes I and II are 20 kb of telomere-associated sequences (TASs) composed of tandem arrays of 86–89 bp repeats which are part of a larger 0.9–1.2-kb repeating unit (109). The TASs of different telomeres appear to have variable numbers and arrangements of the repeats, and their function, if any, is not known. Chromosome III does not have TAS sequences; instead rDNA repeats (10 kb in length) directly abut the telomere repeat tracts (22, 23, 109). The telomeric repeats are packaged into chromatin of unusual structure; there is 300 bp of micrococcal nuclease-resistant chromatin at the ends of chromosomes (28).

Different telomere constructs bearing the *ura4*+ marker gene were used to effect precise chromosome breakage of a non-essential minichromosome to form new chromosome ends with functional telomeres (112). Constructs with telomere repeats and TAS elements, or only the telomere repeats, were able to form functional telomeres. This, and the fact that linear episomes bearing only telomere repeats are stably propagated, indicates that the TASs are not essential for telomere function. The *ura4*+ gene adjacent to the telomere was subject to a reversible position effect. Ura4+ immediately adjacent to telomere repeats was silenced strongly; those separated from telomere repeats by TASs were silenced to a lesser degree. It is not known whether it is simply distance that causes this decrease in silencing, or whether the TAS forms a barrier to the silencing effect. If *ade6*+ is placed at the telomere, colonies with red (repressed state) and white (expressing state) sectors form (see Fig. 2). Pure red or white colonies containing about 2×10^6 cells can be obtained, indicating that a particular state can be stable for at least 20 generations. The stable propagation of states with occasional switching to the opposite state is a feature of position effect variegation. Presumably the repressive telomeric heterochromatin encroaches on adjacent genes in some cells but not others.

Conventional DNA polymerases are unable fully to synthesize telomeres; at each successive round of replication, sequence would be lost from the ends of telomeres. The cell has solved this problem with a specialized polymerase known as telomerase (12, 14). Composed of protein subunits and an RNA molecule, telomerase is a reverse transcriptase enzyme. Telomerase uses the RNA as a template to add nucleotides to the G-rich strand in the 5′ to 3′ direction, leaving a long overhang which is then filled in by conventional polymerases. A telomerase activity has been reported in *S. pombe* extracts and partially purified *S. pombe* telomerase sediments as a 35S particle, suggesting that it exists *in vivo* as part of a large multiprotein complex (111). The putative catalytic subunit of fission yeast telomerase was identified using degenerate PCR based on the sequences of *Euplotes* telomerase and budding yeast *EST2* (113). *trt1*$^+$ (telomerase reverse transcriptase) contains telomerase-specific regions of homology as well as regions shared with the larger family of reverse transcriptases. Deletion of *trt1*$^+$ causes a senescence phenotype: initially cells are healthy but gradually decline as they are propagated. Fewer, ragged-edge colonies form, and cells become elongated and die. These changes are accompanied by a gradual decrease in telomere length. A few 'survivor' colonies form even as the telomeres disappear completely, and it was discovered that the majority of these had circularized their chromosomes (114). Presumably the loss of the protective telomeric cap made the chromosome ends reactive; only cells in which all the chromosome fusions were intramolecular (i.e. avoiding the formation of dicentric chromosomes) would be viable. An alternative mechanism of survival, probably involving recombination, also allowed maintenance of telomeres.

The first telomere-binding protein from *S. pombe* was identified in a one-hybrid screen using the GGTTAC consensus as telomere sequence (115). Taz1 (*t*elomere-*a*ssociated in *Schizosaccharomyces pombe*) bears homology to the DNA-binding domain of the oncoprotein Myb, which is also found in the human telomere-binding proteins TRF1 and TRF2 (115, 116). Budding yeast Rap1p also has structural similarity (117); see below. Deletion of *taz1*$^+$ causes derepression of telomeric silencing and telomeres become vastly elongated, giving a smear of 2.5–4.5 kb when probed with telomere repeat probe, compared with 300 bp in wild-type strains (115). Chromatin structure at telomeres has also been reported to be altered in *taz1*$^-$ strains (115). The phenotypes of *taz1* mutants are consistent with it acting as a negative regulator of telomerase activity: when it is absent, telomeres become elongated. The details of how this regulation might occur are at present unknown, but it is likely that the *S. cerevisiae* model will be applicable (see below).

Loss of telomeric silencing is likely to be the outcome of mutations in a variety of proteins associated with telomeres or important for their function. In a mutant in which telomerase was directly or indirectly activated, there might be insufficient silencing factors to impose silencing on the lengthened telomeres. However, mutants that directly or indirectly cause a reduction in telomerase activity, causing telomere tract shortening, might lack a large enough block of heterochromatin to impose silencing. In addition, mutations in structural components of the telomere cap might allow access to transcription factors. Alternatively, efficient silencing adjacent to telomeres may be mediated by a combination of end-binding factors and telomere repeat-interacting factors, so that mutants that elongate the telomere might alleviate silencing by increasing the distance between the end and the marker gene. A genetic screen for derepression of *ura4*[+] and *his3*[+] genes placed at different telomeres has identified a large collection of mutants (118). *Lot* mutants have *lengthening of telomeres*, whilst other mutants, typified by *rat1* (*repression alleviated at telomeres*), alleviate silencing but have no effect on telomere length. It is encouraging that *lot3* turned out to be an allele of *taz1* (115, 118); it has a mutation within one of the conserved helices of the Myb-like DNA-binding region. *lot2* has similar but more severe phenotypes than *lot3*. Both derepress silencing specifically at telomeres and have hugely elongated telomeres (118). The details of this telomere expansion may be different in the two mutants, based on the pattern of telomere repeats on Southern blots. The products of the *rat1*[+] and *lot2*[+] genes remain to be characterized.

Apart from these phenotypes, the consequences of *lot2* or *taz1* mutation on vegetative cells appear to be insignificant: strains grow healthily for countless generations. However, there are highly deleterious effects on meiosis (115, 118, 119). Asci produced from *lot2* or *taz1* crosses are aberrant, and contain abnormal numbers and sizes of spores. The spores that are produced are often inviable (approximately 7% viability in *lot2* and 25% in *taz1*), suggesting that meiosis is aberrant. In addition, the level of meiotic recombination is reduced in *lot2* × *lot2* and *taz1* × *taz1* crosses compared with wild-type. How do defects in putative telomere-associated proteins have such a profound effect on meiotic division? It is known that large-scale nuclear reorganization occurs at the beginning of meiosis (discussed in detail in Chapter 5). Briefly, in vegetative cells, centromeres are clustered at the spindle pole body (SPB) and telomeres are found elsewhere on the nuclear periphery. Upon entry to the meiotic phase, telomeres become associated at the SPB while centromeres are released (120, 121). The nucleus takes on an elongated 'horsetail' shape and moves rapidly around the cell during meiotic prophase. FISH with telomere probes and live analysis using a GFP–Swi6 fusion protein showed that *lot2* and *taz1* mutants are defective in association of telomeres at the SPB and the nucleus performs only an aberrant horsetail stage (118, 119). It has been hypothesized that telomere clustering (the bouquet stage), observed in several species, may initiate the process of homologous chromosome pairing that must occur as a prelude to the meiotic divisions (reviewed in Reference 122). These observations in fission yeast are probably the best evidence to support such a model. Although the nucleus undergoes horsetail-like movements in *lot2* and *taz1* zygotes, this fails to involve the chromosomes because they are not linked by their telomeres to the SPB. This failure at meiotic prophase appears to lead to reduced recombination and chromosome missegregation during the meiotic divisions (A. L. Pidoux, E. Nimmo, and R. C. Allshire, unpublished observations), probably because of the failure of homologous chromosomes to pair and form crossovers/chiasmata which would ensure their correct orientation on the meiosis I spindle.

There are hints to the involvement of other proteins in telomere function. The

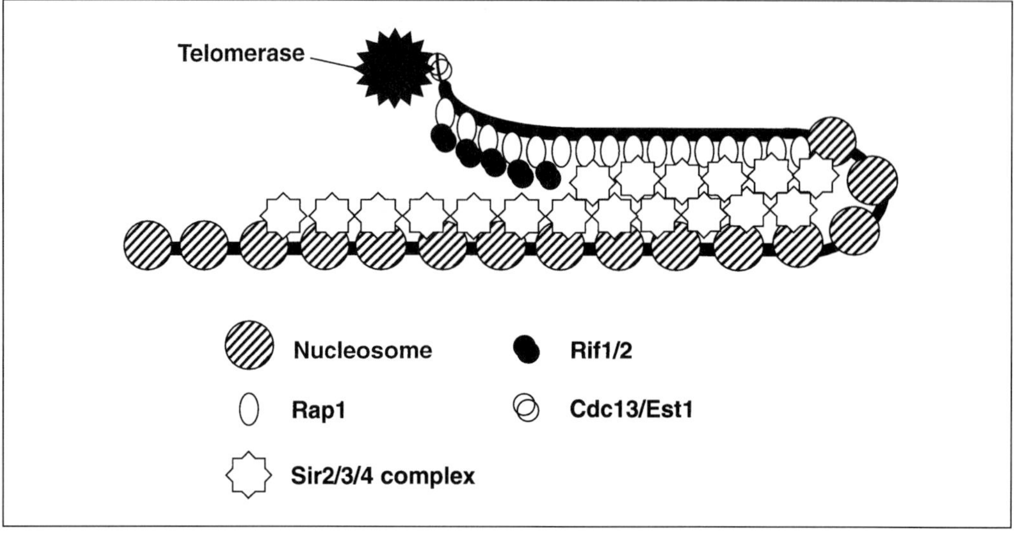

Fig. 6 Model for telomere structure in *S. cerevisiae*. In this model, the Rap1p-containing telomeric repeat region folds back on to the subtelomeric region. This foldback is presumably mediated and stabilized by interactions between the Sir complex and both Rap1p and histones H3 and H4 of the subtelomeric nucleosomes. The activity of telomerase is regulated by some kind of counting mechanism, which monitors the number of Rap1p molecules bound to the telomeric repeats. The Rif proteins influence this mechanism through their binding and negative regulation of Rap1p. Cdc13p and Est1p have been proposed to bind to the single-strand overhangs and may positively modulate telomerase activity, perhaps by serving as a docking site for the catalytic subunits. Ku is likely to bind the ends of the telomere but its precise role is not yet clear. Model adapted from References 13, 14, 126.

proteins Swi6, Clr4, and Rik1, important for silencing at the mating-type loci and centromeres, may also play some role at telomeres since telomeric silencing is alleviated to some extent when these proteins are defective (E. Nimmo and R. C. Allshire, unpublished observations). Telomere shortening is seen in several DNA replication checkpoint mutants (*rad1*, *rad3*, *rad17*, and *rad26*; see Chapter 4), suggesting that the proteins may have a role in telomere length regulation (123). *rad3*$^+$ and *tel1*$^+$ are *S. pombe* homologues of the human ATM gene that is defective in patients with ataxia telangiectasia (124). Deletion of both *rad3* and *tel1* causes an extreme slow growth phenotype, loss of all telomeric repeats, and the production of survivor colonies in which all chromosomes are circular. It is not known whether the telomere-shortening phenotypes result indirectly from defects in other cell cycle events or whether Rad3 and Tel1 are directly associated with the telomeres, perhaps in the formation of a protective cap. Alternatively, they might be involved in coordinating replication of telomere-adjacent sequences with telomerase repeat addition.

A structure involving repressive heterochromatin appears to exist at *S. pombe* telomeres. While few molecular details are known, it is likely that a model such as that described for *S. cerevisiae* in Fig. 6 will be applicable, although some of the components may differ.

8. *S. cerevisiae* **telomeres and associated proteins**

Budding yeast telomeres have been covered comprehensively in several excellent reviews recently, and we therefore give a briefer coverage of this topic (for reviews see References 11, 13, 14, 125, 126). The extreme ends of budding yeast chromosomes are composed of the short sequence repeats C1-3A/TG1-3 (actually C2-3A(CA)1-5/(TG)1-5) that extend for 300±75 bp and are immediately adjacent to blocks of the repetitive sequence element Y' (125). More internal to the Y' elements again there are more repeats and an X element. While all *S. cerevisiae* telomeres bear C1-3A/TG1-3 repeats, they do not all have Y's and the structure of the Y' and X repeat regions vary between chromosomes, making the functional importance of these elements unclear (125).

The telomeres of *S. cerevisiae* are packaged into a nuclease-resistant structure (127) that is heterochromatic in nature, and they are able to impose transcriptional silencing on genes placed within or nearby them (128). This property is known as telomere position effect (TPE) (see Fig. 2) and has many features and components in common with silencing at the mating-type loci (reviewed in References 11, 15). The C1-3A/TG1-3 repeats reveal a non-nucleosomal smear pattern when subjected to micrococcal nuclease digestion, indicating that a chromatin structure different from that of typical regularly spaced nucleosomes exists at the extreme ends of the telomeres (127). In contrast, the X and Y' repeat regions show the regular ladder of nucleosomes upon micrococcal nuclease digestion. Silencing effects are observed up to 3.5 kb from C1-3A/TG1-3 repeats (129). What are the protein components responsible for packaging telomeres into silenced chromatin? Many mutants have been described that alleviate TPE. For some of the genes thus identified, direct association of their gene products with telomeres has been demonstrated biochemically.

Rap1p (repressor-activator protein 1; reviewed in References12–14) is required for the repression or activation of a large number of genes. *rap1* mutants are defective in telomeric silencing and have effects on telomere length. The Rap1 protein binds directly to DNA; its consensus-binding site appears frequently (every 18 bp) in the course of the repeating C1-3A/TG1-3 array (130). Rap1p binding to telomeric DNA *in vivo* has been demonstrated in ChIP experiments (131). The crystal structure of Rap1p complexed to DNA has been solved and reveals that Rap1p bears structural similarity to the Myb protooncogene product (117), although there is little homology at the primary sequence level. Rap1p appears to be associated with DNA close to the ends of telomeres (0.5–4 kb from crosslinking experiments; Reference 131) and with subtelomeric regions tens of kilobases away from chromosome termini. This may indicate that telomeres have a secondary structure that involves looping back on itself in some manner (13, 126, 131) (see Fig. 6).

In addition to their effect on telomeric silencing, *rap1* mutants have interesting effects on telomere length which have led to the exciting proposal that Rap1p is the basis for a counting mechanism that serves to modulate telomere length through regulation of telomerase activity (12, 132). Two other components of the telomere length regulatory pathway, Rif1p and Rif2p, were identified through their interaction with Rap1p in two-hybrid screens (133, 134). Evidence suggests that Rif1p and Rif2p are positive regulators of the interaction of Rap1p with telomeric repeats, and therefore act to keep telomere tracts short (14).

Several factors that were originally identified for their role in silencing at the mating-type loci are also required for telomeric silencing. *SIR2*, *SIR3*, and *SIR4* are non-essential, but their mutation has profound effects on telomere biology (reviewed in References 11, 14, 126). *sir3* and *sir4* deletion mutants have shortened telomeres and derepression of telomeric (and mating-type) silencing (11, 135). Strains with mutations in *sir2*, *3*, or *4* display hyperacetylation of histone H4 at mating-type loci and subtelomeric Y' elements (136). These regions, in common with heterochromatin in other species, normally have hypoacetylated histones (41). Various mutations in the N-terminal tails of histones H3 and H4 also have effects on silencing at telomeres and mating type loci (135, 137). Thus, Sir proteins are important for the structure of repressive chromatin found at telomeres; if they are missing or defective, chromatin structure may be 'loosened', increasing accessibility to transcription factors and other proteins. The physical association of Sir2p, Sir3p, and Sir4p with telomeric chromatin has been demonstrated by ChIP (42, 131).

It appears that Sir3p plays a particularly important role in silencing (126). Over-expression of Sir3p causes 'spreading' of silenced chromatin; marker genes can be silenced up to 16 kb from the telomere in such strains (129), and this is correlated with the physical spreading of Sir3 protein along the chromosome, as shown by chromatin crosslinking studies (42, 131). Several lines of evidence, including *in vitro* studies, two-hybrid experiments, co-immunoprecipitation, and immunolocalization, indicate that silencing components interact physically (11, 13, 14, 126). Both Sir4p and Sir3p interact with Rap1p, and the N-termini of histones H3 and H4, as well as each other and themselves. Immunolocalization experiments using antisera against Rap1p, Sir3p, or Sir4p give a pattern of discrete foci localized to the nuclear periphery and represents clustered telomeres.

The RNA component of budding yeast telomerase was identified through its de-repression of telomeric silencing when overexpressed (138). *TLC1* overexpression leads to loss of telomeric repeats, which is presumably due to titration of other telomerase components away from their task of repeat addition. Telomerase mutants are not expected to die instantly (139): telomeres are gradually 'eroded' away, a few base pairs per division, as the ends of chromosomes fail to be replicated. For many generations this is of no great importance to the cell, but eventually catastrophe occurs as the cell is no longer able to maintain its chromosomes. *est1* (ever shorter telomeres) mutants monitored over 80–100 generations were found to suffer a progressive decrease in telomere length as they were propagated. This is accompanied by decrease in growth rate and increased cell death. Linear plasmids, but not circular ones, are unstable and there is a high frequency of chromosome loss. Four *EST* genes have been identified (139, 140). *TLC1* also has an *est* phenotype as predicted. *EST2* encodes a large protein with homology to reverse transcriptases and telomerases from *Euplotes*, human, and *S. pombe* (141, 142). Est1p and Cdc13p (*CDC13* and *EST4* are allelic) are able to bind telomeric repeats *in vitro* (126, and references therein), and their function is not required for telomerase activity *in vitro*. It is possible that Est1p and Cdc13p bind to single-stranded regions of yeast telomeres and recruit the catalytic components of telomerase including Est2p and *TLC1*. Deletion or over-expression of *EST1* or *EST2* has virtually no effect on telomeric silencing, although overexpression of certain mutant alleles do alleviate silencing (143).

The Ku proteins bind double-strand breaks and are required for DNA non-homologous end-joining pathway and for V(D)J recombination in antibody production (reviewed in Reference 144). Homologues of the Ku heterodimer exist in budding yeast: *HDF1/YKU70* and *HDF2/YKU80*. A two-hybrid screen with *HDF1* showed that it interacts with the C-terminus of Sir4p (145). This has implicated the Sir proteins in DNA end-joining and, indeed, *sir2, 3, 4* mutants are defective in Ku-dependent DNA end-joining, but not homologous recombination (145–147). One view is that they are involved in 'silencing' the broken site so as to protect it from attack by nucleases and to halt transcription. Crosslinking experiments have recently demonstrated that Ku is at telomeres (148). Loss of Ku function in yeast causes shortening of telomeres and loss of telomeric silencing (149, 150). Ku may be a terminus-binding factor, although its function at telomeres would be distinct from its role in non-homologous end-joining. Interestingly, the normal punctate pattern of Sir3p, Sir4p, and Rap1p is lost in Ku deletion strains, and unusually this is accompanied by scattering of Y' FISH signals, suggesting that Ku function is required for telomere clustering (150).

There are likely to be additional telomere components, and some of these will be identified through further genetic screens. *dot* mutants have the same phenotype as *tlc1*, i.e. their overexpression disrupts telomeric silencing (151). They all affect mating-type loci too, suggesting their products may be components or regulators of silent chromatin. There are several mutants that affect telomere length. For instance, like the *S. pombe rad3* and *tel1* mutants, mutations in either *TEL1* (an ATM homologue) or *TEL2* also cause telomere shortening (152–154).

Telomeres are thought to play important roles in meiosis (see above), and it is likely that meiosis-specific telomere-associated proteins exist. One example is the Ndj1 protein, which accumulates at telomeres during meiotic prophase and may act to stabilize homologous DNA interactions at telomeres (155). Meiotic nuclear reorganization has recently been described in budding yeast; telomeres tend to become clustered on one side of the nucleus, at least transitorily, in meiotic prophase (156). This is reminiscent of the bouquet stage seen in other organisms (122).

The wealth of information known about *S. cerevisiae* telomeres suggest models for telomere structure and silencing similar to that shown in Fig. 6.

9. Comparison of *S. pombe* and *S. cerevisiae* telomeres

Our current knowledge suggests that budding yeast and fission yeast telomeres are largely similar to each other and to telomeres in most organisms. The degree of detail to which the similarities extend is not yet known. Both have tracts of simple repeats at the extreme termini though the actual repeat sequence is very different. In addition both have subtelomeric elements (TAS in *S. pombe* and Y'/X in *S. cerevisiae*) which are not conserved in sequence but do have similar features. In neither organism are these subtelomeric elements essential for telomere function since functional chromosomes have been engineered which lack these elements and survive with only the repeat tracts.

Telomeres in budding and fission yeast impose transcriptional silencing on adjacently placed genes, supporting the model that telomeres are packaged into some higher-order heterochromatic structure. Histones are important for telomere function in budding yeast. Interestingly, histones H3 and H4 are hypoacetylated at telomeres as well as centromeres in *S. pombe* (K. Ekwall and R. C. Allshire, unpublished observations).

EST2 and *trt1*$^+$ encode conserved putative telomerase catalytic subunits. In *S. cerevisiae* the RNA component of telomerase, *TLC1*, has been identified, and several genes that might encode subunits or regulators of telomerase are known (*ESTs/CDC13*). No other putative telomerase components have yet been identified in *S. pombe*.

Telomere repeat-binding proteins have been identified in both yeasts. While Taz1 and Rap1p have little similarity in primary sequence, the crystal structure of Rap1p indicates that it shares a Myb-like DNA-binding domain with Taz1 and human TRF1 and TRF2. Loss of *RAP1* or *taz1*$^+$ function has the same effect: massive elongation of telomeres, which suggests that they both act as negative regulators of telomerase. No convincing homologues for *SIR3* or *SIR4* have been identified in *S. pombe* or other organisms. Whilst these proteins clearly play an important role in telomere function in budding yeast, it is possible that this is a detail in which budding yeast differs from other organisms. It is clear that known fission yeast 'silencing' proteins play a role at telomeres; the chromo-domain proteins Swi6 and Chp1 are localized at *S. pombe* telomeres and *clr4*, *rik1*, and *swi6* mutants alleviate telomeric silencing to some extent.

In common with heterochromatin in many organisms, the telomeres of budding and fission yeast are found at the nuclear periphery. It is possible that this location is required for efficient silencing. Fission yeast telomeres become tightly clustered at the SPB as a prelude to the horsetail stage of meiotic prophase. Meiotic nuclear reorganization has recently been described in budding yeast.

10. Conclusions and perspectives

Centromeres and telomeres are specified not only by their DNA sequence but by the formation of specialized chromatin configurations. Silent heterochromatin is central to the function of budding yeast telomeres and fission yeast centromeres and telomeres. It is thought to mediate the formation of higher-order structures, perhaps folds or loops, which the cell recognizes as kinetochore or telomere by its landscape. The importance of chromatin structure is underscored by the TSA experiments in fission yeast in which genetically identical cells manifest different phenotypes as a consequence of the acetylation pattern of their centromeric chromatin.

In the near future it is expected that we will gain an even better understanding of the structure and function of a simple kinetochore through the powerful combination of genetics, biochemistry, and *in vitro* analysis in *S. cerevisiae*. A major challenge in the study of fission yeast centromeres is to understand how a complex centromere works: what is its structure, and which sequences and proteins are required for its specification and maintenance. The knowledge gained from both these systems will also be applicable to understanding the requirements for centromere function in multicellular eukaryotes.

Telomere structure and biology appear to be similar in many important respects in budding yeast, fission yeast, and vertebrates. Telomere structure, particularly with respect to length regulation, is of great relevance to ageing and cancer (157). The advantages of being able to study telomeres in genetically tractable organisms are clear. The next few years are likely to see the identification of further telomere components and a greater understanding of how chromatin structure contributes to telomere function.

Acknowledgements

We thank Britta Borgstrom, Karl Ekwall, Elaine Nimmo and Janet Partridge for allowing us to cite unpublished work. Douglas Stewart is thanked for artwork. Work in Robin Allshire's lab is supported by the Medical Research Council of Great Britain.

References

1. Pluta, A., Mackay, A., Ainsztein, A., Goldberg, I., and Earnshaw, W. (1995) The centromere: hub of chromosomal activities. *Science*, **270**, 1591.
2. Rudner, A. D. and Murray, A. W. (1996) The spindle assembly checkpoint. *Curr. Opin. Cell Biol.*, **8**, 773.
3. Hardwick, K. G. (1998) The spindle checkpoint. *Trends Genet.*, **14**, 1.
4. Blackburn, E. H. and Greider, C. W. (ed) (1995) *Telomeres.* Cold Spring Harbor Laboratory Press, Cold Spring Harbor, NY.
5. Hyman, A. A. and Sorger, P. K. (1995) Structure and function of kinetochores in budding yeast. *Annu. Rev. Cell Dev. Biol.*, **11**, 471.
6. Clarke, L., Baum, M., Marschall, L. G., Ngan, V. K., and Steiner, N. C. (1994) Structure and function of *Schizosaccharomyces pombe* centromeres. *Cold Spring Harbor Symp. Quant. Biol.*, **58**, 687.
7. Allshire, R. C. (1995) Elements of chromosome structure and function in fission yeast. *Semin. Cell Biol.*, **6**, 55.
8. Allshire, R. C. (1996) Transcriptional silencing in the fission yeast: a manifestation of higher order chromosome structure and functions. In *Epigenetic mechanisms of gene regulation*, p443. Cold Spring Harbor Laboratory Press, Cold Spring Harbor, NY.
9. Clarke, L. (1998) Centromeres: proteins, protein complexes, and repeated domains at centromeres of simple eukaryotes. *Curr. Opin. Genet. Dev.*, **8**, 212.
10. Hegemann, J. H. and Fleig, U. N. (1993) The centromere of budding yeast. *Bioessays*, **15**, 451.

11. Shore, D. (1995) Telomere position effects and transcriptional silencing in the yeast *Saccharomyces cerevisiae*. In *Telomeres* (ed E. H. Blackburn and C. W. Greider), p139. Cold Spring Harbor Laboratory Press, Cold Spring Harbor, NY.

12. Shore, D. (1997) Telomerase and telomere-binding proteins: controlling the end-game. *Trends Biochem. Sci.*, **22**, 233.

13. Grunstein, M. (1997) Molecular model for telomeric heterochromatin in yeast. *Curr. Opin. Cell Biol.*, **9**, 383.

14. Lowell, J. E. and Pillus, L. (1998) Telomere tales: chromatin, telomerase and telomere function in *Saccharomyces cerevisiae*. *Cell Mol. Life Sci.*, **54**, 32.

15. Laurenson, P. and Rine, J. (1992) Silencers, silencing and heritable transcriptional states. *Microbiol. Rev.*, **56**, 543.

16. Karpen, G. H. (1994) Position–effect variegation and the new biology of heterochromatin. *Curr. Opin. Genet. Dev.*, **4**, 281.

17. Lohe, A. R. and Hilliker, AJ. (1995) Return of the H-word (heterochromatin). *Curr. Opin. Genet. Dev.*, **5**, 746.

18. Weiler, K. S., and Wakimoto, B. T. (1995) Heterochromatin and gene expression in *Drosophila. Annu. Rev. Genet.*, **29**, 577.

19. Karpen, G. H. and Allshire, R. C. (1997) The case for epigenetic effects on centromere identity and function. *Trends Biochem. Sci.*, **13**, 489.

20. Bloom, K. S. and Carbon, J. (1982) Yeast centromere DNA is in a unique and highly ordered structure in chromosomes and small circular minichromosomes. *Cell*, **29**, 305.

21. Klar, A. J. S. (1992). Molecular genetics of fission yeast cell type: mating type and mating-type interconversion.. In *The molecular and cellular biology of the yeast* Saccharomyces cerevisiae: *gene Expression*, vol. II (ed E. W. Jones, J. R. Pringle, and J. R. Broach), p745. Cold Spring Harbor Laboratory Press, Cold Spring Harbor, NY.

22. Hoheisel, J. D., Maier, E., Mott, R., McCarthy, L., Grigoriev, A. V., Schalkwyk, L. C., Nizetic, D., Francis, F. and Lehrach, H. (1993) High-resolution cosmid and P1-maps spanning the 14-mb genome of the fission yeast *S. pombe. Cell*, **73**, 109.

23. Mizukami, T., Chang, W. I., Garkavtsev, I., Kaplan, N., Lombardi, D., Matsumoto, T., Niwa, O., Kounosu, A., Yanagida, M., Marr, T. G., and Beach, D. (1993) A 13 kb resolution cosmid map of the 14 Mb fission yeast genome by nonrandom sequence-tagged site mapping. *Cell*, **73**, 121.

24. Clarke, L. and Baum, M. P. (1990) Functional analysis of a centromere from fission yeast: a role for centromere-specific repeated sequences. *Mol. Cell. Biol.*, **10**, 1863.

25. Murakami, S., Matsumoto, T., Niwa, O., and Yanagida, M. (1991) Structure of the fission yeast centromere cen3: direct analysis of the reiterated inverted region. *Chromosoma*, **101**, 214.

26. Hahnenberger, K. M., Carbon, J., and Clarke, L. (1991) Identification of DNA regions required for mitotic and meiotic functions within the centromere of *Schizosaccharomyces pombe* chromosome I. *Mol. Cell. Biol.*, **11**, 2206.

27. Takahashi, K.,Murakami, S., Chikashige, Y., Funabiki, H., Niwa, O. , and Yanagida, M. (1992) A low copy number central sequence with strict symmetry and unusual chromatin structure in fission yeast centromere. *Mol. Biol. Cell*, **3**, 819.

28. Chikashige, Y., N. Kinoshita, Y. Nakaseko, T. Matsumoto, S. Murakami, O. Niwa, and M. Yanagida. (1989) Composite motifs and repeat symmetry in *S. pombe* cemtromeres: direct analyses by integration of Not1 restriction sites. *Cell*, **57**, 739.

29. Steiner, N. C., Hahnerberger, K. M., and Clarke, L. (1993) Centromeres of the fission yeast *Schizosaccharomyces pombe* are highly variable genetic loci. *Mol. Cell. Biol.* **13**, 4578.

30. Baum, M., V. K. Ngan, and L. Clarke (1994) The centromeric K-type repeat and the central core are together sufficient to establish a functional *Schizosaccharomyces pombe* centromere. *Mol. Biol. Cell*, **5**, 747.

31. Kuhn, R. M., Clarke, L., and Carbon, J. (1991). Clustered tRNA genes in *Schizosaccharomyces pombe* centromeric DNA sequence repeats. *Proc. Natl. Acad. Sci. U.S.A.*, **88**, 1306.

32. Takahashi, K., Murakami, S., Chikashige, Y., Niwa, O., and Yanagida, M. (1991) A large number of tRNA genes are symmetrically located in fission yeast centromeres. *J. Mol. Biol.*, **218**, 13.

33. Clarke, L., Amstutz, H., Fishel, B., and Carbon, J. (1986) Analysis of centromeric DNA in the fission yeast *Schizosaccharomyces pombe*. *Proc. Natl. Acad. Sci. U.S.A.*, **83**, 8253.

34. Polizzi, C. M. and L. Clarke. (1991) The chromatin structure of centromeres from fission yeast is distinct, with a differentiation of the central core which correlates with function. *J. Cell Biol.*, **112**, 191.

35. Allshire, R. C., Javerzat, J-P., Redhead, N. J., and Cranston, G. (1994) Position effect variegation at fission yeast centromeres. *Cell*, **76**, 157.

36. Allshire, R. C., Nimmo, E. R., Ekwall, K., Javerzat, J-P., and Cranston, G. (1995) Mutations derepressing silent centromeric domains in fission yeast disrupt chromosome segregation. *Genes Dev.*, **9**, 218.

37. Steiner, N. S. and Clarke, L. (1994) A novel epigenetic affect can alter centromere function in fission yeast. *Cell*, **79**, 865.

38. Ekwall, K., Olsson, T., Turner, B. M., Cranston, G., and Allshire, R. C. (1997) Transient inhibition of histone acetylation alters the structural and functional imprint at fission yeast centromeres. *Cell*, **91**, 1021.

39. Ngan, V. K. and Clarke, L. (1997) The centromere enhancer mediates centromere activation in *Schizosaccharomyces pombe*. *Mol. Cell. Biol.*, **17**, 3305.

40. Rodionov, O., Lobocka, M., and Yarmolinsky, M. (1999) Silencing of genes flanking the P1 plasmid centromere. *Science*, **283**, 546.

41. Turner, B. M. and O'Neill, L. P. (1995) Histone acetylation in chromatin and chromosomes. *Semin. Cell Biol.*, **6**, 229.

42. Hecht, A., Strahl-Bolsinger, S., and Grunstein M. (1996). Spreading of transcriptional repressor SIR3 from telomeric heterochromatin. *Nature*, **383**, 92–96.

43. Lorentz, A., Heim, L., and Schmidt, H. (1992) The switching gene *swi6* affects recombination and gene expression in the mating-type region of *Schizosaccharomyces pombe*. *Mol. Gen. Genet.*, **233**, 436.

44. Thon, G. and Klar, A. J.S. (1992) The *clr1* locus regulates the expression of the cryptic mating-type loci in fission yeast. *Genetics*, **131**, 287.

45. Ekwall, K. and Ruusala, T. (1994) Mutations in *rik1, clr2, clr3* and *clr4* genes asymmetrically derepress the silent mating-type loci in fission yeast. *Genetics*, **136**, 53.

46. Thon, G., Cohen, A., and Klar, A. J.S. (1994) Three additional linkage groups that repress transcription and meiotic recombination in the mating-type region of *Schizosaccharomyces pombe*. *Genetics*, **138**, 29.

47. Ekwall, K., Javerzat, J-P., Lorentz, K., Schmidt, H., Cranston, G., and Allshire, R. (1995) The chromo domain protein Swi6: a key component at fission yeast centromeres. *Science*, **269**, 1429.

48. Lorentz, A., Osterman, K., Fleck, O., and Schmidt, H. (1994) Switching gene *swi6*, involved in repression of silent mating type loci in fission yeast, encodes a homologue of chromatin-associated proteins from *Drosophila* and mammals. *Gene*, **143**, 139.

49. Paro, R. and Hogness, D. S. (1991) The polycomb protein shares a homologous domain with a heterochromatin-associated protein of *Drosophila*. *Proc. Natl. Acad. Sci. U.S.A.*, **88**, 263.

50. Funabiki, H., Hagan, I., Uzawa, S., and Yanagida, M. (1993) Cell cycle-dependent specific positioning and clustering of centromeres and telomeres in fission yeast. *J. Cell Biol.*, **121**, 961.

51. Ekwall, K., Nimmo, E. R., Javerzat, J-P., Borgström, B., Egel, R., Cranston, G., and Allshire, R. (1996) Mutations in the fission yeast silencing factors *clr4*[+] and *rik1*[+] disrupt the localisation of the chromo domain protein Swi6p and impair centromere function. *J. Cell Sci.*, **109**, 2637.

52. Ivanova, A. V., Bonaduce, M. J., Ivanov, S. V., and Klar, A. J. (1998) The chromo and SET domains of the Clr4 protein are essential for silencing in fission yeast. *Nature Genet.*, **19**, 192.

53. Jenuwein, T., Labile, G., Dorn, R., and Reuter, G. (1998) SET domain proteins modulate chromatin domains in eu- and heterochromatin. *Cell. Mol. Life Sci.*, **54**, 80.

54. Aagaard, L., Laible, G., Selenko, P., Schmid, M., Dorn, R., Schotta, G., Kuhfittig, S., Wolf, A., Lebersorger, A., Singh, P. B., Reuter, G., and Jenuwein, T. (1999) Functional mammalian homologues of the *Drosophila* PEV-modifier Su(var)3-9 encode centromere-associated proteins which complex with the heterochromatin component M31. *EMBO J.*, **18**, 1923.

55. Doe, C. L., Wang, G., Chow, C., Fricker, M., Singh, P. B., and Mellor, E. J. (1998) The fission yeast chromo domain encoding gene *chp1*[+] is required for chromosome segregation and shows a genetic interaction with alpha tubulin. *Nucleic Acids Res.*, **26**, 4222.

56. Ekwall, K., Cranston, G., and Allshire, R. C. (1999) Novel fission yeast mutants which alleviate transcriptional silencing in centromeric flanking repeats and disrupt chromosome segregation. *Genetics*, **153**, 1153.

57. Javerzat, J.-P., McGurk, G., Cranston, G., Barreau, C., Bernard, P., Gordon, C., and Allshire, R. (1999) Defects in components of the proteasome enhance transcriptional repression at fission yeast centromeres and impair chromosome segregation. *Mol. Cell. Biol.*, **19**, 5155.

58. Allshire, R. C. (1997) Centromeres, checkpoints and chromatid cohesion. *Curr. Opin. Genet. Dev.*, **7**, 264.

59. Murakami, Y., Huberman, J. A., and Hurwitz, J. (1996) Identification, purification, and molecular cloning of autonomously replicating sequence-binding protein 1 from fission yeast *Schizosaccharomyces pombe*. *Proc. Natl. Acad. Sci. U.S.A.*, **93**, 502.

60. Halverson, D., Baum, M., Stryker, J., Carbon, J., and Clarke, L. (1997) A centromere DNA binding protein from fission yeast affects chromosome segregation and has homology to human CENP-B. *J. Cell Biol.*, **136**, 487.

61. Lee, J.-K., Huberman, J. A., and Hurwitz, J. (1997) Purification and characterisation of a CENP-B homologue protein that binds to the centromeric K-type repeat DNA of *Schizosaccharomyces pombe*. *Proc. Natl. Acad. Sci. U.S.A.*, **94**, 8427.

62. Kapoor, M., Montes de Oca Luna, R., Liu, G., Lozano, G., Cummings, C., Mancini, M., Ouspenski, I., Brinkley, B. R., and May, G. S. (1998) The *cenpB* gene is not essential in mice. *Chromosoma*, **107**, 570.

63. Takahashi, K., Yamada, H., and Yanagida, M. (1994) Fission yeast minichromosome loss mutants Mis cause lethal aneuploidy and replication abnormality. *Mol. Biol. Cell*, **5**, 1145.

64. Furuya, K., Takahashi, K., and Yanagida, M. (1998) Faithful anaphase is ensured by Mis4, a sister chromatid cohesion molecule required in S phase and not destroyed in G1 phase. *Genes Dev.*, **12**, 3408.

65. Saitoh, S., Takahashi, K., and Yanagida, M. (1997) Mis6, a fission yeast inner centromere protein, acts during G1/S and forms specialised chromatin required for equal segregation. *Cell*, **90**, 131.

66. Bernard, P., Hardwick, K., and Javerzat, J.-P. (1998) Fission yeast Bub1 is a mitotic

centromere protein essential for the spindle checkpoint and the preservation of correct ploidy through mitosis. *J. Cell Biol.*, **143**, 1775.

67. Ding, R., McDonald, K. L., and McIntosh, J. R. (1993) Three-dimensional reconstruction and analysis of mitotic spindles from the yeast, *Schizosaccharomyces pombe*. *J. Cell Biol.*, **120**, 141.

68. Winey, M., Mamay, C. L., O'Toole, E. T., Mastronarde, D. N., Giddings, T. H., McDonald, K. L., and McIntosh, J. R. (1995) Three-dimensional ultrastructural analysis of the *Saccharomyces cerevisiae* mitotic spindle. *J. Cell Biol.*, **129**, 1601.

69. Doheny, K., Sorger, P., Hyman, A., Tugendreich, S., Spencer, F., and Hieter, P. (1993) Identification of essential components of the *S. cerevisiae* kinetochore. *Cell*, **73**, 761.

70. Kingsbury, J. and Koshland, D. (1991) Centromere-dependent binding of yeast mini-chromosmes to microtubules *in vitro*. *Cell*, **66**, 483.

71. Kingsbury, J. and Koshland, D. (1993) Centromere function on minichromosomes isolated from budding yeast. *Mol. Biol. Cell*, **4**, 859.

72. Hyman, A. A., Middleton, K., Centola, M., Mitchison, T. J., and Carbon, J. (1992) Microtubule-motor activity of a yeast centromere-binding complex. *Nature*, **359**, 5333.

73. Sorger, P. K., Severin, F. F., and Hyman, A. A. (1994) Factors required for the binding of reassembled yeast kinetochores to microtubules *in vitro*. *J. Cell Biol.* **127**, 995.

74. Bram, R. J. and Kornberg, R. D. (1987) Isolation of a *Saccharomyces cerevisiae* centromere DNA-binding protein, its human homologue, and its possible role as a transcription factor. *Mol.Cell. Biol.*, **7**, 403.

75. Mellor, J., Jiang, W., Funk, M., Rathjen, J., Barnes, C., Hinz, T., Hegemann, J. H., and Philippsen, P. (1990) CPF1, a yeast protein which functions in centromeres and promoters. *EMBO J.*, **9**, 4017.

76. Jiang, W. and Philippsen, P. (1989) Purification of a protein binding to the CDEI subregion of *Saccharomyces cerevisiae* centromere DNA. *Mol. Cell. Biol.*, **9**, 5585.

77. Niedenthal, R., Sen-Gupta, M., Wilmen, A., and Hegemann, J. H. (1993) Cpf1 protein induced bending of yeast centromere DNA element I. *Nucleic Acids Res.*, **21**, 4726.

78. Meluh, P. B. and Koshland, D. (1997) Budding yeast centromere composition and assembly as revealed by *in vivo* crosslinking. *Genes Dev.*, **11**, 3401.

79. Ng, R. and Carbon, J. (1987) Mutational and *in vitro* protein binding studies on centromere DNA from *Saccharomyces cerevisiae*. *Mol. Cell. Biol.*, **7**, 4522.

80. Lechner, J. and Carbon, J. (1991) A 240 kD multisubunit protein complex, CBF3, is a major component of the budding yeast centromere. *Cell*, **64**, 717.

81. Kaplan, K. B., Hyman, A. A., and Sorger, P. K. (1997) Regulating the yeast kinetochore by ubiquitin-dependent degradation and Skp1p-mediated phosphorylation. *Cell*, **91**, 491.

82. Jiang, W., Lechner, J., and Carbon, J. (1993) Isolation and characterisation of a gene (*CBF2*) specifying a protein component of the budding yeast kinetochore. *J. Cell Biol.*, **121**, 513.

83. Goh, P. Y. and Kilmartin, J. (1993) *NDC10*: a gene involved in chromosome segregation in *S. cerevisiae*. *J. Cell Biol.*, **121**, 503.

84. Spencer, F., Gerring, S. L., Connelly, C., and Hieter, P. (1990) Mitotic chromosomes transmission fidelity mutants in *Saccharomyces cerevisiae*. *Genetics*, **124**, 237.

85. Strunnikov, A. V., Kingsbury, J., and Koshland, D. (1995) *CEP3* encodes a centromere protein of *Saccharomyces cerevisiae*. *J. Cell Biol.*, **128**, 749.

86. Lechner, J. (1994) A zinc finger protein, essential for chromosome segregation, constitutes a putative DNA binding subunit of the *Saccharomyces cerevisiae* kinetochore complex, CBF3. *EMBO J.*, **13**, 5203.

87. Connelly, C. and Hieter, P. (1996) Budding yeast *SKP1* encodes an evolutionarily conserved kinetochore protein required for cell cycle progression. *Cell*, **86**, 275.

88. Espelin, C. W., Kaplan, K. B., and Sorger, P. K. (1997) Probing the achitecture of a simple kinetochore using DNA–protein crosslinking. *J. Cell Biol.*, **139**, 1383.

89. Stoler, S., Keith, K. C. Curnick, K. E., and Fitzgerald-Hayes, M. (1995) A mutation in *CSE4*, an essential gene encoding a novel chromatin-associated protein in yeast, causes chromosome non-disjunction and cell cycle arrest at mitosis. *Genes Dev.*, **9**, 573.

90. Smith, M. M., Yang, P., Santisteban, M. S., Boone, P. W., Goldstein, A. T., and Megee, P. C. (1996) A novel histone H4 mutant defective for nuclear division and mitotic chromosome transmission. *Mol. Cell. Biol.*, **16**, 1017.

91. Baker, R. E., Harris, K., and Zhang, K. (1998) Mutations synthetically lethal with cep1 target *S. cerevisiae* kinetochore components. *Genetics*, **149**, 73.

92. Sullivan, K. F., Hechenberger, M., and Masri, K. (1994) Human CENP-A contains a histone H3 related histone fole domain that is required for targeting to the centromere. *J. Cell Biol.*, **127**, 581.

93. Meluh, P. B., Yang, P., Glowczewski, L., Koshland, D., and Smith, M. M. (1998) Cse4p is a component of the core centromere of *Saccharomyces cerevisiae*. *Cell*, **94**, 607.

94. Warburton, P. E., Cooke, C. A., Bourassa, S., Vafa, O., Sullivan, B. A., Stetten, G., Gimelli, G., Warburton, D., Tyler-Smith, C., Sullivan, K. F., Poirier, G. G., and Earnshaw, W. C. (1997) Immunolocalistion of CENP-A suggests a distinct nucleosome structure at the inner kinetochore plate of active centromeres. *Curr. Biol.*, **7**, 901.

95. Shelby, R. D. , Vafa, A., and Sullivan, K. F. (1996) Assembly of CENP-A into chromatin requires a cooperative array of nucleosomal contact sites. *J. Cell Biol.*, **136**, 501.

96. Saitoh, H. J., Tomkiel, J., Cooke, C. A., Ratrie, H., Maurer, M., Rothfield, N. F., and Earnshaw, W. (1992) CENP-C, an autoantigen in scleroderma, is a component of the human inner kinetochore plate. *Cell*, **70**, 115.

97. Brown, M. T. (1995) Sequence similarities between the yeast chromosome segregation protein Mif2 and the mammalian centromere protein CENP-C. *Gene*, **160**, 111.

98. Meluh, P. B. and Koshland, D. (1995) Evidence that the *MIF2* gene of *Saccharomyces cerevisiae* encodes a centromere protein with homology to the mammalian centromere protein CENP-C. *Mol. Biol. Cell*, **6**, 793.

99. Meeks-Wagner, D., Wood, J. S., Garvik, B., and Hartwell, L. H. (1986) Isolation of two genes that affect mitotic chromosome segregation in *S. cerevisiae*. *Cell*, **44**, 53.

100. Hyland, K. M., Kingsbury, J., Koshland, D., and Heiter, P. (1999) Ctf19p: a novel kinetochore protein in *Saccharomyces cerevisiae* and a potential link between the kinetochore and mitotic spindle. *J. Cell Biol.*, **145**, 15.

101. Wang, Y. and Burke, D. J. (1995) Checkpoint genes required to delay cell division in response to nocodazole respond to impaired kinetochore function in the yeast *Saccharomyces cerevisiae*. *Mol. Cell. Biol.*, **15**, 6838.

102. Francisco, L., Wang, W., and Chan, C. S. (1994) Type I protein phosphatase acts in opposition to Ipl1 protein kinase in regulating yeast chromosome segregation. *Mol. Cell. Biol.*, **14**, 4731.

103. Biggins, S., Severin, F. F., Bhalla, N., Sassoon, I., Hyman, A. A., and Murray, A. W. (1999) The conserved protein kinase Ipl1 regulates microtubule binding to kinetochores in budding yeast. *Genes Dev.*, **13**, 532.

104. Sassoon, I., Severin, F. F., Andrews, P. D., Taba, M.-R., Kaplan, K. B., Ashford, A. J., Stark, M. J.R., Sorger, P. K., and Hyman, A. A. (1999) Regulation of *Saccharomyces cerevisiae* kinetochores by the type 1 phosphatase Glc7p. *Genes Dev.*, **13**, 545.

105. Bloecher, A. and Tatchell, K. (1999) Defects in *Saccharomyces cerevisiae* protein phosphatase type I activate the spindle/kinetochore checkpoint. *Genes Dev.*, **13**, 517.

106. Middleton, K. and Carbon, J. (1994) *KAR3*-encoded kinesin is a minus end directed motor that functions with centromere binding proteins (CBF3) on an *in vitro* yeast kinetochore. *Proc. Natl. Acad. Sci. U.S.A.*, **91**, 7212.

107. Cadwell, C., Yoon, H. J., Zebarjadian, Y., and Carbon, J. (1997) The yeast nucleolar protein Cbf5p is involved in rRNA biosynthesis and interacts genetically with the RNA polymerase I transcription factor RRN3. *Mol. Cell. Biol.*, **17**, 6175.

108. Woodage, T., Basrai, M. A., Baxevanis, A. D., Hieter, P., and Collins, F. S. (1997) Characterization of the CHD family of proteins. *Proc. Natl. Acad. Sci. U.S.A.*, **94**, 11472.

109. Sugawara, N. F. (1989) DNA sequences at the telomeres of the fission yeast *S. pombe*. PhD thesis, Harvard University, Cambridge, MA.

110. Hiraoka, Y., Henderson, E., and Blackburn, E. H. (1998) Not so peculiar: fission yeast telomere repeats. *Trends Biochem. Sci.*, **23**, 126.

111. Lue, N. F. and Peng, Y. (1997) Identification and characterisation of a telomerase activity from *Schizosaccharomyces pombe*. *Nucleic Acids Res.*, **25**, 4331.

112. Nimmo, E. R., Cranston, G., and Allshire, R. C. (1994) Telomere-associated chromosome breakage in fission yeast results in variegated expression of adjacent genes. *EMBO J.*, **13**, 3801.

113. Nakamura, T. M., Morin, G. B., Chapman, K. B., Weinrich, S. L., Andrews, W. H., Lingner, J., Harley, C. B., and Cech, T. R. (1997) Telomerase catalytic subunit homologues from fission yeast and human. *Science*, **277**, 955.

114. Nakamura, T. M., Cooper, J. P., and Cech, T. R. (1998) Two modes of survival of fission yeast without telomerase. *Science*, **282**, 493.

115. Cooper, J. P., Nimmo, E. R., Allshire, R. C., and Cech, T. R. (1997) Regulation of telomere length and function by a Myb-domain protein in fission yeast. *Nature*, **385**, 744.

116. Broccoli, D., Smogorzewska, A., Chong, L., and de Lange, T. (1997) Human telomeres contain distinct Myb-related proteins, TRF1 and TRF2. *Nature Genet.*, **17**, 236.

117. Konig, P., Giraldo, R., Chapman, L., and Rhodes, D. (1996) The crystal structure of the DNA binding domain of yeast Rap1 in complex with telomeric DNA. *Cell*, **85**, 125.

118. Nimmo, E. R., Pidoux, A. L., Perry, P. E., and Allshire, R. C. (1998) Defective meiosis in telomere-silencing mutants of *Schizosaccharomyces pombe*. *Nature*, **392**, 825.

119. Cooper, J. P., Watanabe, Y., and Nurse, P. (1998) Fission yeast Taz1 protein is required for meiotic clustering and recombination. *Nature*, **392**, 828.

120. Chikashige, Y., Ding, D. Q., Funabiki, H., Haraguchi, T., Mashiko, S., Yanagida, M., and Hiraoka, Y. (1994) Telomere-led premeiotic chromosome movement in fission yeast. *Science*, **264**, 270.

121. Chikashige, Y., Ding, D.-Q., Imai, Y., Yamamoto, M., Haraguchi, T., and Hiraoka, Y. (1997) Meiotic nuclear reorganization: switching the position of centromeres and telomeres in the fission yeast *Schizosaccharomyces pombe*. *EMBO J.*, **16**, 193.

122. Dernberg, A. F., Sedat, J. W., Cande, W. Z., and Bass, H. W. (1995) Cytology of telomeres. In *Telomeres* (ed E. H. Blackburn and C. W. Greider). p. 295. Cold Spring Harbor Laboratory Press, Cold Spring Harbor, NY.

123. Dahlen, M., Olsson, T., Kanter-Smoler, G., Ramne, A., and Sunnerhagen, P. (1998) Regulation of telomere length by checkpoint genes in *Schizosaccharomyces pombe*. *Mol. Biol. Cell*, **9**, 611.

124. Naito, T., Matsuura, A., and Ishikawa, F. (1998) Circular chromosome formation in a fission yeast mutant defective in two ATM homologues. *Nature Genet.*, **20**, 203.

125. Louis, E. J. (1995) The chromosome ends of *Saccharomyces cerevisiae*. *Yeast*, **11**, 1553.

126. Gotta, M. and Cockell, M. (1997) Telomeres, not the end of the story. *Bioessays*, **19**, 367.

127. Wright, J. H., Gottshling, D. E., and Zakian, V. A. (1992) *Saccharomyces* telomeres assume a non-nucleosomal chromatin structure. *Genes Dev.*, **6**, 197.

128. Gottschling, D. E. , Aparicio, O. M., Billington, B. L., and Zakian, V. A. (1990). Position effect at *S. cerevisiae* telomeres: reversible repression of pol II transcription. *Cell*, **63**, 751.

129. Renauld, H., Aparicio, O. M., Zierath, P. D., Billington, B. L., Chablani, S. K., and Gottschling, D. E. (1993) Silent domains are assembled continuously from the telomere and are defined by promoter distance and strength, and SIR3 dosage. *Genes Dev.*, **7**, 1133.

130. Gilson, E., Roberge, M., Giraldo, R., Rhodes, D., and Grunstein, M. (1993) Distortion of the DNA double helix by Rap1 at silencers and multiple telomeric binding sites. *J. Mol. Biol.*, **231**, 293.

131. Strahl-Bolsinger, S., Hecht, A., Luo, K., and Grunstein, M. (1997) Sir2 and Sir4 interactions differ in core and extended telomeric heterochromatin in yeast. *Genes Dev.*, **11**, 83.

132. Marcand, S., Gilson, E., and Shore, D. (1997) A protein counting mechanism for telomere length regulation in yeast. *Science*, **275**, 986.

133. Wooton, D. and Shore, D. (1997) A novel Rap1p-interacting factor, Rif2p, cooperates with Rif1p to regulate telomere length in *Saccharomyces cerevisiae*. *Genes Dev.*, **11**, 748.

134. Hardy, C. F.J., Sussel, L., and Shore, D. (1992) A RAP1-interacting protein involved in transcriptional silencing and telomere length regulation. *Genes Dev.*, **6**, 801.

135. Aparicio, O. M., Billington, B. L., and Gottschling, D. E. (1991) Modifiers of position effect are shared between telomeric and silent mating-type loci in *Saccharomyces cerevisiae*. *Cell*, **66,** 1279.

136. Braunstein, M., Rose, A. B., Holmes, S. G., Allis, C. D., and Broach, J. R. (1993) Transcriptional silencing in yeast is associated with reduced nucleosome acetylation. *Genes Dev.*, **7**, 592.

137. Kayne, P. S., Kim, U.-J., Han, M., Mullen, J. R., Yoshizaki, F., and Grunstein, M. (1988) Extremely conserved histone H4 N terminus is dispensable for growth but essential for repressing the silent mating type loci in yeast. *Cell*, **55**, 27.

138. Singer, M. E. and Gottschling, D. E. (1994) TLC1: template RNA component of *Saccharomyces cerevisiae* telomerase. *Science*, **266**, 404.

139. Lundblad, V. and Szostak, J. W. (1989) A mutant with a defect in telomere elongation leads to senescence in yeast. *Cell*, **57**, 633.

140. Lendvay, T. S., Morris, D. K., Sah, J., Balasubramanian, B., and Lundblad, V. (1996) Senescence mutants of *Saccharomyces cerevisiae* with a defect in telomere replication identify three additional *EST* genes. *Genetics*, **144**, 1399.

141. Lingner, J., Hughes, T. R., Shevchenko, A., Mann, M., Lundblad, V., and Cech, T. R. (1997) Reverse transcriptase motifs in the catalytic subunit of telomerase. *Science*, **276**, 561.

142. Counter, C. M., Meyerson, M., Eaton, E. N., and Weinberg, R. A. (1997) The catalytic subunit of yeast telomerase. *Proc. Natl. Acad. Sci. U.S.A.*, **94**, 9202.

143. Evans, S. K., Sistrunk, M. L., Nugent, C. I., and Lundblad, V. (1998) Telomerase, Ku, and telomeric silencing in *Saccharomyces cerevisiae*. *Chromosoma*, **107**, 352.

144. Critchlow, S. E. and Jackson, S. P. (1998) DNA end-joining: from yeast to man. *Trends Biochem. Sci.*, **23**, 394.

145. Tsukamoto, Y., Kato, J., and Ikeda, H. (1997) Silencing factors participate in DNA repair and recombination in *Saccharomyces cerevisiae*. *Nature*, **388**, 900.

146. Jackson, S. P. (1997) Silencing and DNA repair connect. *Nature*, **388**, 829.

147. Boulton, S. J. and Jackson, S. P. (1998) Components of the Ku-dependent non-homologous end-joining pathway are involved in telomeric length maintenance and telomeric silencing. *EMBO J.*, **17**, 1819.

148. Gravel S., Larrivee, M., Labrecque, P., and Wellinger, R. J. (1998) Yeast Ku as a regulator of chromosomal DNA end structure. *Science*, **280**, 741.

149. Nugent, C. I., Bosco, G., Ross, L. O., Evans, S. K., Salinger, A. P., Moore, J. K., Haber, J. E., and Lundblad, V. (1998) Telomere maintenance is dependent on activities required for end repair of double-strand breaks. *Curr. Biol.*, **8**, 657.

150. Laroche, T., Martin, S. G., Gotta, M., Gorham, H. C., Pryde, F. E., Louis, E. J., and Gasser, S. M. (1998) Mutation of yeast Ku genes disrupts the subnuclear organization of telomeres. *Curr. Biol.*, **8**, 653.

151. Singer, M. S., Kahana, A., Wolf. A. J., Meisinger, L. L., Peterson, S. E., Goggin, C., Mahowald, M., and Gottschling, D. E. (1998) Identification of high-copy disruptors of telomeric silencing in *Saccharomyces cerevisiae*. *Genetics*, **150**, 613.

152. Runge, K. W. and Zakian, V. A. (1996) *TEL2*, an essential gene required for telomere length regulation and telomere position effect in *Saccharomyces cerevisiae*. *Mol. Cell. Biol.*, **16**, 3094.

153. Greenwell, P. W., Kronmal, S. L., Porter, S. E., Gassenhuber, J., Obermaier, B., and Petes, T. D. (1995) *TEL1*, a gene involved in controlling telomere length in *S. cerevisiae*, is homologous to the human ataxia telangiectasia gene. *Cell*, **82**, 823.

154. Morrow, D. M., Tagle, D. A., Shiloh, Y., Collins F. S., and Hieter, P. (1995) *TEL1*, an *S. cerevisiae* homolog of the human gene mutated in ataxia telangiectasia, is functionally related to the yeast checkpoint gene *MEC1*. *Cell*, **82**, 831.

155. Conrad, M. N., Dominguez, A. M., and Dresser, M. E. (1997) Ndj1p, a meiotic telomere protein required for normal chromosome synapsis and segregation in yeast. *Science*, **276**, 1252.

156. Trelles-Sticken, E., Loidl, J., and Scherthan, H. (1999) Bouquet formation in budding yeast: initiation of recombination is not required for meiotic telomere clustering. *J. Cell Sci.*, **112**, 651.

157. Autexier, C., and Greider, C. W. (1996) Telomerase and cancer: revisiting the telomere hypothesis. *Trends Biochem. Sci.*, **21**, 387.

8 | Splicing pre-mRNA introns

ANDREW E. MAYES, JUDITH A. POTASHKIN and JEAN D. BEGGS

1. The occurrence of introns in yeast

In the late 1970s it was discovered that many eukaryotic genes are interrupted by non-coding sequences (introns). Since then, various types of introns have been identified, the mechanisms for their removal have been intensively researched, and the possible reasons for their existence have been the subject of much debate. Introns are normally removed from newly transcribed RNA and the regions present in the mature RNA (exons) are concomitantly joined in a process termed RNA splicing. Three different chemical reactions of RNA splicing are currently known, exemplified by the splicing of 1) group I, 2) group II and nuclear precursor messenger RNA (pre-mRNA), and 3) tRNA introns (1). Genes containing group I introns are found widely in nature, for instance in the chloroplast and mitochondria of plants, in mitochondria of fungi and other lower eukaryotes, in some eubacteria and bacteriophages. These introns have a characteristic, conserved secondary structure, and are spliced in a guanosine co-factor-dependent process that involves two *trans*-esterification reactions (2). Group II introns have been found in chloroplast, mitochondrial and some eubacterial genomes. They have a highly conserved secondary and tertiary structure that is distinct from that of group I introns. Group II introns (3) are spliced in two *trans*-esterification reactions that are very similar to those of pre-mRNA splicing (see below). However, unlike pre-mRNA splicing, which is dependent on a complex RNA- and protein-containing machinery called the spliceosome, some group I and group II introns have the ability to self-splice due to the autocatalytic activity of the RNA itself. In contrast to the first two types of splicing chemistry, tRNA introns are spliced in two distinct protein-catalysed reactions; first the intron is excised by an endonuclease, then the two exons are ligated (4).

Recently, it was found that, unlike most pre-mRNAs, *HAC1* pre-mRNA is spliced by a non-spliceosomal process which resembles the mechanism of tRNA splicing. The *HAC1* gene is involved in the signalling pathway of the unfolded protein response that monitors the accumulation of unfolded proteins in the endoplasmic reticulum. Its intron is removed by a specific endoribonuclease (Ire1p) and the two exons are subsequently ligated by the nuclear tRNA ligase (Rlg1p). At present it is thought that *HAC1* transcript is the unique substrate of this mechanism (5). This review focuses on spliceosome-based nuclear pre-mRNA splicing.[1]

In higher eukaryotes, most protein-encoding genes are interrupted by one or more introns. The presence of multiple introns can allow the joining of different combinations of exons by alternative splicing, to produce distinct mRNAs from identical pre-mRNAs and thereby increase the informational capacity of the genome. Approximately 43% of the known genes in *Schizosaccharomyces pombe* (fission yeast) contain one or more introns. In contrast, introns are found in only 4% of the 6200 or so genes comprising the genome of *Saccharomyces cerevisiae* (referred to as budding yeast in this chapter), and in all but a few cases only one intron is present. Nevertheless, approximately 27% of budding yeast transcripts are spliced, owing to the fact that a large proportion of intron-containing genes are highly expressed (6, 7). Thus the distribution of introns appears to be non-random in the budding yeast genome; about 70% of the highly expressed genes that encode ribosomal proteins contain an intron. Spliceosomal introns are also found in some nuclear genes that encode small RNAs, for example the budding yeast U3 small nucleolar RNAs (snoRNAs) and the U6 small nuclear RNA (snRNA) of fission yeast.

In both *S. cerevisiae* and *S. pombe*, most nuclear introns are relatively small (fewer than 600 nucleotides). In *S. cerevisiae* they are generally found towards the 5′ end of the pre-mRNA, such that the first exon is short. There are exceptions, for example the *DBP2* gene contains an intron that is over 1000 nucleotides in length and lies 1273 nucleotides from the initiating AUG. A plausible hypothesis suggests that a progenitor yeast nuclear genome may have contained more introns, but that reverse transcriptase activity combined with homologous recombination eliminated almost all the introns, leaving mainly those that lie close to the 5′ ends of transcripts (8). Only a few budding yeast nuclear genes are known to have more than one intron, including *MATa1*, *DYN2*, *RPL7A*, *RPL7B*, and *RPS22B*, each of which contains two introns (for databases of budding yeast introns, see http://www.cse.ucsc.edu/research/ compbio/yeast_introns.html or http://www.embl-heidelberg.de/ExternalInfo/ seraphin/yidb.html). By contrast, in *S. pombe* about 50% of the intron-containing genes have two to seven introns that are scattered throughout the transcript (9).

Introns are highly divergent in sequence; however, three short regions of consensus sequence have been identified that define introns and contribute chemically reactive nucleotides to the splicing reaction (Fig. 1). The importance of these sequences has been demonstrated through site-directed mutagenesis and analyses of naturally occurring mutations both *in vivo* and *in vitro* (reviewed in Reference 10, and references therein; see References 6, 11, 12 for recent bioinformatic analyses).

In all eukaryotes studied to date, the nuclear pre-mRNA splicing reaction occurs via an ordered pathway involving two sequential *trans*-esterification reactions (Fig. 2) (13, 14). In the first *trans*-esterification step, the phosphodiester bond at the 5′ splice site (the exon 1–intron boundary) is cleaved as a result of nucleophilic attack by the 2′ hydroxyl group of a conserved adenosine that lies in the intron towards the 3′ splice site, and is referred to as the branchpoint nucleotide. This yields two intermediates: exon 1 with a free 3′ hydroxyl, and intron–exon 2 in a branched-lariat structure, in which the 5′ end of the intron is linked covalently to the branchpoint adenosine via a 2′-5′ phosphodiester bond. In step 2, cleavage at the 3′ splice site

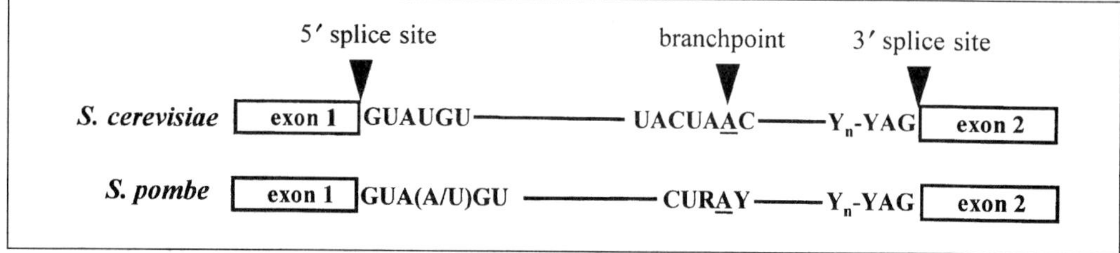

Fig. 1 Schematic representation of typical budding yeast (upper) and fission yeast (lower) introns, indicating the sequences of the three conserved elements: the 5′ splice site, the branch point and the 3′ splice site. Exon sequences (boxes) and intron sequences (lines) are indicated. The distances between the branchpoint and the splice sites are intron specific. The branchpoint adenosine is underlined. R represents purine, Y represents pyrimidine bases.

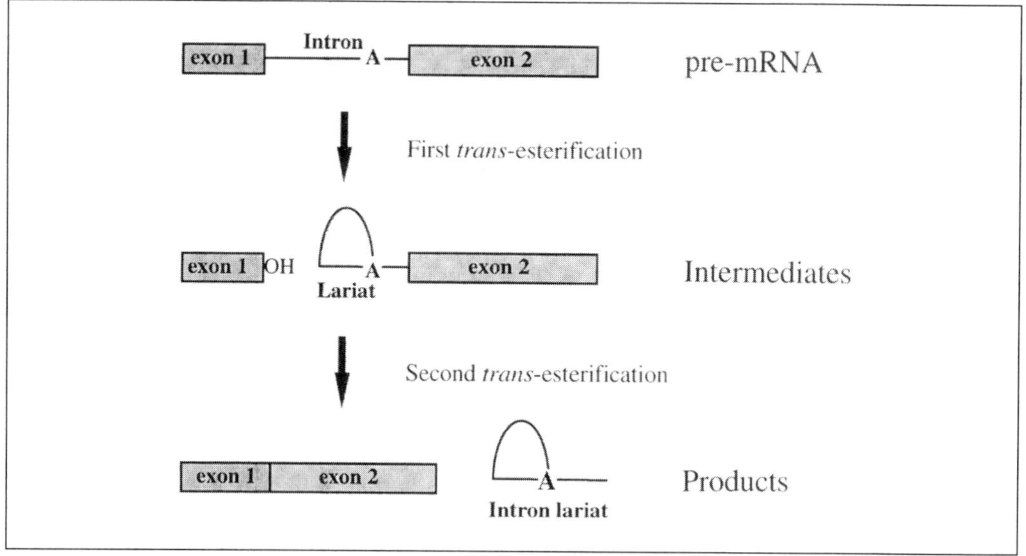

Fig. 2 Pre-mRNA splicing occurs via two *trans*-esterification reactions. Exon sequences are represented by boxes, intron sequences by lines. The branchpoint adenosine is indicated by an A.

occurs as a result of nucleophilic attack by the free 3′ hydroxyl of exon 1. The intron is released, and the exons are simultaneously joined via a 3′-5′ phosphodiester bond, producing the two products of the splicing reaction: a mature messenger RNA (mRNA) and an excised intron-lariat.

Compared with other types of introns, such as group I and group II introns which have highly conserved secondary and tertiary structures, the lack of information content in pre-mRNA intron sequences is compensated by the *trans*-acting components (both RNA and protein) of the spliceosome. Extensive RNA–RNA and RNA–protein interactions are involved in splice-site recognition and the alignment of the splice sites into a conformation suitable for the catalysis of intron removal (10, 15, 16).

2. The spliceosomal machinery

Splicing of pre-mRNAs in the nucleus is dependent upon the formation of a large, dynamic ribonucleoprotein complex, the spliceosome, which is formed by the assembly of multiple RNA and protein factors on to the pre-mRNA transcript. Assembly of the spliceosome occurs in an ordered stepwise manner involving the interconversion of several distinct intermediate complexes (see section 3). The spliceosome is composed of five small nuclear ribonucleoprotein particles (snRNPs) and a number of non-snRNP proteins (reviewed in References 14, 17, 18). The snRNPs are *trans*-acting factors with both RNA and protein components, and are named according to the snRNA each contains (i.e. U1, U2, U4, U5, and U6; Table 1). The snRNA-associated proteins fall into two groups. Seven core or Sm proteins (B, D_1, D_2, D_3, E, F, and G, as originally defined in metazoa) are common components of the U1, U2, U4, and U5 snRNPs, while each snRNP also contains snRNA-specific proteins (18).

The snRNPs play a central role in the recognition and alignment of pre-mRNA splice sites and the snRNA components have been proposed to form the catalytic centre of the spliceosome. The spliceosomal snRNPs are abundant in higher eukaryotes, with approximately 10^5 to 10^6 particles per cell, compared with around 200–500 per cell for *S. cerevisiae* and 10^2 to 10^4 per cell for *S. pombe*. For this reason many of the early biochemical studies were focused on the more abundant higher eukaryotic particles. More recently, much has been learnt from molecular and genetic studies performed with yeast.

2.1 snRNAs: structure, conservation, and interactions

With the exception of U6 snRNA, the lengths and primary sequences of the snRNAs vary greatly from species to species. The limited conservation of primary sequence is mostly restricted to short functionally important regions. U6 snRNA, however, is highly conserved with respect to both size and sequence, which may reflect its central position in the spliceosome and its potentially catalytic role in the splicing reaction. Genetic studies have demonstrated the functional importance of conserved sequences in the U6 snRNA that are absolutely required for both steps of the splicing reaction (10, 19, and references therein). In budding yeast, U5 snRNA exists in two distinct forms, $U5_L$ and $U5_S$ (Table 1), determined by alternative processing events at the 3′

Table 1 *S. cerevisiae* snRNAs

snRNA	Gene	Size (nt)	Polymerase	Structural Features
U1	*SNR19*	568	RNAP II	TMG cap, Sm site
U2	*SNR20*	1171	RNAP II	TMG cap, Sm site
U4	*SNR14*	160	RNAP II	TMG cap, Sm site
$U5_S$	*SNR7*	180	RNAP II	TMG cap, Sm site
$U5_L$	*SNR7*	215		
U6	*SNR6*	112	RNAP III	MP cap, U-rich 3′ end

RNAP, RNA polymerase; TMG, trimethylguanosine; MP, γ-monomethylphosphate end of the primary transcript (20). Both forms appear to be incorporated into active spliceosomes; the reason why two forms exist is not known.

The secondary structures of the snRNAs (as predicted by chemical and enzymatic probing and phylogenetic studies) are largely preserved throughout eukaryotes and are important for the binding of proteins (21). The Sm proteins associate tightly with a conserved structural motif, the Sm site, of the U1, U2, U4, and U5 snRNAs. This is a single-stranded region with consensus sequence $RAU_{3-6}GR$ (where R is a purine base), and is normally flanked by double hairpin loops (22). Deletion or substitution of the Sm site is lethal in budding yeast cells, demonstrating the functional significance of this site (23, 24, 25). However, despite being highly conserved, the Sm site is remarkably tolerant to sequence changes, with only three positions being sensitive to mutation (24).

The biogenesis of snRNPs has been studied in higher eukaryotes, mainly in *Xenopus* oocytes (26). The U1, U2, U4, and U5 snRNAs are transcribed by RNA polymerase II in the nucleus and acquire a 7-methyl guanosine cap structure co-transcriptionally. These U snRNAs are then actively exported to the cytosol (Fig. 3) (21) where they may be chemically modified, primarily by methylation and pseudouridinylation. The RNA associates with the Sm proteins and the cap is hypermethylated (to trimethylguanosine; TMG) and the 3' end is shortened. These core snRNPs are then imported back

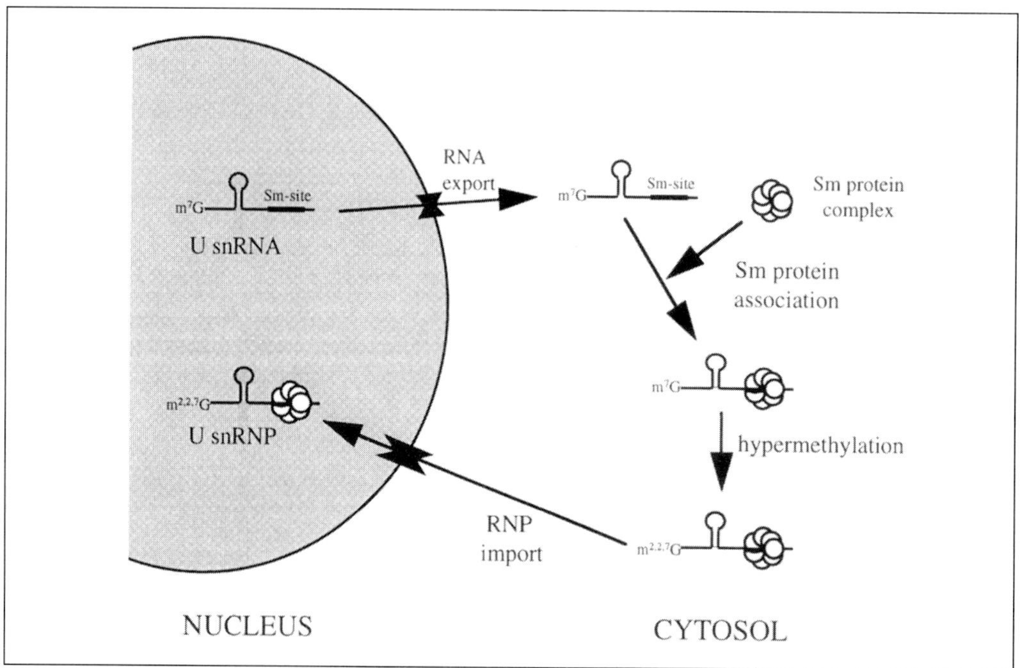

Fig. 3 SnRNP particle biogenesis. Nascent U1, U2, U4, and U5 snRNAs are transported via a receptor-mediated pathway to the cytosol, where they associate with the Sm protein complex. Binding of these proteins triggers hypermethylation of the cap structure and the snRNP is then imported back into the nucleus, again in a receptor-mediated process.

into the nucleus in a process that requires both the TMG cap and the Sm proteins. Proteins that are specific to individual snRNAs then associate with the appropriate core particles to form mature snRNPs. The yeast snRNAs apparently undergo similar maturation steps, although it is not known whether they have a cytoplasmic phase.

U6 snRNA is fundamentally different from the other spliceosomal snRNAs. It is transcribed by RNA polymerase III and, at least in higher eukaryotes, the 5′ triphosphate end is modified by methylation of the γ-phosphate (27) (Table 1), and the 3′ end is elongated by the addition of non-templated uridines and converted to a 2′,3′ cyclic phosphate (28). Experiments in *Xenopus* oocytes suggest that it is largely retained in the nucleus, but the localization in yeast has not been studied.

The U6 snRNA does not associate directly with the Sm proteins, although a distinct set of seven structurally related Sm-like (Lsm; see below) proteins associate with U6 snRNA (29, 30). The U4 and U6 snRNAs contain extensive sequence complementarity, and most or all of the U4 snRNA is complexed with U6 snRNA through Watson–Crick base-pairing in a U4/U6 di-snRNP (31). The U6 snRNA exists in several-fold excess over the other spliceosomal snRNAs. Whether this free U6 particle has a function other than to drive the formation of U4/U6 di-snRNPs is not known.

2.2 Proteins

2.2.1 Identification of proteins involved in splicing

Several complementary genetic methods have been used to identify splicing factors in yeast (reviewed in Reference 32, and references therein). Since splicing is an essential cellular process and, as yeast splicing factors are encoded by single-copy genes, mutations in these genes are often lethal. A common genetic strategy is to isolate mutations with conditional phenotypes, for example conferring heat or cold sensitivity on haploid cells. In this way many of the *PRP* (*Precursor RNA Processing*) genes were identified as encoding proteins involved in splicing. Other splicing genes have been identified by their ability to suppress these conditional *prp* mutations at the restrictive temperature. Such extragenic suppressors can be either mutations in other genes which alleviate the original defect, for example a mutation in *PRP21* suppresses the *prp24-1* heat-sensitive growth defect (33), or wild-type genes that confer suppression when over-expressed, for example *SPP2* suppression of *prp2-1* (34). Another very successful approach has been to search for mutations that are lethal in combination with (i.e. enhance the defect of) a mutation in another splicing factor. Such screens have identified *MUD* (*Mutant U1 Die*) (35), *SLU* (*Synthetic Lethal with U5*) (36), and *SLT* (*Synthetic Lethal with Utwo*) (37) genes in budding yeast.

The publication of the complete genome sequence for *S. cerevisiae* has led to the rapid identification of further yeast splicing proteins through homology searches with spliceosomal protein sequences from other eukaryotes, for example Snu114p and the Sm proteins (see section 2.2.2). Increasingly, a combination of biochemical purification and mass spectrometry is also being used to identify the components of protein complexes, again facilitated by the availability of the budding yeast genome sequence (see Chapter 1). This approach was used to identify the proteins of the U1 snRNP (38,

39), the U2 snRNP (40), and the U4/U6.U5 tri-snRNP (41, 42) with novel snRNP-associated proteins being identified in each case.

The use of two-hybrid protein interaction screens, starting with characterized splicing factors as baits, has detected interactions between known splicing factors, and also identified candidates for novel splicing factors. In addition, exhaustive and iterative screens have revealed extensive networks of protein interactions and suggested connections between the splicing machinery and other metabolic pathways (43; M. Fromont-Racine, A. E. Mayes, J. D. Beggs, and P. Legrain, unpublished results).

2.2.2 Sm and Sm-like core snRNP proteins

Yeast homologues of the metazoan snRNP Sm proteins were identified only relatively recently. They were not detected in conventional genetic screens for splicing mutants, possibly because the genes are small (less than 600 bp) and the proteins are quite resilient to mutagenesis (44). Also, until recently, the low abundance of yeast snRNP particles hampered the identification of snRNP proteins by traditional biochemical techniques that were successfully used in HeLa studies.

Several lines of evidence suggested that Sm proteins also exist in budding yeast: (1) the snRNAs have consensus sequences for Sm sites (45), (2) certain Sm-site mutations in the snRNAs are lethal or inhibitory to growth (24), (3) anti-Sm antisera precipitate budding yeast snRNPs (46, 47), and (4) a set of six to eight small proteins was found in biochemically purified U1 and U4/U6.U5 snRNPs (48).

The existence of budding yeast Sm proteins was confirmed by the identification and characterization of Smd1p, Smd3p, and Sme1p, which correspond to the mammalian Sm proteins D1, D3, and E respectively (49, 50, 51). The strongest evidence that they are genuine Sm proteins comes from the ability of the human D1 and E proteins functionally to complement null alleles of their respective budding yeast homologues (51, 52).

Sequence comparisons of the Sm proteins from a range of species led to the identification of a conserved motif, the Sm, or snRNP core protein motif (53, 54, 55). This motif is composed of two conserved blocks of 32 and 14 amino acids, separated by a non-conserved spacer region of variable length. There is only one invariant residue, an aspartate in Sm motif 1. Conservation of the physiochemical nature of residues at a number of other sites indicates that the structural fold of the Sm motif may be more important for the function of these proteins than is the primary sequence. This is supported by the observation that a number of mutations that map to the Sm motif of the *S. cerevisiae* Sme protein do not significantly affect the interactions made with Smf or Smg, *in vivo* or *in vitro* (44). However, truncation of the Sm motif of either human SmB or SmD3 prevents these proteins forming a complex, suggesting that both conserved regions are required for intermolecular interactions (54).

With the definition of the Sm motif and the completion of the *S. cerevisiae* genome-sequencing project, putative yeast homologues for the remaining Sm proteins became apparent (Table 2). The biochemical purification and sequencing of the proteins of the budding yeast U1 snRNP confirmed the identity of these homologues, and revealed that yeast snRNPs contain a set of seven Sm proteins analogous to the

Table 2 *S. cerevisiae* snRNP proteins

Protein	ORF	snRNP	Size (kDa)	Structural features†
Smb1	YER029c	U1,2,4,5	22	Sm motif
Smd1	YGR074w	U1,2,4,5	16	Sm motif
Smd2	YLR275w	U1,2,4,5	13	Sm motif
Smd3	YLR147c	U1,2,4,5	11	Sm motif
Sme1	YOR159c	U1,2,4,5	10	Sm motif
Smx3 (Smf)	YPR182w	U1,2,4,5	10	Sm motif
Smx2 (Smg)	YFL017w-A	U1,2,4,5	8	Sm motif
Luc7	YDL087c	U1	30	Two zinc finger motifs
Mud1	YBR119w	U1	33	Two RRMs
Nam8	YHR086w	U1	57	Three RRMs
Prp39	YML046w	U1	75	TPR repeats
Prp40	YKL012w	U1	69	Two WW repeats
Prp42	YDR235w	U1	65	TPR repeats
Snp1	YIL061c	U1	34	RRM
Snu56	YDR240c	U1	57	Basic, serine-rich
Snu71	YGR013w	U1	71	Acidic, SR/RD/RE
Yhc1	YLR298c	U1	27	Zinc finger
Cus1	YMR240c	U2	50	Proline-rich
Cus2	YNL286w	U2	32	Two RRMs
Hsh49	YOR319w	U2	25	Two RRMs
Lea1	YPL213w	U2	27	Leucine-rich
Msl1	YIR009w	U2	13	Two RRMs
Rse1	YML049c	U2	154	
Snu17‡/1st3	YIR005w	U2	17	RRM
Prp9	YDL030w	U2*	63	Zinc finger
Prp11	YDL043c	U2*	30	Zinc finger
Prp21	YJL203w	U2*	33	Surp
Prp3	YDR473c	U4/U6	56	
Prp4	YPR178w	U4/U6	52	WD repeats
Prp6	YBR055c	U4/U6.U5	104	TPR repeats, Zinc finger
Prp31	YGR091w	U4/U6.U5	56	
Prp38	YGR075c	U4/U6.U5	28	Serine-rich, acidic C-term
Snu13	YEL026w	U4/U6.U5	13	
Snu23	YDL098c	U4/U6.U5	23	Zinc finger
Snu66	YOR308c	U4/U6.U5	66	
Spp381	YBR152w	U4/U6.U5	34	PEST domain
Brr2	YER172c	U5	246	DExH
Dib1	YPR082c	U5	17	
Prp8	YHR165c	U5	280	Proline-rich, acidic N-terminal
Prp18	YGR006w	U5	28	
Snu114	YKL137w	U5	114	EF2-like
Lsm2	YBL026w	U6	11	Sm motif
Lsm3	YLR438c-A	U6	10	Sm motif
Lsm4	YER112w	U6	21	Sm motif
Lsm5	YER146w	U6	10	Sm motif
Lsm6	YDR378c	U6	14	Sm motif
Lsm7	YNL147w	U6	12	Sm motif
Lsm8	YJR022w	U6	15	Sm motif

*Required for the addition of the U2 snRNP to the commitment complex. †Structural features: DEAD or DExH,members of superfamily of ATP-dependent RNA helicases; PEST, protein degradation sequence; RD, arginine, aspartate dipeptide repeats; RE, arginine, glutamate dipeptide repeats; RRM, RNA recognition motif; SR, serine, arginine, dipeptide repeats; surp, potential RNA-binding motif; TPR, tetratricopeptide repeats; WD, protein interaction motif characteristic of β subunit of G proteins; WW, protein motif believed to interact with proline-rich regions; zinc finger/knuckle motif, clustered histidine and/or cysteine residues that often contribute to zinc binding. ‡A. Gottschalk, G. Neubauer, R. Lührmann and P. Fabrizio, personal communication.

situation in humans (38, 39). Given the similarity in the sequences and associations, it seems reasonable to assume that the yeast proteins will assemble and function in a manner analogous to the characterized human proteins, although this remains to be demonstrated.

The use of electron microscopy to study the ultrastructure of the HeLa snRNP core complex suggested that the Sm proteins form a globular structure, which may have a central hole or depression (56). This proposal is now reinforced by crystallographic data of pairs of Sm proteins, from which a model of a heptameric, ring-like structure was produced (57). The central hole of the predicted ring would have a positive charge, and a diameter of 20 Å, large enough to accommodate a single-stranded RNA. Although the Sm protein complex associates with the snRNAs, none of the Sm proteins contains recognizable RNA-binding motifs. It is currently believed that the RNA-binding properties of the Sm proteins result from the formation of a novel RNA recognition motif present only when the proteins form a heteromeric complex.

In addition to the canonical Sm proteins, nine other sequences containing the Sm motif have been identified in budding yeast (43, 53, 55), and named (or renamed in the cases of Uss1p and Smx4p) the Lsm proteins (*Like Sm*), (29, 30).

The construction of conditional alleles of *LSM2* to *LSM8* demonstrated that they are important for the accumulation of normal levels of U6 snRNA, and for pre-mRNA splicing. Two-hybrid and immunoprecipitation studies indicated that the Lsm2–8 proteins associate with one another and form a U6-associated complex. In addition, crosslinking and co-precipitation experiments demonstrated that several Lsm proteins bind directly to the U6 snRNA, most likely at the 3′ end (58). Despite the obvious similarities between the Sm and Lsm proteins, the Sm and Lsm protein complexes may not be functionally analogous. As U6 snRNA may not leave the nucleus (and therefore should not need to be re-imported) and its 5′ end is not hypermethylated, neither of the two primary roles of the Sm proteins appears necessary for this RNA. The Lsm protein complex may therefore perform a different function. One proposal, based on two-hybrid and genetic interactions with Prp24p, is that the Lsm complex facilitates the recycling of the U6 snRNA into U4/U6 di-snRNPs following completion of the splicing reactions and dissociation of the spliceosome (29). In addition, the association of Lsm8p with Lhp1p, the equivalent of human La protein (59), suggests a role for Lhp1p in the biogenesis of the U6 snRNA (and possibly other polymerase III transcripts), acting as a chaperone for nascent transcripts produced by RNA polymerase III.

Human homologues of the Lsm proteins have been isolated as a protein complex that exhibits a toroidal shape under the electron microscope. This complex was found to associate with the 3′ end of U6 RNA *in vitro* and was shown to facilitate the formation of U4/U6 di-snRNPs (60).

2.2.3   snRNP particle-specific proteins

Each of the snRNP core particles associates with a set of proteins specific for that snRNP. Tables 2 and 3 list the snRNP proteins for *S. cerevisiae* and *S. pombe* respect-

Table 3 *S. pombe* putative snRNP-associated splicing factors

Homologues*	Protein†	ORF/accession no.‡	Size (kDa)	Structural features§
Smb1		SPAC26A3.08	16	Sm motif
Smd1		SPAC27D7.07C	13	Sm motif
Smd2		SPAC2C4.03C	13	Sm motif
Smd3		SPBC19C2.14	11	Sm motif
Sme1		SPBC11G11.06C	10	Sm motif
Smx3		SPAC2F3.17C	8	Sm motif
Smx2		SPBC4B4.05	9	Sm motif
Prp39		SPBC4B4.09	72	TPR repeats
Prp40		SPAC4D7.13	76	WW repeats
Snp1/U1-70K		SPAC19A8.13	30	RRM
SAP155	Prp10	SPAC27F1.09c	130	RWDETP, PP2A
SAP130	Prp12	BAA86918	135	
Prp21/SAP114		013900	48	Two surp, proline-rich
Cus1/SAP145		SPAC22F8.10c	69	Proline-rich
Cus2	Uap2	SPBC1289.02C	42	Two RRMs
Lea1/U2A'		SPBC1861.08C	27	
Msl1/U2B"		SPBC4B4.07C	28	
Prp3/SAP90		SPAC30.06	63	
Prp4/SAP60		SPAC227.12	52	WD repeats
Prp6	Prp1/Zer1	SPBC6B1.07	103	TPR repeats
Prp31		SPBC119.13C	58	
Brr2		SPAC9.03C	249	DExH
Prp8/p220	Spp42	SPAC30.06	275	
Prp18		SPCC126.14	40	
U5-40kD		SPBC1289.11	47	WD repeats
Lsm2		SPCC1620.01C	11	Sm motif
Lsm3		SPAC9B6.05C	11	Sm motif
Lsm4		SPBC30D10.06	14	Sm motif
Lsm5		SPBC20F10.09	9	Sm motif
Lsm6		SPAC2F3.17C	8	Sm motif
Lsm7		SPCC285.12	12	Sm motif
Lsm8		SPCC1840.10	11	Sm motif

*Proteins of *S. cerevisiae* and/or *Homo sapiens* that show significant sequence similarity. Functional complementation has not yet been demonstrated in most cases.
†At this time, only Prp10, Prp12, Prp1/Zer1, and Spp42 have been demonstrated to play a role in splicing.
‡Links to references may be found at http://www.sanger.ac.uk/Projects/S_pombe/FUNCAT/splicing.shtml
§Structural features as for Table 2; PP2A see Reference 112.

ively. (It should be noted that many of the *S. pombe* proteins listed in Tables 3 and 5 have been identified only tentatively as splicing factors based on sequence identities.) Although these lists are up to date at the time of writing, they are unlikely to be complete, as new snRNP-associated factors are still being discovered. Further information and references concerning each protein are available through the Saccharomyces Genome Database (SGD; http://genome-www.stanford.edu/Saccharomyces/), Yeast Proteome Database (YPD; http://www.proteome.com), or the *S. pombe* web site (http://www.sanger.ac.uk/Projects/S_pombe/FUNCAT/splicing.shtml.). The functions of some of these proteins are discussed in sections 4 and 5 below.

2.2.4 Non-snRNP proteins

In addition to those factors found to be tightly associated with snRNP particles, several classes of non-snRNP proteins are required. These include proteins involved in snRNP biogenesis, spliceosome assembly, the molecular rearrangements of splice-osomal components, spliceosome disassembly, intron debranching and degradation, and the recycling of spliceosomal components for further rounds of splicing. Some of these factors associate only transiently with the snRNPs and/or spliceosome machinery, apparently being released after the step at which they are required. Tables 4 and 5 list the non-snRNP proteins identified in *S. cerevisiae* and *S. pombe* respectively. Further information on individual proteins is available via the SGD and Proteome databases. The roles of several of these proteins are discussed in more detail in sections 4 and 5.

Amongst the non-snRNP associated proteins are several members of the DEAD- or DExH-box superfamily of putative ATP-dependent RNA helicases. These proteins are believed to play important roles in facilitating the unwinding or melting of RNA duplexes within snRNPs and spliceosomes, thereby controlling the molecular re-arrangements that take place during the spliceosome cycle (see section 5 for a fuller description).

2.2.5 Conservation of splicing factors between yeasts and humans

For a large number of these proteins, yeast and human homologues have been identi-fied (reviewed in References 14, 61, and references therein). For example, as pre-viously discussed, budding yeast counterparts of the human Sm proteins exist, and recently human versions of the Lsm proteins have also been identified (60).

The U5 snRNP protein Prp8p is the most highly conserved splicing factor identi-fied to date (62). It is highly conserved with respect to its primary sequence, large size, and pre-mRNA binding properties. Both budding yeast Prp8p and its HeLa counterpart, p220, bind to the pre-mRNA at the 5′ splice site prior to step 1, and to the branchpoint-3′ splice-site region after the first step of splicing. Prp8p is believed to play a crucial role in the stabilization of the U5 snRNA–exon interactions (see section 4).

Although a large number of splicing factors have clear structural and functional homologues in yeasts and humans, several significant differences exist. For example, the budding yeast Prp28 protein shows no snRNP association, whereas the corres-ponding human protein, p100, is a U5 snRNP component. Perhaps more striking are the differences that exist between the U1 snRNPs from budding yeast and verte-brates. The yeast snRNP particle, as biochemically purified, is significantly larger than the vertebrate U1 snRNP (18S compared with 12S). The yeast U1 snRNA is 568 nucleotides in length, compared with 164 nucleotides for the human equivalent and, in addition to the core Sm proteins and homologues of the three U1-specific proteins, U1A (Mud1p), U1C (Yhc1p), and U1-70K (Snp1p), the yeast particle contains a fur-ther seven proteins that have no currently characterized vertebrate homologues (38, 39, 63). This may reflect mechanistic differences of 5′ splice-site recognition in these two organisms. As 5′ splice site sequences are more stringently conserved in budding

Table 4 *S. cerevisiae* non-snRNP proteins

Protein	ORF	Size (kDa)	Activity	Structural features†
Cef1	YMR213w	68	Prp19 associated	
Dbr1	YKL149c	48	Debranchase	
Ecm2	YBR065c	41		Two zinc fingers, lysine-rich
Isy1	YJR050w	28		
Msl5	YLR116w	53	Binds BPS in CC*	Two zinc knuckles, proline-rich C-terminal, maxi-KH domain
Mud2	YKL074c	58	U2*	Three RRMs
Prp2	YNR011c	100	ATPase	DEAH
Prp5	YBR237w	96	U2*	DEAD
Prp16	YKR086w	120	ATPase	DEAH
Prp17	YDR364c	52		WD repeats
Prp19	YLL036c	57	Recombination	Myb-like DNA binding
Prp22	YER013w	130	ATPase and mRNA release	DEAH
Prp24	YMR268c	51	U4/U6 recycling	Three RRMs
Prp28	YDR243c	67	ATPase	DEAD
Prp43	YGL120c	88	Intron release	DEAH
Sad1	YFR005c	52	U4/U6 biogenesis	Zinc finger
Slu7	YDR088c	45	Long BP-3′ ss recognition	Zinc knuckle
Snt309	YPR101w	21	Prp19p associated	
Spp2	YOR148c	21	Prp2p addition	Lysine-rich
Syf1	YDR416w	100		TPR repeats
Syf2	YGR129w	25		
Syf3	YLR117c	82		TPR repeats

*Binds branchpoint sequence (BPS) in commitment complex (CC2); †KH, hnRNP K homology; other structural features as for Table 2.

Table 5 *S. pombe* non-snRNP proteins

Homologue*	Protein†	Cosmid/accession no. ‡	Size (kDa)	Structural features§
Mud2p/U2AF65	Prp2	SPBC146.07	59	RS, threeRRMs
U2AF35	U2AF23	SPAP8A3.06	25	
	Prp4	SPCC777.14	55	Kinase
Prp46/PRL1	Prp5	AB004535	52	WD repeat
	Prp8/Cdc28	SPBC21B10.01c	112	DEAH
	Srp1¶	SPBC11C11.08	31	RRM, glycine hinge, RS
	Srp2¶	SPAC16.02C	40	Two RRMs, RS
Msl5/SF1		SPCC962	64	Maxi-KH, two zinc knuckles, proline-rich
Prp43		SPBC16H5.10C	84	DEAH
Prp22		SPAC10F6.02C	130	DEAH
Prp16		SPBC17G9.01	113	DEAH
Prp28		SPCC63.11	74	DEAH
Slu7		SPBC365.05C	43	Zinc knuckle
Dbr1		SPAC17A5.02C	53	
Syf3		SPBC31F10.11C	81	TPR repeats
Cef1	Cdc5	P39964	87	

*Proteins of *S. cerevisiae* and/or *H. sapiens* that show significant sequence similarity. Functional complementation has not yet been demonstrated in most cases.
†At this time, only Prp2, Prp4, Prp5, Prp8/Cdc28, and Cdc5 have been shown genetically to play a role in splicing.
‡Links to references may be found at http://www.sanger.ac.uk/Projects/S_pombe/FUNCAT/splicing.shtml
§Structural features as for Tables 2 and 4.
¶Similar to mammalian SR proteins, but a function in splicing has not yet been demonstrated.

yeast and there are no examples of alternative 5′ splice site usage, recognition of these sequences is presumably a simpler process than in humans.

In higher eukaryotes, SR proteins (rich in arginine, serine repeats) play key roles in splice-site selection and regulating alternative splicing, partly by acting as components of protein bridges that link splice sites across introns. In addition, SR proteins bind to exonic enhancer sequences and form protein bridges that activate splicing of neighbouring weak introns (those that have short pyrimidine stretches between the branch point sequence and the 3′ end of intron). SR proteins appear to be unnecessary for budding yeast splicing, and it is conceivable that the extra proteins in the budding yeast U1 snRNP may carry out protein linkage and/or RNA recognition functions which the many SR proteins perform in higher eukaryotes. Thus, the need for alternative splice-site usage, exon skipping, and tissue-specific splicing may mean that higher eukaryotes require a more flexible splicing machinery, in which the U1 snRNP is modified by *trans*-acting factors that regulate splice site selection.

Nevertheless, two SR proteins, Srp1 and Srp2, have been identified in *S. pombe* (64, 65). Overexpression of Srp1 in a wild-type strain of fission yeast results in accumulation of pre-mRNA, but the role of Srp2 in splicing is presently unknown. Although the existence of natural splicing enhancers in yeast has not yet been investigated, mammalian exonic splicing enhancers can activate splicing of weak upstream introns in fission yeast (J. A. Wise, personal communication). This is intriguing in view of the large number of introns in fission yeast genes and the potential for alternative patterns of splicing.

Budding yeast cells apparently have only one type of spliceosomal machinery, unlike higher eukaryotes which contain U12-specific (or AT-AC) spliceosomes in addition to the classical U2-specific (GT-AG) spliceosomes. The U12 spliceosome processes a distinct class of introns that often have non-canonical terminal dinucleotide sequences (i.e. AT-AC), although other conserved sequences are more critical in defining these introns (66). The U12 spliceosome is so named as it contains an alternative set of snRNPs, U11, U12, U4^{ATAC} and U6^{ATAC}, but apparently shares the U5 snRNP with U2 spliceosomes (67). These alternative snRNPs do not appear to be present in budding yeast.

3. An overview of spliceosome assembly

The development of *in vitro* splicing assays opened up the field, allowing a biochemical dissection of the process. *In vitro* studies have been performed mainly with HeLa nuclear extracts (68) or whole cell extracts of *S. cerevisiae* (69). To date, an *in vitro* system for *S. pombe* has proved elusive. Following the addition of substrate pre-mRNA to a splicing extract, there is a lag period before the intermediates and products of splicing appear. During this incubation period the pre-mRNA is bound by snRNPs and other protein factors in the highly ordered spliceosome assembly pathway, certain stages of which require ATP hydrolysis. In a budding yeast *in vitro* splicing reaction the products are detectable within 1 min or less of the addition of the substrate RNA. The process is slower with HeLa nuclear extract, requiring 5–30 min of incubation, depending on the pre-mRNA substrate used. Spliceosome complex assembly is a highly dynamic process, involving the addition of many components and conformational changes. For analytical purposes the process is divided into several stages as defined by the detection of distinct complexes following fractionation by gel electrophoresis, gel filtration, density gradient centrifugation, or affinity chromatography. In yeast, different fractionation conditions have resulted in different nomenclatures for splicing complexes. The following pathway will be discussed CC→B→A2-1→A1→A2-2→A2-3→I; in which CC, B, A2-1, and A1 contain pre-mRNA, A2-2 and A2-3 contain intermediates and products, and I contains the excised intron (Fig. 4).

The earliest splicing-specific complex formed on the pre-mRNA is the commitment complex (CC), the stage in spliceosome assembly when a pre-mRNA is no longer competed out of the splicing pathway by excess competitor pre-mRNA. It results from the ATP-independent association of the U1 snRNP with the 5′ splice site through base pairing between a conserved sequence in U1 snRNA and a complementary sequence at the 5′ end of the intron (70, 71). The primary function of the U1 snRNP seems to be in defining the 5′ splice site.

CC is converted to complex B (or pre-spliceosome) with the ATP-dependent association of U2 snRNP at the branchpoint sequence of the pre-mRNA through Watson–Crick base-pairing. This leaves the branchpoint adenosine unpaired and therefore free to act as a nucleophile in the first step of the splicing reaction. To this pre-spliceosome a U4/U6.U5 tri-snRNP complex and numerous non-snRNP factors are added, to produce complex A2-1, a pre-catalytic form of spliceosome. The U4/U6 and U5 snRNPs associate to form a tri-snRNP complex before entry into the spliceosome.

Conversion of complex A2-1 to the active spliceosome, or complex A1, is accompanied by a loosening of the U4 snRNP association, and neither the U4 snRNA nor the U1 snRNA seems to be required for catalysis (72, 73). Interaction of the U6 snRNA with the 5′ end of the intron may displace the U1 snRNA. As the U4:U6 base-pairing is destabilized, interactions between the U2 and U6 snRNAs now contribute to the formation of the catalytic centre. Simultaneously, an invariant loop sequence in U5 snRNA becomes closely aligned with exon sequences just upstream of the 5′ splice site. These dynamic RNA–RNA interactions are illustrated schematically in Fig. 5. The roles of individual proteins in these and subsequent rearrangements are discussed in the next section (for reviews see References 10, 15, 16).

4. Who does what

Although a large number of components of the spliceosome have been identified, the functions of many of the proteins have yet to be determined. In this section the roles of the snRNAs and some of the proteins are discussed within the context of the spliceosome cycle (see Fig. 4).

The association of U1 snRNP with the 5′ splice site during CC formation is enhanced by the presence of the nuclear cap-binding complex (consisting of Cbp20p

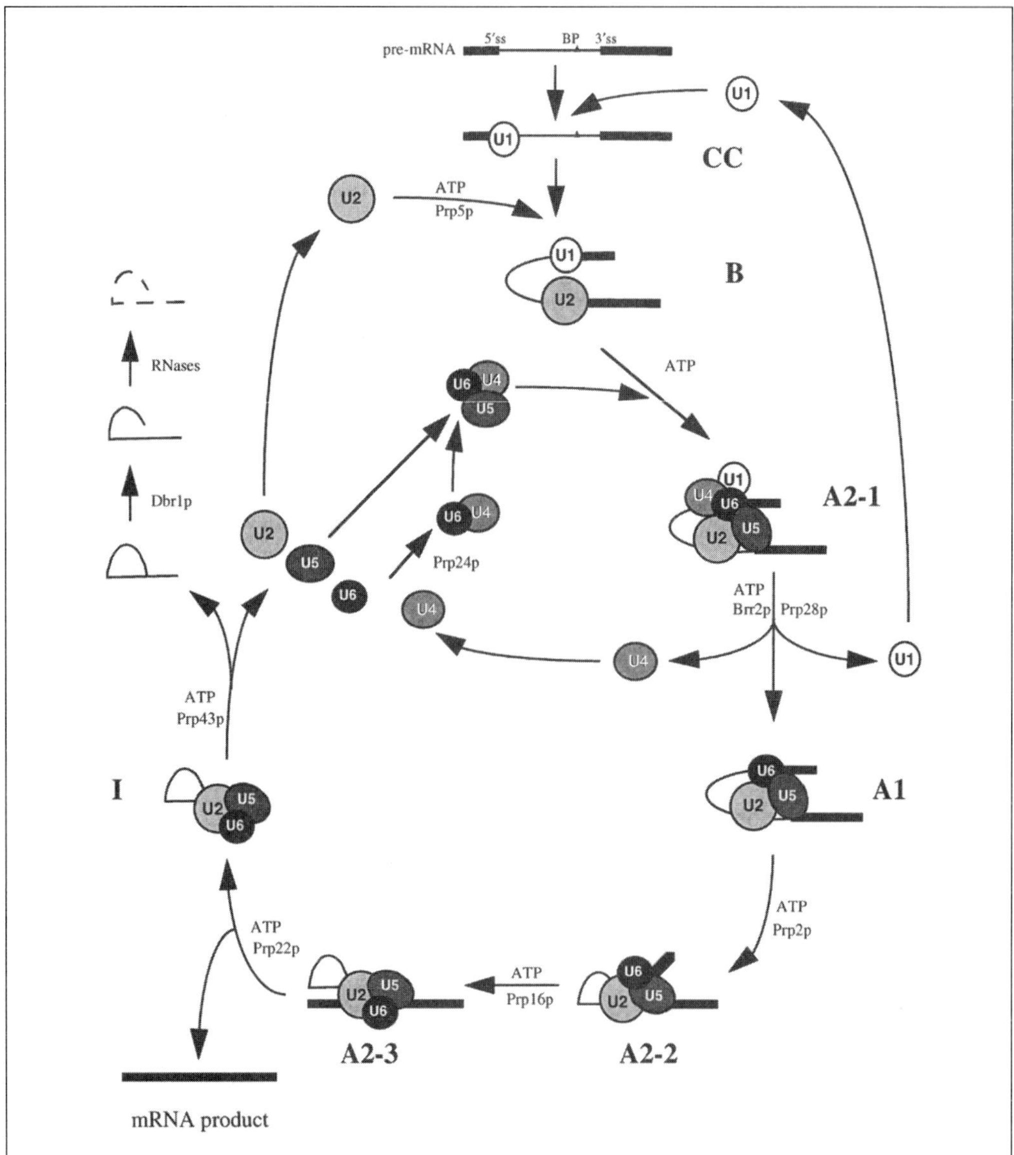

Fig. 4 The spliceosome cycle. The pathway of spliceosome assembly and the transitions necessary for the splicing reactions and snRNP recycling are shown. Macromolecular complexes identified biochemically and genetically are indicated, as are the sites of action for some of the proteins involved in the molecular rearrangements. Modified from Reference 14.

and Cbp80p) on the m^7G cap of the pre-mRNA. The nuclear cap-binding complex is proposed to interact with U1 snRNP, stabilizing the U1 snRNP–pre-mRNA interaction, and thereby promoting CC formation (74, 75). A *cbp80* null mutant exhibits reduced usage of the cap-proximal 5′ splice site in a reporter construct containing

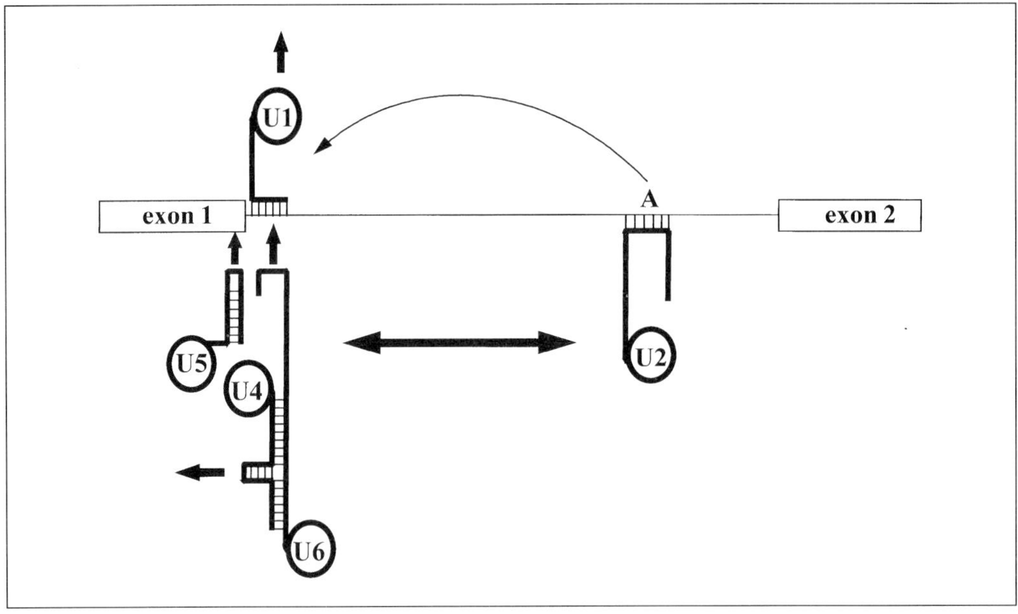

Fig. 5 Schematic representation of RNA–RNA interactions during spliceosome assembly. For simplicity no proteins are shown. The snRNAs are represented by thick lines and named circles, and the snRNA structures are highly simplified. The intron is represented by a thin line; base-pairing interactions are indicated by rows of fine lines. See section 3 for details of the interactions.

two identical competing 5′ splice sites. Thus, as for HeLa splicing, cap-binding complex influences usage of cap-proximal splice sites.

CC formation is also promoted by Luc7p, Nam8p, Prp39p, and Yhc1p, all of which are U1 snRNP-specific proteins. At least seven U1 snRNP proteins (Nam8p, Snp1p, Snu56p, Yhc1p, and three Sm core proteins) bind to the pre-mRNA (76) and therefore may stabilize the U1 snRNP-5′ splice-site interaction. Yhc1p contacts the 5′ splice site in the intron, whereas Snp1p binds to exon sequences upstream of the 5′ splice site, and Nam8p binds non-conserved sequences in the intron, downstream of the 5′ splice site (77). Nam8p is not normally essential, but is required for uncapped pre-mRNAs and for the efficient recognition of non-canonical 5′ splice sites. Nam8p apparently stabilizes commitment complexes. The *LUC7* gene was identified by a mutation that causes lethality in a budding yeast strain lacking the nuclear cap-binding complex (78). Although the splicing of wild-type introns is affected only mildly in a *luc7* mutant strain, non-consensus introns are more sensitive to this mutation and, as for Cbp80p-depleted cells, there is reduced usage of the cap-proximal 5′ splice site in a reporter construct containing duplicated 5′ splice sites (63). The genetic data, together with the observation that the cap-binding complex could be co-precipitated with U1 snRNP from wild-type but not from *luc7* mutant extracts, suggests a possible role for Luc7p in directly or indirectly mediating the interaction of U1 snRNP and cap-binding complex. Thus, the association of U1 snRNP with the pre-mRNA is probably influ-

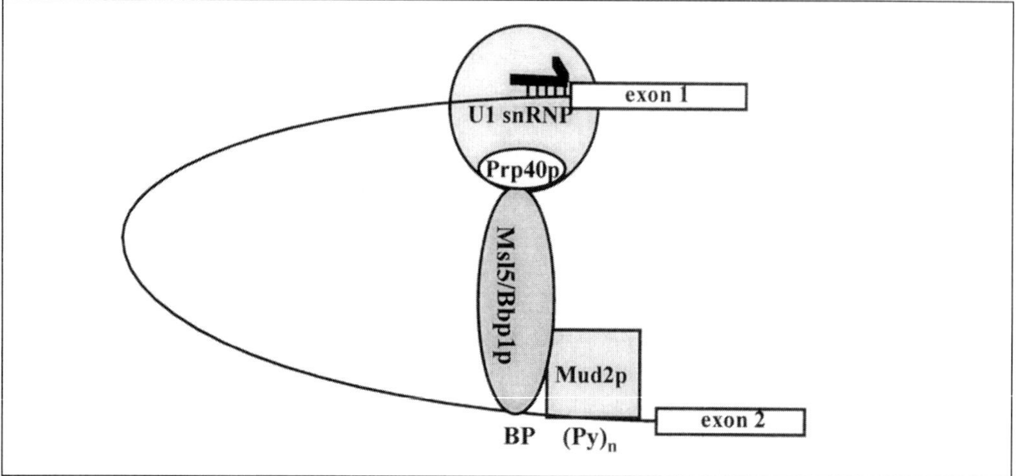

Fig. 6 Molecular interactions in the commitment complex 2. See section 4 for details.

enced by both protein–protein and protein–RNA interactions and, even in budding yeast, different U1 snRNP components differentially affect the splicing of different introns.

Under appropriate electrophoretic conditions, two commitment complexes, CC1 and CC2, can be resolved. The U1 snRNP and a 5′ splice site are required for CC1 formation, whereas CC2 forms only when there is also a branchpoint sequence in the pre-mRNA. The transition from CC1 to CC2 involves branchpoint bridging protein (Msl5p/Bbp1p) and Mud2p, which respectively bind to the branchpoint sequence and to the adjacent pyrimidine-rich tract prior to the addition of U2 snRNP. Msl5p is part of a cross-intron protein bridge. It binds directly to Prp40p in the U1 snRNP and to Mud2p. Thus, a link is formed between the 5′ and 3′ splice sites prior to U2 snRNP binding. (Fig. 6) (35, 79–81).

The formation of complex B, the prespliceosome, is the first ATP-dependent step in spliceosome assembly, and involves addition of the U2 snRNP. Four proteins, Prp5p, Prp9p, Prp11p, and Prp21p, are particularly important for the branchpoint localization of the U2 snRNP on the pre-mRNA, with the putative splicing factor Sub2p also implicated at this stage (16). The Prp5p and Sub2p proteins are putative ATP-dependent RNA helicases, and presumably account for the ATP requirement of this event. Prp5p in particular has been proposed to facilitate the molecular rearrangements required to allow the base pairing of U2 with the branchpoint sequence. Indeed, the ATPase activity of Prp5p was preferentially stimulated by U2 snRNA *in vitro* (82).

Conversion of complex B to complex A2-1 also requires ATP, and at least two proteins, Prp31p and Syf3p/Clf1p, have been implicated in the association of the tri-snRNP with the prespliceosome. Concurrently with and/or immediately after tri-snRNP addition, several intramolecular and intermolecular rearrangements take place, one being the association of U6 snRNA with the 5′ splice site. As the interactions of the U1 and U6 snRNAs at the 5′ splice site are mutually exclusive, it is thought that U1 dissociates at this point, with the putative RNA helicase Prp28p implicated in disrupting the U1:5′ splice-site interaction (83, 84).

The transition from complex A2-1 to complex A1, the catalytically active spliceosome, requires the destabilization of the U4:U6 duplex. Prp8p may monitor the exchange of U6 for U1 at the 5′ splice site and regulate the unwinding of the U4:U6 duplex (85). U6 snRNA now base pairs with the intron at the 5′ splice site and with the U2 snRNA. Again this is an energy-requiring process, and Prp38p and Brr2p are important for these changes. Brr2p is another member of the putative ATP-dependent RNA helicase family and is an integral component of the U5 snRNP. Brr2p has been shown to unwind U4:U6 duplexes *in vitro* and within the context of snRNP particles (16, 86).

Curiously, a high molecular weight protein complex that contains at least eight polypeptides, including Prp19p, Cef1p, and Snt309p, joins the spliceosome at about the time of U4 dissociation. The role of this complex in splicing is not known; however, proteins of this complex may mediate a link between splicing and the cell cycle (see below).

At this stage, the 5′ splice site is bound on the exon side by a conserved single-stranded loop in the U5 snRNA and on the intron side by U6 snRNA. A further rearrangement is required in order to produce the active spliceosome. Again it is ATP dependent and, once more, a putative helicase protein is involved. Prp2p associates transiently with spliceosomes at this point (an interaction that requires Spp2p) and probably remains associated until after the first *trans*-esterification reaction is complete. The Prp2p-dependent rearrangement performed at this point represents the most proximal event prior to step 1, and brings the 2′ hydroxyl of the branchpoint into proximity with the phosphodiester bond at the 5′ splice site.

In budding yeast splicing extracts, the formation of the active spliceosome requires the 5′ splice site and the branchpoint sequence, but not the 3′ splice site, and only the U2, U5, and U6 snRNPs are required for the first step of the splicing reaction. The invariant loop I of the U5 snRNA is not essential for step 1 *in vitro* (87); however, the interactions made by U5 snRNA with the 5′ exon appear to be important in positioning the end of exon 1 correctly for the second *trans*-esterification reaction (88). Given that U5 loop I is dispensable for step 1, it has been reasoned that the catalytic site of the spliceosome must be composed of U2 and/or U6 with or without associated proteins. Interestingly however, Prp8p (an integral U5 snRNP component) interacts with the 5′ splice site, the 3′ splice site, and the branchpoint sequence, implying a direct role in the catalysis of the reaction. Alternatively, Prp8p may act as a scaffold for the assembly of other factors at the active site, including anchoring the ends of both exons (89), while regulating or monitoring conformational changes (85, 90, 91). At present no information is available on what actually triggers the catalytic reaction, or how the target phosphodiester bond is identified.

Following step 1 (the first point at which intermediates can be detected), the completion of the splicing reaction is a rapid process; however, before step 2 can proceed, a reorganization is required in which the catalytic site is remodelled. This involves a conformational change in which the 3′ splice site is brought into juxtaposition with the 5′ splice site, apparently anchored and aligned by the U5 snRNP. These pre-step 2 changes are ATP dependent and the DExH box protein Prp16p has been demonstrated to affect rearrangements around the 3′ splice site. It has been proposed that Prp16p performs a proofreading function, monitoring the accuracy of branched nucleotide formation such that only appropriate intermediate species proceed to step 2 (92).

The U5 snRNP component Snu114p has also been implicated at this stage. Snu114p displays striking structural similarity to the ribosomal elongation factor EF-2 (ribosomal translocase), including a G domain that is essential, thus suggesting that GTP hydrolysis maybe important for Snu114p function (93). This is, to date, the first evidence that a G domain-containing protein plays an essential role in the pre-mRNA splicing process. By analogy with the function of EF-2 in translation, this may indicate a role for Snu114p in a GTP-dependent translocation event in the spliceosome.

Prp17p is also required at this ATP-dependent stage, although its function has not been determined. Two other proteins, Prp18p and Slu7p, act after the ATP-dependent step and interact with the 3′ splice site. Prp18p associates loosely with the U5 snRNP and has also been demonstrated to interact genetically with Prp8p, Prp16p, Prp17p, and Slu7p, presumably at this stage within the spliceosome (94). Although Slu7p is an integral spliceosome component, it is essential only for the splicing of introns in which the branchpoint to 3′ splice site distance is greater than seven nucleotides. In addition to these proteins, a biochemical activity designated SSF1 (*Second Step Factor 1*) is required for the second *trans*-esterification reaction, but the factor(s) responsible for this activity remains unidentified (95).

Although Prp8p can apparently influence 3′ splice-site usage (91), the mechanism by which the 3′ splice site is recognized has yet to be fully determined. The conserved YAG motif is known to be the only 3′ splice-site sequence required for step 2 to proceed; however, the presence of a polypyrimidine tract between the branchpoint and the 3′ splice site is also known to contribute to splice-site recognition. A simple 5′–3′ scanning mechanism has been proposed to explain the preference to use the first YAG sequence downstream of the branchpoint as the 3′ splice site. However, several lines of evidence argue against this. For example, any downstream AG sequence can (at least to some extent) compete with an upstream AG, and the splicing machinery can still cope even if a structural block is placed between the branchpoint and the 3′ splice site.

An alternative 'diffusion–collision' model has also been proposed. In this scenario the 3′ splice site closest to the branchpoint sequence would be most likely to interact with the splicing machinery and would therefore be used more frequently than more distal splice sites. It remains possible that a combination of these models may prove to be the true mechanism of 3′ splice-site recognition and selection.

Completion of the second *trans*-esterification sees the formation of complex A2-3, which contains the products of splicing (i.e. a spliced mRNA and an excised intron). Two further putative RNA helicases are involved in the release of these products from the splicing complex. First, Prp22p (and possibly other factors) facilitates the release of the spliced mRNA in an energy-dependent reaction, and Prp43p mediates the release of the intron. The intron's lariat form is subsequently debranched by Dbr1p.

Following spliceosome disassembly, the snRNP components are recycled. This requires the reformation of the U4/U6 di-snRNP and the U4/U6.U5 tri-snRNP. The reannealing of the U4 and U6 snRNAs is performed (at least in part) by the RNA recognition motif (RRM)-containing protein Prp24p. The re-annealing reaction catalysed by Prp24p proceeds more efficiently with snRNPs than with deproteinized snRNAs, suggesting that other endogenous factors are also required for optimal re-annealing (96). The U6-specific Lsm proteins were recently proposed to be such factors (29). The reassociation of U4/U6 with U5 requires at least Prp6p and Prp8p to form a functional tri-snRNP, which is then able to assemble with complex B to produce a new spliceosome.

5. DEAD/DExH-box proteins

It should be apparent from the description of the splicing pathway in section 4 that the putative ATP-dependent RNA helicase proteins play crucial roles in the molecular rearrangements that are necessary to complete the splicing reactions. These proteins have been proposed to be the mechanical devices of the spliceosome, responsible for the rearrangements of mutually exclusive RNA–RNA and RNA–protein interactions. The proteins of this family are characterized by the presence of seven conserved amino acid sequence elements, and are often referred to as DEAD- or DExH-box proteins after one of the most highly conserved motifs. Although neither the mechanism of action nor the substrates for these proteins are defined, all display RNA-activated ATPase activity. From the crystal structure of members of this family, it appears that the core region, containing the conserved motifs, may function as an ATP-dependent motor, with the non-conserved terminal extensions conferring substrate specificity. For example, the unique N-terminal domain of Prp16p is responsible for its transient binding to the spliceosome prior to step 2 (97).

Given that these proteins modulate the rearrangement of only short duplex regions of RNA that lack extensive similarities, it is possible that proteins stabilizing such regions may be the true targets for the DEAD/DExH-box proteins. In this model, the conformational rearrangements would be facilitated by the DEAD- or DExH-box protein disrupting the interaction between the RNA duplex and the stabilizing protein. For example, it has been proposed that Prp5p, which is important for the binding of the U2 snRNP to the branchpoint region, may disrupt an interaction with Prp9p thereby allowing the molecular rearrangement of the U2 snRNA to produce a structure capable of associating with the branchpoint sequence.

The specificity and timing of action of DEAD- or DExH-box proteins may be controlled via interactions with other proteins. For example, in the case of Prp2p, which acts immediately before the first *trans*-esterification, Spp2p is required for its binding to the spliceosome (98). In this way Spp2p may control the action of Prp2p by providing a spliceosome-binding site only at this unique step. Brr2p, the helicase proposed to be responsible for the unwinding of the U4:U6 duplex, appears to have an antagonistic relationship with Prp24p (the protein proposed to re-anneal the U4 and U6 snRNAs following spliceosome disassembly). Whether U4:U6 is the direct target for Brr2p *in vivo* and whether it unwinds this duplex during spliceosome assembly remain to be determined.

The Prp16 protein, which is required before step 2 of splicing, has been proposed to function in a kinetic proofreading model to ensure splicing fidelity. Prp16p can be strongly crosslinked at the 3′ splice site, and mutants of Prp16p which show reduced RNA-dependent ATPase activity and reduced branchpoint specificity also have reduced splicing fidelity. It has been proposed that the ATP hydrolysis performed by Prp16p can act as a 'timer' such that if a mutant branchpoint is used, causing splicing to be slowed, the Prp16 protein will detect this and activate a discard pathway. However, if the correct branchpoint sequence is used, the Prp16 protein would facilitate the changes necessary to allow step 2 to proceed (92).

6. The regulation of splicing

6.1 Regulated splicing of specific introns

As the splicing of introns is an essential step in gene expression, it represents a potential point for regulation. To date, only a small number of regulated splicing events has been reported for budding yeast.

The ribosomal protein L30 (formerly L32) autoregulates the splicing of its own transcript, *RPL30* (formerly *RPL32*). Sequences within the 5′ exon interact with intronic nucleotides to produce a structure that binds the L30 protein (99). This binding does not preclude U1 snRNP association, but prevents the U2 snRNP from entering the assembling spliceosome. Thus, in the presence of excess L30 protein, the transcript is not spliced, protein production is reduced, and the level is thereby regulated. The L30 protein also directly affects the regulation of *RPL30* translation in the cytoplasm.

Mer2p is essential for meiosis to proceed and is required for the formation of proper chiasmata and for meiotic chromosome dysjunction. Its primary role is thought to be in the generation of double-stranded breaks in meiotic genetic recombination. Although the *MER2* pre-mRNA is transcribed in both mitosis and meiosis, it is spliced efficiently only during meiosis. The MER2 pre-mRNA has a non-canonical 5′ splice site and an unusually large 5′ exon. Sequences in the 5′ exon play a role in the regulation of splicing. Splicing is regulated by the requirement for Mer1p, which itself is expressed only during meiosis. Mer1p contains a KH, RNA-binding motif and associates with *MER2* pre-mRNA at several sites in the 5′ exon and within the intron. Thus, by restricting the expression of Mer1p to meiosis, the cell can control the splicing and therefore the expression of Mer2p (100).

6.2 Phosphorylation

There are many indications that phosphorylation of splicing factors, especially of SR proteins, is important for splicing regulation in higher eukaryotes (for example, see

Reference 101). Little is known about phosphorylation of yeast splicing factors, and, although the discovery of two SR proteins in *S. pombe* (64, 65) and of the splicing-related kinase Prp4 (102) is provocative, so far there are no data to indicate that phosphorylation of SR proteins is important for splicing in yeast.

7. Links between splicing and other processes

7.1 Splicing, capping, and polyadenylation

Gene expression is controlled at many levels and recent reports have suggested links between pre-mRNA splicing and other cellular processes. As mentioned in section 4, pre-mRNA splicing is influenced by capping. The cap-binding complex plays a highly conserved role in the recognition of pre-mRNA substrates at an early step in the splicing process and, in particular, enhances splicing of the first intron of a pre-mRNA.

Removal of the last intron of metazoan pre-mRNAs is promoted by the proximity of a polyA-cleavage site downstream in the RNA. This is apparently mediated by an interaction between the polyadenylation-cleavage complex and splicing factors associated with the proximal 3' splice site, probably bridged by SR proteins. In budding yeast few transcripts contain more than a single intron, so this is not a major factor.

7.2 Transcription and splicing

Although it has not been demonstrated in yeast, transcription is thought to be coupled to splicing in higher eukaryotes, with components of both systems co-localizing in 'speckles' within the nucleus. A physical association between mammalian RNA polymerase II (RNAP II) and capping, splicing, and polyadenylation factors has also been reported (reviewed in Reference 103). Following initiation of transcription, RNAP II pauses, and the C-terminal domain (CTD) of the large subunit becomes hyperphosphorylated before the elongation phase of transcription (see Chapter 6, section 3.2). The presence of the CTD is important for efficient capping, splicing and cleavage/polyadenylation, and the phosphorylated form of the CTD has affinity for capping enzyme, SR proteins, and polyadenylation factors. It is difficult to distinguish direct effects of transcription on splicing from related effects of capping and polyadenylation, however, conceivably, RNAP II may pause to recruit processing machinery to the emerging nascent transcript and/or there may be a checkpoint to ensure the coordination of transcription and RNA processing events. This represents a potential means of regulating transcription through RNA processing factors, or of controlling patterns of splicing through factors binding specifically at individual promoters.

7.3 Splicing and mRNA stability

Potential links have been found between the spliceosome machinery and other cellular processes through the use of exhaustive and iterative yeast two-hybrid screens. In this way, the Lsm proteins, which are integral components of the U6 snRNP, have been implicated in mRNA decapping and stability. These small proteins have been shown to interact with the machinery of the decapping complex (Dcp1p and Dcp2p) and also with proteins implicated for the stability of mRNA (Pat1p/Mrt1p and Xrn1p) (M. Fromont-Racine, A. E. Mayes, P. Legrain and J. D. Beggs, unpublished results; 104). In addition, depletion of the Lsm proteins has been shown to cause increased stability of mRNAs, indicating a direct role in the control of degradation. It appears that distinct Lsm protein complexes are involved in these different functions, a complex containing Lsm1–7p being involved in mRNA turnover, whereas Lsm2–8p are U6 associated (104).

7.4 Splicing and the cell cycle

Over the years, a number of links have been established between the splicing of pre-mRNA and the cell cycle in budding yeast. As early as 1982, Johnston *et al.* identified mutations in several *PRP* genes which were defective in DNA replication and which produced a dumbbell-forming, or *dbf*, morphology. These authors found alleles of *PRP3* in a genetic screen for *dbf* mutants and subsequently showed that DNA synthesis was rapidly reduced in mutants of *PRP2*, *PRP3*, *PRP4*, *PRP8*, and *PRP11* (105, 106). In addition, a *prp22* mutation was isolated in a genetic screen for mitotic arrest mutants designed to find proteins involved in cyclin proteolysis (107), and *PRP17*, which encodes a second step splicing factor, was independently identified as a cell cycle gene, *CDC40*. Deletion of *PRP17/CDC40* is not lethal, but causes heat-sensitive growth. At the restrictive temperature, cells display a splicing defect and fail to progress through the G2–M transition. Genetic screens for factors interacting with a *prp17/cdc40Δ* allele identified suppressor mutations in *PRP8* (108) and synthetic lethal mutations in *PRP8* plus five other known splicing factor genes (*PRP16*, *PRP22*, *SLU7*, *SLT11*, and *SNR20/U2* snRNA) as well as three genes (*SYF1*, *SYF2*, and *SYF3*) subsequently shown also to encode spliceosome components (109, 110). It has been proposed that these proteins may together provide a checkpoint or monitoring function that prevents progression from the G2 to the M phase in the event of a general splicing defect, or a defect in the splicing of specific mRNAs encoding cell cycle proteins (108).

In *S. pombe* also, mutations in a number of splicing factor genes, including *prp1/zer1*, *prp5*, *prp6*, *prp8/cdc28*, *prp11*, *prp12*, *prp13*, and *prp14*, result in cell cycle defects. The observation that *S. pombe cdc5-120* is synthetic lethal with *prp5-1* led to experiments that showed that *cdc5* cell cycle mutants are defective in splicing (111).

The nature of the functional or mechanistic link between splicing and the cell cycle has yet to be determined, although it seems likely that a set of specific introns may be responsible. If such introns are in some way suboptimal or more sensitive to changes within the cell, they may have evolved to act like a cell cycle checkpoint. Failure to splice these introns could result in a block at the particular point in the cell cycle at which their protein product is required. As the Dsk1 kinase, which regulates mitosis, also phosphorylates SR proteins *in vitro*, this could play a role in coordinating splicing and the cell cycle in *S. pombe*.

8. What remains to be done?

Despite the wealth of knowledge gained about the splicing reactions and the factors involved, many important questions have yet to be resolved. The current list of splicing-related proteins is unlikely to represent the full complement of factors involved, with new members being identified and added on an almost monthly basis. However, with the application of highly efficient affinity purification techniques and mass spectrometric analyses, full characterization of the splicing machinery should not be far off. As discussed in section 4, the precise functions of very few splicing factors are actually known, and it will require a major undertaking to fill this gap in the knowledge. The factor(s) directly responsible for the first cleavage reaction at the 5′ end of the intron remains one of the most intriguing mysteries: is catalysis RNA or protein driven, or both? What directs the remodelling of the spliceosome for the second catalytic step?

Major questions also remain regarding the regulation and control of splicing. What drives the splicing reactions, and how are the rearrangements within the spliceosome coordinated? Related to this is the need for accurate regulation of the DEAD/ DExH-box proteins; what controls their specificity and timing? Does phosphorylation play a role in the regulation of splicing in yeast? What is the functional significance of links between splicing and other cellular processes? Some possible answers to these questions have been suggested, but remain to be tested.

It seems likely that the current view of how introns are defined will prove to be simplistic. As with splicing enhancers in higher eukaryotes, subtle structural elements within the sequence of yeast introns will probably add to the complexity of the story, with subsets of transcripts requiring specific co-factors for their efficient splicing. A paradigm for this already exists in the requirement for Slu7p when the branchpoint to 3′ splice-site distance is greater than seven nucleotides (112), and the subtle effects of U1 snRNP proteins on 5′ splice-site usage hint at regulation. Above all, why does *S. cerevisiae* tolerate the continued presence of introns when an estimated 100 or so splicing factors have to be produced in order to splice only about 250 pre-mRNAs in a highly ATP-dependent process?

In summary, the past two decades have seen the accumulation of an enormous pool of data, and many novel phenomena have been revealed, but our understanding of the mechanisms and means of regulating pre-mRNA splicing remains far from complete. The expectation is that at least some of the regulatory aspects will be elucidated by functional genomics early in the twenty-first century.

Acknowledgements

The authors thank Paul Smith and David Page for helpful comments on this manuscript, the Wellcome Trust for a Prize Studentship to A. E. M., the National Institutes of Health for a grant to J. A. P., and The Royal Society for a Cephalosporin Fund Senior Research Fellowship to J. D. B.

References

1. Dix, I. and Beggs, J. D. (2000) Introns. In *Encyclopedia of genetics* (ed E. C. R. Reeve). Fitzroy Dearborn, London, in press.
2. Cech, T. R. (1990) Self-splicing of group I introns. *Ann. Rev. Biochem.,* **59**, 543.
3. Michel, F. and Ferat, J. L. (1995) Structure and activities of group II introns. *Ann. Rev. Biochem.,* **64,** 435.
4. Phizicky, E. M. and Greer, C. L. (1993) Pretransfer RNA splicing—variation on a theme or exception to the rule. *Trends Biochem. Sci.,* **18**, 31.
5. Gonzalez, T. N., Sidrauski, C., Dorfler, S., and Walter, P. (1999) Mechanism of non-spliceosomal mRNA splicing in the unfolded protein response pathway. *EMBO J.,* **18**, 3119.
6. Lopez, P. J. and Séraphin, B. (1999) Genomic-scale quantitative analysis of yeast pre-mRNA splicing: implications for splice-site recognition. *RNA,* **5**, 1135.
7. Ares, M., Jr., Grate, L., and Pauling, M. H. (1999) A handful of intron-containing genes produces the lion's share of yeast mRNA. *RNA,* **5**, 1138.
8. Fink, G. R. (1987) Pseudogenes in yeast? *Cell,* **49**, 5.
9. Wentz-Hunter, K. and Potashkin, J. (1995) The evolutionary conservation of the splicing apparatus between fission yeast and man. *Nucleic Acids Symp.,* **33**, 226.
10. Nilsen, T. W. (1998) RNA-RNA interactions in nuclear pre-mRNA splicing. In *RNA structure and function* (ed R. Simons and M. Grunberg-Manago), p279. Cold Spring Harbor Laboratory Press, Cold Spring Harbor, NY.
11. Long, M., de Souza, S. J., and Gilbert, W (1998) The yeast splice site revisited: new exon consensus from genomic analysis. *Cell,* **91**, 739.
12. Spingola, M., Grate, L., Haussler, D., and Ares, M., Jr. (1999) Genome-wide bioinformatic and molecular analysis of introns in *Saccharomyces cerevisiae*. *RNA,* **5**, 221.
13. Ruby, S. W. and Abelson, J. (1991) Pre-messenger RNA splicing in yeast. *Trends Genet.,* **7**, 79.
14. Burge, C. B., Tuschl, T., and Sharp, P. A. (1999) Splicing of precursors to mRNAs by the spliceosome. In *The RNA world,*2nd edn (ed R. F. Gesteland, T. R. Cech, and J. F. Atkins), p525. Cold Spring Harbor Laboratory Press, Cold Spring Harbor, NY.
15. Nilsen, T. W. (1994) RNA–RNA interactions in the spliceosome: unraveling the ties that bind. *Cell,* **78**, 1.
16. Staley, J. P. and Guthrie, C. (1998) Mechanical devices of the spliceosome: motors, clocks, springs, and things. *Cell,* **92**, 315.
17. Hodges, P. E., Plumpton, M., and Beggs, J. D. (1997) Pre-mRNA splicing factors in the yeast *Saccharomyces cerevisiae*. In *Eukaryotic mRNA processing* (ed A. R. Krainer), p213. IRL Press, Oxford.
18. Will, C. L. and Lührmann, R. (1997) Protein functions in pre-mRNA splicing. *Curr. Opin. Cell Biol.,* **9**, 320.
19. Brow, D. A. and Guthrie, C. (1988) Spliceosomal RNA U6 is remarkably conserved from yeast to mammals. *Nature,* **334**, 213.
20. Chanfreau, G., AbouElela, S., Ares, M., and Guthrie, C. (1997) Alternative 3'-end processing of U5 snRNA by RNase III. *Genes Dev.,* **11**, 2741.
21. Reddy, R. and Busch, H. (1988) Small RNAs: RNA sequences, structure, and modifications. In *Structure and function of major and minor small nuclear ribonucleoprotein particles* (ed M. L. Birnstiel), p1. Springer, Berlin.
22. Branlant, C., Krol, A., Ebel, J. P., Lazar, E., Bernard, H., and Jacob, M. (1982) U2 snRNA shares a structural domain with U1, U4 and U5 snRNAs. *EMBO J.,* **1**, 1259.

23. McPheeters, D. S., Fabrizio, P., and Abelson, J. (1989) *In vitro* reconstitution of functional yeast U2 snRNPs. *Genes Dev.*, **3**, 2124.

24. Jones, M. H. and Guthrie, C. (1990) Unexpected flexibility in an evolutionarily conserved protein–RNA interaction: genetic analysis of the Sm binding site. *EMBO J.*, **9**, 2555.

25. Siliciano, P. G., Kivens, W. J., and Guthrie, C. (1991) More than half of yeast U1 snRNA is dispensable for growth. *Nucl.Acids Res.*, **19**, 6367.

26. Mattaj, I. W. (1988) UsnRNP assembly and transport. In *Structure and function of major and minor small nuclear ribonucleoprotein particles* (ed M. L. Birnstiel), p100. Springer, Berlin.

27. Singh, R. and Reddy, R. (1989) γ-Monomethyl phosphate: a cap structure in spliceosomal U6 small nuclear RNA. *Proc. Natl. Acad. Sci. U.S.A.*, **86**, 8280.

28. Lund, E. and Dahlberg, J. E. (1992) Cyclic 2′,3′-phosphates and nontemplated nucleotides at the 3′ end of spliceosomal U6 small nuclear RNAs. *Science,* **255**, 327.

29. Mayes, A. E., Verdone, L., Legrain, P., and Beggs, J. D. (1999) Characterization of Sm-like proteins in yeast and their association with U6 snRNA. *EMBO J.*, **18**, 4321.

30. Salgado-Garrido, J., Bragado-Nilsson, E., Kandels-Lewis, S., and Séraphin, B. (1999) Sm and Sm-like proteins assemble in two related complexes of deep evolutionary origin. *EMBO J.*, **18**, 3451.

31. Fortner, D. M., Troy, R. G., and Brow, D. A. (1994) A stem/loop in U6 RNA defines a conformational switch required for pre-messenger RNA splicing. *Genes Dev.*, **8**, 221.

32. Beggs, J. D. (1995) Yeast splicing factors and genetic strategies for their analysis. In *Pre-mRNA processing* (ed A. I. Lamond), p79. R. G. Landes, Texas.

33. Vaidya, V. C., Seshadri, V., and Vijayraghavan, U. (1996) An extragenic suppressor of prp24-1 defines genetic interaction between PRP24 and PRP21 gene products of *Saccharomyces cerevisiae. Mol. Gen. Genet.*, **250**, 267.

34. Last, R. L., Maddock, J. R., and Woolford, Jr. J. L. (1987) Evidence for related functions of the *RNA* genes of *Saccharomyces cerevisiae. Genetics*, **117**, 619.

35. Abovich, N., Liao, X. L. C., and Rosbash, M. (1994) The yeast MUD2 protein—an interaction with PRP11 defines a bridge between commitment complexes and U2 snRNP addition. *Genes Dev.*, **8**, 843.

36. Frank, D., Patterson, B., and Guthrie, C. (1992) Synthetic lethal mutations suggest interactions between U5 small nuclear RNA and four proteins required for the 2nd step of splicing. *Mol. Cell. Biol.*, **12**, 5197.

37. Xu, D., Field, D. J., Tang, S-J., Moris, A., Bobechko, B. P., and Friesen, J. D. (1998) Synthetic lethality of yeast *slt* mutations with U2 small nuclear RNA mutations suggests functional interactions between U2 and U5 snRNPs that are important for both steps of pre-mRNA splicing. *Mol. Cell. Biol.*, **18**, 2055.

38. Neubauer, G., Gottschalk, A., Fabrizio, P, Séraphin, B., Lührmann, R., and Mann, M. (1997) Identification of the proteins of the yeast U1 small nuclear ribonucleoprotein complex by mass spectrometry. *Proc. Natl. Acad. Sci. U.S.A.*, **94**, 385.

39. Gottschalk, A., Tang, J., Puig, O., Salgado-Garido, J., Neubauer, G., Colot, H. V., Mann, M., Séraphin, B., Rosbash, M., Lührmann, R., and Fabrizio, P. (1998) A comprehensive biochemical and genetic analysis of the yeast U1 snRNP reveals five novel proteins. *RNA*, **4**, 374.

40. Caspary, F., Shevchenko, A., Wilm, M., and Séraphin, B. (1999) Partial purification of the yeast U2 snRNP reveals a novel yeast pre-mRNA splicing factor required for pre-spliceosome assembly. *EMBO J.*, **18**, 3463.

41. Bridge, E., Riedel, K. U., Johansson, B. M., and Pettersson, U. (1996) Spliced exons of adenovirus late RNAs colocalize with snRNP in a specific nuclear domain. *J. Cell Biol.*, **135**, 303.

42. Gottschalk, A., Neubauer, G., Banroques, J., Mann, M., Lührmann, R., and Fabrizio, P. (1999) Identification by mass spectrometry and functional analysis of novel proteins of the yeast [U4/U6.U5] tri-snRNP. *EMBO J.*, **18**, 4535.

43. Fromont-Racine, M., Rain, J-C., and Legrain, P. (1997) Towards a functional analysis of the yeast genome through exhaustive two-hybrid screens. *Nature Genet.*, **16**, 277.

44. Camasses, A., Bradado-Nilsson, E., Martin, R., Séraphin, B., and Bordonne, R. (1998) Interactions within the yeast Sm core complex: from proteins to amino acids. *Mol. Cell. Biol.*, **18**, 1956.

45. Siliciano, P. G., Brow, D. A., Roiha, H., and Guthrie, C. (1987) An essential snRNA from *S. cerevisiae* has properties predicted for U4, including interaction with a U6-like snRNA. *Cell*, **50**, 585.

46. Tollervey, D. and Mattaj, I. W. (1987) Fungal small nuclear ribonucleoproteins share properties with plant and vertebrate U snRNPs. *EMBO J.*, **6**, 469.

47. Riedel, N., Wolin, S., and Guthrie, C. (1987) A subset of yeast snRNAs contains functional binding-sites for the highly conserved Sm antigen. *Science*, **235**, 328.

48. Fabrizio, P., Esser, S., Kastner, B., and Lührmann, R. (1994) Isolation of *S. cerevisiae* snRNPs—comparison of U1 and U4/U6.U5 to their human counterparts. *Science*, **264**, 261.

49. Rymond, B. C. (1993) Convergent transcripts of the yeast *PRP38–SMD1* locus encode 2 essential splicing factors, including the D1 core polypeptide of small nuclear ribonucleoprotein particles. *Proc. Natl. Acad. Sci. U.S.A.*, **90**, 848.

50. Roy, J., Zheng, B. H., Rymond, B. C., and Woolford, J. L., Jr. (1995) Structurally related but functionally distinct yeast SmD core small nuclear ribonucleoprotein particle proteins. *Mol. Cell. Biol.*, **15**, 445.

51. Bordonne, R. and Tarassov, I. (1996) The yeast *SME1* gene encodes the homologue of the human E core protein. *Gene*, **176**, 111.

52. Rymond, B. C., Rokeach, L. A., and Hoch, S. O. (1993) Human snRNP polypeptide-D1 promotes pre-messenger RNA splicing in yeast and defines nonessential yeast Smd1p sequences. *Nucleic Acids Res.*, **21**, 3501.

53. Cooper, M., Parkes, V., Johnston, L. H., and Beggs, J. D. (1995) Identification and characterisation of Uss1p (Sdb23p): a novel U6 snRNA-associated protein with significant similarity to core proteins of small nuclear ribonucleoproteins. *EMBO J.*, **14**, 2066.

54. Hermann, H., Fabrizio, P., Raker, V. A., Foulaki, K., Hornig, H., Brahms, H., and Lührmann, R. (1995) snRNP Sm proteins share two evolutionarily conserved sequence motifs which are involved in Sm protein–protein interactions. *EMBO J.*, **14**, 2076.

55. Séraphin, B. (1995) Sm and Sm-like proteins belong to a large family: identification of proteins of the U6 as well as the U1, U2, U4 and U5 snRNPs. *EMBO J.*, **14**, 2089.

56. Plessel, G., Lührmann, R., and Kastner, B. (1997) Electron microscopy of assembly intermediates of the snRNP core: morphological similarities between the RNA-free (E.F.G) protein heteromer and the intact snRNP core. *J. Mol. Biol.*, **265**, 87.

57. Kambach, C., Walke, S., Young, R., Avis, J. M., de la Fortelle, E., Raker, V. A., Lührmann, R., Li, J., and Nagai, K. (1999) Crystal structures of two Sm protein complexes and their implications for the assembly of the spliceosomal snRNPs. *Cell*, **96**, 375.

58. Vidal, V. P. I., Verdone, L., Mayes, A. E., and Beggs, J. D. (1999) Characterization of U6 snRNA–protein interactions. *RNA*, **5**, 1470.

59. Pannone, B. K., Xue, D., and Wolin, S. L. (1998) A role for the yeast La protein in U6 snRNP assembly: evidence that the La protein is a molecular chaperone for RNA polymerase III transcripts. *EMBO J.*, **17**, 7442.

60. Achsel, T., Brahms, H., Kastner, B., Bachi, A., Wilm, M., and Lührmann, R. (1999) A doughnut-shaped heteromer of human Sm-like proteins binds to the 3'-end of U6 snRNA, thereby facilitating U4/U6 duplex formation *in vitro*. *EMBO J.,* **18**, 5789.

61. Caceres, J. F. and Krainer, A. R. (1997) Mammalian pre-mRNA splicing factors. In *Eukaryotic mRNA processing* (ed A. R. Krainer), p174. IRL Press, Oxford.

62. Hodges, P. E., Jackson, S. P., Brown, J. D., and Beggs, J. D. (1995) Extraordinary sequence conservation of the PRP8 splicing factor. *Yeast,* **11**, 337.

63. Fortes, P., Bilbao-Cortes, D., Fornerod, M., Rigaut, G., Raymond, W., Séraphin, B., and Mattaj, I. W. (1999) Luc7p, a novel yeast U1 snRNP protein with a role in 5' splice site recognition. *Genes Dev.,* **13**, 2425.

64. Gross, T., Richert, T., Mierke, C., Lützelberger, M., and Käufer, N. F. (1999) Identification and characterization of *srp1*, a gene of fission yeast encoding a RNA binding domain and a RS domain typical of SR splicing factors. *Nucleic Acids Res.,* **26**, 505.

65. Lützelberger, M., Gross, T., and Käufer, N. F. (1999) Srp2, an SR protein family member of fission yeast: *in vivo* characterization of its modular domains. *Nucleic Acids Res.,* **27**, 2618.

66. Burge, C. B. and Karlin, S. (1997) Prediction of complete gene structures in human genomic DNA. *J. Mol. Biol.,* **268**, 78.

67. Luo, H. B. R., Moreau, G. A., and Moore, M. J. (1999) The human Prp8 protein is a component of both U2- and U12-dependent spliceosomes. *RNA,* **5**, 893.

68. Krainer, A. R., Maniatis, T., Ruskin, B., and Green, M. (1984) Normal and mutant human β-globin pre-mRNAs are faithfully and efficiently spliced *in vitro*. *Cell,* **36**, 993.

69. Lin, R.-J., Newman, A. J., Cheng, S.-C., and Abelson, J. (1985) Yeast mRNA splicing *in vitro*. *J. Biol. Chem.,* **260**, 14780.

70. Séraphin, B. and Rosbash, M. (1989) Identification of functional U1 snRNA pre-messenger RNA complexes committed to spliceosome assembly and splicing. *Cell,* **59**, 349.

71. Rosbash, M. and Séraphin, B. (1991) Who's on 1st? The U1 snRNP-5' splice site interaction and splicing. *Trends Biochem. Sci.,* **16**, 187.

72. Yean, S-L. and Lin, R-J. (1991) U4 small nuclear RNA dissociates from a yeast spliceosome and does not participate in the subsequent splicing reaction. *Mol. Cell. Biol.,* **11**, 5571.

73. Yean, S-L. and Lin, R-J. (1996) Analysis of small nuclear RNAs in a precatalytic spliceosome. *Gene Expression,* **5**, 301.

74. Colot, H. V., Stutz, F., and Rosbash, M. (1996) The yeast splicing factor Mud13p is a commitment complex component and corresponds to CBP20 the small subunit of the nuclear cap-binding complex. *Genes Dev.,* **10**, 1699.

75. Lewis, J. D., Gorlich, D., and Mattaj, I. W. (1996) A yeast cap binding protein complex (yCBC) acts at an early step in pre-mRNA splicing. *Nucleic Acids Res.,* **24**, 3332.

76. Zhang, D. and Rosbash, M. (1999) Identification of eight proteins that cross-link to pre-mRNA in the yeast commitment complex. *Genes Dev.,* **13**, 581.

77. Puig, O., Gottschalk, A., Fabrizio, P., and Seraphin, B. (1999) Interaction of the U1 snRNP with nonconserved intronic sequences affects 5' splice site selection. *Genes Dev.,* **13**, 569.

78. Fortes, P., Kufel, J., Fornerod, M., Polycarpou-Schwartz, M., Lafontaine, D., Tollervey, D., and Mattaj, I. W. (1999) Genetic and physical interactions involving the yeast nuclear cap-binding complex. *Mol. Cell. Biol.,* **19**, 6543.

79. Abovich, N. and Rosbash, M. (1997) Cross-intron bridging interactions in the yeast commitment complex are conserved in mammals. *Cell,* **89**, 403.

80. Berglund, J. A., Chua, K., Abovich, N., Reed, R., and Rosbash, M. (1997) The splicing factor BBP interacts specifically with the pre-mRNA branchpoint sequence UACUAAC. *Cell,* **89**, 781.

81. Rain, J-C., Rafi, Z., Rhani, Z., Legrain, P., and Krämer, A. (1998) Conservation of functional domains involved in RNA binding and protein–protein interactions in human and *Saccharomyces cerevisiae* pre-mRNA splicing factor SF1. *RNA*, **4**, 551.
82. O'Day, C. L., Dalbadie-McFarland, G., and Abelson, J. (1996) The *Saccharomyces cerevisiae* Prp5 protein has RNA-dependant ATPase activity with specificity for U2 small nuclear RNA. *J. Biol. Chem.*, **271**, 33261.
83. Staley, J. P. and Guthrie, C. (1999) An RNA switch at the 5′ splice site requires ATP and the DEAD box protein Prp28p. *Mol. Cell*, **3**, 55.
84. Murray, H. L. and Jarrell, K. A. (1999) Flipping the switch to an active spliceosome. *Cell*, **96**, 599.
85. Kuhn, A. N., Li, Z., and Brow, D. A. (1999) Splicing factor Prp8 governs U4/U6 RNA unwinding during activation of the spliceosome. *Mol. Cell*, **3**, 65.
86. Raghunathan, P. L. and Guthrie, C. (1998) RNA unwinding in U4/U6 snRNPs requires ATP hydrolysis and DEIH-box splicing factor Brr2. *Curr. Biol.*, **8**, 847.
87. O'Keefe, R. T., Norman, C., and Newman, A. J. (1996) The invariant U5 snRNA loop 1 sequence is dispensable for the first catalytic step of pre-mRNA splicing in yeast. *Cell*, **86**, 679.
88. O'Keefe, R. T. and Newman, A. J. (1998) Functional analysis of the U5 snRNA loop 1 in the second catalytic step of yeast pre-mRNA splicing. *EMBO J.*, **17**, 565.
89. Dix, I., Russell, C. S., O'Keefe, R., Newman, A. J., and Beggs, J. D. (1998) Protein–RNA interactions in the U5 snRNP of *Saccharomyces cerevisiae*. *RNA*, **4**, 1675.
90. Teigelkamp, S., Newman, A. J., and Beggs, J. D. (1995) Extensive interactions of PRP8 protein with the 5′ and 3′ splice sites during splicing suggest a role in stabilization of exon alignment by U5 snRNA. *EMBO J.*, **14**, 2602.
91. Umen, J. G. and Guthrie, C. (1995) A novel role for a U5 snRNP protein in 3′ splice site selection. *Genes Dev.*, **9**, 855.
92. Burgess, S. M. and Guthrie, C. (1993) A mechanism to enhance mRNA splicing fidelity: the RNA-dependent ATPase Prp16 governs usage of a discard pathway for aberrant lariat intermediates. *Cell*, **73**, 1377.
93. Fabrizio, P., Laggerbauer, J. L., Lane, W. S., and Lührmann, R. (1997) An evolutionarily conserved U5 snRNP-specific protein is a GTP-binding factor closely related to the ribosomal translocase EF-2. *EMBO J.*, **16**, 4092.
94. Umen, J. G. and Guthrie, C. (1995) The second catalytic step of pre-mRNA splicing. *RNA*, **1**, 869.
95. Ansari, A. and Schwer, B. (1995) SLU7 and a novel activity, SSF1, act during the PRP16-dependent step of yeast pre-mRNA splicing. *EMBO J.*, **14**, 4001.
96. Raghunathan, P. L. and Guthrie, C. (1998) A spliceosomal recycling factor that reanneals U4 and U6 small nuclear ribonucleoprotein particles. *Science*, **279**, 857.
97. Wang, Y. and Guthrie, C. (1998) PRP16, a DEAH-box RNA helicase, is recruited to the spliceosome primarily via its nonconserved N-terminal domain. *RNA*, **4**, 1216.
98. Roy, J., Kim, K., Maddock, J. R., Anthony, J. G., and Woolford, J. L., Jr. (1995) The final stages of spliceosome maturation require Spp2p that can interact with the DEAH box protein Prp2p and promote step 1 of splicing. *RNA*, **1**, 375.
99. Li, B. J., Vilardell, J., and Warner, J. R. (1996) An RNA structure involved in feedback regulation of splicing and of translation is critical for biological fitness. *Proc. Natl. Acad. Sci. U.S.A.*, **93**, 1596.
100. Nandabalan, K. and Roeder, G. S. (1995) Binding of a cell-type-specific RNA splicing factor to its target regulatory sequence. *Mol. Cell. Biol.*, **15**, 1953.

101. Prasad, J., Colwill, K., Pawson, T., and Manley, J. L. (1999) The protein kinase Clk/Sty directly modulates SR protein activity: both hyper- and hypophosphorylation inhibit splicing. *Mol. Cell. Biol.,* **19**, 6991.

102. Gross, T., Lützelberger, M., Wiegmann, H., Klingenhoff, A., Shenoy, S., and Käufer, N. F. (1997) Functional analysis of the fission yeast Prp4 protein kinase involved in pre-mRNA splicing and isolation of a putative mammalian homologue. *Nucleic Acids Res.,* **25**, 1028.

103. Steinmetz, E. J. (1997) Pre-mRNA Processing and the CTD of RNA polymerase II: the tail that wags the dog. *Cell,* **89**, 491.

104. Tharun, S., He, W., Mayes, A. E., Lennertz, P., Beggs, J. D., and Parker, R. (2000) Sm-like proteins function in mRNA decapping and decay. *Nature,* **404**, 515.

105. Johnston, L. H. and Thomas, A. P. (1982) The isolation of new DNA-synthesis mutants in the yeast *Saccharomyces cerevisiae. Mol. Gen. Genet.,* **186**, 439.

106. Shea, J. E., Toyn, J. H., and Johnston, L. H. (1994) The budding yeast U5 snRNP PRP8 is a highly conserved protein which links RNA splicing with cell cycle progression. *Nucleic Acids Res.,* **22**, 5555.

107. Hwang, L. H. and Murray, A. W. (1997) A novel yeast screen for mitotic arrest mutants identifies *DOC1*, a new gene involved in cyclin proteolysis. *Mol. Biol. Cell,* **8**, 1877.

108. Ben-Yehuda, S., Russell, C. S., Dix, I., Beggs, J. D., and Kupiec, M. (2000) Extensive genetic interactions between *PRP8* and *PRP17/CDC40*, two genes involved in pre-mRNA splicing and cell cycle progression. *Genetics,* **154**, 61.

109. Chung, S., McLean, M. R., and Rymond, B. C. (1999) Yeast ortholog of the *Drosophila* crooked neck protein promotes spliceosome assembly through stable U4/U6.U5 snRNP addition. *RNA,* **5**, 1042.

110. Ben-Yehuda, S., Dix, I., Russell, C. S., McGarvey, M., Beggs, J. D., and Kupiec, M. (2000) Genetic and physical interactions between factors involved in both cell cycle progression and pre-mRNA splicing in *Saccharomyces cerevisiae. Genetics*, in press.

111. McDonald, W., Ohi, H. R., Smelkova, N., Frendewey, D., and Gould, K. L. (1999) Myb-related fission yeast cdc5p is a component of a 40S snRNP-containing complex and is essential for pre-mRNA splicing. *Mol. Cell. Biol.* **19**, 5352.

112. Brys, A. and Schwer, B. (1996) Requirement for SLU7 in yeast pre-mRNA splicing is dictated by the distance between the branchpoint and the 3′ splice site. *RNA,* **2**, 707.

113. Wang, C., Chua, K., Seghezzi, W., Lees, E., Gozani, O., and Reed, R. (1998) Phosphorylation of spliceosomal protein SAP155 coupled with splicing catalysis. *Genes Dev.,* **12**, 1409.

Notes

1. The literature on splicing is vast, and only a selection of references can be cited here. References for specific yeast proteins can be readily obtained from the Yeast Proteome Database at http://www.proteome.com, while references for *S. pombe* proteins can be obtained at http://www.sanger.ac.uk/Projects/S_pombe/FUNCAT/rna_proc.shtml

9 | Yeast nuclear transport

CHARLES N. COLE

1. Introduction

The presence of a nucleus is one of the defining characteristics of eukaryotic cells. Nuclei are bounded by a double-membraned nuclear envelope. This separation of the cytoplasm from the nucleus compartmentalizes the genetic material in an environment that is physically and biochemically distinct from the remainder of the cell. Therefore, cellular components that function within the nucleus but are produced in the cytoplasm must enter the nucleus, and RNA molecules required in the cytoplasm must exit the nucleus. The double-membraned nuclear envelope is perforated by nuclear pore complexes (NPCs), which are the only channels for nucleocytoplasmic transport. The importance of nucleocytoplasmic transport to eukaryotic organisms is underscored by the extremely high degree of conservation, from yeasts to mammals, in the nuclear transport machinery (for reviews see References 1–4). Rapid progress has been made in recent years in defining the proteins involved in nuclear transport and in elucidating their mechanisms of action. In addition to NPCs, nuclear transport requires specific signals in the macromolecules (cargo) to be transported, soluble receptor proteins that recognize these signals and move across the nuclear envelope with the cargo, Gsp1p/Ran, a small ras-like GTPase, and additional factors.

This chapter summarizes current understanding of nuclear transport in yeasts and describes the variety of methods that have been used to advance understanding. Studies using metazoans are described in some cases where they provided important insights into nuclear transport. Because the preponderance of work on yeast nuclear transport has been in the budding yeast, *Saccharomyces cerevisiae*, the term 'yeast' will be used to refer to studies with *S. cerevisiae*. *Schizosaccharomyces pombe* will be referred to by its species name or as fission yeast. The term 'yeasts' will be used to refer to both of these organisms.

2. Nuclear pore complexes

Each yeast cell contains approximately 150 NPCs, distributed at random over the surface of the nuclear envelope and with a density of about 15 NPCs/μm^2 (5). New NPCs are formed throughout the cell cycle. Very little is known about the assembly of NPCs or how their number is controlled.

2.1 Organization and structure of NPCs

NPCs are more than 20 times the size of a ribosome but are assembled from fewer components. When examined by high-resolution electron microscopy (EM), NPCs from all organisms display an eightfold rotational symmetry perpendicular to the plane of the nuclear envelope. Three-dimensional image reconstruction suggests that NPCs contain nuclear and cytoplasmic rings with spokes radiating inwardly to define a central channel through which macromolecular transport takes place. Figure 1A shows a structural diagram of a yeast NPC and Fig. 1B shows a thin-section electron micrograph of a portion of the yeast nuclear envelope, containing several NPCs. Metazoan NPCs have a mass of approximately 125 MDa. Yeast NPCs appear to have a slightly lower mass (6) and are slightly smaller in both thickness and diameter. However, all NPCs are thought to have a central channel with a diameter of about 25 nm, and peripheral channels approximately 9 nm in diameter. The central channel is the site of receptor-mediated, signal-dependent nuclear transport, while the smaller channels permit passive diffusion of small molecules. On the nuclear side is the nuclear basket or 'fishtrap'. This structure is composed of protein fibres that end in a smaller nuclear basket ring; in metazoans, additional protein fibers extend from the basket deep into the nuclear interior (7), but it is not yet known whether these are present in the much smaller yeast nucleus. On the cytoplasmic face of the NPC, short fibres extend into the cytoplasm. Figure 1C shows high-resolution electron micrographs of individual yeast NPCs, revealing the nuclear basket and cytoplasmic fibrils.

A central difference between fungal and metazoan cells is the closed mitotic cycle of fungi (see Chapter 5). While NPCs disassemble and reassemble during the cell cycle of metazoan cells, yeast NPCs remain intact. Yeast NPCs could be static structures and, once formed, might persist for many generations. Alternatively, NPCs might be dynamic, with components frequently associating and dissociating. Perhaps the central core portions of the NPC are stable while more peripheral components are dynamic.

2.2 Nucleoporins

The polypeptides comprising NPCs are called nucleoporins. Budding yeast has been an organism of choice for identification of nucleoporins (for a review see Reference 1). Two major factors contribute to this. First, many nucleoporins were identified genetically, either by screening temperature-sensitive (ts) mutants for defects in nuclear transport or by conducting synthetic lethal screens, starting with mutant alleles of previously identified nucleoporins. Second, the absence of a nuclear lamina in yeast facilitated purification of yeast NPCs (8). Proteins in the purified NPC fraction were separated and their identity determined by microsequencing. A related approach has been to isolate subcomplexes of the yeast NPC from strains producing a protein A-tagged nucleoporin (9). Various extraction procedures permit disruption of NPCs so that subcomplexes are released. Proteins in complexes with the tagged nucleoporin of interest can be isolated and either sequenced or identified immunochemically.

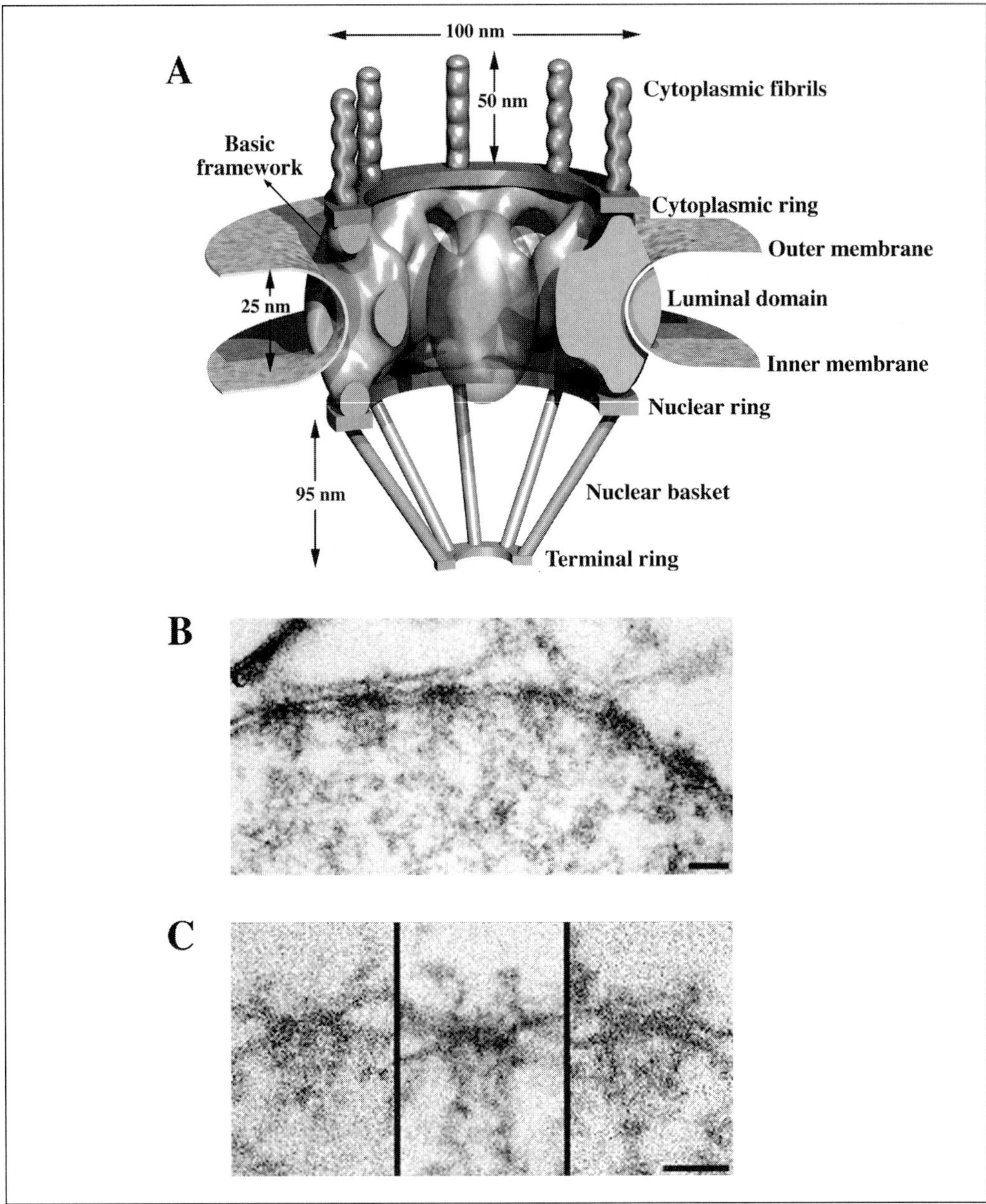

Fig. 1 (A) Diagram of the yeast nuclear pore showing structural components and dimensions (yeast nuclear pores are about 15% smaller in their linear dimensions than vertebrate nuclear pores). (B) Cross-section through the nuclear envelope of a yeast cell reveal the double-membraned nuclear envelope and NPCs. Scale bar, 100 nm. (C) Enlargements of thin-section electron micrographs of individual NPCs showing peripheral structures, such as the cytoplasmic fibrils and the nuclear baskets. Scale bar, 100 nm. Panel A was adapted from a review in *The International Review of Cytology*, 1995, **162B**, 225–255 (Reference 138), by copyright permission of Academic Press. Panels B and C were adapted from *The Journal of Cell Biology* 1998, **143,** 577–588 (Reference 13), by copyright permission of The Rockefeller University Press. The composite figure was provided by B. Fahrenkrog and N. Panté.

Table 1 Nucleoporins of *S. cerevisiae*

Protein name*	Other names	Size (kDa)	Essential at 23°C	Essential at 37°C	Repeats	Other features	Import defect	mRNA export defect	NE abnormal	NPC clusters	Closest metazoan nucleoporin	*S. pombe* homologue
Nup1p		130	N	Y	FXFG		Y	Weak	Y			
Nup2p		95	N	N	FXFG	Ran-binding domain	N	N	N	N	NUP358	
Nsp1p		110	Y	Y	FXFG	Coiled-coil	Y	N	Y		p62	
Nup49p	Nsp49p	49	Y	Y	GLFG		Some alleles	Some alleles	N		p54	
Nup53p		53	N	N	No		Y					
Nup57p	Nsp57p	57	Y	Y	GLFG	Coiled-coil					p58	
Nup59p	Asm4p	59	N	N	No		N					
Nup82p		82	Y	Y	No	Coiled-coil	N	Y	N		NUP84/NUP88	
Nup84p		84	N	Y	No		N	Y	Y	N	NUP107	
Nup85p	Rat9p	85	N	Y	No		N	Y	N	Y		
Nic96p		96	Y	Y	No	Coiled-coil	Y		Y		NUP93	Npp106
Nup100p	Nsp100p	100	N	N	GLFG	RNP motif	N	N	N		NUP98	
Nup116p	Nsp116p	116	N	Y	GLFG	RNP motif	Y	Y	Herniations	NPCs sealed	NUP98	
Nup120p	Rat2p	120	N	Y	No		N	Y	Y	Y		
Nup133p	Rat3p	133	N	Y	No		N	Y	Y	Y		
Nup145Np	Rat10p	60	N	N	GLFG	RNP motif	N	N	Y	Y		
Nup145Cp	Rat10p	85	N	Y	No		N	Y	Y	Y		
Pom152p		152	N	N	No	Transmembrane	N	N	N	N	gp210?	
Nup157p		157	N	N	No		N	N	N			
Nup159p	Rat7p	159	Y	Y	XXFG	Coiled-coil	N	Y	N	at 23°C	CAN/NUP214	
Nup170p		170	N	N	No		N	N	N		NUP155	
Nup188p		188	N	N	No		N	N	Y			
Nup192p	YJL039c	192			No						NUP205	
Seh1p†		39	N	Y	No	WD repeats; KO cs‡	Very weak	N	N			
Ndc1p		74			No	Transmembrane					gp210?	
Npl4p		66	Y	Y	Degenerate		Y	Very weak	Herniations	Y		
Gle2p		41	Y	Y	No		N	Y	Y	At 37°C		Rae1
Rip1p	Nup42p	42	N	N	XXFG		N	N	N	N	hRIP/Rab	

*Numbers were assigned to nucleoporins based on molecular weight or identification through genetic screens
†Seh, sec13 homologue; ‡KO cs, knockout cold-sensitive.

Based on their amino acid sequences, nucleoporins fall into two major classes. Approximately half contain repeats of distinctive short peptides, called 'nucleoporin repeats'. Repeat-containing nucleoporins were initially identified in mammalian cells through the use of antibodies prepared against rat liver nuclear envelopes (10, 11). These antibodies were subsequently used to identify yeast nucleoporins. Three types of amino acid repeats have been identified: GLFG, FG, and FXFG. However, in many nucleoporins, some of the repeats are closely related but not identical. Repeats are separated by non-conserved spacers. Examination of the yeast genome sequence indicates that all repeat-containing nucleoporins have been identified but nucleoporins that lack repeats continue to be identified as of early 1999.

2.2.1 Essential and non-essential nucleoporins

Several analyses are commonly conducted when studying nucleoporins. These include essentiality, functionality, and synthetic lethality. Approximately one-third of yeast nucleoporins are not essential at either 23°C or 37°C (Table 1). Another one-third confer a ts phenotype when disrupted. The final one-third are essential under all conditions. Nup145p is novel in that it is cleaved proteolytically into two nucleoporins, Nup145Np and Nup145Cp (12).

2.2.2 Nucleoporin localization and NPC clustering

Determination that a protein is a nucleoporin requires that it be localized to the nuclear periphery by indirect immunofluorescence or immunogold EM. When the morphology of the NPC is well preserved, immunogold EM also permits localization of nucleoporins to specific structures or domains of NPCs (13). Fig. 2A shows immunogold electron micrographs that indicate the locations of five yeast nucleoporins. The diagram in Fig. 2B summarizes where several yeast nucleoporins are located. At present, 28 nucleoporins of *S. cerevisiae* have been identified, and are listed in Table 1 (for a review see Reference 1). It is likely that most yeast nucleoporins are now known. Only two nucleoporins from *S. pombe* have been identified.

The distribution of NPCs is frequently altered in strains lacking a non-essential nucleoporin, or carrying a ts allele of an essential one. Clustering of NPCs to one or a few regions of the nuclear envelope is often observed (14). Fig. 3 shows the distribution of NPCs in wild-type cells (panel A), in cells that lack Nup120p (panel E), or where one of the two coiled-coil regions of Nup159p is absent due to truncation of the protein (panel C). The degree and patterns of clustering vary among different strains that have clustered NPCs. This is seen most clearly in thin-section electron micrographs (panels G–J). In some strains there are also dramatic perturbations of the nuclear envelope. In other mutant strains herniations in the nuclear envelope are observed and membrane seals cover and block transport through NPCs (15). In most cases, clustering is constitutive. This means that the NPCs within clusters remain sufficiently functional so that cells can grow at the permissive temperature. In two cases, clustering patterns are altered dramatically by temperature shifts (16, 17).

An interesting series of experiments to examine NPC clustering involved mating two yeast cells, one with clustered NPCs, due to disruption of *NUP133*, and the other

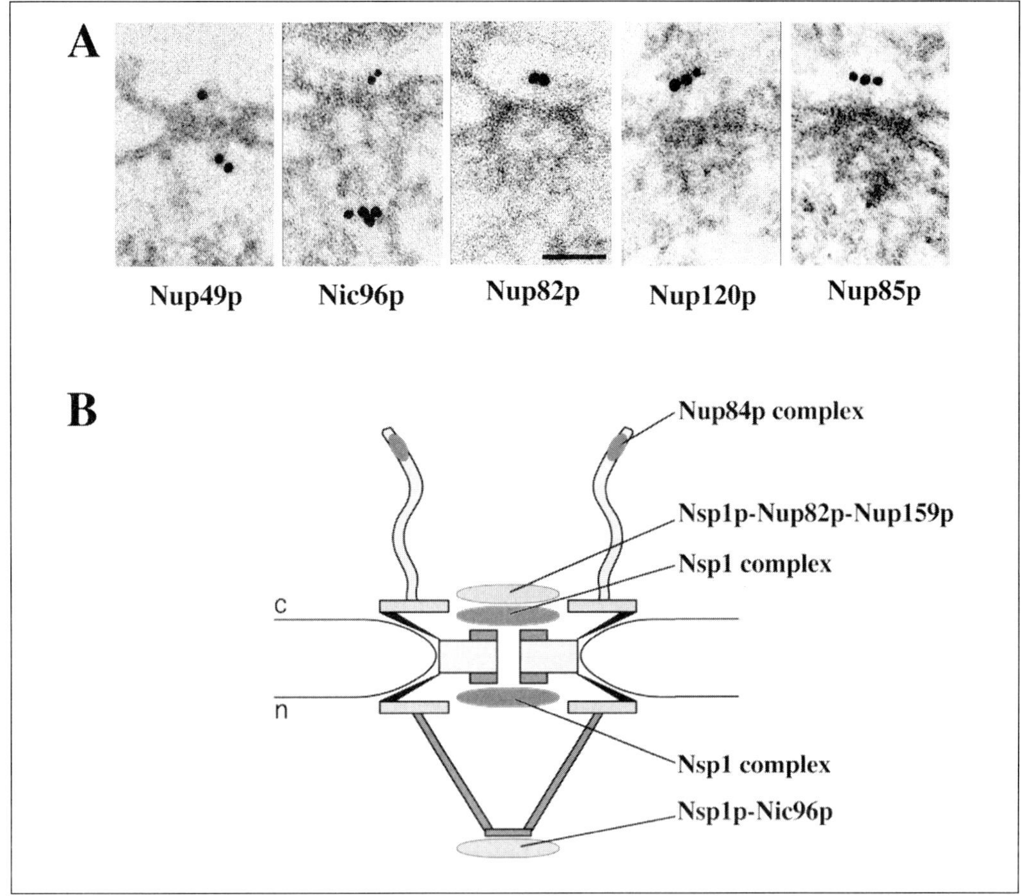

Fig. 2 Immunogold EM permits individual nucleoporins to be localized within NPCs. Immunolocalization was performed in Triton X100-extracted spheroplasts from yeast strains expressing different protein A-tagged nucleoporin. A polyclonal anti-protein A antibody directly conjugated to 8-nm colloidal gold was used for pre-embedding labelling. Scale bar, 100 nm. n, nucleus; c, cytoplasm. (A) Each panel shows the localization of a different nucleoporin. (B) Diagram summarizing the immunolocalization of several yeast nucleoporins. Adapted from *The Journal of Cell Biology* 1998, **143,** 577–588 (Reference 13); provided by B. Fahrenkrog and N. Panté, and used by copyright permission of The Rockefeller University Press.

expressing a GFP-tagged form of Nup133p (18, 19). The GFP-tagged protein entered clustered NPCs rapidly. In addition, NPCs could be seen to move very rapidly within the nuclear envelope, indicating that they are highly mobile within the envelope and are not anchored to cellular structures that would prevent mobility.

Membrane-spanning nucleoporins have been identified in both yeast and metazoan organisms. Yeast Pom152p belongs to this class (20). This nucleoporin is not essential, suggesting that other membrane-spanning nucleoporins are probably present. One candidate is Ndc1p, a component of spindle pole bodies also found in NPCs (21).

A sizeable fraction of yeast nucleoporins contain sequences predicted to form coiled-

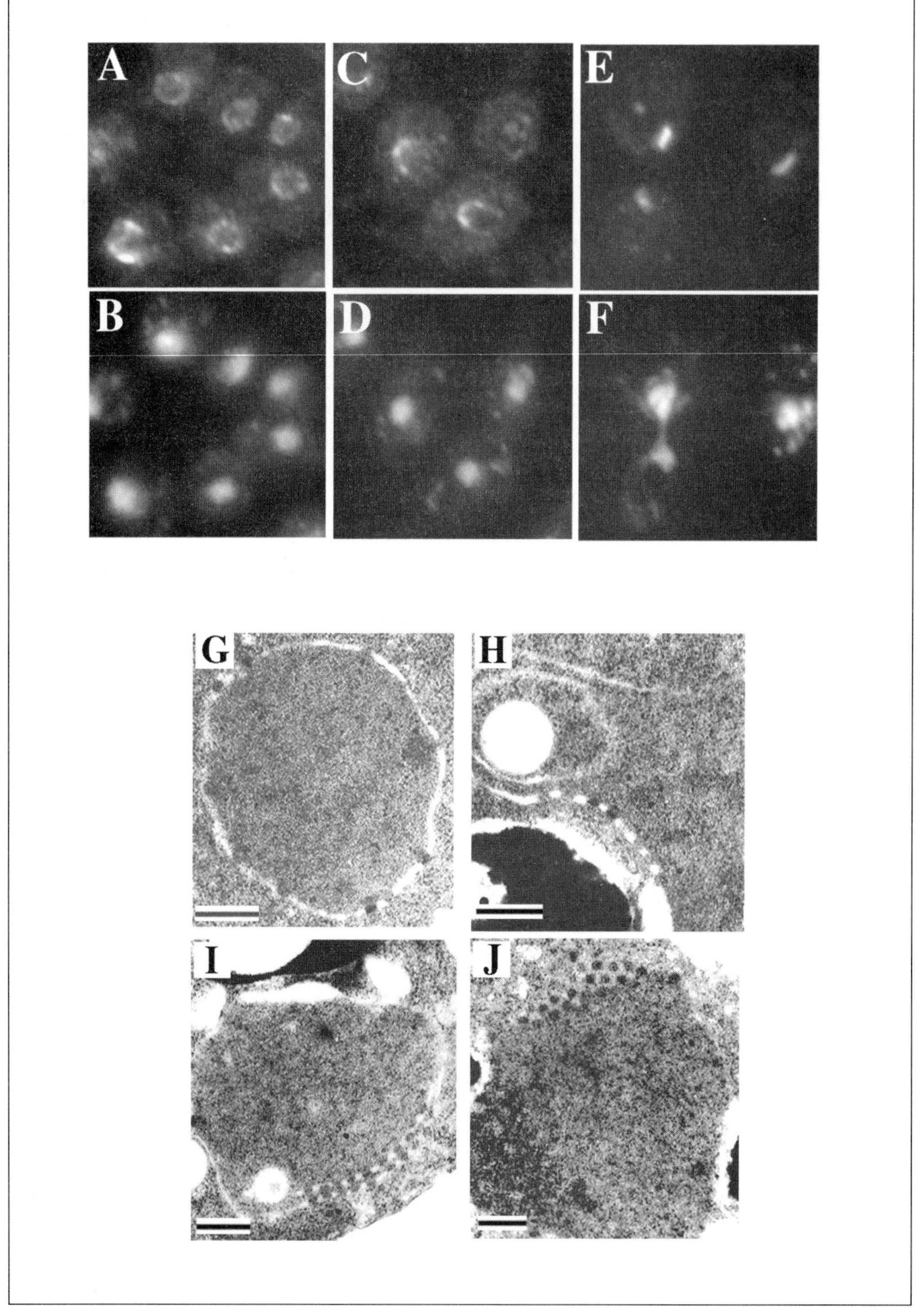

Fig. 3 Clustering of NPCs visualized by indirect immunofluorescence using an anti-nucleoporin antibody. Standard indirect immunofluorescence was performed using the RL1 monoclonal antibody raised against rat liver nuclear envelopes (11), which also recognizes yeast NPCs. (A) Wild-type cells; (C) *nup159-1* cells; (E) *nup120Δ* cells. Panels B, D, and F indicate the location of DNA within the same cells as in panels A, C, and E respectively. (G–J) Thin-section electron micrographs reveal clustering of NPCs. Scale bars, 250 nm. (G) Distribution of NPCs in a wild-type cell. (H) Truncation of Nup159p results in clustering, with all NPCs remaining in the plane of the nuclear envelope. (I) The absence of Nup85p also results in clustering of NPCs within the plane of the nuclear envelope. In addition, nuclear envelope abnormalities are also seen. Two layers of nuclear envelope are seen in some areas and NPCs present in different layers are somehow aligned with one another. (J) The absence of Nup120p results in clustering, with most or all of the NPCs in one region of the nuclear envelope, often within a protuberance. Figure provided by C. Snay-Hodge, Dartmouth Medical School, Hanover, NH.

---

coils, a common motif mediating protein–protein interactions. Nup100p, Nup116p, and Nup145Np all contain an RNP motif, and this motif from Nup145p has been shown to bind RNA homopolymers *in vitro* (22). Many RNA-binding proteins contain multiple RNP motifs, raising the possibility that nucleoporin RNP motifs may mediate a direct interaction between RNA and nucleoporins.

2.2.3 Structural and functional roles of nucleoporins

Nucleoporins can be involved in one or more of the following: protein import, protein export, RNA export, NPC assembly, or NPC stability. Assays have been developed to examine most of these functions.

In situ hybridization assays are used to determine whether mutant strains are able to maintain export of mRNA. Indirect immunofluorescence, or use of GFP-tagged reporter proteins carrying a nuclear localization signal (NLS) or nuclear export signal (NES), is used to examine nuclear protein transport. Assays have been developed recently to examine tRNA export (23) and rRNA/ribosome export (24), but only a few mutant strains have been screened for their ability to export tRNA or rRNA.

A role in NPC assembly can be inferred by determining how the number of NPCs per cell changes following a shift of a ts strain to 37°C. In strains carrying a mutant allele of *NIC96* or *NSP1*, cells gradually reduce their growth rate and eventually cease dividing (after 6–12 h) (25, 26). Freeze-fracture EM indicates that these mutant strains cease NPC assembly following the temperature shift. Direct assays to measure NPC stability have not yet been developed.

Disruption or mutation of many nucleoporins results in an mRNA export defect. A defect in nuclear protein import occurs in fewer mutant strains. In none of the nucleoporin mutant strains examined to date has any defect in protein export been observed. The reasons for these different nucleoporin requirements are uncertain. Poly(A)$^+$ RNA is exported as a complex containing RNA-binding proteins, and these hnRNPs are much larger than even the largest nuclear protein. Therefore, mRNA (and ribosome) export may be more complex and more sensitive to disruptions of NPC structure than is protein import or protein export. RNA export could well involve more interactions between the RNA–protein complex (hnRNP) and nucleoporins, perhaps requiring the activity of more receptors and NPC-associated transport factors.

Another variable among mutant strains concerns the speed with which transport defects develop when cells are shifted to the non-permissive temperature. In some strains, transport defects can be detected within a few minutes of a shift to 37°C, and accumulation of protein or RNA occurs synchronously in all cells of the population. This suggests that the mutated nucleoporin normally plays a direct role in RNA export. In other cases, transport defects appear over an extended period of time, and only some of the cells show an export block, suggesting an indirect role in transport for these nucleoporins.

Nucleoporins are expected to perform multiple functions. Some are likely to play primarily or exclusively structural roles. Others are likely to perform primarily functional roles by interacting directly with transport factors. Some nucleoporins may have important roles in both NPC structure and function. Some may function primarily in NPC assembly by ensuring that other nucleoporins are assembled properly into NPCs, while others may play critical structural roles by interacting with the nuclear envelope.

2.2.4 Nucleoporin subcomplexes

Biochemical approaches have been used to identify subcomplexes of the NPC. Hurt's laboratory identified a subcomplex containing Nup84p, Nup85p, Nup120p, Nup145Cp, Seh1p, and a minor fraction of Sec13p (27). Sec13p plays an essential role in secretion, and the function it may play within the NPC remains uncertain. Interestingly, none of these nucleoporins (except Sec13p) is essential, although deletion of more than one generally results in synthetic lethality.

A second subcomplex consists of Nsp1p, Nup159p, and Nup82p (28). Each of these nucleoporins is essential. Upon a shift to 37°C, Nup159p is lost from nuclear pores in strains expressing either Nup159p or Nup82p whose coiled-coil domains have been truncated. Nup159p and Nup82p have been localized by immunogold EM to the cytoplasmic face of the NPC (29, 30), and the data suggest that Nup159p is the more peripheral of these proteins.

A third subcomplex contains Nsp1p, Nic96p, Nup49p, and Nup57p (9). This complex has been localized to both sides of the central transport channel of the NPC (13). In addition, Nup188p, Pom152p, and Nic96p can be co-immunoprecipitated, suggesting that they are adjacent within NPCs. Together, Nic96p, Pom152p, Nup188p, Nup157p, Nup170p, and Nup192p comprise about 25% of the mass of the yeast NPC (31). Based on immuno-EM, these nucleoporins appear to be part of the central core of the NPC attached to the nuclear envelope.

Fewer nucleoporins from metazoan cells have been identified but several of those identified are related to yeast nucleoporins and may be functional homologs (Table 1). However, in only one case has it been possible to substitute a mammalian nucleoporin for a yeast nucleoporin: mammalian NUP155 can replace yeast Nup170p (31). Nevertheless, analysis of subcomplexes of metazoan NPCs also suggests that specific NPC subcomplexes are conserved between yeasts and metazoans. An important question about nucleoporin subcomplexes is whether they are assembly intermediates, as they could arise as a consequence of preferred NPC breakage patterns.

2.2.5 Assembly of NPCs

At present we have very limited knowledge about how NPCs are assembled. With only 150 NPCs per cell, yeast cells must produce on average only a single new NPC per minute during growth.

Bucci and Wente developed a genetic screen to identify mutants defective in this process (32). Cells expressing GFP-Nup49p were mutagenized, sorted using a fluorescence-activated cell sorter (FACS), and cells that failed to incorporate GFP-Nup49p into NPCs were analysed further. Some of the strains identified carried mutations of Nup57p, suggesting that assembly of Nup49p into NPCs requires Nup57p. This approach has the potential to shed light on the assembly pathway for NPCs, and could also identify any non-nucleoporin assembly factors involved in NPC biogenesis.

3. Protein import

Over the past several years considerable efforts from many laboratories have gone into defining the requirements and mechanisms for nuclear protein import. Genetic, cytological, and biochemical approaches have contributed in important ways to our current understanding. Studies of protein import in yeast paralleled studies in metazoan organisms.

3.1 Early studies on protein import

Early studies on protein import were conducted primarily in *Xenopus* oocytes. Feldherr and colleagues showed that colloidal gold particles are imported into nuclei by a diffusional process, since import occurs as well at 4°C as at 25°C and appears to have no energy or ATP requirement (33). Larger colloidal gold particles also enter nuclei but only when they are coated with a nuclear protein or NLSs. Import appears to be active or facilitated, as it requires physiological temperatures and ATP. If ATP is omitted or if the studies are done at 4°C, coated colloidal gold particles accumulate at the nuclear periphery. These and other studies suggest that nuclear proteins first bind to the cytoplasmic side of the NPC in an energy- and temperature-independent step, and are subsequently translocated through the NPC (34).

3.2 Defining the signals for protein import

Standard approaches in the 1980s, using molecular biology and recombinant DNA technology, showed that most nuclear proteins contain one of two types of closely related NLSs. Unipartite or simple NLSs were defined first in *Simain virus 40* (SV40) large T antigen, whose NLS has the sequence PKKKRKV (35). Bipartite NLSs, defined in nucleoplasmin, consist of two shorter clusters of basic residues (two or three basic amino acids) separated by approximately 10 amino acids (36). These signals can be located anywhere within a protein as long as protein folding or complex formation does not render them inaccessible. By injecting radiolabelled protein into oocytes, it was shown that the import process was saturable and could be competed by other macromolecules that also carried NLSs. This is another property that distinguishes facilitated or active transport from diffusion. These signals can also direct reporter proteins into yeast nuclei.

A major advance in studying protein import came from the development of '*in vitro*' systems. Cells whose plasma membranes have been perforated, but whose nuclear envelopes are intact, can be produced by treating metazoan cells with the detergent digitonin (37), or by subjecting yeast cells to cycles of freezing and thawing (38). When proteins bearing NLSs or gold particles coated with NLS peptides are added to semi-intact cells, they are imported into nuclei. Import requires cytosol from the permeabilized cells and physiological temperatures, and is inhibited by non-hydrolysable analogues of ATP (39).

3.3 Identification of the NLS receptor

Fractionation of the cytosol was performed to identify the components involved in protein import. In this way, the receptor that recognizes and binds NLSs was discovered. This receptor is a heterodimer of 60- and 95-kDa proteins (40, 41). Since they were identified in many laboratories, there are several names for each. In budding yeast, the 60-kDa protein is called Srp1p because it was initially discovered as a suppressor of a ts allele affecting *RNA polymerase I* (42). Srp1p/importin α recognizes NLSs directly. The crystal structure of mammalian importin α complexed with an NLS peptide was recently solved and indicates how this protein can recognize diverse single and bipartite NLSs (43).

The 95-kDa importin β subunit in yeast, Kap95p, was identified by biochemical fractionation and through a synthetic lethal screen starting with a ts allele of *RNA1* (44–46). As would have been expected, yeast strains carrying ts alleles affecting either Srp1p or Kap95p display strong, rapidly-occurring defects in nuclear protein import following a shift to the non-permissive temperature (46, 47). Both proteins are normally found in the cytoplasm and are concentrated at NPCs. Kap95p appears to target the trimeric α/β/cargo complex to NPCs by interacting directly with nucleoporins, most likely through nucleoporin repeats (44, 48). A recent review, based mainly on studies in metazoans, summarizes the details of the interactions and mechanisms of function of importin α and importin β (49).

3.4 The Gsp1p/Ran GTPase system

Table 2 lists the soluble non-receptor factors that appear to play roles in nuclear transport in yeast.

3.4.1 Gsp1p/Ran

A central player in nucleocytoplasmic exchange is the small ras-like GTPase, Gsp1p. In yeast, this protein was identified genetically through a screen for suppressors of ts mutations in the *prp20* gene (50). The metazoan homologue of Gsp1p, called Ran, was first noticed as an open reading frame with homology to mammalian ras, and is

Table 2 *Nuclear transport factors*

Protein name	Size (kDa)	Essential?	Structural features	Functions in					Metazoan homologues	*S. pombe* homologue
				Protein import	Protein export	mRNA export	tRNA export	rRNA export		
Gsp1p	25	Y		Y	Y	Y	Y	Y	Ran	Spi1
Gsp2p	25	N		N	N	N	N	N	Ran	Spi1
Rna1p	46	Y		Y	Y	Y	Y	Y	RanGAP1	Rna1
Prp20p	53	Y		Y	Y	Y	Y	Y	RCC1	Pim1
Yrb1p	23	Y	RanBD	Y		N			RanBP1	
Yrb2p	36	CS	RanBD, XXFGs	N	CS	N				Hap1
Ntf2p	14	Y	RanBD	Y	N	N			p10 (p15,NTF2)	
Rss1p/Gle1p/Brr3p	62	Y	Coiled-coil; NES?	N	N	Y			GLE1	
Mex67p	67	Y	RNA binding	N		Y			TAP	
Mtr2p	21	Y				Y				
Dbp5p/Rat8p	54	Y	DEAD-box prot	N	N	Y			mDEAD5, hDBP5	

RanBD, Ran binding domain; CS, cold-sensitive

called RAN/TC4. Its critical role in nuclear transport was demonstrated using biochemical approaches at about the same time that yeast Gsp1p was identified (39, 51). Gsp1p/Ran is a highly conserved protein found in all eukaryotic organisms examined for its presence. The *S. pombe* homologue of Gsp1p/Ran is Spi1.

Like other small GTPases, Gsp1p/Ran goes through a cycle of GTP hydrolysis followed by exchange of bound GDP for GTP (for a review see Reference 52). The importance for nuclear transport of GTP hydrolysis by Ran/Gsp1p was first shown *in vivo* in *S. cerevisiae* (53). Based on its homology to ras, a mutant of Gsp1p was prepared (Gsp1pG21V) that was locked in the GTP-bound state (no GTPase activity). Overexpression of Gsp1pG21V resulted in cell death following inhibition of protein import and RNA export (53).

As with other small GTPases, Gsp1p/Ran has very low GTPase activity when acting alone (54), but a GTPase-activating protein (RanGAP1; Rna1p in *S. cerevisiae* and *S. pombe*) increases its hydrolysis rate by several orders of magnitude. Yeast Prp20p (RCC1 in metazoans, Pim1 in *S. pombe*) is the factor that exchanges bound GDP with GTP following GTP hydrolysis (55).

3.4.2 Rna1p, the GTPase-activating protein for Gsp1p

The yeast *RNA1* gene was discovered in the 1960s in a screen for mutants defective in protein synthesis. Subsequent analysis showed that the mutant allele, *rna1-1*, had pleiotropic defects in RNA metabolism, including a defect in export of poly(A)$^+$ RNA (56, 57). The gene was cloned and antibodies prepared, which showed that Rna1p is located in the cytoplasm and enriched at the nuclear periphery (58).

Additional studies demonstrated that *rna1-1* cells are also defective in nuclear protein import (59). Since defects in import might lead to defects in RNA export, and vice versa, these studies could not determine whether or not Rna1p participated directly in both RNA export and protein import. *In vitro* assays indicate a direct role for Rna1p in protein import (59). Although assays to examine RNA export *in vitro* have not been developed, it is believed that Rna1p plays an essential function in all types of nuclear transport.

3.4.3 Prp20p, the guanine nucleotide exchange factor for Gsp1p

Prp20p was first identified in a screen for mutants with defects in pre-mRNA processing. The same gene was also identified, and called *SRM1*, in a screen for mutants that could suppress a mating defect caused by mutation of *STE3* (60). Finally, the gene was identified (and called *MTR1*) using an *in situ* hybridization screen to identify mutants defective in RNA export (61).

Although genetic approaches are rarely possible using tissue culture cells, the mammalian homologue of Prp20p was identified genetically in Chinese hamster ovary cells because of its ability to complement the ts defect of tsBN2 cells. When tsBN2 cells are shifted to the non-permissive temperature, premature condensation of chromosomes occurs (62, 63). Therefore, the cloned gene was called *RCC1* (regulator of chromosome condensation). This gene was also identified genetically in *S. pombe* where mutation of the gene caused premature induction of mitosis (providing its name, Pim1) (64). Immunohistochemical and fractionation studies indicated that the RCC1 protein (and homologues from other organisms) is restricted to the nucleus and associated with chromatin.

Because of the diversity of phenotypes associated with mutation of the gene and its identification in diverse genetic screens, the actual function of Prp20p remained uncertain for quite some time. Biochemical studies demonstrated that RCC1 was an exchange factor for Ran (55). Discovery of the key role Ran plays in nuclear transport suggested that its regulators, RanGAP1 and RCC1, would also be involved directly in nuclear transport. This was consistent with defects in protein import and RNA export seen in *prp20* cells and tsBN2 hamster cells (61, 65, 66).

3.4.4 Are components of the Ran/Gsp1p system involved in processes other than nuclear transport?

Several studies have suggested that Ran/Gsp1p plays roles in processes other than nuclear transport. Ran and its regulators were identified in *S. pombe* because of their effects on the cell cycle (for a review see Reference 67). The *srm1-1* allele of *PRP20* was isolated in a screen for genes able to suppress a mating defect (60). The tsBN2 cell line was isolated because of its defects in chromsome condensation and cell cycle progression (63). Mutation of these genes in yeasts and metazoans causes defects in multiple aspects of RNA metabolism. Complex genetic interactions among these genes have been observed in both budding and fission yeast. Are all of these other phenotypes consequences of defects in nuclear transport? We do not yet know. However, it is easy to see how defects in transport could have diverse consequences for the cell. More experimentation will be required to resolve this important question.

3.4.5 Gsp1p/Ran-binding proteins

Yeasts and metazoans contain several proteins that interact with Ran/Gsp1p. These proteins share a related Ran-binding domain and include Kap95p/importin β. Yrb1p (yeast Ran-binding protein 1) is a homologue of metazoan RanBP1. Yeast strains carrying a mutant allele of *yrb1* are defective in both protein import and RNA export, but not in protein export (68). Yeast Yrb1p and metazoan RanBP1 stimulate hydrolysis of GTP bound to Ran and catalysed by Rna1p/RanGAP1. Yrb1p is located in the cytoplasm and enriched near the nuclear envelope, consistent with an interaction with Rna1p.

A second yeast protein with a Ran-binding domain is Yrb2p (69). Like Yrb1p, Yrb2p binds only to Gsp1p–GTP. This protein is not essential, but strains lacking Yrb2p are cold sensitive. At lower temperatures, defects were observed in protein export but not in protein import or RNA export. Additional studies will be required to determine Yrb2p's *in vivo* function.

Ntf2p is a 15-kDa Ran-binding protein first identified biochemically as one of the components required for protein import in digitonin-permeabilized mammalian cells (70, 71). Unlike Yrb1p and Yrb2p, Ntf2p interacts with RanGDP and not with RanGTP. In budding yeast, Ntf2p was identified by cloning the gene mutated in a strain ts for protein import (72). This protein is found at the nuclear rim and interacts with several nucleoporins. In metazoan systems, NTF2 has been shown to mediate import into the nucleus of RanGDP (73, 74), and most likely plays the same role in yeasts as in metazoan systems.

In both yeasts and metazoans, Ran-binding domains are found in some nucleoporins. Yeast Nup2p contains a Ran-binding domain (75) while the related metazoan Nup358 contains four (76). Since Nup2p is not essential, and cells disrupted for it are phenotypically wild type, the importance of its Ran-binding domain remains uncertain.

3.5 Other receptors for protein import

Yeast and metazoan cells contain several proteins that are related to Kap95p, primarily in their Ran-binding domains (for reviews see References 77–79). All interact with Gsp1p/Ran and all appear to be directly involved in nuclear transport. Those of *S. cerevisiae* are listed in Table 3, along with the identity of metazoan homologues, if known. In contrast to heterodimeric Srp1p/Kap95p, these other receptors are believed to function as monomers, interacting directly with cargo. Those whose cargo has been identified recognize a much smaller number of cargoes than does Srp1p, and this may explain the lack of a requirement for an adaptor. Most function in protein import. Interestingly, many are not essential, suggesting that different receptors have overlapping cargo-recognition ability.

Mammalian transportin was the second protein import receptor identified (80). Its cargo is hnRNPA1, one of approximately 20 very abundant proteins that bind hnRNA in the nucleus. hnRNP proteins (including yeast Npl3p) are thought to participate in RNA processing and to package the mRNA for export and subsequent function (for a review see Reference 81). Some mammalian hnRNP proteins are removed before export, while most accompany the mRNA through the nuclear pore, are removed following transport, and shuttle back into the nucleus. Kap104p is the yeast homologue

Table 3 Receptors for nuclear transport

Protein	Alternative names	Size (kDa)	Essential?	Import or export	Cargo	Metazoan homologues
Kap95p	Rsl1p	95	Y	Import	NLS proteins	Importin β/karyopherin β
Srp1p		60	Y	Import	NLS proteins	Importin α/karyopherin α
Kap104p		104	N	Import	RNA-binding proteins	Transportin 1
Kap121p	Pse1p	121	N	Import	Pho4p	
Kap123p	Yrb4p	123	N	Import	Ribosomal proteins	
Sxm1p		108	N	Import	Lhp1p (La protein)	
Nmd5p		102	N	Import	Hog1p	
Mtr10p	Kap111p	111	N	Import	Npl3p	
Lph2p		116	N			
Sok1p		101	N			
Crm1p	Xpo1p	132	Y	Export	NES proteins	CRM1
Cse1p	Hrc135p	109	Y	Export	Srp1p	CAS
Los1p		127	N	Export	tRNA	Exportin-t; Exportin-tRNA
Msn5p	Ste21p	142	N	Export	Pho4p	

of human transportin and mediates nuclear import of Nab2p and Hrp1p/Nab4p, two yeast hnRNP proteins that shuttle between the nucleus and the cytoplasm (48).

The most abundant yeast hnRNP protein, Npl3p, uses another Kap95p-related protein, Mtr10p, as its import receptor (82, 83). The domain of Npl3p thought to contain its NLS shows no homology to either class of canonical NLSs. Some ribosomal proteins use Kap123p as an import receptor (84). The NLSs of ribosomal proteins also do not resemble the canonical NLSs.

Evidence that these receptors play important roles in import come from examining the location of their cargoes in strains carrying disruptions or ts alleles of the receptors. Under these conditions, cytoplasmic accumulation of cargo proteins occurs. Although many of these receptors are not essential, their normal cargoes are transported less efficiently in their absence and accumulate in the cytoplasm. Poly(A)$^+$ RNA accumulates in nuclei in cells carrying a mutant allele of *mtr10, kap104,* or *kap123*, but none of these is thought to play a direct role in mRNA export. This indicates that the actual phenotypes seen in mutant strains may result from downstream consequences of the primary defect. Thus, simple phenotypic assays are not capable of revealing whether a protein is involved directly or indirectly in import or export.

3.6 Mechanism of protein import

Determining the mechanism of protein import and the functions of each of these factors has been one of the major goals of the nuclear transport field. Studies from both yeast and metazoans have played important parts in arriving at the understanding we have today. In the most widely held model, the asymmetrical distribution of cytoplasmic Rna1p and nuclear Prp20p (the effectors for Gsp1p) plays a fundamental role in nuclear transport. These locations, along with their opposite effects on nucleotides bound to Ran, can define biochemically whether a substrate and its receptor are on the nuclear or cytoplasmic side of the nuclear envelope.

We now know that proteins with signals for nuclear import can interact with their receptor in the absence of Ran. RanGDP does not disrupt these complexes and it can associate with them. However, these complexes are readily disassembled by RanGTP. Therefore, NLS protein–receptor complexes will form in the cytoplasm in the absence of RanGTP, move through NPCs by an unknown mechanism, and arrive in the nucleus where they will be disassembled by RanGTP (for reviews see References 2–4). This delivers cargoes to the nucleus.

By this model, GTP hydrolysis by Ran is not used to perform physical work, such as moving proteins through the NPC. Rather, hydrolysis is part of the mechanism to maintain the asymmetrical distribution of RanGTP versus RanGDP, and to permit recycling of transport components.

4. Protein shuttling and protein export

Clearly, protein export is essential if receptors that bring cargo into the nucleus are to function in multiple rounds of transport. Thus, the first proteins observed to move bidirectionally across the nuclear envelope were transport receptors, Gsp1p/Ran, and transport factors. However, studies in diverse systems indicate that many proteins shuttle between the nucleus and the cytoplasm (for a review see Reference 4). If protein export is to function in ways similar to protein import, we expect that proteins that leave the nucleus will contain NESs.

4.1 Assays for studying protein export and shuttling

Several assay systems have been developed to study protein export. For proteins that are concentrated in the nucleus, heterokaryon assays have proved particularly useful in both yeast and mammalian cells. These assays require that the protein to be analysed be absent from one of the nuclei or that a distinguishable form of the protein (epitope or GFP tag) be used. Cells are fused in the presence of cycloheximide to block continued production of the protein during the assay. Heterokaryons are fixed at various times after mating and examined by fluorescence microscopy. If the protein of interest is detected in both nuclei, then it must have been exported from one nucleus and imported into the other.

An alternative approach, which can be used only in yeasts, takes advantage of the existence of mutant strains that are temperature sensitive for protein import. The *nup49-313* allele permits protein import at 23°C but not at 37°C (85). Cells carrying this mutation and a detectable protein of interest are shifted to 37°C and examined at subsequent times (86). If the protein of interest shuttles, it will accumulate in the cytoplasm when protein import is prevented by temperature shift. The *nup49-313* strain appears to be defective for import mediated by any of the importin β family members involved in import.

4.2 Nuclear export signals (NESs)

Protein export requires NESs, signals within the cargo analogous to NLSs, but specifying export rather than import. The first protein studied in detail for export signals and requirements was the human immunodeficiency virus 1 (HIV-1) Rev protein. Rev is able to leave the nucleus alone or bound to RNA, and therefore contains information needed for its export. Mutation of Rev demonstrated that a specific leucine-rich region constituted the actual export signal (87, 88). This class of NESs appears to be the most common, but other types are known to exist (89). The finding that many RNA-binding proteins contain NESs and shuttle between nucleus and cytoplasm suggests that export of RNA molecules is actually mediated by NES-containing proteins bound to the RNA. Studies from the Rosbash lab indicate that HIV-1 Rev functions in yeast to mediate the export of RNAs to which Rev is able to bind (90). This suggested that the machinery involved in protein export is conserved among eukaryotes.

4.3 Receptors for protein export

As might be predicted, proteins related to Kap95p function as nuclear export receptors. Interestingly, receptors are involved either in import or export; none has yet been shown to be involved in both processes. The receptor that mediates export of proteins carrying leucine-rich NESs is called Crm1p or Xpo1p (91). This protein is found primarily in the nucleus, but can also be detected in the cytoplasm. Several different threads of investigation converged on Crm1p as the NES receptor. Many years ago, *crm1* mutant strains of *S. pombe* were isolated because of their defects in higher-order chromosome structure (chromosome region maintenance; *crm*) (92). From a separate line of investigation, it was known that the fungal metabolite, lepto-mycin B (LMB), caused cell cycle arrest in *S. pombe* (93). A strain resistant to LMB was isolated and the mutation was found to lie in the *CRM1* gene (94). From a third line of investigation, LMB was identified as an inhibitor of Rev's export activity in mam-malian cells, (95). Therefore, a budding yeast strain mutant for *crm1* was analysed for the distribution of a GFP–NES–NLS reporter protein normally found primarily in the cytoplasm. In the *crm1* mutant strain, this protein accumulated in nuclei (96). Finally, it was shown that injection of mammalian CRM1 into *Xenopus laevis* oocytes stimulated export from the nucleus of NES-containing proteins.

While the interaction of Crm1p with leucine-rich NESs is analogous to the inter-action of Srp1p with canonical NLSs, there is a very important difference between the two. The trimeric Srp1p–Kap95p–cargo complex forms without a Ran requirement, but can associate with RanGDP. In contrast, RanGTP, Crm1p, and cargo interact cooperatively (97).

A second yeast protein export receptor identified is Cse1p, another member of the Kap95p family. Cse1p serves as an export receptor for Srp1p (98, 99). As with Crm1p, formation of Cse1p–Srp1p complexes occurs cooperatively with RanGTP.

Note how the asymmetrical distribution of Ran effectors and of RanGTP versus RanGDP accounts for the direction of transport (Fig. 4). Import cargo–receptor com-plexes form in the cytoplasm and are unstable in the presence of RanGTP. Once import cargo–receptor complexes arrive in the nucleus, they are disassembled by RanGTP. Export cargo–receptor complexes include RanGTP and are unstable in the presence of RanGDP. Following export, GTP hydrolysis by Rna1p results in dis-sociation of the complex and delivery of the cargo to the cytoplasm.

4.4 Other protein export factors

Studies identical to those performed to identify protein import factors have been conducted in order to identify protein export factors. In all cases so far examined, no nucleoporin mutant strains have been found to be defective in export of proteins carrying Leu-rich NESs. The only known transport factor required for protein export is Yrb2p, which is essential only at lower temperatures (16°C) (69).

4.5 Recycling of transport receptors

After delivering cargo to the appropriate compartment, transport receptors must be recycled so they can participate in additional rounds of transport. These receptors are all related to Kap95p/importin β, except for Srp1p, which uses Csc1p as a conven-

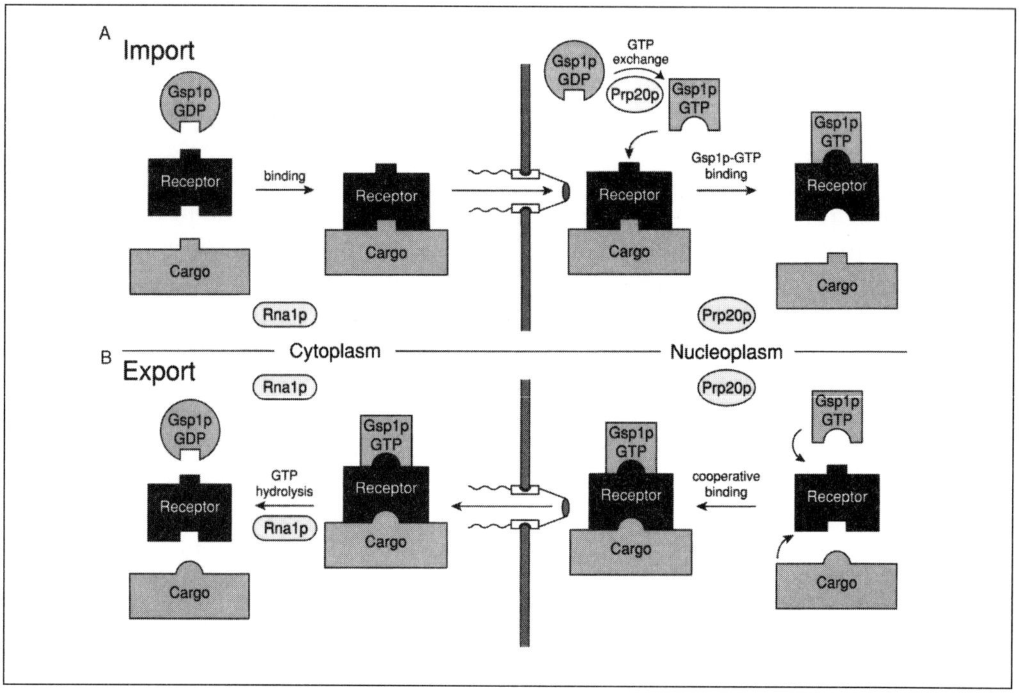

Fig. 4 Nucleocytoplasmic transport pathways. Notice that the effectors for Gsp1p, Prp20p (exchange factor), and Rna1p (GTPase-activating protein) are located on opposite sides of the nuclear envelope. This is expected to result in Gsp1p–GTP primarily in the nucleus and Gsp1p–GDP primarily in the cytoplasm. (A) Nuclear import. Each macromolecule targeted to the nucleus must carry a nuclear localization signal (NLS), which is recognized by a nuclear import receptor. Import complexes are dissociated by Gsp1p–GTP when they reach the nucleus. The receptor then returns to the cytoplasm. (B) Nuclear export. Macromolecules that leave the nucleus must carry a nuclear export signal (NES) which mediates interaction with an export receptor. A cooperative interaction occurs among Gsp1p–GTP, the receptor, and the cargo. This complex is dissociated in the cytoplasm through the action of Rna1p, which hydrolyzes GTP bound to Gsp1p, causing dissociation of the cargo/receptor/Gsp1 complex.

tional protein export receptor. Although the export signal for Srp1p has not been identified, the export signal of a mammalian homologue, RCH1, has been localized to a 10 amino acid region, and does not resemble Leu-rich NESs (100). Some studies suggest that transport receptors of the importin β family may be able to be recycled without requiring an interaction with Ran or other soluble proteins (101, 102). Understanding receptor recycling will require further study.

5. RNA export

RNAs are among the most complex molecules transported across the nuclear envelope. mRNAs are exported as complexes containing processed RNA and a diverse set of RNA-binding proteins called hnRNP proteins (for a review see Reference 81). Ribosomal subunits are assembled in the nucleolus and contain dozens of ribosomal proteins in addition to rRNAs. tRNAs are exported following extensive processing and base modification. In metazoans, most U snRNAs are exported to the cytoplasm for maturation and processing before being reimported. There is no evidence that yeast snRNAs ever leave the nucleus.

5.1 mRNA export

Export of mRNAs is prevented until all processing steps have been completed. The requirement to complete splicing could reflect the entry of pre-mRNA into spliceosomal complexes, which prevent export of unspliced pre-mRNAs. The development of *in situ* hybridization assays to detect nuclear accumulation of poly(A)$^+$ RNA permitted sets of ts mutants to be screened to identify those defective for RNA export (56, 61). Collections of yeast ts mutants were analysed using this assay to identify strains that showed nuclear accumulation of poly(A)$^+$ RNA when shifted to 37°C (Fig. 5B). In this way, several nucleoporins required for RNA export were identified, as were various transport factors. Possible transport factors were also identified through high-copy suppression, synthetic lethal, and two-hybrid screens (for a review see Reference 103).

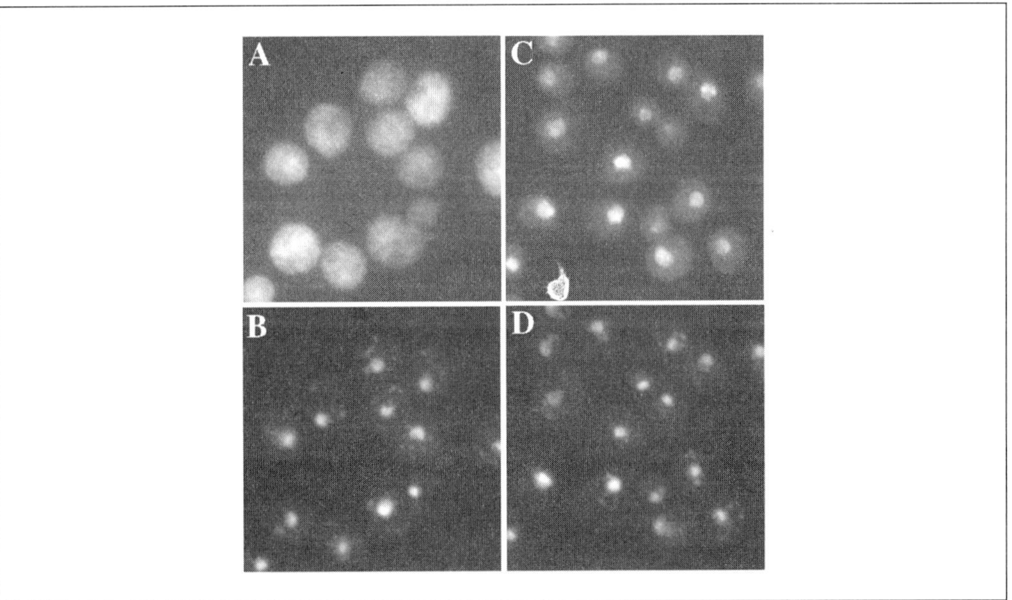

Fig. 5 *In situ* hybridization permits localization of poly(A)$^+$ RNA in yeast cells. An oligo-(dT)$_{50}$ probe, tagged with digoxigenin, hybridizes to the poly(A) tails of mRNA molecules. A fluorescein-labelled anti-digoxigenin antibody indicates where the probe is bound to mRNA. (A) Wild-type cells. (B) Staining of the DNA in the same cells shown in (A). (C) Cells in which Nup159p has been truncated, removing the third of the protein N-terminal to its central repeat domain. These cells are defective in mRNA export and accumulate poly(A)$^+$ RNA in their nuclei. (D) Staining of the DNA in the same cells shown in (C). Figure provided by C. Snay-Hodge, Dartmouth Medical School, Hanover, NH.

5.1.1 Nucleoporins required for mRNA export

Mutation of any one of several nucleoporins prevents or reduces mRNA export (see Table 1). The nucleoporins required for RNA export are Nup159p, Nup82p, Nup84p, Nup85p, Nup120p, Nup133p, and Nup145Cp. Using an identical screen, the fission yeast Rae1 nucleoporin was shown to be required for mRNA export (104), as is its budding yeast homologue, Gle2p (17). Cells lacking Nup116p are also defective for mRNA export. Since deletion of *NUP116* causes a seal that forms over NPCs after a shift to 37°C, RNA export is prevented by a physical barrier (15). Therefore, Nup116p may not be directly involved in mRNA export. Interestingly, mutations to these nucleoporins do not appear to affect protein import or export, suggesting that these nucleoporins form NPC structures specifically designed for RNA export.

Most nucleoporins important for mRNA export are found in two subcomplexes of the yeast NPC: the Nup159p–Nup82p–Nsp1p subcomplex and the Nup84p–Nup85p–Nup120p–Nup145Cp–Seh1p subcomplex. This suggests that these subcomplexes play specific roles in RNA export. Immuno-EM indicates that both subcomplexes are on the cytoplasmic side of the nuclear envelope.

5.1.2 Additional mRNA export factors

Several additional factors required for mRNA export have been identified (Table 3) through study of mutants that accumulate nuclear poly(A)$^+$ mRNA. However, the actual functions of these factors is not yet known.

Nuclear accumulation of poly(A)$^+$ RNA can occur in mutant strains affecting proteins that are unlikely to play a direct role in transport. For example, a mutant allele of acetyl-CoA-carboxylase (*ACC1*) was identified as *MTR7* by screening ts mutant strains for nuclear accumulation of poly(A)$^+$ RNA (105). Cells carrying this mutation have dramatic alterations to their nuclear envelopes, reflecting perturbation of lipid metabolism. Perhaps these envelopes are less able to support construction of functional NPCs, leading to transport defects as an indirect consequence of a defect in lipid metabolism. Determining which of these factors play direct roles in mRNA export will require a more detailed understanding of the mechanism of mRNA export.

Proteins that bind pre-mRNAs in the nucleus to form hnRNPs are thought to be mediators of mRNA export. The observation that that mammalian hnRNP A1 contains both import and export signals is consistent with the idea that signals on RNA-binding proteins are the actual information responsible for mRNA export. In yeast, Npl3p is the protein that appears to be most similar to hnRNP A1, and it has been shown to shuttle between the nucleus and cytoplasm (86). Nuclear accumulation of poly(A)$^+$ RNA occurs in strains carrying ts alleles of *npl3* (53) and in cells where Npl3p is overexpressed. Mutant alleles of *npl3* also show genetic interactions with many other proteins involved in mRNA export. Finally, export of Npl3p does not occur if it cannot bind to RNA, if mRNA itself is not being exported due to mutations in other factors, or if RNA is not being synthesized. Npl3p may play an essential role of packaging the mRNA into an exportable configuration. RNA-binding proteins from several organisms, including yeast Npl3p, are known to be methylated on arginine residues (106). Although this process is not essential, disruption of the gene encoding the methyltransferase Hmt1p shows synthetic lethality and other genetic interactions with mutant alleles of proteins involved in RNA export.

Another RNA-binding protein that appears to play a direct role in mRNA export is Mex67p (107), which was identified in a synthetic lethal analysis starting with a disruption of the non-essential *NUP85* gene. The protein can be UV crosslinked to poly(A)$^+$ RNA, and is found in a complex with Mtr2p (108), a protein identified by screening ts mutants for nuclear accumulation of poly(A)$^+$ RNA (109). In addition, some Mex67p is associated with NPCs, and this association is lost when ts *mex67-5* cells are shifted to 37°C (107). Immuno-EM demonstrates that the Mtr2p–Mex67p complex can be found at both sides of NPCs, and it is thought to interact with NPCs through Nup85p (108).

Another probable export factor is the Rss1p/Gle1p/Brr3p. This gene was identified through a screen for high-copy suppressors of the *rat7-1* allele of Nup159p (110), by screening cold-sensitive mutants for nuclear accumulation of poly(A)$^+$ RNA (111), and through a synthetic lethal screen starting from a strain lacking Nup100p (112). Rss1p/Gle1p/Brr3p has an extended putative coiled-coil domain, as do several nucleoporins including Nup159p, and contains a sequence that can function as leucine-rich NES when attached to a reporter protein (112). A mutant lacking this putative NES is non-functional. However, Gle1p has never been shown to shuttle and does not interact with Crm1p in a two-hybrid assay (112, 113). Protein import and export are unaffected in strains carrying mutant alleles of *rss1/gle1/brr3*.

Screening a ts mutant collection by *in situ* hybridization identified the essential DEAD-box protein, Dbp5p/Rat8p, as an important mRNA export mediator (114). This protein was also identified through genetic interactions with Rss1p/Gle1p/Brr3p (115) and by analysis of DEAD-box proteins identified through a PCR strategy (116). This abundant protein is located in the cytoplasm but concentrated in the perinuclear region where it associates with NPCs (114). Budding yeast contains more than two dozen genes encoding DEAD-box proteins. Among these, Dbp5p is most closely related to Tif1p/Tif2p (yeast eIF4A), an RNA helicase that functions in initiation of protein synthesis. Purified Dbp5p/Rat8p binds and hydrolyses ATP, but the purified protein has not so far shown RNA-unwinding activity (116). *dbp5* mutants do not display defects in pre-mRNA splicing or translation, processes that require the activity of multiple DEAD-box proteins.

The role of Dbp5p in RNA export remains to be determined. Since some of the proteins that accompany mRNAs out of the nucleus are removed from the RNA and shuttle back to the nucleus, Dbp5p could participate in stripping these proteins from mRNA. The mRNA would then associate with other RNA-binding proteins and with ribosomal subunits, leading to translation. Alternatively, Dbp5p could assist in the association of RNA–protein complexes with NPCs, which may require remodelling their structure so that they can be exported efficiently.

5.1.3 Receptors for mRNA export

Four yeast importin β family members have been shown to be export receptors. Crm1p/Xpo1p plays a role in protein export but has not yet been shown to play a direct role in mRNA export. Cse1p is involved in recycling of Srp1p to the cytoplasm (98, 99). Los1p appears to be the export receptor for tRNA, but there must be another tRNA receptor since cells lacking Los1p are viable. Msn5p is involved in export of the transcription factor, Pho4p.

When cells carrying the ts *xpo1-1* allele of *CRM1* are shifted to 37°C, export of NES-containing reporter proteins is blocked rapidly. Somewhat later, nuclear accumulation of poly(A)$^+$ mRNA can be detected (117). This could reflect an indirect role for Crm1p in mRNA export or could reflect different sensitivities for the *in situ* assay to detect poly(A)$^+$ RNA and the direct observation of the location of GFP–NES reporter proteins. However, many transport proteins, including Dbp5p, shuttle and some use Crm1p as their export receptor (114). Nuclear accumulation of Dbp5p in *xpo1-1 cells* appears central to the mRNA export block in *xpo1-1* cells, as overexpression of Dbp5p prevents nuclear accumulation of poly(A)$^+$ RNA in these cells and maintains some Dbp5p in the cytoplasm (114).

This lack of evidence supporting a direct role for any of the importin β-family members can be explained in at least two ways. An as yet unidentified protein could play a role in mRNA export mechanistically identical to the roles played by known importin β-family members in export of other classes of macromolecules. However, it is also possible that mRNA export is mechanistically distinct from export of smaller macromolecules. These smaller cargoes (e.g. proteins, tRNA) are of a size similar to their receptors. In contrast, mRNA (as ribonucleoprotein) and ribosomal subunits are very large and quite possibly these assemblies make multiple contacts with NPCs simultaneously during export. Dbp5p associates with NPCs through Nup159p, located at the terminal extension of the cytoplasmic filaments. This places Dbp5p at a location where it could play an important role at a late stage of mRNA export, by removing hnRNP proteins (e.g. Npl3p) that subsequently return to the nucleus. In addition, the ATPase activity of Dbp5p could permit Dbp5p to 'pull' or 'ratchet' the mRNP out of the nucleus.

5.2 tRNA export

tRNA export is mediated by Los1p, another member of the Kap95p family of transport receptors (23). Los1p is not essential, so there must be another receptor also able to mediate tRNA export. The *los1-1* allele was identified almost 20 years ago in a screen to identify 'loss of suppression' mutants (*LOS*) defective for processing of suppressor tRNAs (118). It was also isolated in a synthetic lethal screen beginning with a mutant allele of the nucleoporin gene *NSP1* (119). Los1p and several proteins involved in tRNA maturation localize to NPCs.

The observation that Los1p is related to other transport receptors first suggested that it might be involved in tRNA export. An *in situ* hybridization assay was developed using probes to tRNA families (23). In both *los1-1* and *los1Δ* cells, nuclear accumulation of tRNA was seen, indicating an export defect. tRNA export was also defective in cells lacking Nup116p or carrying the *rna1-1* allele, but not in cells carrying the *rat7-1* allele of Nup159p. It will be interesting to extend these *in situ* hybridization analyses to strains carrying mutations in other genes important for RNA export.

Although not yet shown to be the case in yeast systems, tRNA export in *X. laevis* requires that the RNA be fully processed and is much more efficient if the tRNA is charged with amino acids (120). This kind of surveillance would prevent export of defective tRNA molecules, since amino-acylation requires that the tRNA be processed completely and correctly. Intron-containing pre-tRNAs also accumulate in the nuclei of *los1* mutant cells (23). Metazoan exportin-t has been shown to bind tRNA directly, but this has not yet been shown for yeast Los1p.

5.3 Export of rRNA and ribosomal subunits

Ribosomal RNAs must be processed and assembled into ribosomal subunits before they can be exported (121). Cryofixation and cryoelectron microscopy were used to examine ribosome biogenesis in *S. pombe* (122). These studies suggested that ribosomal subunits might move from the nucleolus to the nuclear periphery along specific track-like pathways, and also showed that ribosomal subunits can be exported through NPCs located anywhere within the nuclear envelope, not just NPCs located closest to the nucleolus. Hurt and colleagues developed an assay to monitor ribosomal export *in vivo* by producing a functional GFP fusion to ribosomal protein L25 (24). They showed that the Gsp1p/Ran system is required for ribosome export, since induction of expression of dominant-negative Gsp1p/Ran (locked in the GTP-bound state) blocks ribosomal subunit export. Export of ribosomal subunits was also inhibited by mutations affecting some nucleoporins.

This assay will permit screening of ts mutant banks to identify other genes important for ribosome export. One concern is with the long times of incubation required to detect defects in ribosome export, particularly in the case of nucleoporin mutants. Mutants defective in assembly of NPCs are likely to require many hours at the non-permissive temperature before transport defects can be detected, since transport defects may require a dramatic reduction in the number of functional NPCs before the rate of transport is compromised. Therefore, it will be necessary to determine that mutants defective in ribosome export after 5–15 h at 37°C do not have primary defects in NPC biogenesis.

The final step in production of the 18S small rRNA involves processing from a 20S precursor in the cytoplasm. Using a probe to the sequences removed from the 20S pre-rRNA, Moy and Silver were able to identify mutants defective for export of the 20S precursor to the small ribosomal subunit, since they accumulated the 20S rRNA in nuclei (123). Mutants known to be defective in ribosome biogenesis accumulated the 20S rRNA in the nucleolus. Nucleolar accumulation was also seen in strains with mutations affecting Srp1p, Kap95p, Pse1p, Cse1p, and Mtr10p. Since some of these proteins are known to mediate import of ribosomal proteins, defects in ribosome biogenesis can be distinguished from defects in export, where nucleoplasmic accumulation of the pre-rRNA occurs. They found that small subunit export requires the Ran GTPase system and certain nucleoporins, but did not require many of the other factors involved in nuclear transport. No importin β-family member has been connected with rRNA export. As with export of mRNA, export of rRNA may occur by a mechanism distinct from that of the importin family of transporters.

6. Regulation of nucleocytoplasmic transport

Regulation is known to occur at essentially every level of the gene expression pathway, from modification of chromatin to permit transcription through transcription and RNA processing to translation, RNA turnover, and protein turnover. Several studies indicate that nuclear transport can also be regulated. It is likely that many additional examples of regulated transport will be identified.

6.1 Regulated protein transport

Previous studies of metazoan systems have provided many examples where protein import is regulated. These include the glucocorticoid receptor, which forms a cytoplasmic complex with hsp90 that is disrupted by the receptor's ligand, dexamethasone, permitting import of the receptor. Many laboratories have studied regulated transport of transcription factor NFκB (for a review see Reference 124). This protein is retained in the cytoplasm in a complex with an inhibitory protein, IκB. When the cell receives an appropriate signal, a signal transduction cascade occurs that results in phosphorylation of IκB, dissociation of the NFκB–IκB complex, and nuclear import of NFκB. Other studies indicate that NLSs are often located close to phosphorylation sites and that phosphorylation can enhance the import of the nuclear protein. In all of these systems, regulation occurs by modifying the substrate protein or protein complex, rather than modification of the nuclear transport machinery. Protein phosphorylation and dephosphorylation may play major roles in regulating nuclear transport of specific cargoes.

Regulated protein transport also occurs in yeasts. In budding yeast, high osmolarity results in activation of a signal transduction cascade mediated in part by the mitogen-activated protein (MAP) kinase, Hog1p (125). Activation requires that Hog1p be phosphorylated by a MAP kinase kinase. Using a GFP–Hog1p fusion protein, Silver and colleagues (125) discovered that Hog1p moved from the cytoplasm to the nucleus following osmotic stress. Translocation required that Hog1p be phosphorylated. The kinase that phosphorylates Hog1p remained in the cytoplasm, as did all other components of this signal transduction pathway. Interestingly, nuclear import of Hog1p was independent of Srp1p/Kap95p but required a previously uncharacterized member of the importin β family, Nmd5p, which was transiently translocated to the nucleus after stress. After 30 min of stress, most Hog1p was dephosphorylated and returned to the cytoplasm, using Crm1p as the export receptor. Similar studies have been conducted in *S. pombe* (126).

The budding yeast transcription factor Yap1p is normally cytoplasmic, but translocates to the nucleus following oxidative stress. Normally, Yap1p shuttles between the nucleus and the cytoplasm, and its export is mediated by Crm1p (127). Oxidative stress prevents the interaction between Crm1p and the NES of Yap1p, and this requires at least one of three cysteines close to the NES. Through *in vitro* studies, it could be shown that the interaction between Crm1p and Yap1p was affected dramatically by oxidation/reduction. Including dithiothreitol in the binding buffer resulted in 20-fold more binding of Yap1p to Crm1p.

A third example of this type of regulation in yeast concerns the Pho4p transcription factor, which is cytoplasmic and phosphorylated when phosphate levels are high, but nuclear and unphosphorylated when phosphate levels are low (128). This transcription factor is also imported by a novel importin β-family member, Pse1p. Phosphorylation of Pho4p inhibits its interaction with this receptor.

Another interesting case of regulated export concerns the cell cycle in *S. pombe*. In this case, DNA damage activates a checkpoint to prevent cell division. This occurs because Cdc25p is phosphorylated as a consequence of DNA damage, creating a binding site for the 14-3-3 protein, Rad24p. It appears that Rad24p acts as an attachable nuclear export signal that controls the subcellular distribution of Cdc25p (129).

Protein import is also regulated following heat or ethanol shock in budding yeast (130). After stress, a GFP–NLS–NES reporter protein was unable to enter the nucleus. Many additional examples of regulated protein transport are likely to be discovered in yeast systems.

6.2 Regulation of mRNA export

RNA export is also a target of stress regulation. The rapid and strong induction of heat-shock genes following heat stress is seen in all organisms. This occurs at many levels, although induction of transcription is one of the most important. By adapting *in situ* hybridization so that individual species of mRNA could be detected in yeast, it was shown that *SSA4* mRNA (encoding an inducible Hsp70) is produced and efficiently exported after heat or ethanol shock (130). Stress (heat or ethanol) was also applied to a set of mutant yeast strains to determine which gene products required for regular mRNA export were also needed for export of heat-shock mRNAs after stress. The same nucleoporins are required for both processes. Export of heat-shock mRNAs also requires several factors involved in export of mRNA under non-stress conditions, including Dbp5p, Rss1p, and Crm1p, but not Npl3p (131). The finding that Npl3p can no longer be crosslinked to mRNA following stress suggests that dissociation of Npl3p (and perhaps other RNA-binding proteins) from nuclear poly(A)$^+$ mRNA may play an important role in blocking export of most mRNAs following stress (132).

An interesting question concerns what novel gene products facilitate export of heat-shock mRNAs following stress. Following up on studies to investigate the ability of HIV-1 Rev to function in yeast cells (90), a two-hybrid screen was conducted and identified Rip1p (Rev-interacting protein) (133). It is now known that the Rev–Rip1p interaction is indirect and is bridged by Crm1p (96), which interacts directly with the NES of Rev. Rip1p contains many nucleoporin repeats and is located at the nuclear rim and in the nucleoplasm. Quite possibly, Rip1p is a nucleoporin; it has also been named Nup42p. Rip1p is required for export of heat-shock mRNAs, but is completely dispensable for export under normal growth conditions (131). However, Rip1p becomes essential in cells carrying mutant alleles of *RSS1/GLE1/BRR3* (134) suggesting that Rip1p may also function in normal mRNA export. Identifying the mechanisms underlying selective mRNA export in yeast will require additional studies.

7. Coupling nucleocytoplasmic transport to other biological processes

Several recent studies indicate that the processes of transcription and pre-mRNA processing are linked in both yeast and metazoan systems (see Chapter 6, section 3.2 and Chapter 8, section 7.2). Several enzymes and factors required for pre-mRNA processing associate with the C-terminal domain (CTD) of the largest subunit of RNA polymerase II during transcription (for a review see Reference 135). These include capping enzymes, splicing factors, and polyadenylation factors. The CTD from yeast contains 26 repeats of the sequence YSPTSPS, and up to 50 copies of a related sequence are found in metazoan RNA polymerase II CTDs.

There is also evidence that suggests a link between mRNA 3′-end processing and mRNA export. Completion of 3′ processing and polyadenylation is required for efficient mRNA export. Sherman and co-workers found that deletion of critical 3′ processing signals in *CYC1* mRNA (the *cyc1-512* allele) resulted in the production of extended transcripts that were cleaved and polyadenylated at sites downstream from the normal processing site (136). These extended transcripts accumulate in nuclei (C. A. Hammell and C. N. Cole, unpublished results). The mechanism for generation of these longer transcripts is unknown. The normal 3′ processing machinery could recognize cryptic sites downstream of the normal one. Alternatively, cleavage could occur through the action of other endonucleases, and these extended transcripts might then become substrates for poly(A) polymerase. Mutations affecting 3′ processing factors Rna14p and Rna15p also result in production of extended transcripts which end at downstream sites (137). Poly(A)$^+$ RNA can be detected in nuclei from cells of some strains carrying mutant alleles affecting 3′ processing factors (C. A. Hammell and C. N. Cole, unpublished results). Thus, mutations in either *cis*-acting sequences or *trans*-acting factors required for 3′ processing can result in mRNA export defects. What might be the important difference between polyadenylated 3′ ends formed by the normal mechanism and those formed in these aberrant situations? Perhaps one or more 3′-processing factors remain on RNAs processed through the normal pathway, but might never associate with mRNAs whose 3′ ends and poly(A) tails are produced by an abnormal mechanism. Alternatively, 3′ processing through the normal pathway could be required for association of one or more export factors with the mRNA.

Conversely export factors may play a role in mRNA 3′-end processing. Shifting *prp20-1* or *rna1-1* cells to 37°C results in rapidly-occurring defects in 3′-end formation and the production of extended transcripts (66). In both *rna1-1* and *prp20-1* mutant cells, *CUP1* mRNAs were initiated at multiple sites upstream from the normal site and 3′ processed at multiple sites downstream from the normal site. In both strains, the same sites were used. *CUP1* mRNAs with aberrant 3′ ends can be detected within 2 min of the shift to 37°C. Proteins of the Ran system have been suggested to play a role in nuclear structure, so it is possible that these mutations result in rapidly-occurring defects in the ability of mutant cells to maintain proper nuclear organization, which may be a requirement for proper and efficient 3′-end processing and export. Another possibility is that processing factors function in mRNA export and export factors function in 3′ processing, thereby linking these processes to each other and to transcription.

8. Questions for future research

Much of the progress towards understanding nuclear transport has occurred in the past decade. It is probable that a thorough understanding of transport will be achieved within the next 10 years. What are the key questions that remain to be addressed?

- What is the mechanism by which macromolecules are translocated from one side of the NPC to the other? How is energy (ATP or GTP) used for nuclear transport? Is export of very large assemblies (messenger ribonucleoprotein and ribosomal subunits) mechanistically distinct from that of smaller cargoes (proteins and tRNA)?
- Do Gsp1p/Ran and other proteins of the Ran system play direct roles in cellular processes other than nuclear transport, or are they involved only in transport? If these proteins play direct roles only in nuclear transport, how are defects in transport related to the observed cell cycle defects seen in some mutant strains, and RNA synthesis and processing defects seen in others?
- How are NPCs assembled? To what degree are they dynamic?
- What mechanisms are used to regulate nuclear transport?

Future studies about nuclear transport should provide answers to these questions, and will certainly raise others.

Acknowledgements

The author thanks Nelly Panté and Christine Hodge for contributing figures for this chapter, Reed Detar and Chris Maute for assistance with the figures, and members of his laboratory for helpful discussions and critical reading of the manuscript.

References

1. Doye, V. and Hurt, E. (1997) From nucleoporins to nuclear pore complexes. *Curr. Opin. Cell Biol.*, **9**, 401
2. Izaurralde, E. and Adam, S. (1998) Transport of macromolecules between the nucleus and the cytoplasm. *RNA*, **4**, 351.

3. Mattaj, I. W. and Englmeier, L. (1998) Nucleocytoplasmic transport: the soluble phase. *Annu. Rev. Biochem.*, **67**, 265.

4. Corbett, A. H. and Silver, P. A. (1997) Nucleocytoplasmic transport of macromolecules. *Microbiol. Mol. Biol. Rev.*, **61**, 193.

5. Winey, M., Yarar, D., Giddings T. H., Jr., and Mastronarde, D. N. (1997) Nuclear pore complex number and distribution throughout the *Saccharomyces cerevisiae* cell cycle by three-dimensional reconstruction from electron micrographs of nuclear envelopes. *Mol. Biol. Cell*, **8**, 2119.

6. Yang, Q., Rout, M. P., and Akey, C. W. (1998) Three-dimensional architecture of the isolated yeast nuclear pore complex: functional and evolutionary implications. *Mol. Cell*, **1**, 223.

7. Cordes, V. C., Reidenbach, S., Rackwitz, H. R., and Franke, W. W. (1997) Identification of protein p270/Tpr as a constitutive component of the nuclear pore complex-attached intranuclear filaments. *J. Cell Biol.*, **136**, 515.

8. Rout, M. P. and Blobel, G. (1993) Isolation of the yeast nuclear pore complex. *J. Cell Biol.*, **123**, 771.

9. Grandi, P., Doye, V., and Hurt, E. C. (1993) Purification of NSP1 reveals complex formation with 'GLFG' nucleoporins and a novel nuclear pore protein NIC96. *EMBO J.*, **12**, 3061.

10. Davis, L. I. and Blobel, G. (1986) Identification and characterization of a nuclear pore complex protein. *Cell*, **45**, 699.

11. Snow, C. M., Senior, A., and Gerace, L. (1987) Monoclonal antibodies identify a group of nuclear pore complex glycoproteins. *J. Cell Biol.*, **104**, 1143.

12. Teixeira, M. T., Siniossoglou, S., Podtelejnikov, S., Benichou, J. C., Mann, M., Dujon, B., and Hurt, E. (1997) Two functionally distinct domains generated by in vivo cleavage of Nup145p: a novel biogenesis pathway for nucleoporins. *EMBO J.*, **16**, 5086.

13. Fahrenkrog, B., Hurt, E. C., Aebi, U., and Pante, N. (1998) Molecular architecture of the yeast nuclear pore complex: localization of Nsp1p subcomplexes. *J. Cell Biol.*, **143**, 577.

14. Heath, C. V., Copeland, C. S., Amberg, D. C., Del Priore, V., Snyder, M., and Cole, C. N. (1995) Nuclear pore complex clustering and nuclear accumulation of poly(A)$^+$ RNA associated with mutations of the *Saccharomyces cerevisiae RAT2 / NUP120* gene. *J. Cell Biol.*, **131**, 1677.

15. Wente, S. R. and Blobel, G. (1993) A temperature-sensitive *NUP116* null mutant forms a nuclear envelope seal over the yeast nuclear pore complex thereby blocking nucleocytoplasmic traffic. *J. Cell Biol.*, **123**, 275.

16. Gorsch, L. C., Dockendorff, T. C., and Cole, C. N. (1995) A conditional allele of the novel repeat-containing yeast nucleoporin *RAT7/NUP159* causes both rapid cessation of mRNA export and reversible clustering of nuclear pore complexes. *J. Cell Biol.*, **129**, 939.

17. Murphy, R., Watkins, J. L., and Wente, S. R. (1996) *GLE2*, a *Saccharomyces cerevisiae* homologue of the *Schizosaccharomyces pombe* export factor *RAE1*, is required for nuclear pore complex structure and function. *Mol. Biol. Cell*, **7**, 1921.

18. Belgareh, N. and Doye, V. (1997) Dynamics of nuclear pore distribution in nucleoporin mutant yeast cells. *J. Cell Biol.*, **136**, 747.

19. Bucci, M. and Wente, S. R. (1997) *In vivo* dynamics of nuclear pore complexes in yeast. *J. Cell Biol.*, 136:,1185.

20. Wozniak, R. W., Blobel, G., and Rout, M. P. (1994) POM152 is an integral protein of the pore membrane domain of the yeast nuclear envelope. *J. Cell Biol.*, **125**, 31.

21. Chial, H. J., Rout, M. P., Giddings, T. H., and Winey, M. J. (1998) *Saccharomyces cerevisiae* Ndc1p is a shared component of nuclear pore complexes and spindle pole bodies. *J. Cell Biol.*, **143**, 1789.

22. Fabre, E., Boelens, W. C., Wimmer, C., Mattaj, I. W., and Hurt, E. C. (1994) Nup145p is required for nuclear export of mRNA and binds homopolymeric RNA *in vitro* via a novel conserved motif. *Cell*, **78**, 275.

23. Sarkar, S. and Hopper, A. K. (1998) tRNA nuclear export in *Saccharomyces cerevisiae*: *in situ* hybridization analysis. *Mol. Biol. Cell*, **9**, 3041.

24. Hurt, E., Hannus, S., Schmeizl, B., Lau, D., Tollervey, D., and Simos, G. (1999) A novel *in vivo* assay reveals inhibition of ribosomal nuclear export in ran-cycle and nucleoporin mutants. *J. Cell Biol.*, **133**, 389.

25. Mutvei, A., Dihlmann, S., Herth, W., and Hurt, E. C. (1992) NSP1 depletion in yeast affects nuclear pore formation and nuclear accumulation. *Eur. J. Cell Biol.*, **59**, 280.

26. Zabel, U., Doye, V., Tekotte, H., Wepf, R., Grandi, P., and Hurt, E. C. (1996) Nic96p is required for nuclear pore formation and functionally interacts with a novel nucleoporin, Nup188p. *J. Cell Biol.*, **133**, 1141.

27. Siniossoglou, S., Wimmer, C., Rieger, M., Doye, V., Tekotte, H., Weise, C., Emig, S., Segref, A., and Hurt, E. C. (1996) A novel complex of nucleoporins which includes Sec13p and a Sec13p homolog, is essential for normal nuclear pores. *Cell*, **84**, 265.

28. Belgareh, N., Snay-Hodge, C., Pasteau, F., Dagher, S., Cole, C. N., and Doye, V. (1998) Functional characterization of a Nup159p-containing nuclear pore subcomplex. *Mol. Biol. Cell*, **9**, 3475.

29. Hurwitz, M. E., Strambio-de-Castilla, C., and Blobel, G. (1998) Two yeast nuclear pore complex proteins involved in mRNA export form a cytoplasmically oriented subcomplex. *Proc. Natl. Acad. Sci. U.S.A.*, **95**, 11241.

30. Kraemer, D. M., Strambio-de-Castillia, C., Blobel, G., and Rout, M. P. (1995) The essential yeast nucleoporin NUP159 is located on the cytoplasmic side of the nuclear pore complex and serves in karyopherin-mediated binding of transport substrate. *J. Biol. Chem.*, **270**, 19017.

31. Aitchison, J. D., Rout, M. P., Marelli, M., Blobel, G., and Wozniak, R. W. (1995) Two novel related yeast nucleoporins Nup170p and Nup157p: complementation with the vertebrate homologue Nup155p and functional interactions with the yeast nuclear pore-membrane protein Pom152p. *J. Cell Biol.*, **131**, 1133.

32. Bucci, M. and Wente, S. R. (1998) A novel fluorescence-based genetic strategy identifies mutants of *Saccharomyces cerevisiae* defective for nuclear pore complex assembly. *Mol. Biol. Cell*, **9**, 2439.

33. Feldherr, C. M., Kallenbach, E., and Schultz, N. (1984) Movement of a karyophilic protein through the nuclear pores of oocytes. *J. Cell Biol.*, **99**, 2216.

34. Newmeyer, D. D. and Forbes, D. J. (1988) Nuclear import can be separated into two distinct steps *in vitro*: nuclear pore binding and translocation. *Cell*, **52**, 641.

35. Kalderon, D., Roberts, B. L., Richardson, W. D., and Smith, A. E. (1984) A short amino acid sequence able to specify nuclear location. *Cell*, **39**, 499.

36. Dingwall, C., Sharnick, S. V., and Laskey, R. A. (1982) A polypeptide domain that specifies migration of nucleoplasmin into the nucleus. *Cell*, **30**, 449.

37. Adam, S. A., Marr, R. S., and Gerace, L. (1990) Nuclear protein import in permeabilized mammalian cells requires soluble cytoplasmic factors. *J. Cell Biol.*, **111**, 807.

38. Schlenstedt, G., Hurt, E., Doye, V., and Silver, P. A. (1993) Reconstitution of nuclear protein transport with semi-intact yeast cells. *J. Cell Biol.*, **123**, 785.

39. Moore, M. S. and Blobel, G. (1992) The two steps of nuclear import, targeting to the nuclear envelope and translocation through the nuclear pore, require different cytosolic factors. *Cell*, **69**, 939.

40. Chi, N. C., Adam, E. J., and Adam, S. A. (1995) Sequence and characterization of cytoplasmic nuclear protein import factor p97. *J. Cell Biol.*, **130**, 265.

41. Görlich, D., Prehn, S., Laskey, R. A., and Hartmann, E. (1994) Isolation of a protein that is essential for the first step of nuclear protein import. *Cell*, **79**, 767.

42. Yano, R., Oakes, M., Yamaghishi, M., Dodd, J. A., and Nomura, M. (1992) Cloning and characterization of SRP1, a suppressor of temperature-sensitive RNA polymerase I mutations, in *Saccharomyces cerevisiae*. *Mol. Cell. Biol.*, **12**, 5640.

43. Conti, E., Uy, M., Leighton, L., Blobel, G., and Kuriyan, J. (1998) Crystallographic analysis of the recognition of a nuclear localization signal by the nuclear import factor karyopherin alpha. *Cell*, **94**, 193.

44. Iovine, M. K., Watkins, J. L., and Wente, S. R. (1995) The GLFG repetitive region of the nucleoporin Nup116p interacts with Kap95p, an essential yeast nuclear import factor. *J. Cell Biol.*, **131**, 1699.

45. Enenkel, C., Blobel, G., and Rexach, M. (1995) Identification of a yeast karyopherin heterodimer that targets import substrate to mammalian nuclear pore complexes. *J. Biol. Chem.*, **270**, 499.

46. Koepp, D. M., Wong, D. H., and Corbett, A. H. (1996) Dynamic localization of the nuclear import receptor and its interactions with transport factors. *J. Cell Biol.*, **133**, 1163.

47. Loeb, J., Schlenstedt, G., Pellman, D., Kornitzer, D., Silver, P. A., and Fink, G. R. (1995) The yeast nuclear import receptor is required for mitosis. *Proc. Natl. Acad. Sci. U.S.A.*, **92**, 7647.

48. Aitchison, J. D., Blobel, G., and Rout, M. P. (1996) Kap104p: a karyopherin involved in the nuclear transport of messenger RNA binding proteins. *Science*, **274**, 624.

49. Görlich, D. (1998) Transport into and out of the cell nucleus. *EMBO J.*, **17**, 2721.

50. Belhumeur, P., Lee, A., Tam, R., DiPaolo, T., Fortin, N., and Clark, M. W. (1993) GSP1 and GSP2, genetic suppressors of the prp20-1 mutant in *Saccharomyces cerevisiae*: GTP-binding proteins involved in the maintenance of nuclear organization. *Mol. Cell. Biol.*, **13**, 2152.

51. Melchior, F., Paschal, B., Evans, J., and Gerace, L. (1993) Inhibition of nuclear protein import by nonhydrolyzable analogues of GTP and identification of the small GTPase Ran/TC4 as an essential transport factor. *J. Cell Biol.*, **123**, 1649.

52. Moore, M. S. (1998) Ran and nuclear transport. *J. Biol. Chem.*, **273**, 22857.

53. Schlenstedt, G., Saavedra, C., Loeb, J. D. J., Cole, C. N., and Silver, P. A. (1995) The GTP-bound form of the yeast Ran/TC4 homologue blocks nuclear protein import and appearance of poly (A)$^+$ RNA in the cytoplasm. *Proc. Natl. Acad. Sci. U.S.A.*, **92**, 225.

54. Bischoff, F. R., Krebber, H., Kempf, T., Hermes, I., and Ponstingl, H. (1994) RanGAP1 induces GTPase activity of nuclear Ras-related Ran. *Proc. Natl. Acad. Sci. U.S.A.*, **91**, 2587.

55. Bischoff, F. R. and Ponstingl, H. (1991) Catalysis of guanine nucleotide exchange on Ran by the mitotic regulator RCC1. *Nature*, **354**, 80.

56. Amberg, D. A., Goldstein, A. L., and Cole, C. N. (1992) Isolation and characterization of *RAT1*, an essential gene of *Saccharomyces cerevisiae* required for the efficient nucleo-cytoplasmic trafficking of mRNA. *Genes Dev.*, **6**, 1173.

57. Shiokawa, K. and Pogo, K. O. (1974) The role of cytoplasmic membranes in controlling the transport of nuclear messenger RNA and initiation of protein synthesis. *Proc. Natl. Acad. Sci. U.S.A.*, **71**, 2658.

58. Hopper, A. K., Traglia, H. M., and Dunst, R. W. (1990) The yeast RNA1 gene product necessary for RNA processing is located in the cytoplasm and apparently excluded from the nucleus. *J. Cell Biol.*, **111**, 309.

59. Corbett, A. H., Koepp, D. M., Schlenstedt, G., Lee, M. S., Hopper, A. K., and Silver, P. A. (1995) Rna1p, a Ran/TC4 GTPase activating protein, is required for nuclear import. *J. Cell Biol.*, **130**, 1017.

60. Clark, K. L. and Sprague, G. F. (1989) Yeast pheromone response pathway: characterization of a suppressor that restores mating to receptorless mutants. *Mol. Cell. Biol.*, **9**, 2682.

61. Kadowaki, T., Zhao, Y., and Tartakoff, A. M. (1992) A conditional yeast mutant deficient in mRNA transport from nucleus to cytoplasm. *Proc. Natl. Acad. Sci. U.S.A.*, **89**, 2312.

62. Ohtsubo, M., Kai, R., Furuno, N., Sekiguchi, T., Sekiguchi, M., Hayashida, H., Kuma, K., Miyata, T., Fukushige, S., Murotsu, T., Matsubara, K., and Nishimoto, T. (1987) Isolation and characterization of the active cDNA of the human cell cycle gene (RCC1) involved in the regulation of onset of chromosome condensation. *Genes Dev.*, 1. 585.

63. Nishimoto, T., Eilen, E., and C., B (1978) Premature chromosome condensation in a ts DNA-mutant of BHK cells. *Cell*, **15**, 475.

64. Sazer, S. and Nurse, P. (1994) A fission yeast RCC1-related protein is required for the mitosis to interphase transition. *EMBO J.*, **13**, 606.

65. Amberg, D. C., Fleischmann, M., Stagljar, I., Cole, C. N., and Aebi, M. (1993) Nuclear PRP20 protein is required for mRNA export. *EMBO J.*, **12**, 233.

66. Forrester, W., Stutz, F., Rosbash, M., and Wickens, M. (1992) Defects in mRNA 3′-end formation, transcription initiation, and mRNA transport associated with the yeast mutation prp20: possible coupling of mRNA processing and chromatin structure. *Genes Dev.*, **6**, 1914.

67. Sazer, S. (1996) The search for the primary function of the Ran GTPase continues. *Trends Cell Biol.*, **6**, 81.

68. Schlenstedt, G., Wong, D. H., Koepp, D. M., and Silver, P. A. (1995) Mutants in a yeast Ran binding protein are defective in nuclear transport. *EMBO J.*, **14**, 5367.

69. Taura, T., Krebber, H., and Silver, P. A. (1998) A member of the Ran-binding protein family, Yrb2p, is involved in nuclear protein export. *Proc. Natl. Acad. Sci. U.S.A.*, **95**, 7427.

70. Moore, M. S. and Blöbel, G. (1994) Purification of a Ran-interacting protein that is required for protein import into the nucleus. *Proc. Natl. Acad. Sci. U.S.A.*, **91**, 10212.

71. Paschal, B. M. and Gerace, L. (1995) Identification of NTF2, a cytosolic factor for nuclear import that interacts with nuclear pore complex protein p62. *J. Cell Biol.*, **129**, 925.

72. Corbett, A. H. and Silver, P. S. (1996) The *NTF2* gene encodes an essential, highly conserved protein that function in nuclear transport *in vivo*. *J. Biol. Chem.*, **271**, 18477.

73. Smith, A., Brownawell, A., and Macara, I. G. (1998) Nuclear import of Ran is mediated by the transport factor NTF2. *Curr. Biol.*, **8**, 1403.

74. Ribbeck, K., Lipowsky, G., Kent, H. M., Stewart, M., and Görlich, D. (1998) NTF2 mediates nuclear import of Ran. *EMBO J.*, **17**, 6587.

75. Hartmann, E. and Görlich, D. (1995) A Ran-binding motif in nuclear pore proteins? *Trends Cell Biol.*, **5**, 192.

76. Wu, J., Matunis, M. J., Kraemer, D., Blobel, G., and Coutavas, E. (1995) Nup358, a cytoplasmically exposed nucleoporin with peptide repeats, Ran-GTP binding sites, zinc fingers, a cyclophilin A homologous domain, and a leucine-rich region. *J. Biol. Chem.*, **270**, 14209.

77. Gorlich, D., Dabrowski, M., Bischoff, F. R., Kutay, U., Bork, P., Hartmann, E., Prehn, S., and Izaurralde, E. (1997) A novel class of RanGTP binding proteins. *J. Cell Biol.*, **138**, 65.

78. Moroianu, J. (1998) Distinct nuclear import and export pathways mediated by members of the karyopherin beta family. *J. Cell. Biochem.*, **70**, 231.

79. Ohno, M., Fornerod, M., and Mattaj, I. W. (1998) Nucleocytoplasmic transport: the last 200 nanometers. *Cell*, **92**, 327.

80. Pollard, V. W., Michael, W. M., Nakielny, S., Siomi, M., Wang, F., and Dreyfuss, G. (1996) A novel receptor-mediated nuclear protein import pathway. *Cell*, **86**, 985.

81. Siomi, H. and Dreyfuss, G. (1997) RNA-binding proteins as regulators of gene expression. *Curr. Opin. Genet. Dev.*, **7**, 345.

82. Senger, B., Simos, G., Bischoff, F. R., Podtelejnikov, A., Mann, M., and Hurt, E. (1998) Mtr10p functions as a nuclear import receptor for the mRNA-binding protein Npl3p. *EMBO J.*, **17**, 2196.

83. Pemberton, L. F., Rosenblum, J. S., and Blobel, G (1997) A distinct and parallel pathway for the nuclear import of an mRNA-binding protein. *J. Cell Biol.*, **139**, 1545.

84. Rout, M. P., Blobel, G., and Aitchison, J. D. (1997) A distinct nuclear import pathway used by ribosomal proteins. *Cell*, **89**, 715.

85. Doye, V., Wepf, R., and Hurt, E. C. (1994) A novel nuclear pore protein Nup133p with distinct roles in poly(A)$^+$ RNA transport and nuclear pore distribution. *EMBO J.*, **13**, 6062.

86. Lee, M., Henry, M., and Silver, P. A. (1996) A protein that shuttles between nucleus and cytoplasm, is an important mediator of mRNA export. *Genes Dev.*, **10**, 1233.

87. Bogerd, H. P., Fridell, R. A., Madore, S., and Cullen, B. R. (1995) Identification of a novel cellular cofactor for the Rev/Rex class of retroviral regulatory proteins. *Cell*, **82**, 485.

88. Fischer, U., Huber, J., Boelens, W. C., Mattaj, I. W., and Lührmann, R. (1995) The HIV-1 Rev activation domain is a nuclear export signal that accesses an export pathway used by specific cellular RNAs. *Cell*, **82**, 475.

89. Michael, W. M., Choi, M., and Dreyfuss, G. (1995) A nuclear export signal in hnRNP A1: a signal-mediated, temperature-dependent nuclear protein export pathway. *Cell*, **83**, 415.

90. Stutz, F. and Rosbash, M. (1994) A functional interaction between Rev and yeast pre-mRNA is related to splicing complex formation. *EMBO J.*, **17**, 4096.

91. Stade, K., Ford, C. S., Guthrie, C., and Weis, K. (1997) Exportin 1 (Crm1p) is an essential nuclear export factor. *Cell*, **90**, 1041.

92. Adachi, Y. and Yanagida, M. (1989) Higher order chromosome structure is affected by cold-sensitive mutations in a *Schizosaccharomyces pombe* gene crm1$^+$ which encodes a 115-kD protein preferentially localized in the nucleus and its periphery. *J. Cell Biol.*, **108**, 1195.

93. Yoshida, M., Nishikawa, M., Nishi, K., Abe, K., Horinouchi, S., and Beppu, T. (1990) Effects of leptomycin B on the cell cycle of fibroblasts and fission yeast cells. *Exp. Cell Res.*, **187**, 150.

94. Nishi, K., Yoshida, M., Fujiwara, D., Nishikawa, M., Horinouchi, S., and Beppu, T. (1994) Leptomycin B targets a regulatory cascade of crm1, a fission yeast nuclear protein, involved in control of higher order chromosome structure and gene expression. *J. Biol. Chem.*, **269**, 6320

95. Wolff, B., Sanglier, J. J., and Wang, Y. (1997) Leptomycin B is an inhibitor of nuclear export: inhibition of nucleo-cytoplasmic translocation of the human immunodeficiency virus type 1 (HIV-1) Rev protein and Rev-dependent mRNA. *Chem. Biol.*, **4**, 1391.

96. Neville, M., Stutz, F., Lee, L., Davis, L. I., and Rosbash, M. (1997) The importin-beta family member Crm1p bridges the interaction between Rev and the nuclear pore complex during nuclear export. *Curr. Biol.*, **7**, 767.

97. Fornerod, M., Ohno, M., Yoshida, M., and Mattaj, I. W. (1997) CRM1 is an export receptor for leucine-rich nuclear export signals. *Cell*, **90**, 1051.

98. Solsbacher, J., Maurer, P., Bischoff, F. R., and Schlenstedt, G. (1998) Cse1p is involved in export of yeast importin alpha from the nucleus. *Mol. Cell. Biol.*, **18**, 6805.

99. Hood, J. K. and Silver, P. A. (1998) Cse1p is required for export of Srp1p/Importin-alpha from the nucleus in *Saccharomyces cerevisiae*. *J. Biol. Chem.*, **273**, 35142.

100. Boche, I. and Fanning, E. (1997) Nucleocytoplasmic recycling of the nuclear localization signal receptor alpha subunit *in vivo* is dependent on a nuclear export signal, energy, and RCC1. *J. Cell Biol.*, **139**, 313.

101. Kose, S., Imamoto, N., Tachibana, T., Shimamoto, T., and Yoneda, Y. (1997) Ran-unassisted nuclear migration of a 97-kD component of nuclear pore-targeting complex. *J. Cell Biol.*, **139**, 841.

102. Nakielny, S. and Dreyfuss, G. (1998) Import and export of the nuclear protein import receptor transportin by a mechanism independent of GTP hydrolysis. *Curr. Biol.*, **8**, 89.

103. Stutz, F. and Rosbash, M. (1998) Nuclear RNA export. *Genes Dev.*, **12**, 3033.

104. Brown, J. A., Bharathi, A., Ghosh, A., Whalen, W., Fitzgerald, E., and Dhar, R. (1995) A mutation in the *Schizosaccharomyces pombe* rae1 gene causes defects in poly(A)$^+$ RNA export and in the cytoskeleton. *J. Biol. Chem.*, **270**, 7411.

105. Schneiter, R., Hitomi, M., Ivessa, A. S., Fasch, E. V., Kohlwein, S. D., and Tartakoff, A. M. (1996) A yeast acetyl coenzyme A carboxylase mutant links very-long-chain fatty acid synthesis to the structure and function of the nuclear membrane-pore complex. *Mol. Cell. Biol.*, **16**, 7161.

106. Henry, M. and Silver, P. (1996) A novel arginine methyltransferase (HMT1) modifies a poly(A) RNA-binding protein. *Mol. Cell. Biol.*, **16**, 3668.

107. Segref, A., Sharma, K., Doye, V., Hellwig, A., Huber, J., Luhrmann, R., and Hurt, E. (1997) Mex67p, a novel factor for nuclear mRNA export, binds to both poly(A)$^+$ RNA and nuclear pores. *EMBO J.*, **16**, 3256.

108. Santos-Rosa, H., Moreno, H., Simos, G., Segref, A., Fahrenkrog, B., Pante, N., and Hurt, E. (1998) Nuclear mRNA export requires complex formation between Mex67p and Mtr2p at the nuclear pores. *Mol. Cell. Biol.*, **18**, 6826.

109. Kadowaki, T., Hitomi, M., Chen, S., and Tartakoff, A. M. (1994) Nuclear mRNA accumulation causes nucleolar fragmentation in yeast mtr2 mutant. *Mol. Biol. Cell*, **5**, 1253.

110. Del Priore, V., Snay, C. A., Bahr, A., and Cole, C. N. (1996) Identification of *RSS1*, a high-copy extragenic suppressor of the *rat7-1* temperature-sensitive allele of the *Saccharomyces cerevisiae RAT7/NUP159* nucleoporin. *Mol. Biol. Cell*, **7**, 1601.

111. Noble, S. M. and Guthrie, C. (1996) Identification of novel genes required for yeast pre-mRNA splicing by means of cold-sensitive mutants. *Genetics*, **143**, 67.

112. Murphy, R. and Wente, S. R. (1996) An RNA-export mediator with an essential nuclear export signal. *Nature*, **383**, 357.

113. Watkins, J. L., Murphy, R., Emtage, J. L., and Wente, S. R. (1998) The human homologue of *Saccharomyces cerevisiae* Gle1p is required for poly(A)$^+$ RNA export. *Proc. Natl. Acad. Sci. U.S.A.*, **95**, 6779.

114. Hodge, C. A., Colot, H. V. Stafford, P. and Cole, C. N. (1999) Rat8p/Dbp5p is a shuttling transport factor that interacts with Rat7p/Nup159p and Gle1p and suppresses the mRNA export defect of *xpo1–1* cells. *EMBO J.*, **18**, 5778.

115. Strahm, Y., Fahrenkrog, B., Zenklusen, D., Rychner, E., Kantor, J., Rosbash, M., and Stutz, F. (1999) The RNA export factor Gle1p localizes on the cytoplasmic side of the NPC and physically interacts with the nucleoporin Rip1p, the DEAD-box protein Rat8p/Dbp5p and a new protein Ymr255p/Gfd1p. *EMBO J.*, **18**, 5761.

116. Tseng, S. S. I., Weaver, P. L., Liu, Y., Hitomi, M., Tartakoff, A. M., and Chang, T. H. (1998) Dbp5p, a cytosolic RNA helicase, is required for poly(A)$^+$ RNA export. *EMBO J.*, **17**, 2651.

117. Weis, K., Dingwall, C., and Lamond, A. I. (1996) Characterization of the nuclear protein import mechanism using Ran mutants with altered nucleotide binding specificities. *EMBO J.*, **15**, 7120.

118. Hopper, A. K., Schultz, L. D., and Shapiro, R. D. (1980) Processing of intervening sequences: a new yeast mutant which fails to excise intervening sequences from precursor tRNAs. *Cell*, **19**, 741.

119. Simos, G., Tekotte, H., Grosjean, H., Segref, A., Sharma, K., Tollervey, D., and Hurt, E. C. (1996) Nuclear pore proteins are involved in the biogenesis of functional tRNA. *EMBO J.*, **15**, 2270.

120. Lund, E. and Dahlberg, J. E. (1998) Proofreading and aminoacylation of tRNAs before export from the nucleus. *Science*, **282**, 2082.

121. Warner, J. R. (1990) The nucleolus and ribosome formation. *Curr. Opin. Cell Biol.*, **2**, 521.

122. Léger-Silvestre, I., Noaillac-Depeyre, J., Faubladier, M., and Gas, N. (1997) Structural and functional analysis of the nucleolus of the fission yeast *Schizosaccharomyces pombe*. *Eur. J. Cell Biol.*, **72**, 13.

123. Moy, T. I. and Silver, P. A. (1999) Nuclear export of small ribosomal subunits requires the Ran-GTPase cycle and certain nucleoporins. *Genes Dev.*, 2118.

124. Baldwin, A. S., Jr. (1996) The NF-kB and IkB proteins: new discoveries and insights. *Annu. Rev. Immunol.*, **14**, 649.

125. Ferrigno, P., Posas, F., Koepp, D., Saito, H., and Silver, P. A. (1998) Regulated nucleo/cytoplasmic exchange of HOG1 MAPK requires the importin beta homologs *NMD5* and *XPO1*. *EMBO J.*, **17**, 5606.

126. Toone, W. M., Kuge, S., Samuels, M., Morgan, B. A., Toda, T., and Jones, N. (1998) Regulation of the fission yeast transcription factor Pap1 by oxidative stress: requirement for the nuclear export factor Crm1 (Exportin) and the stress-activated MAP kinase Sty1/Spc1. *Genes Dev.*, **12**, 1453.

127. Yan, C., Lee, L. H., and Davis, L. I. (1998) Crm1p mediates regulated nuclear export of a yeast AP-1-like transcription factor. *EMBO J.*, **17**, 7416.

128. Kaffman, A., Rank, N. M., ONeill, E. M., Huang, L. S., and O'Shea, E. K. (1998) The receptor Msn5 exports the phosphorylated transcription factor Pho4 out of the nucleus. *Nature*, **396**, 482.

129. Lopez-Girona, A., Furnari, B., Mondesert, O., and Russell, P. (1999) Nuclear localization of Cdc25 is regulated by DNA damage and a 14-3-3 protein. Natuire, **397**, 172.

130. Saavedra, C., Tung, K.-S., Amberg, D. C., Hopper, A. K., and Cole, C. N. (1996) Regulation of mRNA export in response to stress in *Saccharomyces cerevisiae*. *Genes Dev.*, **10**, 1608.

131. Saavedra, C. A., Hammell, C. M., Heath, C. V., and Cole, C. N. (1997) Yeast heat shock mRNAs are exported through a distinct pathway defined by Rip1p. *Genes Dev.*, **11**, 2845.

132. Krebber, H., Taura, T., Lee, M. S., and Silver, P. A. (1999) Uncoupling of the hnRNP Npl3p from mRNAs during the stress-induced block in mRNA export. *Genes Dev.*, **13**, 1994.

133. Stutz, F., Neville, M., and Rosbash, M. (1995) Identification of a novel nuclear pore-associated protein as a functional target of the HIV-1 Rev protein in yeast. *Cell*, **82**, 495.

134. Stutz, F., Kantor, J., Zhang, D., McCarthy, T., Neville, M., and Rosbash, M. (1997) The yeast nucleoporin Rip1p contributes to multiple export pathways with no essential role for its FG-repeat region. *Genes Dev.*, **11**, 2857.

135. Steinmetz, E. J. (1997) Pre-mRNA processing and the CTD of RNA polymerase II: the tail that wags the dog. *Cell*, **89**, 491.
136. Zaret, K. S. and Sherman, F. (1982) DNA sequences required for efficient transcription termination in yeast. *Cell*, **28**, 563.
137. Colgan, D. F. and Manley, J. L. (1997) Mechanism and regulation of mRNA poly-adenylation. *Genes Dev.*, **11**, 2755.
138. Panté, N. and Aebi, U. (1995) Toward a molecular understanding of the structure and function of the nuclear pore complex. *Int. Rev. Cytol.*, 162B, 225.

Index